Chemistry

2

Alyn G. McFarland
Nora Henry

AQA material is reproduced by permission of AQA.

Orders: please contact Bookpoint Ltd, 130 Milton Park, Abingdon, Oxon OX14 4SB. Telephone: +44 (0)1235 827720. Fax: +44 (0)1235 400454. Lines are open 9.00 a.m.–5.00 p.m., Monday to Saturday, with a 24-hour message answering service. Visit our website at www.hoddereducation.co.uk

First published in 2015 by
Hodder Education,
An Hachette UK Company
Carmelite House, 50 Victoria Embankment, London EC4Y ODZ

Impression number 5 4 3 2 1

Year 2019 2018 2017 2016 2015

Cover photo © SSilver–Fotolia
Illustrations by Aptara
Typeset in 11/13 pt ITC Berkeley Oldstyle Std by Aptara, Inc.
Printed in Italy

A catalogue record for this title is available from the British Library.

ISBN 978 1471 807701

Activities

These practical-based activities will help consolidate your learning and test your practical skills.

In this edition the authors describe many important experimental procedures as "Activities" to conform to recent changes in the A-level curriculum. Teachers should be aware that, although there is enough information to inform students of techniques and many observations for exam-question purposes, there is not enough information for teachers to replicate the experiments themselves or with students without recourse to CLEAPSS Hazcards or Laboratory worksheets which have undergone a thorough risk assessment procedure.

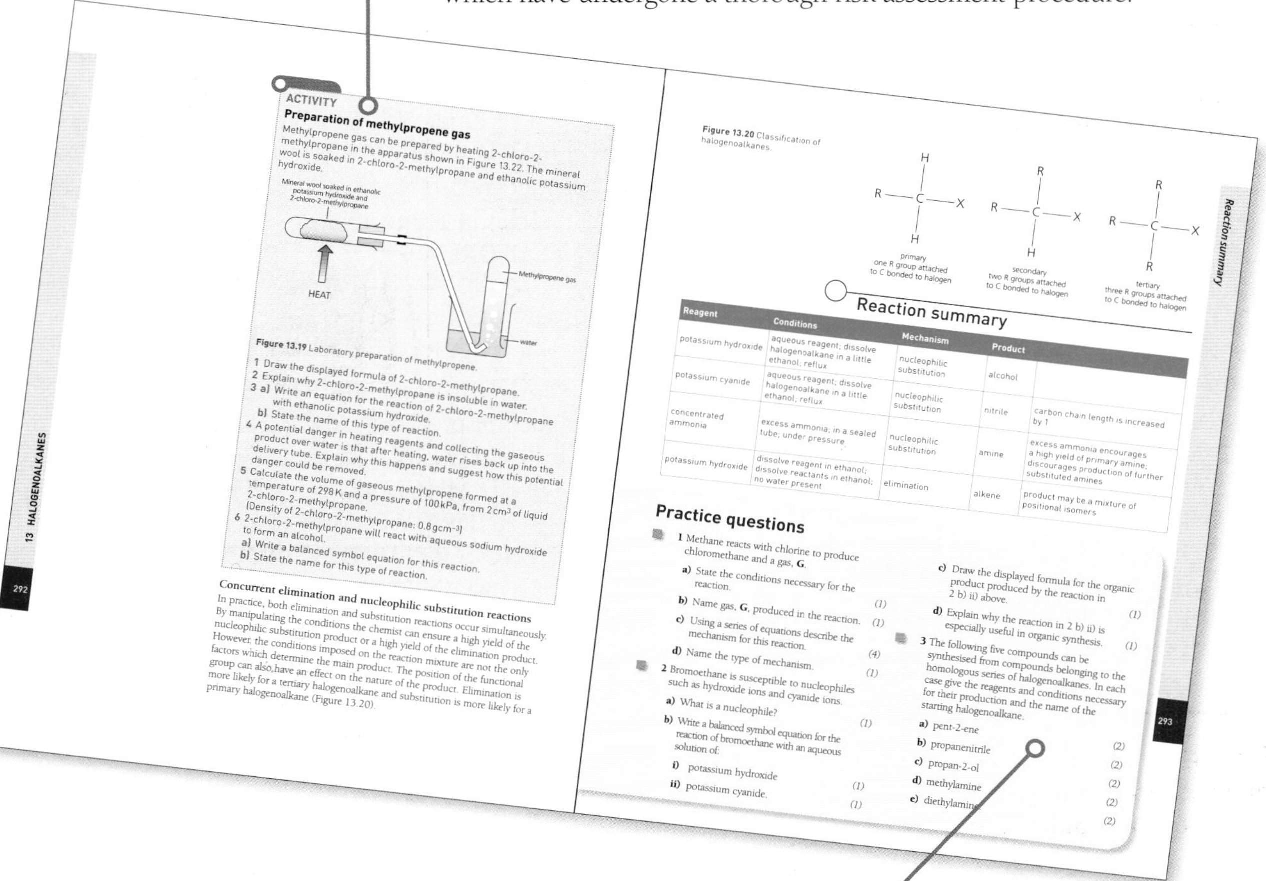

Practice questions

You will find practice questions, including multiple-choice questions, at the end of every chapter. These follow the style of the different types of questions with short and longer answers that you might see in your examination, and they are colour coded to highlight the level of difficulty.

- Green – Basic questions that everyone should be able to answer without difficulty.
- Orange – Questions that are a regular feature of exams and that all competent candidates should be able to handle.
- Purple – More demanding questions which the best candidates should be able to do.

A dedicated chapter for developing your **Maths** can be found at the back of this book.

Photo credits

The Publisher would like to thank the following for permission to reproduce copyright material:

p. 16 Courtesy of Morgan Riley – Wikipedia Commons (http://creativecommons.org/licenses/by/3.0/deed.en); p. 27 *both* © Hodder Education; p. 27 © Hodder Education; p. 32 © CHARLES D. WINTERS/SCIENCE PHOTO LIBRARY; p. 32 *both* © ANDREW LAMBERT PHOTOGRAPHY/SCIENCE PHOTO LIBRARY; p. 35 *left* © Pumehana – Fotolia.com; p. 35 *right* © WLADIMIR BULGAR/SCIENCE PHOTO LIBRARY; p. 45 © CHARLES D. WINTERS/SCIENCE PHOTO LIBRARY; p. 43 and p.46 © minifilm – Fotolia.com; p. 56 © MARTYN F. CHILLMAID/SCIENCE PHOTO LIBRARY; p. 68 and p.76 © CHARLES D. WINTERS/SCIENCE PHOTO LIBRARY; p. 78 and p.79 © SCIENCE PHOTO LIBRARY; p. 80 © ANDREW LAMBERT PHOTOGRAPHY/SCIENCE PHOTO LIBRARY; p. 87 © ANDREW LAMBERT PHOTOGRAPHY/SCIENCE PHOTO LIBRARY; p. 90 © DR P. MARAZZI/SCIENCE PHOTO LIBRARY; p. 100 and 107 © ANDREW LAMBERT PHOTOGRAPHY/SCIENCE PHOTO LIBRARY; p. 104 © somjring34 – Fotolia.com; p.106 © Max Tactic – Fotolia.com; p. 113 © vlad61_61 – Fotolia.com; p. 116 © photongpix –Fotolia.com; p. 135 and p.137 © MARTYN F. CHILLMAID/SCIENCE PHOTO LIBRARY; p. 138 © ANDREW LAMBERT PHOTOGRAPHY/SCIENCE PHOTO LIBRARY; p. 141 © MIKE WALKER, VISUALS UNLIMITED /SCIENCE PHOTO LIBRARY; p. 142 © fotosuper - iStock via Thinkstock; p. 143 © ANDREW LAMBERT PHOTOGRAPHY/SCIENCE PHOTO LIBRARY; p. 150 © ANDREW LAMBERT PHOTOGRAPHY/SCIENCE PHOTO LIBRARY; p. 153 © airborne77 – Fotolia.com; p. 154 © janeness – Fotolia.com; p. 160 © LAGUNA DESIGN/SCIENCE PHOTO LIBRARY; p. 167 © ANDREW LAMBERT PHOTOGRAPHY/SCIENCE PHOTO LIBRARY; p. 149 and p.168 © JERRY MASON/SCIENCE PHOTO LIBRARY; p. 169 © ANDREW LAMBERT PHOTOGRAPHY/SCIENCE PHOTO LIBRARY; p. 171 © ANDREW LAMBERT PHOTOGRAPHY/SCIENCE PHOTO LIBRARY; p. 175 and 186 © adam88xx – Fotolia.com; p. 176 © ANDREW LAMBERT PHOTOGRAPHY/SCIENCE PHOTO LIBRARY; p. 179 © SCIENCE PHOTO LIBRARY; p. 187 © uckyo – Fotolia.com; p. 188 © SCIENCE PHOTO LIBRARY; p. 198 and p.200; © DAVID MUNNS/SCIENCE PHOTO LIBRARY; p. 202 © SATURN STILLS/SCIENCE PHOTO LIBRARY; p. 206 *top* © ANDREW LAMBERT PHOTOGRAPHY/SCIENCE PHOTO LIBRARY; p. 206 *bottom* © ANDREW LAMBERT PHOTOGRAPHY/SCIENCE PHOTO LIBRARY; p. 215 © MARTYN F. CHILLMAID/SCIENCE PHOTO LIBRARY; p. 224 © ANDREW LAMBERT PHOTOGRAPHY/SCIENCE PHOTO LIBRARY; p. 233 and p.235 © valdis torms – Fotolia.com; p. 236 © ANDREW LAMBERT PHOTOGRAPHY/SCIENCE PHOTO LIBRARY; p. 245 © Nutthaphol – iStock via Thinkstock; p. 263 © CHARLES D. WINTERS/SCIENCE PHOTO LIBRARY; p. 276 © ST. BARTHOLOMEW'S HOSPITAL/SCIENCE PHOTO LIBRARY; p. 278 © geogphotos / Alamy; p. 261 and p.279 © RUSSELL KIGHTLEY/SCIENCE PHOTO LIBRARY; p. 293 © DENNIS SCHROEDER, NREL/US DEPARTMENT OF ENERGY/SCIENCE PHOTO LIBRARY; p. 298 © A. BENOIST/BSIP/SCIENCE PHOTO LIBRARY; p. 302 © Photos contributed by Natrij, Wikimedia (http://creativecommons.org/licenses/by-sa/3.0/); p. 287 and p.303 © MAXIMILIAN STOCK LTD/SCIENCE PHOTO LIBRARY; p. 305 © DR JURGEN SCRIBA/SCIENCE PHOTO LIBRARY; p. 309 © SL-66 – Fotolia.com; p. 318 © Maurice Crooks / Alamy.

Every effort has been made to trace all copyright holders, but if any have been inadvertently overlooked, the Publisher will be pleased to make the necessary arrangements at the first opportunity.

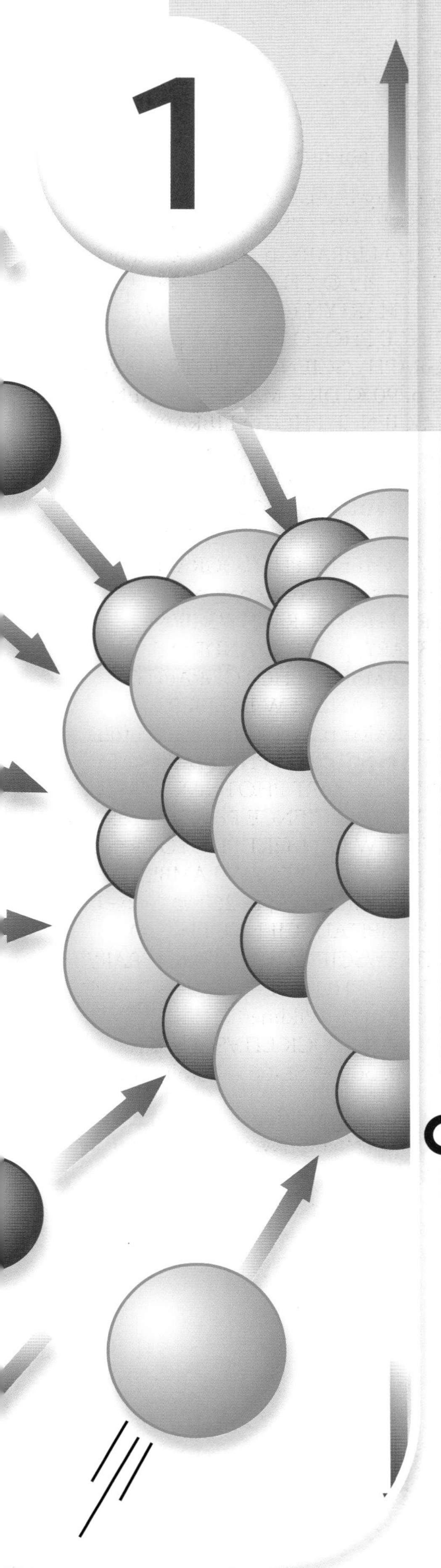

1 Thermodynamics: Born-Haber cycles

PRIOR KNOWLEDGE

It is expected that you are familiar with all of the content of the **Energetics** unit in Year 1 of the A-level Chemistry course. The following are some key points of Prior Knowledge:

- The enthalpy change in a chemical reaction is represented by ΔH.
- ΔH is the heat energy change measured under conditions of constant pressure.
- For exothermic reactions, ΔH has a negative value.
- For endothermic reactions, ΔH has a positive value.
- The standard enthalpy change of combustion is the enthalpy change when 1 mole of a substance is burned completely in excess oxygen with all substances in standard states under standard conditions.
- The standard enthalpy change of formation is the enthalpy change when 1 mole of a compound is formed from its constituent elements in their standard states under standard conditions.
- Hess's Law states that the enthalpy change for a chemical reaction is independent of the route taken and depends only on the initial and final states.
- Hess's Law can be used to calculate enthalpy changes for chemical reactions from the enthalpy changes of other reactions.
- Temperature changes may be used to calculate enthalpy changes using $q = mc\Delta T$.
- The mean bond enthalpy is a measure of the energy required to break one mole of a covalent bond measured in $kJ\,mol^{-1}$ averaged across compounds containing that bond.
- Mean bond enthalpy values may be used to calculate the enthalpy change for a chemical reaction by considering the covalent bonds broken (endothermic) and the covalent bonds made during the reaction.

TEST YOURSELF ON PRIOR KNOWLEDGE 1

1 Write equations, including state symbols, to represent the standard enthalpies of formation given below:

a) Formation of magnesium oxide, $MgO(s)$.

b) Formation of hydrogen chloride, $HCl(g)$.

2 Write equations, including state symbols to represent the standard enthalpies of combustion given below:

a) Combustion of propane, $C_3H_8(g)$.

b) Combustion of methanol, $CH_3OH(l)$.

3 The standard enthalpies of formation of ethane, carbon dioxide and water are −85, −394 and −286 kJ mol^{-1}, respectively.

a) State Hess's Law.

b) Write an equation, including state symbols, to represent the standard enthalpy of combustion of ethane.

c) Calculate a value for the standard enthalpy of combustion of ethane.

4 In an experiment the temperature of 200 g of water rose by 22.4 °C when 0.015 mole of ethanol were burned in air and the heat used to warm the water. Calculate a value for the enthalpy change when one mole of ethanol is burned. The specific heat capacity of water is 4.18 J K^{-1} g^{-1}).

5 The mean bond energy values for certain covalent bonds are given below:

Bond	Mean bond enthalpy (kJ mol^{-1})
C–H	412
C–C	348
O–H	463
O=O	496
C=O	803

Calculate a value for the following reaction using these values:

$C_3H_8(g) + 5O_2(g) \rightarrow 3CO_2(g) + 4H_2O(g)$

Enthalpy changes in ionic compounds

The changes that occur during the formation of an ionic compound are considered in this unit. The dissociation or formation of an ionic compound may be calculated from other values using Hess's Law in the form of a Born-Haber cycle.

Notations

- ΔH represents an enthalpy change.
- ΔH is measured under stated conditions, i.e. ΔH_{298} is the enthalpy change at 298 K and the pressure is 100 kPa.
- The **enthalpy change** is the heat energy change at constant pressure.
- Standard Enthalpy Values are the ΔH value for enthalpy changes of specific reactions measured under standard conditions.
- Standard conditions are represented by the symbol $\ominus$ which is used after ΔH to indicate that an enthalpy changes occurs under standard conditions.
- Standard conditions are 298 K, 100 kPa pressure, all solutions of concentration 1 mol dm^{-3} and all substances are present in their standard states.

The definitions of the enthalpy changes associated with ionic compounds must be understood to be able to link the enthalpy changes together, particularly in terms of what the 'per mol' in $kJ\,mol^{-1}$ means.

TIP
This is an enthalpy profile for an endothermic reaction. The enthalpy change is positive. Make sure you can draw and recognise the enthalpy profile diagram for an exothermic reaction as well from the Energetics topic in AS Year 1.

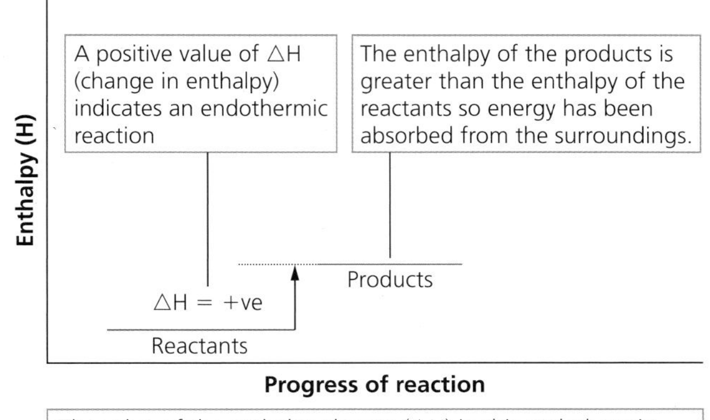

The value of the enthalpy change (ΔH) in this endothermic reaction is positive. This is because there has been an increase in enthalpy from reactants to products. This is a standard feature of **endothermic** reactions and it must be remembered that **ΔH is positive.**

Figure 1.1 Enthalpy profile for an endothermic reaction.

The word enthalpy is based on the Greek noun *enthalpos*, which means *heating*. It comes from the classical Greek prefix ἐν-, meaning 'to put into', and the verb θάλπειν, *thalpein*, meaning 'to heat'. The word enthalpy first appeared in scientific literature in 1909. Many different symbols were used to denote enthalpy and it was not until 1922 that the symbol '*H*' was accepted as standard.

Lattice enthalpy

Lattice enthalpy can be represented as $\Delta_L H^{\ominus}$ or $\Delta_{latt} H^{\ominus}$.

The **enthalpy of lattice dissociation** is the enthalpy change when one mole of an ionic compound is separated into its component gaseous ions. The phrase 'at infinite separation' is often used in the definition because all bonding between the ions must be broken so the ions must be infinitely separated.

The enthalpy of lattice dissociation is also called the lattice dissociation enthalpy. It is an endothermic process and has a positive value measured in $kJ\,mol^{-1}$. Energy is required to overcome the strong attraction between the positive and negative ions. The 'per mol' refers to one mole of the ionic compound.

The **enthalpy of lattice formation** is the enthalpy change when one mole of an ionic compound is formed from its constituent ions in the gaseous state. It is an exothermic process and has a negative value measured in $kJ\,mol^{-1}$. Again, the 'per mol' refers to one mole of the ionic compound.

Equations for lattice dissociation and formation

Equations, including state symbols, are often asked for the process which occurs when the lattice of an ionic compound dissociates or forms.

EXAMPLE 1

Write an equation, including state symbols, for the reaction that has an enthalpy change equal to the lattice dissociation enthalpy of sodium chloride.

Answer

This question is asking for an equation for the dissociation of **one mole of sodium chloride** into **its constituent gaseous ions** as that is the definition of the enthalpy of lattice dissociation.

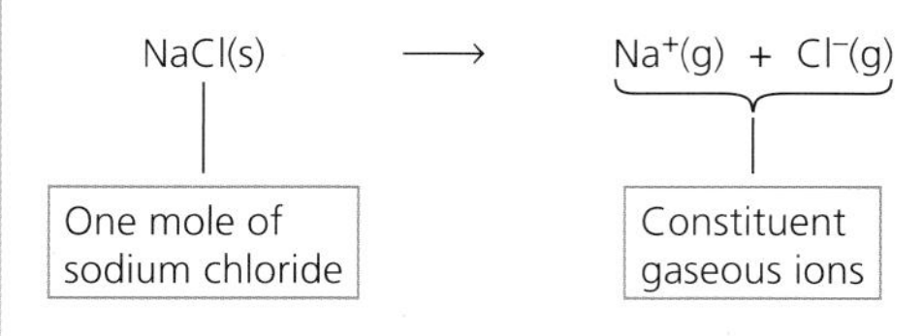

TIP

Make sure you include the state symbols. The notes below the equation are used for clarity and are not expected in the answer.

EXAMPLE 2

Write an equation, including state symbols, for the reaction that has an enthalpy change equal to the enthalpy change of lattice formation of calcium chloride.

Answer

This question is asking for an equation for the formation of **one mole of calcium chloride** from **its constituent gaseous ions** as that is the definition of the enthalpy of lattice formation.

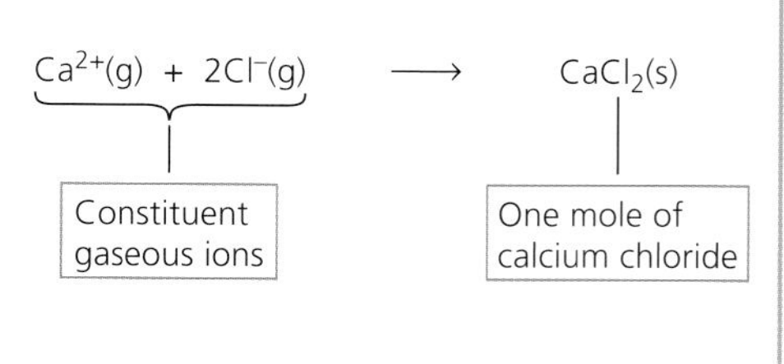

Enthalpy of formation

Enthalpy of formation can be represented as $\Delta_f H^{\ominus}$. The enthalpy (change) of formation is the enthalpy change when one mole of a compound is formed from its elements when all reactants and products are in their standard states under standard conditions.

Again equations for the process of formation are asked with state symbols.

TIP

You will have written equations for enthalpies of formation in AS.

EXAMPLE 3

Write an equation, including state symbols, for the process that has an enthalpy change equal to the standard enthalpy of formation of silver(I) fluoride.

Answer

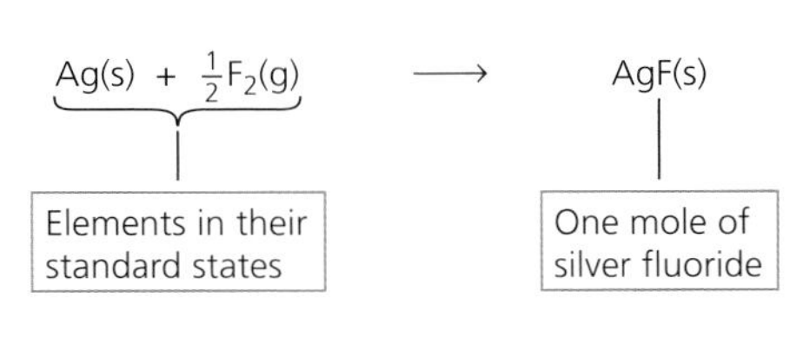

This question asks for an equation for the formation of **one mole of silver fluoride** from its **constituent elements when all reactants and products are in their standard states under standard conditions**.

EXAMPLE 4

Write an equation, including state symbols, for the process that has an enthalpy change equal to the standard enthalpy of formation of calcium chloride.

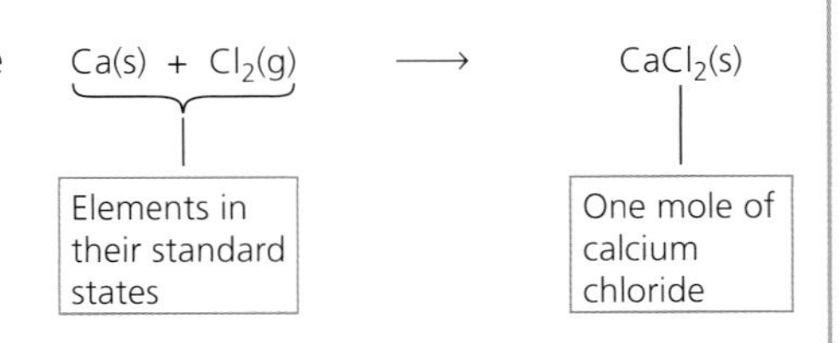

Answer

This question asks for an equation for the formation of **one mole of calcium chloride** from its **constituent elements when all reactants and products are in their standard states under standard conditions**.

TIP

Do not confuse the enthalpy of lattice formation with enthalpy of formation of a compound as these can also be asked. For calcium chloride, as shown in the examples above, these equations are different. Do not forget the diatomic elements in enthalpy change of formation equations.

EXAMPLE 5

Write an equation, including state symbols, for the process that has an enthalpy change equal to the standard enthalpy of formation of magnesium oxide.

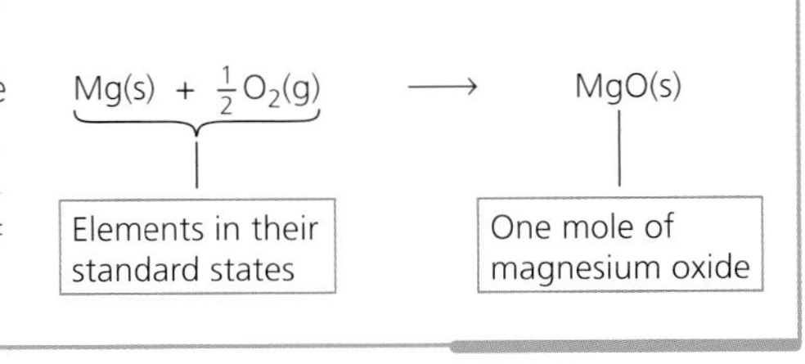

Enthalpy of atomisation

Enthalpy of atomisation can be represented as $\Delta_a H^\ominus$ or $\Delta_{at} H^\ominus$. The **enthalpy of atomisation of an element** is the enthalpy change for the formation of one mole of gaseous atoms from the element in its standard state.

Again equations for the process of formation may be asked with state symbols but it is more likely you have to understand the process in a Born-Haber cycle. For diatomic elements, the equation must produce one mole of gaseous atoms so $\frac{1}{2}X_2$ is used where $\frac{1}{2}X_2$ represents half a mole of the diatomic element.

EXAMPLE 6

Write an equation, including state symbols, for the process that has an enthalpy change equal to the enthalpy of atomisation of sodium.

$Na(s) \rightarrow Na(g)$

TIP

A common mistake here is to list Br_2 as a gas but the element must be in its standard state. The atomisation of iodine would be $\frac{1}{2}I_2(s) \rightarrow I(g)$.

EXAMPLE 7

Write an equation, including state symbols, for the process that has an enthalpy change equal to the enthalpy of atomisation of bromine.

$\frac{1}{2}Br_2(l) \rightarrow Br(g)$

EXAMPLE 8

Write an equation, including state symbols, for the process that has an enthalpy change equal to the enthalpy of atomisation of oxygen.

$\frac{1}{2}O_2(g) \rightarrow O(g)$

Bond dissociation enthalpy (bond enthalpy)

Bond dissociation enthalpy can be represented as $\Delta_{BDE}H^{\ominus}$ or $\Delta_{dis}H^{\ominus}$ or $\Delta_{dissoc}H^{\ominus}$. The bond dissociation enthalpy is a value for the enthalpy change when one mole of a covalent bond is broken under standard conditions in the gaseous state. It takes a positive value as one mole of the covalent bond is broken (endothermic).

For example, the bond dissociation enthalpy for chlorine is $+242\,kJ\,mol^{-1}$. This means that 242 kJ of energy is required to convert 1 mol of chlorine molecules (containing 1 mol of the covalent bond) into 2 mol of gaseous chlorine atoms.

The equation for this process is: $Cl_2(g) \rightarrow 2Cl(g)$

TIP

The definition of bond dissociation enthalpy is usually asked for with reference to a specific bond or diatomic molecule. It is used in Born-Haber cycles with diatomic elements.

- It is important to note that mean bond enthalpy values were met in AS and are a measure of the energy required to break one mole of a covalent bond measured in mol^{-1} (the 'per mole' is per mole of the covalent bond) averaged across many compounds containing the bond.
- For diatomic molecules, the bond enthalpy is **not** averaged across many compounds containing the bond as the bond only occurs in the diatomic molecules.
- Also for mean bond enthalpy calculations the enthalpy of bond breaking is endothermic and the enthalpy of bond making is exothermic as will be considered again later in this unit, but bond dissociation enthalpy is always endothermic as it is for the dissociation of the bond.

TIP

The value for the bond dissociation enthalpy for the diatomic molecules is twice the value for the enthalpy of atomisation of these elements. This is important in deciding what value to use in a Born-Haber cycle later in this unit.

EXAMPLE 9

Write an equation, including state symbols, for the process that has an enthalpy change equal to the bond dissociation enthalpy of fluorine.

$F_2(g) \rightarrow 2F(g)$

This equation shows the dissociation of one mole of diatomic fluorine, F_2, into two moles of gaseous fluorine atoms.

Sometimes the definition of bond dissociation enthalpy is asked as applied to a particular diatomic element and the answer must be in terms of this element. The initial element should be in the gaseous state even if it is bromine or iodine.

EXAMPLE 10

Define *bond dissociation enthalpy* as applied to fluorine.

Answer

The enthalpy change to break the bond in one mole of fluorine to form two moles of gaseous fluorine atoms.

Ionisation enthalpies

Ionisation enthalpies can be represented as $\Delta_{IE}H^{\ominus}$. A number after IE can be used to indicate if it is a first or second, etc., ionisation enthalpy, for example $\Delta_{IE1}H^{\ominus}$ represents the first ionisation enthalpy. At AS, ionisation enthalpies (or they were called ionisation energies in Atomic Structure) were examined as evidence for the existence of energy levels.

The **first ionisation enthalpy** is the enthalpy change when one mole of electrons is removed from one mole of gaseous atoms to form one mole of gaseous ions with a single positive charge.

The **second ionisation enthalpy** is the enthalpy change when one mole of electrons is removed from one mole of gaseous ions with a single positive charge to form one mole of gaseous ions with a 2+ charge.

The **third ionisation enthalpy** is the enthalpy change when one mole of electrons is removed from one mole of gaseous ions with a 2+ charge to form one mole of gaseous ions with a 3+ charge.

For the purposes of Born-Haber cycles, you would be expected to be able to define and carry out calculations using first, second and sometimes third ionisation enthalpies.

EXAMPLE 11

Write an equation, including state symbols, for the process that has an enthalpy change equal to the first ionisation enthalpy of magnesium.

Answer

$Mg(g) \rightarrow Mg^{+}(g) + e^{-}$

Remember that both the atoms and ions have to be gaseous. The value for the above first ionisation enthalpy is +736 kJ mol^{-1}. Remember also that first ionisation enthalpies decrease down a group. The first ionisation enthalpy of calcium is +590 kJ mol^{-1}.

TIP

To go from $Ca(g)$ to $Ca^{2+}(g)$ requires a total of the first and second ionisation enthalpies.
The first ionisation enthalpy of calcium is +590 kJ mol^{-1}.
The second ionisation enthalpy of calcium is +1150 kJ mol^{-1}.
The change in enthalpy required for $Ca(g) \rightarrow Ca^{2+}(g) + 2e^{-}$ is + 590 + 1150 = +1740 kJ.

EXAMPLE 12

Write an equation, including state symbols, for the process that has an enthalpy change equal to the second ionisation enthalpy of calcium.

$Ca^{+}(g) \rightarrow Ca^{2+}(g) + e^{-}$

Electron affinity

Electron affinity can be represented as $\Delta_{EA}H^{\ominus}$. Again a number after EA can be used to represent the first or second electron affinities.

The first electron affinity is the enthalpy change when one mole of gaseous atoms forms one mole of negative ions with a single negative charge. An equation, including state symbols, representing the first electron affinity of fluorine would be:

$F(g) + e^{-} \rightarrow F^{-}(g)$

The second electron affinity is the enthalpy change when one mole of gaseous ions with a single negative charge forms one mole of gaseous ions with a double negative charge.

EXAMPLE 13

Write an equation, including state symbols, for the process that has an enthalpy change equal to the second electron affinity of oxygen.

$O^{-}(g) + e^{-} \rightarrow O^{2-}(g)$

TIP

When the definition of *electron affinity* is asked give the definition of first electron affinity.

Values for second electron affinities are positive as the process is endothermic. This is due to the repulsion of the negative ion for the negative electron being added.

TEST YOURSELF 2

1 Write equations, including state symbols, for the process that has an enthalpy change equal to the following processes:
  a) Enthalpy of lattice formation of calcium chloride.
  b) Bond dissociation enthalpy of fluorine.
  c) First electron affinity of oxygen.
  d) Enthalpy of formation of silver fluoride.
  e) First ionisation enthalpy of potassium.

2 Name the enthalpy changes represented by the following equations:
  a) $Ca^{+}(g) \rightarrow Ca^{2+}(g) + e^{-}$
  b) $Na(s) \rightarrow Na(g)$
  c) $Cl(g) + e^{-} \rightarrow Cl^{-}(g)$
  d) $Na(s) + \frac{1}{2}F_2(g) \rightarrow NaF(s)$
  e) $Mg^{2+}(g) + 2Cl^{-}(g) \rightarrow MgCl_2(s)$

3 Define the following:
  a) Standard enthalpy of formation.
  b) Enthalpy of lattice formation.
  c) Lattice dissociation enthalpy.
  d) Enthalpy of atomisation of chlorine.
  e) First ionisation enthalpy of lithium.

Born-Haber cycles

The Born-Haber cycle is a technique for applying Hess's Law to the standard enthalpy changes that occur when an ionic compound is formed.

The Born-Haber cycle for NaCl can be drawn simply as shown in the diagram below. It should be noted that this cycle is simply being used to explain the processes but would not be accepted as a Born-Haber cycle in an examination question.

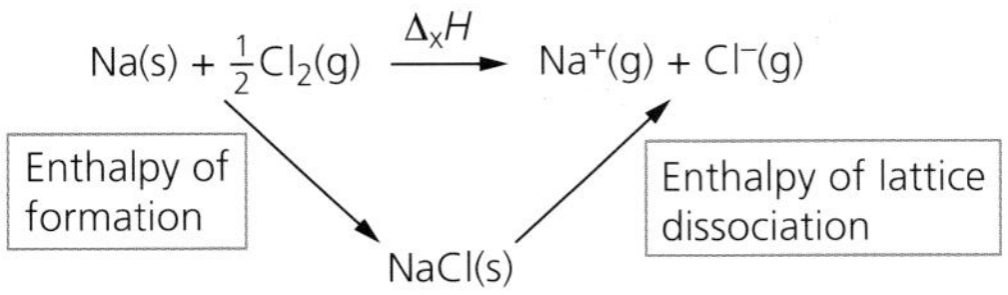

$$Na(s) + \tfrac{1}{2}Cl_2(g) \xrightarrow{\Delta_x H} Na^{+}(g) + Cl^{-}(g)$$

Enthalpy of formation
NaCl(s)
Enthalpy of lattice dissociation

$\Delta_x H$ is the sum of all the changes required to convert Na(s) to $Na^{+}(g)$ and $\frac{1}{2}Cl_2(g)$ to $Cl^{-}(g)$.

These can be summarised in the table below.

Process	Name of enthalpy change	$\Delta H^{\ominus}$/kJ mol^{-1}
$Na(s) \rightarrow Na(g)$	Enthalpy of atomisation of sodium	+109
$Na(g) \rightarrow Na^{+}(g) + e^{-}$	First ionisation enthalpy of sodium	+494
$\frac{1}{2}Cl_2(g) \rightarrow Cl(g)$	Enthalpy of atomisation of chlorine	+121
$Cl(g) + e^{-} \rightarrow Cl^{-}(g)$	First electron affinity of chlorine	−364

$$\Delta_x H^{\ominus} = \Delta_a H^{\ominus} + \Delta_{IE1} H^{\ominus} + \Delta_a H^{\ominus} + \Delta_{EA1} H^{\ominus}$$

$$\Delta_x H^{\ominus} = +109 + 494 + 121 - 364 = +360\,\text{kJ}$$

Hess's Law states that the enthalpy change for a chemical reaction is independent of the route taken and depends only on the initial and final states.

This means that $\Delta_x H$ = Enthalpy of formation + Enthalpy of lattice dissociation

This can be written as: $\Delta_x H = \Delta_f H + \Delta_L H$

The standard symbols are often left out when writing these expressions.

The value for the enthalpy of formation of sodium chloride is $-411\,\text{kJ mol}^{-1}$.

The enthalpy of lattice dissociation ($\Delta_L H$) can be calculated.

$$\Delta_x H = \Delta_f H + \Delta_L H$$

$$+360 = -411 + \Delta_L H$$

$$\Delta_L H = +360 + 411 = +771\,\text{kJ mol}^{-1}$$

The enthalpy of lattice dissociation for sodium chloride is $+771\,\text{kJ mol}^{-1}$. The enthalpy of lattice formation is $-771\,\text{kJ mol}^{-1}$.

Drawing a Born-Haber cycle

The diagram below shows a Born-Haber cycle for a Group 1 halide, in this case for sodium chloride, NaCl.

For all Born-Haber cycles, the arrows for the enthalpy changes are not to scale.

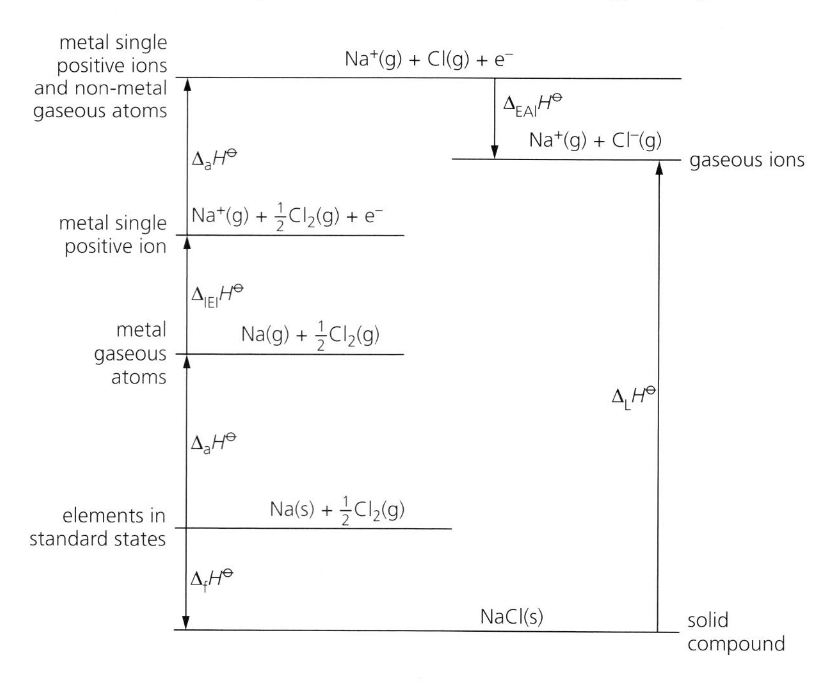

Features of the cycle

- The arrows that are going up represent endothermic processes.
- The arrows that are going down represent exothermic processes.
- $\Delta_L H^\ominus$ is the enthalpy of lattice dissociation in this Born-Haber cycle.
- The labels at the left and right of the different enthalpy levels show the species which should be present on these levels. These labels are there

only to help you understand the cycle but are not essential if you are asked to draw one.

- You may be expected to complete some of the levels in a cycle with the species present or even to complete part of the cycle as shown in the examples which follow.

EXAMPLE 14

Complete the Born-Haber cycle shown on the right for potassium fluoride by drawing the missing energy levels, symbols and arrows.

Answer

In this type of question you would be expected to complete the levels in the diagram as shown in the previous example, with endothermic changes having levels above the previous one and exothermic changes having levels below.

The symbols for the species (atoms, ions and electrons) on each enthalpy level with appropriate state symbols would also be expected.

The correct direction of the arrows representing the changes should be given.

The completed cycle with annotations is shown below.

$K^+(g) + \frac{1}{2}F_2(g) + e^-$

+420 $K(g) + \frac{1}{2}F_2(g)$

+90 $K(s) + \frac{1}{2}F_2(g)$

−569 KF(s)

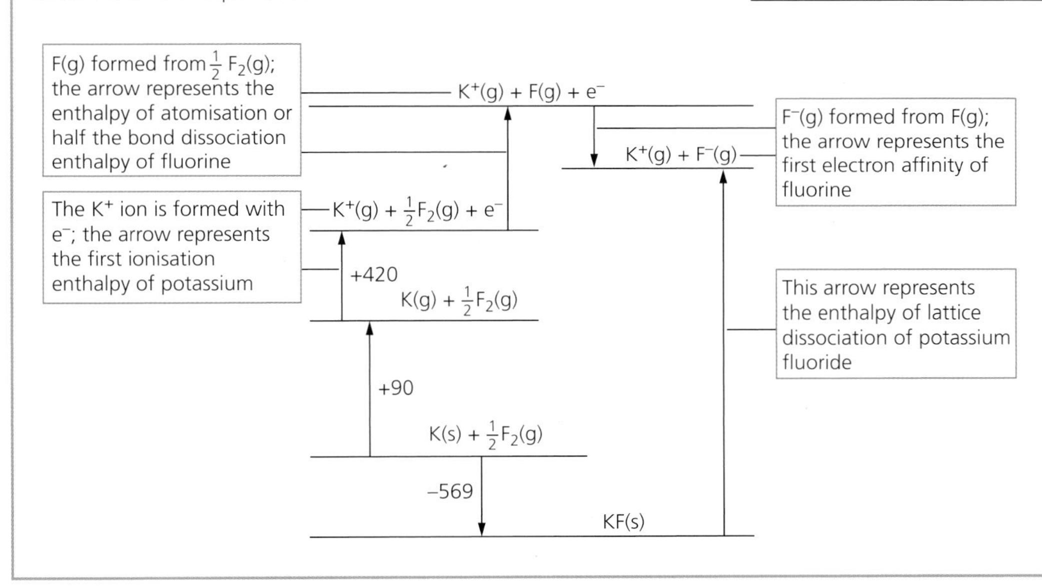

Calculations from a Born-Haber cycle

In calculations, the enthalpy of atomisation of chlorine may be given or, alternatively, the bond dissociation enthalpy. It is vital that you understand that for **gaseous** diatomic elements like the first two halogens, the bond dissociation enthalpy is twice the enthalpy of atomisation. For Group 1 halides and other halides where the oxidation number of the metal is +1, like the one shown in the cycle below, the change required is $\frac{1}{2}Cl_2(g) \rightarrow Cl(g)$. This is the value of the enthalpy of atomisation of chlorine or half the value for the bond dissociation enthalpy of chlorine.

Calculations may be of other values in the Born-Haber cycle, if the lattice enthalpy is known.

EXAMPLE 15

Using the Born-Haber cycle shown on the right, calculate the enthalpy of formation of silver(I) fluoride using the enthalpy changes given. The diagram is not to scale.

Name of enthalpy change	$\Delta H^\ominus$/kJ mol^{-1}
Enthalpy of lattice dissociation of silver fluoride	+967
Bond dissociation enthalpy of fluorine	+158
Electron affinity of fluorine	−348
First ionisation enthalpy of silver	+732
Enthalpy of atomisation of silver	+289

$Ag^+(g) + F^-(g) + e^-$

$Ag^+(g) + F^-(g)$

$Ag^+(g) + \frac{1}{2}F_2^-(g) + e^-$

$Ag(g) + \frac{1}{2}F_2^-(g)$

$Ag(s) + \frac{1}{2}F_2^-(g)$

$AgF^-(s)$

Answer

The most important point in this Born-Haber cycle is the type of value given for the conversion of fluorine to fluorine atoms.

The value given is the bond dissociation enthalpy which is +158 kJ mol^{-1} and this is the enthalpy value for the change $F_2(g) \rightarrow 2F(g)$.

However in this cycle the change required is:

$\frac{1}{2}F_2(g) \rightarrow F(g)$

The enthalpy value for this change is half of the value for the bond dissociation enthalpy.

It is good practice to add the values for the enthalpy changes to the Born-Haber cycle as this makes the calculations simpler. The diagram below shown the Born-Haber cycle for silver fluoride with the known enthalpy changes added.

From the cycle, the enthalpy of formation of silver fluoride is calculated by going round the cycle in the other direction.

Start at the start of the arrow for the enthalpy change you want to find and go around the cycle in the other direction.

If you go through an arrow in the wrong direction change the sign of the enthalpy value.

For example, the enthalpy of lattice dissociation of AgF

$$= -\Delta_f H^\ominus + \Delta_a H^\ominus + \Delta_{IE1} H^\ominus + \frac{1}{2}\Delta_{BDE} H^\ominus + \Delta_{EA1} H^\ominus$$

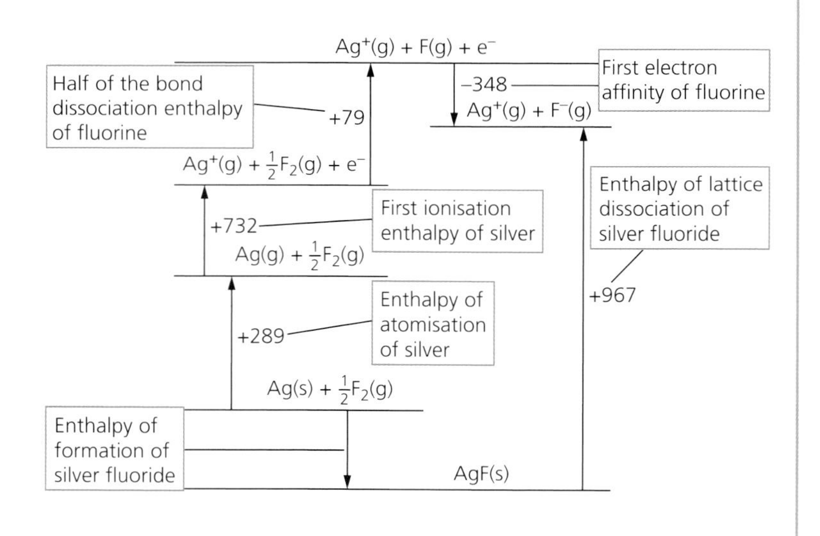

Calculating $\Delta_f H^\ominus$

$$\Delta_f H^\ominus = +289 + 732 + 79 - 348 - 967$$

$$\Delta_f H^\ominus = -215\ \text{kJ mol}^{-1}$$

The calculations are sometimes carried out without a cycle. The values may be given and you would be expected to remember the sequence of enthalpy changes as shown in the next example.

EXAMPLE 16

The table below gives some values for standard enthalpy changes. Use these values to calculate the standard enthalpy of lattice dissociation of potassium bromide.

Name of enthalpy change	$\Delta H^{\ominus}$/kJ mol^{-1}
Enthalpy of formation of potassium bromide ($\Delta_f H^{\ominus}$)	-392
Enthalpy of atomisation of bromine ($\Delta_a H^{\ominus}$(Br))	+112
Electron affinity of bromine ($\Delta_{EA1} H^{\ominus}$)	-342
First ionisation enthalpy of potassium ($\Delta_{IE1} H^{\ominus}$)	+420
Enthalpy of atomisation of potassium ($\Delta_a H^{\ominus}$(K))	+90

Answer

The enthalpy of lattice dissociation ($\Delta_L H^{\ominus}$) can be calculated from the other values:

$$\Delta_L H^{\ominus} = -\Delta_f H^{\ominus}(\text{KBr}) + \Delta_a H^{\ominus}(\text{K}) + \Delta_{IE1} H^{\ominus}(\text{K}) + \Delta_a H^{\ominus}(\text{Br}) + \Delta_{EA1} H^{\ominus}(\text{Br})$$

$$\Delta_L H = +392 + 90 + 420 + 112 + (-342)$$

$$\Delta_L H = +672\ \text{kJ mol}^{-1}$$

Born-Haber cycles for metals with oxidation state +2

Born-Haber cycles for compounds of metals where the metal has an oxidation state of +2 are slightly more complex. The cycle below represents the Born-Haber cycle for magnesium chloride.

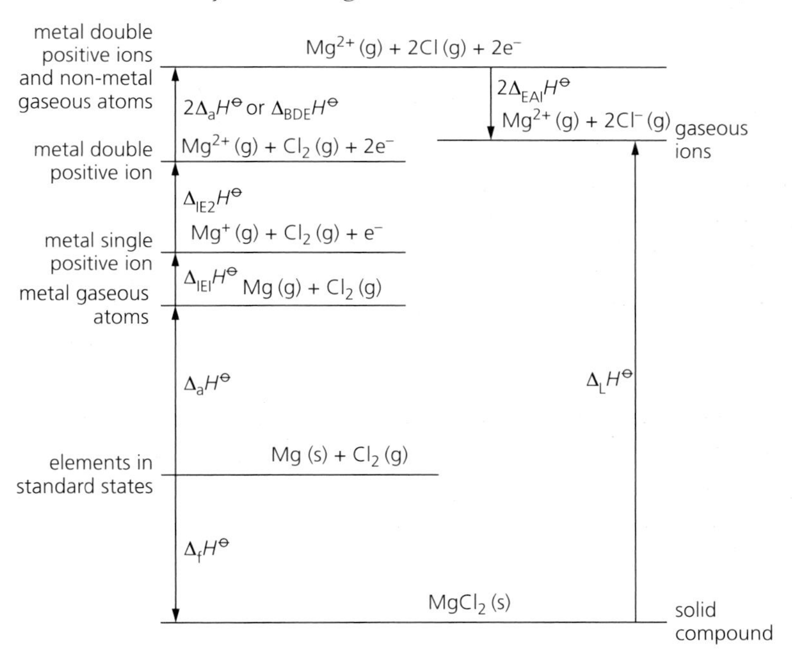

For all these diagrams the enthalpy changes are not to scale. The labels at the left and right of the different enthalpy levels show the species which should be present on these levels. These labels are there only to help you understand the cycle but are not essential if you are asked to draw one.

Metals such as Group 2 metals and many transition metals will have the metal in the oxidation state of +2. This means that an additional enthalpy change is required, the second ionisation enthalpy of the metal ($\Delta_{IE2} H^{\ominus}$).

As the compounds require two moles of halide ions per mole of the metal ion, the enthalpy of atomisation for the halogen is multiplied by 2 ($2\Delta_{at} H^{\ominus}$)

or the bond dissociation enthalpy is used ($\Delta_{BDE}H^{\ominus}$) and the first electron affinity is multiplied by 2 ($2\Delta_{EA1}H^{\ominus}$) as seen on the cycle.

Calculations using this type of cycle

The calculations using this Born-Haber cycle are similar to the calculations with the previous one.

Name of enthalpy change	$\Delta H^{\ominus}$/kJ mol^{-1}
Enthalpy of formation of magnesium chloride ($\Delta_f H^{\ominus}$)	−642
Bond dissociation enthalpy of chlorine ($\Delta_{BDE}H^{\ominus}$)	+242
Electron affinity of chlorine ($\Delta_{EA1}H^{\ominus}$)	−364
First ionisation enthalpy of magnesium ($\Delta_{IE1}H^{\ominus}$)	+736
Second ionisation enthalpy of magnesium ($\Delta_{IE2}H^{\ominus}$)	+1450
Enthalpy of atomisation of magnesium ($\Delta_a H^{\ominus}$(K))	+150

$$\Delta_L H^{\ominus} = -\underset{(MgCl_2)}{\Delta_f H^{\ominus}} + \underset{(Mg)}{\Delta_a H^{\ominus}} + \underset{(Mg)}{\Delta_{IE1} H^{\ominus}} + \underset{(Mg)}{\Delta_{IE2} H^{\ominus}} + \underset{(Cl_2)}{\Delta_{BDE} H^{\ominus}} + \underset{(Cl)}{2\Delta_{EA1} H^{\ominus}}$$

$$= +642 + 150 + 736 + 1450 + 242 - 2(364) = +2492\ \text{kJ mol}^{-1}$$

Enthalpy of lattice formation

The Born-Haber cycle for magnesium chloride where enthalpy of lattice formation is used is exactly the same, except the arrow for lattice enthalpy goes down as it is an exothermic process.

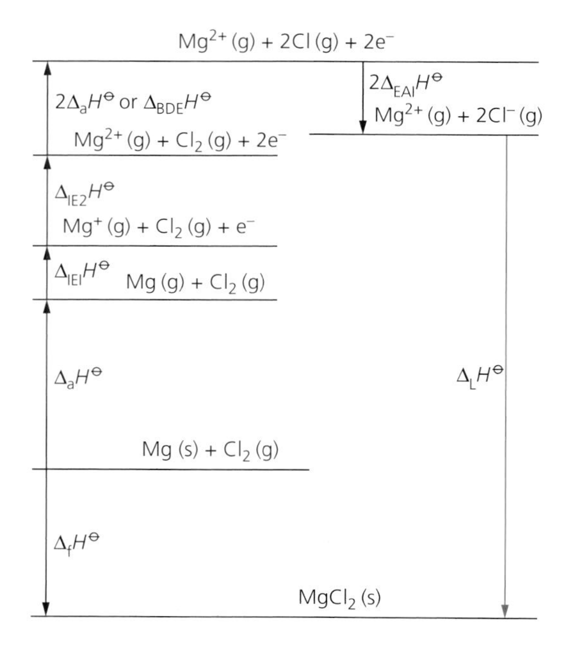

The calculation of the lattice enthalpy can be carried out as before and the sign changed to – at the end to show the exothermic enthalpy of lattice formation.

Alternatively the following expression may be used:

$$\underset{(MgCl_2)}{\Delta_f H^\ominus} = \underset{(Mg)}{\Delta_a H^\ominus} + \underset{(Mg)}{\Delta_{IE1} H^\ominus} + \underset{(Mg)}{\Delta_{IE2} H^\ominus} + \underset{(Cl_2)}{\Delta_{BDE} H^\ominus} + \underset{(Cl)}{2\Delta_{EA1} H^\ominus} + \underset{(MgCl_2)}{\Delta_L H^\ominus}$$

$$-642 = +150 + 736 + 1450 + 242 + 2(-364) + \Delta H^\ominus_L$$

$$\Delta_L H^\ominus = -642 - 150 - 736 - 1450 - 242 + 728$$

$$\Delta_L H^\ominus = -2492\ \text{kJ mol}^{-1}$$

TIP
The use of enthalpy of lattice formation means that the enthalpy of formation is simply a total of all the other changes in the cycle. This can be done for previous examples as well but beware if enthalpy of lattice dissociation is to be calculated.

This is using the cycle in a slightly different way. The expression may use $2\Delta_a H^\ominus$ in place of $\Delta_{BDE} H^\ominus$ for chlorine but $\Delta_{BDE} H^\ominus = 2\Delta_a H^\ominus$ for diatomic molecules.

Any unknown in the expression may be calculation from given data.

Born-Haber cycle for magnesium oxide

When drawing a Born-Haber cycle for the oxide of a metal with oxidation state +2, there is an additional enthalpy change. The second electron affinity of oxygen is required.

- The first electron affinity ($\Delta_{EA1} H^\ominus$) of oxygen is represented by the equation:

  $O(g) + e^- \longrightarrow O^-(g)$

- This is an exothermic process due to the net attraction between the nucleus of the oxygen atom and the incoming electron.
- The second electron affinity ($\Delta_{EA2} H^\ominus$) of oxygen is represented by the equation:

  $O^-(g) + e^- \longrightarrow O^{2-}(g)$

- The second electron affinity of oxygen is endothermic because of the repulsion between the negatively charged $O^-(g)$ ion and the incoming negative electron.

The values required to calculate the enthalpy of lattice dissociation for magnesium oxide are given below.

Name of enthalpy change	$\Delta H^\ominus$/kJ mol^{-1}
Enthalpy of formation of magnesium oxide ($\Delta_f H^\ominus$)	−602
Enthalpy of atomisation of oxygen ($\Delta_a H^\ominus$)	+248
First electron affinity of oxygen ($\Delta_{EA1} H^\ominus$)	−142
Second electron affinity of oxygen ($\Delta_{EA2} H^\ominus$)	+844
First ionisation enthalpy of magnesium ($\Delta_{IE1} H^\ominus$)	+736
Second ionisation enthalpy of magnesium ($\Delta_{IE2} H^\ominus$)	+1450
Enthalpy of atomisation of magnesium ($\Delta_a H^\ominus$(K))	+150

Figure 1.2 A refractory material is one that is physically and chemically stable at high temperatures and is used to line furnaces. Magnesium oxide is a commonly used refractory material due to its high lattice enthalpy.

When drawing the Born-Haber cycle for magnesium oxide, it is important to realise that the second electron affinity of oxygen is endothermic.

The Born-Haber cycle below is for magnesium oxide. The values for the enthalpy changes have been added to the cycle.

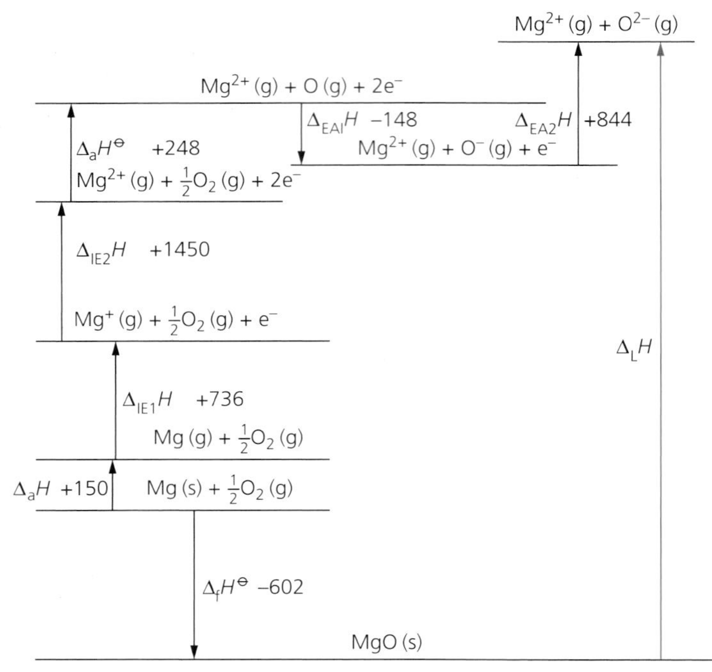

The cycle is different from previous ones as it includes the endothermic second electron affinity of oxygen so there is a level above $Mg^{2+}(g) + O^-(g) + e^-$. Also as the second electron affinity is +844 and the first electron affinity is −142, the level with $Mg^{2+}(g) + O^{2-}(g)$ must be above the level for $Mg^{2+}(g) + O(g) + 2e^-$.

The calculation of the enthalpy of lattice dissociation of magnesium oxide is:

$$\Delta_LH^\ominus = -\underset{(MgO)}{\Delta_fH^\ominus} + \underset{(Mg)}{\Delta_aH^\ominus} + \underset{(Mg)}{\Delta_{IE1}H^\ominus} + \underset{(Mg)}{\Delta_{IE2}H^\ominus} + \underset{(O)}{\Delta_aH^\ominus} + \underset{(O)}{\Delta_{EA1}H^\ominus} + \underset{(O)}{\Delta_{EA2}H^\ominus}$$

$$= +602 + 150 + 736 + 1450 + 248 - 142 + 844$$

$$= +3888\ \text{kJ mol}^{-1}$$

TIP

Remember that the enthalpy of lattice formation is simply the negative value of the enthalpy of lattice dissociation. The enthalpy of lattice formation of magnesium oxide is −3888 kJ mol^{-1}.

Understanding enthalpy changes in a Born-Haber cycle

Enthalpy change	Exothermic (ΔH negative)	Endothermic (ΔH positive)
Enthalpy of lattice dissociation		✓
Enthalpy of formation	✓ (mostly)	
Enthalpy of atomisation		✓
Bond dissociation enthalpy		✓
First ionisation energy		✓
Second ionisation energy		✓
First electron affinity	✓	
Second electron affinity		✓

The enthalpy of lattice dissociation is endothermic whereas the enthalpy of lattice formation would be exothermic and have the same numerical value but be a negative enthalpy change.

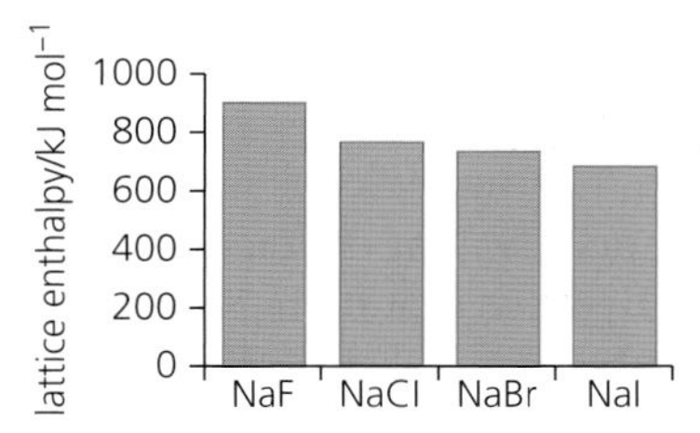

Figure 1.3 The bar chart shows the effect of ion size on lattice enthalpy of sodium halides as Group 7 is descended. The negative ions get bigger and so the distance between the centres of the oppositely charged ions is greater and the attraction less, resulting in a decrease in lattice enthalpy.

Comparison of lattice enthalpy values

The value of the lattice enthalpy gives a measure of the strength of the ionic bond and this depends on the charge on the ions and the size of the ions. The greater the value of the lattice enthalpy, the stronger the ionic bond. Smaller ions and higher charge ions give stronger ionic bonding.

- Smaller ions are more closely packed in the lattice and so are more attracted to each other.
- Ions with a high charge are more attracted to each other as well.
- For example, the lattice enthalpy for NaCl is $+771\,\text{kJ mol}^{-1}$ but that for NaF is $+915\,\text{kJ mol}^{-1}$.
- The ions have the same + and – charges but the F^- ion is smaller than the Cl^- ion so the ions in NaF can pack closer together in the lattice and so lattice enthalpy is greater as the attraction between the ions is greater in NaF compared to NaCl.
- The lattice enthalpy for $MgCl_2$ is $+2492\,\text{kJ mol}^{-1}$ but that for MgO is $+3888\,\text{kJ mol}^{-1}$.
- The lattice enthalpy for $MgCl_2$ is greater than that for MgO due to the 2⁻ charge on the O^{2-} ion compared to the – charge on the Cl^- ion. The O^{2-} ion is also smaller than the Cl^- ion so there is a stronger attraction between Mg^{2+} and O^{2-} ions than between the Mg^{2+} and Cl^- ions.
- Note the lattice enthalpy for MgO is very high (and hence it is very stable and has a very high melting point) due to the 2+ and 2– charge on the small ions.

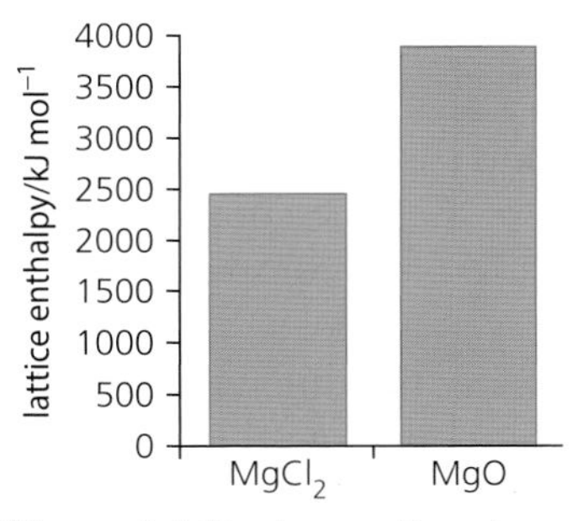

Figure 1.4 The larger the charge on the ion, the stronger the electrostatic force of attraction between the ions which results in high lattice enthalpy.

The balance of enthalpy values in a Born-Haber cycle may be examined. You may be asked to comment on the difference in lattice enthalpy values based on other enthalpy values in the cycle. For the difference between NaCl and NaF, it is the halogen bond enthalpy values that are different as well as the enthalpy of formation. The enthalpy of atomisation of sodium and first ionisation energy of sodium are the same in both cycles.

TEST YOURSELF 3

1 The enthalpy changes below can be used to calculate the enthalpy of lattice dissociation of rubidium chloride.

Step	Process	$\Delta H^\ominus$/kJ mol⁻¹
1	$Rb(s) + \frac{1}{2}Cl_2(g) \rightarrow RbCl(s)$	–435
2	$Rb(s) \rightarrow Rb(g)$	+81
3	$Rb(g) \rightarrow Rb^+(g) + e^-$	+403
4	$\frac{1}{2}Cl_2(g) \rightarrow Cl(g)$	+121
5	$Cl(g) + e^- \rightarrow Cl^-(g)$	–364

a) Name the enthalpy change represented by step 1.

b) Calculate the enthalpy of lattice dissociation of rubidium chloride.

2 The enthalpy of lattice dissociation of potassium chloride is $+715\,\text{kJ mol}^{-1}$.

a) Explain the difference between the enthalpy of lattice dissociation of rubidium chloride compared to the enthalpy of lattice dissociation of potassium chloride.

b) The enthalpy of formation of potassium chloride is $-437\,\text{kJ mol}^{-1}$ and the enthalpy of atomisation of potassium is $+90\,\text{kJ mol}^{-1}$. Calculate the first ionisation enthalpy of potassium from this data and the data given in the table in question 1.

3 The bond dissociation enthalpy of fluorine is $+158\,\text{kJ mol}^{-1}$.

a) Write an equation, including state symbols, representing the bond dissociation of fluorine.

b) What is meant by the enthalpy of atomisation of fluorine.

c) The first electron affinity of fluorine is $-348\,\text{kJ mol}^{-1}$. Calculate the enthalpy change for the process $\frac{1}{2}F_2(g) + e^- \rightarrow F^-(g)$.

Perfect ionic model

The distinction between ionic and covalent compounds is not black and white.

- Some covalent compounds or molecules are 100% covalent as they are non-polar.
- Covalent compounds can have some ionic character for example polar covalent compounds have some ionic character, i.e. the molecules are polar.
- Ionic compounds can be perfectly ionic with no covalent character.
- A small, large charge cation can distort a large anion so that there is covalent character to the 'ionic compound'. This means that ions are distorted and not spherical as would be assumed in the perfect ionic model.

It is a sliding scale from covalent to ionic. The diagram on the left shows the scale between 100% covalent and 100% ionic.

100% covalent (non-polar)

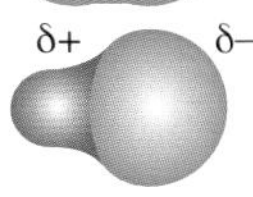

Polar covalent (covalent with ionic character)

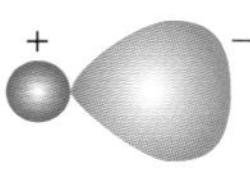

Ionic with covalent character (ions are not spherical)

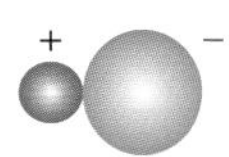

100% ionic (perfectly spherical ions)

The value of lattice enthalpy calculated using a Born-Haber cycle is the real value (i.e. that found by experiment).

The perfect ionic model of ionic compounds allows the calculation of the lattice enthalpy based on two assumptions:

1 the bonding in the compound is 100% ionic.

2 the ions are regarded as point charges or perfect spheres and are not distorted.

If the values of the experimental lattice enthalpy and the lattice enthalpy calculated from the perfect ionic model are different then the compound has some covalent character. The forces of attraction in the compound are found to be greater than predicted by the perfect ionic model so there is covalent character to the ionic compound. The greater the difference in the values, the greater the covalent character in the compound.

TIP

Make sure that you fully understand the perfect ionic model. It is the last two categories which are of interest in this section as 100% ionic compounds will have an experimental lattice enthalpy (calculated from a Born-Haber cycle) roughly equal to the lattice enthalpy calculated from the perfect ionic model.

EXAMPLE 17

The following lattice dissociation enthalpy values were determined.

	NaBr(s)	AgBr(s)
$\Delta_L H^\ominus$(experimental)/kJ mol^{-1}	+732	+891
$\Delta_L H^\ominus$(theoretical)/kJ mol^{-1}	+731	+760

1 State what you would deduce about the bonding in NaBr using the data in the table.
2 State what you would deduce about the bonding in AgBr using that data in the table.

If the values for the experimental (determined from a Born-Haber cycle) and the theoretical (determined from calculations using the perfect ionic model) are almost the same, the bonding in the compound is 100% ionic.

If the experimental value is greater than the experimental value, there is additional covalent bonding in the lattice which means that the ionic compound has some covalent character.

Answers

1 NaBr is ionic as the values are almost identical.
2 AgBr is ionic with some covalent character due to the additional strength of the covalent bonding.

EXAMPLE 18

The experimental lattice enthalpy of magnesium iodide is –2327 kJ mol^{-1}. The theoretical value calculated using the perfect ionic model is –1944 kJ mol^{-1}.

Explain the difference in the values.

Answer

The experimental value is greater than the theoretical value.

Magnesium iodide has covalent character.

There is additional covalent bonding in magnesium iodide.

Enthalpy of solution

Ionic compounds dissolve in water when the ionic lattice breaks up (enthalpy of lattice dissociation) and the polar water molecules form bonds with the ions (enthalpy of hydration).

Enthalpy of solution = Enthalpy of lattice dissociation + Enthalpy of hydration

For sodium chloride:

$$\Delta_{sol}H^{\ominus}_{(NaCl)} = \Delta_{L}H^{\ominus}_{(NaCl)} + \Delta_{hyd}H^{\ominus}_{(Na^+)} + \Delta_{hyd}H^{\ominus}_{(Cl^-)}$$

The balance of the break-up of the ionic lattice and the bonds forming with water determines the enthalpy of solution.

The **enthalpy of solution** is defined as the enthalpy change when one mole of a solute dissolves in water. It can be determined from other enthalpy values. For example:

$$NaCl(s) \longrightarrow Na^+(aq) + Cl^-(aq)$$

The enthalpy of lattice dissociation for sodium chloride is represented by the equation

$$NaCl(s) \longrightarrow Na^+(g) + Cl^-(g) \quad \Delta_L H^{\ominus} = +776\,kJ\,mol^{-1}$$

The enthalpy of hydration is defined as the enthalpy change when one mole of gaseous ions is converted into one mole of aqueous ions.

$$Na^+(g) \longrightarrow Na^+(aq) \quad \Delta_{hyd}H^{\ominus} = -407\,kJ\,mol^{-1}$$

$$Cl^-(g) \longrightarrow Cl^-(aq) \quad \Delta_{hyd}H^{\ominus} = -364\,kJ\,mol^{-1}$$

TIP

Some texts may use '+(aq)' when writing the equation for solution of a substance to represent the addition of water on the left-hand side of the equation. This is acceptable but do not write '+ H_2O'. This also applies to equations for hydration of ions. It is best to keep the equations as shown in the example for NaCl.

These enthalpy changes fit together in the following cycle.

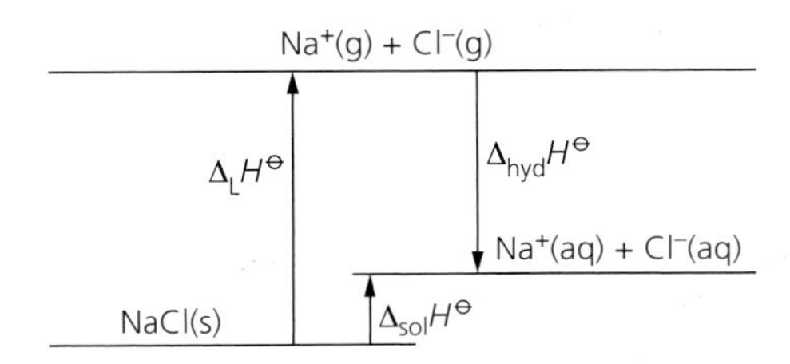

$$\Delta_{sol}H^{\ominus} = (+776) + (-407) + (-364) = +5\,kJ\,mol^{-1}$$

5 kJ of energy are released when 1 mole of NaCl dissolves in water to form a solution. The enthalpy of solution may be exothermic depending on the balance between the lattice enthalpy and the enthalpies of hydration.

It is not suggested that when NaCl dissolves in water, the ionic lattice breaks up into scattered gaseous ions which then dissolve. However, it allows an alternative route for calculation enthalpy of solution values which according to Hess's Law should give the same overall enthalpy change as the process the ions really undergo.

A simpler diagram for calculation of solution may be like the diagram below.

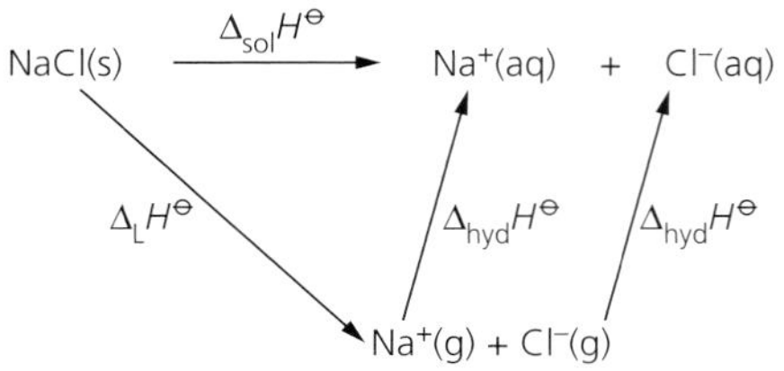

$NaCl(s) \xrightarrow{\Delta_{sol}H^{\ominus}} Na^{+}(aq) + Cl^{-}(aq)$

$\Delta_{L}H^{\ominus}$: $NaCl(s) \rightarrow Na^{+}(g) + Cl^{-}(g)$

$\Delta_{hyd}H^{\ominus}$: $Na^{+}(g) \rightarrow Na^{+}(aq)$

$\Delta_{hyd}H^{\ominus}$: $Cl^{-}(g) \rightarrow Cl^{-}(aq)$

Be careful with enthalpy of solution for Group 2 halides, as 2× enthalpy of hydration for the halide ion is required as there are two moles of halide ion in one mole of the ionic compound.

When ionic compounds dissolve in water the polar water molecules are attracted to the charged ions. The δ– O atoms in H_2O molecules are attracted to the positive ions and the δ+ H atoms in H_2O molecules are attracted to the negative ions.

O δ– Na^{+} δ– O

δ+ H Cl^{-}

TIP

Remember to show the polarity of the water molecules and the charge on the ions if you are asked for this type of diagram. The diagram shows the interaction of 4 water molecules with the ions. There would normally be 6 for each ion. The ones above and below the plane of the paper are not shown so that we can see the ions.

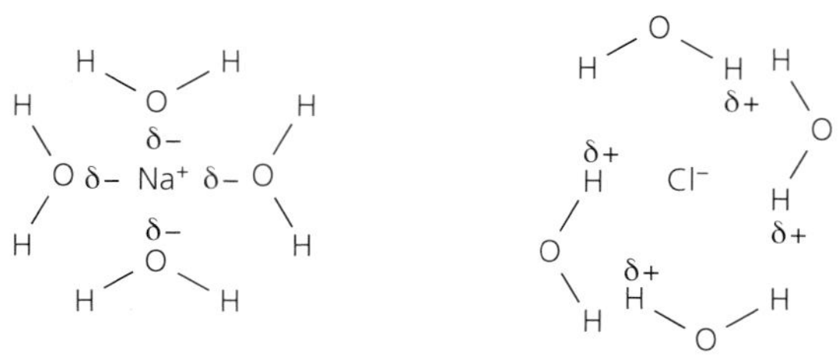

Values for enthalpy of hydration

Enthalpy of hydration values are exothermic as energy is released when the ions are attracted to the polar water molecules.

The enthalpy of hydration for fluoride ions is $-506\,kJ\,mol^{-1}$ which is more negative than the enthalpy of hydration for chloride ions ($-364\,kJ\,mol^{-1}$).

This is because a smaller ion with the same charge has a higher charge density so the negative charge on the ion attracts the δ+ H of the water more strongly.

The same applies to cations: The enthalpy of hydration of the Mg^{2+} ion is $-1920\,kJ\,mol^{-1}$ whereas the enthalpy of hydration of Ca^{2+} ions is $-1650\,kJ\,mol^{-1}$. The Mg^{2+} ion is smaller and has a higher charge density than the Ca^{2+} ion so the ions attracts the δ– O of the water more strongly.

Halides of metals with oxidation state +2

For halides of Group 2 and other metals with an oxidation state of +2, 2× the enthalpy of hydration of the halide ion is required.

For calcium chloride:

Name of enthalpy change	$\Delta H^\ominus$/kJ mol^{-1}
Enthalpy of solution of calcium chloride	−141
Enthalpy of lattice dissociation of calcium chloride	+2237
Enthalpy of hydration of calcium ions	−1650

$$\Delta_{sol}H^\ominus = \Delta_L H^\ominus + \Delta_{hyd}H^\ominus(Ca^{2+}) + 2\Delta_{hyd}H^\ominus(Cl^-)$$

$$-141 = +2237 + (-1650) + 2\Delta_{hyd}H^\ominus(Cl^-)$$

$$-141 - 2237 + 1650 = 2\Delta_{hyd}H^\ominus(Cl^-)$$

$$2\Delta_{hyd}H^\ominus(Cl^-) = -728$$

$$\Delta_{hyd}H^\ominus(Cl^-) = \frac{-728}{2} = -364\,\text{kJ mol}^{-1}$$

The enthalpy of hydration of the chloride ions is −364 kJ mol^{-1}.

The enthalpy of solution of calcium chloride is −141 kJ mol^{-1}.

The process $CaCl_2(s) \rightleftharpoons Ca^{2+}(aq) + 2Cl^-(aq)$ is exothermic. When treating this process as a reversible reaction, increasing the temperature would move the position of equilibrium in the direction of the reverse endothermic process. This would theoretically decrease the amount of calcium chloride which would dissolve as the position of equilibrium would be moved to the left.

TIP

The dissolution of calcium chloride also depends on entropy considerations so this information may be used to predict how the amount of calcium chloride which dissolves changes as temperature changes but it is not the whole picture.

ACTIVITY

Experimental determination of the enthalpy of solution of sodium hydroxide and ammonium nitrate

100 cm^3 of deionised water was measured into a polystyrene cup and the temperature recorded. A mass of solid was quickly dissolved in the water, stirred with the thermometer and the final temperature recorded. The results are shown in the table below.

Substance	Mass/g	Initial temperature/°C	Final temperature/°C	Temperature change/°C
Sodium hydroxide	3.99	24.9	30.0	
Ammonium nitrate	2.02	25.0	23.6	

1 Describe how approximately 2 g of ammonium nitrate was weighed out accurately in this experiment.

2 Explain any safety precautions which should be taken when using solid sodium hydroxide pellets.

3 Both of the substances used in this experiment consist of ions. Give the chemical formulae of the two ions that make up each of the compounds.
4 For each of the solutes, calculate the change in temperature of the water.
5 Calculate the mass of water in the polystyrene cup. The density of water at 25 °C is 0.998 g/cm^3.
6 Calculate the enthalpy change that occurred in each solution. The specific heat capacity of water is 4.18 J K^{-1} g^{-1}.
7 Calculate the number of moles of each solute and use this to calculate the molar enthalpy of solution for each solute.
8 Data books give the enthalpy of solution of NaOH as −44.20 kJ mol^{-1} and for NH_4NO_3, it is +25.40 kJ mol^{-1}. Calculate the percentage error, in this experiment for each solute.
9 State one source of error in this experiment and describe how this source of error can be minimised.
10 If excess solute was used in this experiment, a saturated solution would be formed, with the excess solute in equilibrium with the solution.

$NH_4NO_3(s) + aq \rightleftharpoons NH_4^+(aq) + NO_3^-(aq)$

By applying Le Chatelier's principle to this equilibrium determine if the solubility of ammonium nitrate will increase or decrease when the temperature is raised.

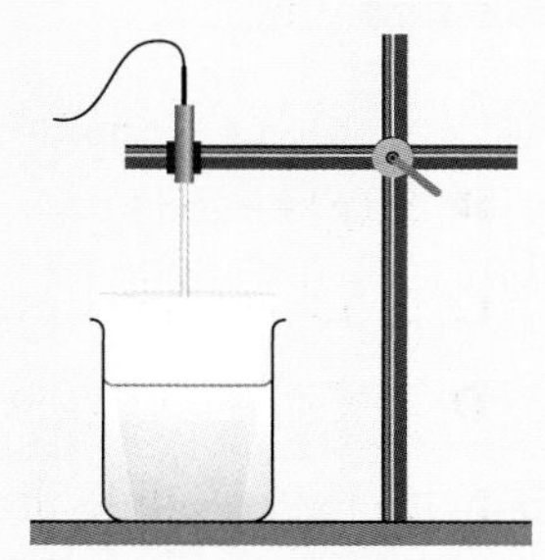

Figure 1.8 A temperature probe may be used to measure temperature changes accurately.

TEST YOURSELF 4

1 The experimental enthalpy of lattice dissociation of silver chloride is +905 kJ mol^{-1}. The theoretical value of the enthalpy of lattice dissociation of silver chloride is +770 kJ mol^{-1}. Explain the difference in these values.
2 The table below gives some enthalpy changes. Calculate the enthalpy of solution of potassium chloride.

Name of enthalpy change	$\Delta H^{\ominus}$/kJ mol^{-1}
Enthalpy of lattice dissociation of potassium chloride	+715
Enthalpy of hydration of potassium ions	−320
Enthalpy of hydration of chloride ions	−364

3 The enthalpy of hydration of fluoride ions is −506 kJ mol^{-1}. Explain why the enthalpy of hydration of fluoride ions is more negative than the enthalpy of hydration of chloride ions given in question 2.

Practice questions

1 Which one of the following enthalpy changes is exothermic?

A enthalpy of atomisation of magnesium

B enthalpy of bond dissociation of chlorine

C first electron affinity of chlorine

D enthalpy of lattice dissociation of magnesium chloride. *(1)*

2 Which one of the following equations represents the second ionisation energy of strontium?

A $Sr(s) \rightarrow S^{2+}(g) + 2e^-$

B $Sr(g) \rightarrow Sr^{2+}(g) + 2e^-$

C $Sr^+(s) \rightarrow Sr^{2+}(g) + e^-$

D $Sr^+(g) \rightarrow Sr^{2+}(g) + e^-$ *(1)*

3 Nitrogen triiodide decomposes on contact according to the equation. The reaction is exothermic.

$$2NI_3(g) \rightarrow N_2(g) + 3I_2(g)$$

Calculate the enthalpy change for the reaction using the mean bond enthalpy values given in the table below. *(3)*

Bond	Mean bond enthalpy ($kJ\,mol^{-1}$)
N–I	159
N≡N	944
I–I	151

4 Lithium fluoride is formed from lithium and fluorine. The following data may be used to calculate the enthalpy of lattice formation of lithium fluoride.

$\Delta H^{\ominus}/kJ\,mol^{-1}$

First ionisation enthalpy of lithium +520

Enthalpy of atomisation of lithium +159

Enthalpy of formation of lithium fluoride –612

Enthalpy of atomisation of fluorine +79

Electron affinity of fluorine atoms –348

a) Write equations, including state symbols, to represent the following enthalpy changes.

i) The first ionisation energy of lithium *(1)*

ii) The enthalpy of formation of lithium fluoride *(1)*

iii) The first electron affinity of fluorine *(1)*

iv) The enthalpy of lattice formation of lithium fluoride. *(1)*

b) Calculate the enthalpy of lattice formation of lithium fluoride. *(2)*

5 The Born-Haber cycle below is for sodium chloride.

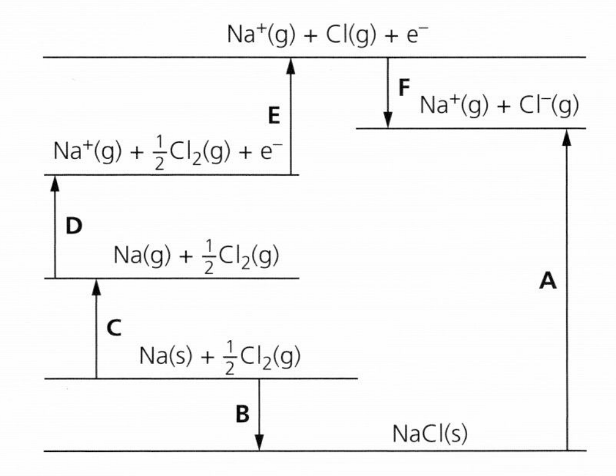

a) The table below shows the enthalpy changes A to F. Complete the table by giving the names of the missing enthalpy changes. *(4)*

Step	Name of enthalpy change	$\Delta H^{\ominus}/kJ\,mol^{-1}$
A	Enthalpy of lattice dissociation of sodium chloride	to be calculated
B		–411
C		+109
D		+494
E	Enthalpy of atomisation of chlorine	+121
F		–364

b) Calculate a value for enthalpy change A using the data in the table. *(2)*

c) The enthalpy of lattice dissociation for potassium chloride is $+710\,kJ\,mol^{-1}$. Explain why the value for sodium chloride calculated in (b) is greater than the value for potassium chloride. *(2)*

6 The Born-Haber cycle shown below is for magnesium chloride. The table gives the values for the enthalpy changes involved in the cycle below.

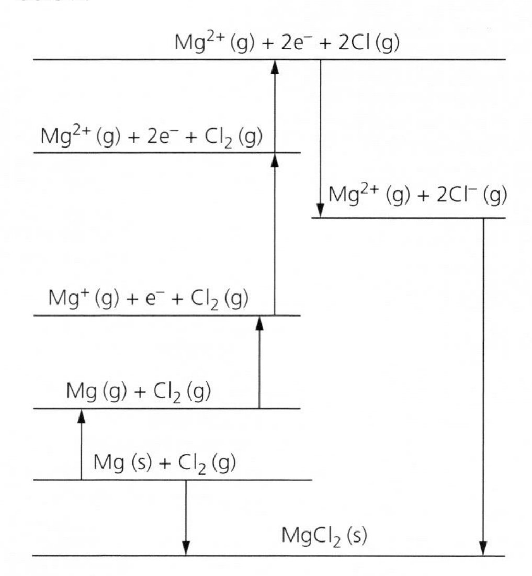

Name of enthalpy change	$\Delta H^{\ominus}/kJ\,mol^{-1}$
Bond dissociation enthalpy of chlorine	+242
Electron affinity of chlorine	−364
First ionisation enthalpy of magnesium	+736
Second ionisation enthalpy of magnesium	+1450
Enthalpy of atomisation of magnesium	+150
Enthalpy of lattice formation of magnesium chloride	−2492

a) Copy out the cycle and put the correct values on the arrows to show the enthalpy change represented by the arrows in the Born-Haber cycle. *(4)*

b) Use the information in the table and the Born-Haber cycle to calculate the enthalpy of formation of magnesium chloride. *(3)*

7 The following data gives the enthalpy of hydration of some ions.

Ion	Enthalpy of hydration ($kJ\,mol^{-1}$)
Chloride	to be calculated
Bromide	−335
Iodide	−293
Lithium	−519
Sodium	−406
Potassium	−320
Iron(II)	−1950

a) Explain why the enthalpy of hydration of the bromide ion is more exothermic than the enthalpy of hydration of an iodide ion. *(2)*

b) The enthalpy of solution for potassium iodide is $+20.5\,kJ\,mol^{-1}$. Calculate the enthalpy of lattice formation for potassium iodide. *(3)*

c) The enthalpy of lattice dissociation of iron(II) chloride calculated from a Born-Haber cycle is $+2631\,kJ\,mol^{-1}$ and the enthalpy of solution of iron(II) chloride is $-47\,kJ\,mol^{-1}$ Calculate the enthalpy of hydration of a chloride ion. *(3)*

d) The theoretical value for the enthalpy of lattice dissociation for iron(II) chloride calculated from the perfect ionic model is $+2525\,kJ\,mol^{-1}$.

i) Explain the meaning of perfect ionic model. *(1)*

ii) Explain why the enthalpy of lattice formation of iron(II) chloride calculated from the Born-Haber cycle, given in (c), is greater than the theoretical value calculated from the perfect ionic model. *(2)*

Stretch and challenge

8 The diagram below gives part of the Born-Haber cycle for aluminium oxide. Examine the data given in the table and complete the cycle.

Calculate the second electron affinity of oxygen based on the data given below.

Compare the value you obtain for the second electron affinity for the given value for the first electron affinity. (If you have not been able to obtain a value for the second electron affinity, use +701 kJ mol^{-1}. This is not the correct value). *(10)*

Name of enthalpy change	$\Delta H^{\ominus}$/kJ mol^{-1}
Enthalpy of atomisation of oxygen	+249.2
First electron affinity of oxygen	−141.0
First ionisation enthalpy of aluminium	+580
Second ionisation enthalpy of aluminium	+1800
Third ionisation enthalpy of aluminium	+2700
Enthalpy of atomisation of aluminium	+324.3
Enthalpy of lattice formation of aluminium oxide	−15 178.9
Enthalpy of formation of aluminium oxide	−1675.7

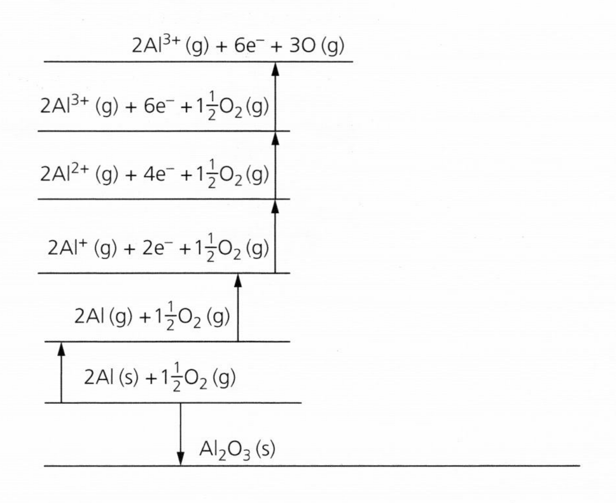

2 Thermodynamics: Gibbs free energy change, ΔG, and entropy change, ΔS

PRIOR KNOWLEDGE

It is expected that you are familiar with all of the content of Chapter 4 **Energetics** in the Year 1 book of the A-level Chemistry course. The following are some key points of Prior Knowledge:

- A positive value for ΔH indicates an endothermic reaction.
- A negative value for ΔH indicates an exothermic reaction.
- Enthalpy change for a reaction may be calculated using standard enthalpy changes of combustion or formation.
- Hess's Law states that the enthalpy change for a chemical reaction is independent of the route taken and depends only on the initial and final states.
- Enthalpy changes may be calculated from other enthalpy changes using Hess's Law.

TEST YOURSELF ON PRIOR KNOWLEDGE 1

1 Calcium carbonate decomposes on heating to form calcium oxide and carbon dioxide. Using the standard enthalpies of formation, calculate the enthalpy change for this reaction.

$CaCO_3(s) \rightarrow CaO(s) + CO_2(g)$

	$\Delta_f H^\ominus/kJ\,mol^{-1}$
$CaCO_3(s)$	−1207
$CaO(s)$	−636
$CO_2(g)$	−394

2 Calculate the standard enthalpy of formation of ammonia, where the enthalpy change for the reaction $N_2(g) + 3H_2(g) \rightarrow 2NH_3(g)$ is $-92\,kJ\,mol^{-1}$.

3 Ammonia reacts with oxygen to form nitrogen monoxide and steam according to the equation.

$4NH_3(g) + 5O_2(g) \rightarrow 4NO(g) + 6H_2O(g)$

a) Using the standard enthalpies of formation, calculate the enthalpy change for this reaction.

	$\Delta_f H^\ominus/kJ\,mol^{-1}$
$NH_3(g)$	−46
$O_2(g)$	0
$NO(g)$	+ 90
$H_2O(g)$	−242

b) Explain why oxygen has a value of zero.

Feasible reactions

A reaction that can occur at a particular temperature is called a feasible reaction. The enthalpy change on its own is not enough to determine whether or not a reaction will be feasible as some feasible reactions are endothermic. These include:

$KCl(s) + (aq) \rightarrow K^+(aq) + Cl^-(aq)$ $\Delta H^\ominus = +19\ kJ\,mol^{-1}$

$H_2O(s) \rightarrow H_2O(l)$ $\Delta H^\ominus = +6\ kJ\,mol^{-1}$

$H_2O(l) \rightarrow H_2O(g)$ $\Delta H^\ominus = +44\ kJ\,mol^{-1}$

$(NH_4)_2CO_3(s) \rightarrow 2NH_3(g) + CO_2(g) + H_2O(g)$ $\Delta H^\ominus = +68\ kJ\,mol^{-1}$

All four of these reactions above have one thing in common – an increase in disorder. This is the disorder of the particles in the system (reaction). Gases are more disordered than liquids and solution which are more disordered than solids.

All of them show a change from an ordered system to a more random system. The degree of disorder or randomness in a system is called its entropy.

Entropy

Entropy is a measure of disorder within a system.

Entropy is given the symbol S and **standard or absolute entropy** $S^\ominus$.

Entropy is measured in $J\,K^{-1}\,mol^{-1}$.

Figure 2.1 Tossing a soft drink into the air is more likely to give a disorganised pile when it falls, than a neatly ordered stack. Disorder is more probable than order.

Absolute entropy values (standard entropy values)

Absolute or standard entropy values are calculated for one mole of a substance based on a scale where the substance has an entropy value of 0 (zero) at 0 K. This assumes that the crystalline form of the substance at 0 K is a perfect crystal.

Some absolute entropy ($S^\ominus$) values are given in the table:

Solids	$S^\ominus/J\,K^{-1}\,mol^{-1}$	Liquids	$S^\ominus/J\,K^{-1}\,mol^{-1}$	Gases	$S^\ominus/J\,K^{-1}\,mol^{-1}$
C(s) (diamond)	2.4	$Br_2(l)$	75.8	$O_2(g)$	205
Si(s)	19.0	Hg(l)	76.1	$Cl_2(g)$	165
Mg(s)	32.7	$C_2H_5OH(l)$	161	$N_2(g)$	192
Al(s)	28.3	$H_2O_2(l)$	110	$H_2(g)$	131
NaCl(s)	72.4	$CS_2(l)$	151	$NH_3(g)$	193
CaO(s)	39.7	$NH_4NO_3(aq)$	190	$CO_2(g)$	214
$I_2(s)$	58.4	NaCl(aq)	116	CO(g)	198
$H_2O(s)$	48.0	$H_2O(l)$	70.0	$H_2O(g)$	189

The higher the absolute entropy value, the higher the degree of disorder of a substance.

From the table it is clear to see that solids are more ordered than liquids and solutions, which are more ordered than gases. This can be used as a general rule to explain changes in entropy later.

Entropy linked to changes in state

A crystalline solid with a highly ordered arrangement has low entropy. When it melts, the system becomes less ordered (it is impossible to tell exactly where each ion or atom or molecule is relative to each other) so the entropy has increased.

In Year 1 of the course (AS) the four types of crystalline solids studied are:

1 metallic
2 ionic
3 molecular
4 macromolecular

From the previous table, examples of these were given.

Type of crystalline solid	Solids	$S^\ominus$/J K^{-1} mol^{-1}
Macromolecular	C(s) (diamond)	2.4
Macromolecular	Si(s)	19.0
Metallic	Mg(s)	32.7
Metallic	Al(s)	28.3
Ionic	NaCl(s)	72.4
Ionic	CaO(s)	40
Molecular	I_2 (s)	58.4
Molecular	H_2O(s)	41

From the table it is clear that macromolecular substances have a very low entropy value and this property results from their very regular macromolecular structure.

Metallic substances also have low entropy values based on them having a very ordered metallic lattice, and similarly with ionic compounds and their ionic lattice. Even molecular crystals such as iodine and ice have low entropy values as they are regular ordered crystalline substances.

There is an increase in entropy from solid to liquid, but the increase in entropy from liquid to gas is much larger as gases are very disordered. This is shown most clearly by the $S^\ominus$ values of water as a solid, liquid and gas below.

Solid	$S^\ominus$/J K^{-1} mol^{-1}	Liquid	$S^\ominus$/J K^{-1} mol^{-1}	Gas	$S^\ominus$/J K^{-1} mol^{-1}
H_2O(s)	48	H_2O(l)	70	H_2O(g)	189

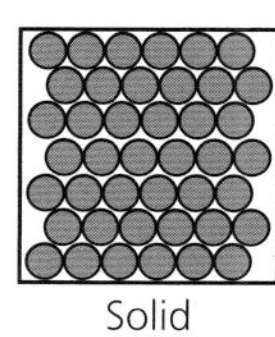

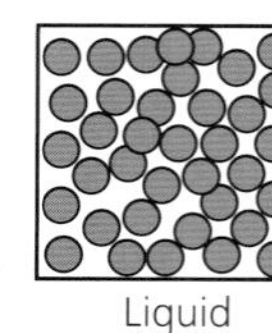

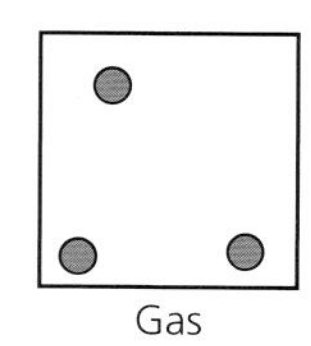

Figure 2.2 The changes in the arrangement of water molecules when ice melts, to form a liquid and then evaporates, show an increase in disorder.

A graph of the entropy values of water against temperature is shown below.

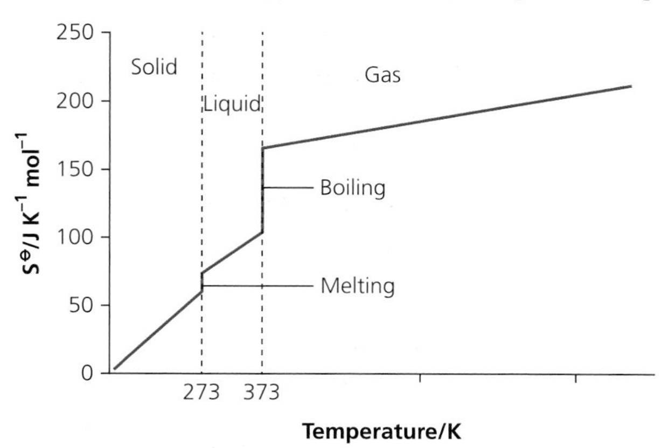

- At 0 K the entropy of a perfect crystalline substance is 0 (zero).
- The graph shows a general increase in entropy as temperature increases.
- Within a solid the entropy increases as temperature increases as the particles gain energy and vibrate more.
- As the solid melts and forms a liquid there is a large increase in disorder at a particular temperature. This temperature is called the melting point.
- Again within a liquid there is an increase in entropy (disorder) increases as temperature increases.
- There is a very large increase in entropy at the boiling point when a liquid changes into a gas as gases are very disordered.

Entropy related to M_r

Some absolute entropy values of gases are given below with their M_r values.

	$H_2(g)$	$N_2(g)$	$CO(g)$	$O_2(g)$	$CO_2(g)$
M_r	2	28	28	32	44
$S^\ominus$/J K^{-1} mol^{-1}	131	192	198	205	214

In general the higher the M_r value of substances in the same state, the higher the entropy value.

This is a general rule but can be used to predict the entropy values of substances in the same state.

Entropy change ($\Delta S^\ominus$)

The entropy change in a reaction can be calculated from absolute entropy values.

Absolute entropy values are quoted with units J K^{-1} mol^{-1} (joules per Kelvin per mole).

The entropy change in a reaction can be calculated from the absolute entropy values.

$\Delta S^\ominus$ = Sum of entropy values of products – Sum of entropy values of reactants

This can be summarised mathematically as:

$$\Delta S^\ominus = \Sigma S^\ominus_{\text{products}} - \Sigma S^\ominus_{\text{reactants}}$$

where Σ represents the 'sum of'.

The 'per mole' must be considered here as absolute entropy values are per mole of the substance.

If two moles of substance appear in the reaction, the absolute entropy value must be multiplied by two before being used in the expression.

EXAMPLE 1

Calcium carbonate decomposes on heating according to the equation:

$$CaCO_3(s) \rightarrow CaO(s) + CO_2(g)$$

Some entropy values are given below.

	$CaCO_3$ (s)	CaO(s)	CO_2(g)
$S^{\ominus}$/J K^{-1} mol^{-1}	93	40	214

Calculate a value for the standard entropy change for this reaction.

Answer

$$\Delta S^{\ominus} = \Sigma S^{\ominus}_{products} - \Sigma S^{\ominus}_{reactants}$$

$$\Delta S^{\ominus} = (214 + 40) - 93 = +161\ J\ K^{-1}\ mol^{-1}$$

It would be expected that this reaction would have a positive entropy change.

The decomposition of a solid to produce a solid and a gas would show an increase in entropy (disorder) as the gas would have a greater level of disorder than any solid.

1 mol of solid produces 1 mol of solid and 1 mol of gas.

TIP

In the next section you will be expected to judge whether a reaction has a positive or negative entropy change based on the reactants and products and their states.

EXAMPLE 2

The equation for the thermal decomposition of $NaHCO_3$ is:

$$2NaHCO_3(s) \rightarrow Na_2CO_3(s) + CO_2(g) + H_2O(l)$$

Absolute entropy values are given in the table below.

	$NaHCO_3$ (s)	Na_2CO_3 (s)	CO_2(g)	H_2O(l)
$S^{\ominus}$/J K^{-1} mol^{-1}	102	135	214	70

Calculate a value of $\Delta S^{\ominus}$ for this reaction.

Answer

$$\Delta S^{\ominus} = \Sigma S^{\ominus}_{products} - \Sigma S^{\ominus}_{reactants}$$

$$\Delta S^{\ominus} = (135 + 214 + 70) - (2 \times 102) = +215\ J\ K^{-1}\ mol^{-1}$$

This reaction shows an increase in entropy as system becomes more disordered. There are 2 mol of solid producing 1 mol of solid and 1 mol of liquid and 1 mol of gas. The liquid and gas are more disordered and so have a higher entropy value. Entropy increases in this reaction.

EXAMPLE 3

Nitrogen reacts with hydrogen to form ammonia according to the equation:

$N_2(g) + 3H_2(g) \rightarrow 2NH_3(g)$

	$N_2(g)$	$H_2(g)$	$NH_3(g)$
$S^\ominus$/J K^{-1} mol^{-1}	192	131	193

Calculate a value of the entropy change for this reaction.

Answer

$\Delta S^\ominus = \Sigma S^\ominus_{products} - \Sigma S^\ominus_{reactants}$

$\Delta S^\ominus = (2 \times 193) - (192 + 3 \times 131) = -199\ J\ K^{-1}\ mol^{-1}$

This reaction shows a decrease in entropy as it becomes more ordered, i.e. 4 mol of gas becomes 2 mol of gas. Fewer moles of gas are more ordered so the reaction shows a decrease in entropy.

TIP
The units of $S^\ominus$ (absolute entropy) and $\Delta S^\ominus$ (entropy change) are J K^{-1} mol^{-1}.

Entropy change predicted from an equation

From the equation of a reaction, it is often possible to tell if $\Delta S^\ominus$ is positive or negative.

A reaction is often called a system, so often we say that the entropy of the system has decreased or increased meaning that $\Delta S^\ominus$ is negative or positive, respectively.

EXAMPLE 4

Predict whether the entropy change of the reaction is positive or negative and explain your answer.

$NH_4NO_3(s) \rightarrow N_2O(g) + 2H_2O(g)$

Answer

1 mol of solid ammonium nitrate forms 1 mol of dinitrogen oxide gas and 2 mol of water vapour (gas).

This will have a positive $\Delta S^\ominus$ value as the gases are much more disordered than the crystalline solid (ammonium nitrate).

TIP
Under standard conditions the water formed in the equation above would be a liquid which is more ordered than a gas, so the $\Delta S^\ominus$ would be less positive.

EXAMPLE 5

Predict whether the entropy change of the reaction is positive or negative and explain your answer.

$CaO(s) + H_2O(l) \rightarrow Ca(OH)_2(s)$

Answer

$\Delta S^\ominus$ is negative as 1 mol of solid and 1 mol of liquid produce 1 mol of solid so the reaction becomes more ordered. Entropy decreases.

TIP
Only $\Delta S^\ominus$ values need to have a positive or negative sign. $S^\ominus$ values are always positive in the same way as $H^\ominus$ values are always positive. Values for $S^\ominus$ may be quoted to different numbers of decimal places or as whole number values.

TIP

When all the reactants and products are in the same state, the increase in entropy can be caused by an increase in the number of moles in the same state.

EXAMPLE 6

Predict whether the entropy change of the reaction is positive or negative and explain your answer.

$4NH_3(g) + 5O_2(g) \rightarrow 4NO(g) + 6H_2O(g)$

Answer

$\Delta S^{\ominus}$ is positive as 9 mol of gas produces 10 mol of gas so the reaction becomes more disordered. Entropy would be expected to increase.

Figure 2.3 Hydrated crystals of copper sulfate are shown above, with a copper sulfate solution made from the same mass of crystals. Which has the highest entropy?

TEST YOURSELF 2

1 Predict whether the entropy of the following reactions will increase or decrease?
  a) $2H_2(g) + O_2(g) \rightarrow 2H_2O(l)$
  b) $CaCO_3(s) + 2HCl(aq) \rightarrow CaCl_2(aq) + CO_2(g) + H_2O(l)$
  c) $2Na(s) + Cl_2(g) \rightarrow 2NaCl(s)$
  d) $CaO(s) + 2NH_4Cl(s) \rightarrow CaCl_2(s) + 2NH_3(g) + H_2O(l)$
2 Calculate the entropy change of the following reaction using the standard entropy data given below.

Substance	$S^{\ominus}$ value/J K^{-1} mol^{-1}
$KNO_3(s)$	133
$KNO_2(s)$	152
$O_2(g)$	205
$Fe(s)$	27
$Fe_3O_4(s)$	146
$(NH_4)_2SO_4(s)$	220
$Ca(OH)_2(s)$	83
$CaSO_4(s)$	107
$NH_3(g)$	193
$H_2O(l)$	70

  a) $2KNO_3(s) \rightarrow 2KNO_2(s) + O_2(g)$
  b) $3Fe(s) + 2O_2(g) \rightarrow Fe_3O_4(s)$
  c) $(NH_4)_2SO_4(s) + Ca(OH)_2(s) \rightarrow CaSO_4(s) + 2NH_3(g) + 2H_2O(l)$

Feasibility of a reaction

The feasibility of a chemical reaction depends on the enthalpy change of the reaction, the entropy change of the reaction and the temperature at which the reaction is occurring.

Gibbs free energy or the free energy change is written as $\Delta G^{\ominus}$ and this links these three properties.

$\Delta G^{\ominus}$ is calculated using the expression:

$$\Delta G^{\ominus} = \Delta H^{\ominus} - T\Delta S^{\ominus}$$

- $\Delta G^\ominus$ is used to predict the feasibility of a chemical reaction.
- For a reaction to be feasible, ΔG must be equal to or less than 0 (zero).
- You can determine if a reaction is feasible at a certain temperature by calculating $\Delta G^\ominus$.
- Also you can calculate the temperature at which a reaction becomes feasible using $\Delta G^\ominus$ and $\Delta H^\ominus$ assuming that $\Delta G^\ominus = 0$ for a feasible reaction.
- $\Delta G^\ominus$ and $\Delta H^\ominus$ have units of $kJ\,mol^{-1}$. $\Delta S^\ominus$ has units of $J\,K^{-1}\,mol^{-1}$.
- The value given or calculated for $\Delta S^\ominus$ in $J\,K^{-1}\,mol^{-1}$ must be divided by 1000 to change into $kJ\,K^{-1}\,mol^{-1}$ before it is used to calculate $\Delta G^\ominus$ or used in the $\Delta G^\ominus$ expression.

TIP
A reaction for which $\Delta G^\ominus = 0$ is in equilibrium so both the forward and reverse reactions are feasible. $\Delta G \leq 0$ is normally used for a reaction to be feasible.

TIP
This is the most often forgotten step in these calculations. $\Delta S^\ominus$ must be converted to $kJ\,K^{-1}\,mol^{-1}$ by dividing the given value by 1000 before using it in the $\Delta G^\ominus$ expression.

Feasible and spontaneous

There is a difference between a reaction being feasible and a reaction being spontaneous. A feasible reaction may not occur at a particular temperature as the activation energy may be too high for it to proceed. A spontaneous reaction is one that is feasible and occurs under standard conditions. For a reaction to be spontaneous, $\Delta G^\ominus$ must be less than zero and the activation energy for the reaction must be low enough for the reaction to proceed under standard conditions.

TIP
The terms feasible and spontaneous are often used interchangeably but it is important to realise that the activation energy has a part to play in the spontaneous nature of a reaction. If a reaction is feasible but does not occur spontaneously, the activation energy is high.

EXAMPLE 7

The reaction below shows the possible decomposition of sodium carbonate into sodium oxide and carbon dioxide.

$$Na_2CO_3(s) \rightarrow Na_2O(s) + CO_2(g)$$

For this reaction, $\Delta H^\ominus = +323.0\,kJ\,mol^{-1}$ and $\Delta S^\ominus = +153.7\,J\,K^{-1}\,mol^{-1}$

a) Show that the thermal decomposition of sodium carbonate is not feasible at 1200 K.
b) Calculate the temperature at which the value of $\Delta G = 0$ for this reaction. Give your answer to 1 decimal place.

Answer

a) $\Delta G^\ominus$ has units of $kJ\,mol^{-1}$ exactly like $\Delta H^\ominus$, so the units of ΔS must be changed to $kJ\,K^{-1}\,mol^{-1}$. This is achieved by dividing the $\Delta S^\ominus$ value by 1000 to convert to $kJ\,K^{-1}\,mol^{-1}$.
So $\Delta S^\ominus = +0.1537\,kJ\,K^{-1}\,mol^{-1}$
$\Delta G^\ominus = \Delta H^\ominus - T\Delta S^\ominus$
$\Delta G^\ominus = 323.0 - (1200 \times 0.1537) = +138.6\,kJ\,mol^{-1}$
Because $\Delta G^\ominus$ is positive at this temperature the reaction is not feasible.
b) $\Delta G^\ominus = \Delta H^\ominus - T\Delta S^\ominus$
When $\Delta G^\ominus = 0$
Then $\Delta H^\ominus - T\Delta S^\ominus = 0$
$+323.0 - T(0.1537) = 0$
$T(0.1537) = 323$
$T = \frac{323.0}{0.1537} = 2101.4964\,K.$
$T = 2101.5\,K$ (to 1 decimal place)
T must be equal to or greater than 2101.5 K for the reaction to be feasible.

MATHS
2101.5 K is the same as 1828.5 °C; to convert between the Kelvin and Celsius temperature scale, simply subtract 273. To convert between Celsius and Kelvin, add 273.

To determine the temperature at which a reaction changes from not feasible to feasible divide $\Delta H^{\ominus}$ by $\Delta S^{\ominus}$ but remember to make sure ΔS is in $kJ\,K^{-1}\,mol^{-1}$.

EXAMPLE 8

The table below gives some enthalpy and entropy data for some elements and compounds.

	$N_2(g)$	$H_2(g)$	$O_2(g)$	$NO(g)$	$NH_3(g)$	$H_2O(g)$
$\Delta_f H^{\ominus}/kJ\,mol^{-1}$	0	0	0	+90	−46	−242
$S^{\ominus}/J\,K^{-1}\,mol^{-1}$	192	131	205	211	193	189

1 Explain why the entropy value of oxygen is greater than that of hydrogen.
2 Explain why the standard enthalpy of formation of oxygen is 0 (zero).
3 The equation for the catalytic oxidation of ammonia is:

$$4NH_3(g) + 5O_2(g) \rightarrow 4NO(g) + 6H_2O(g)$$

a) Calculate the enthalpy change for the reaction using the data in the table.
b) Calculate the entropy change for the reaction. State the units.
c) Explain why this reaction is feasible at any temperature.

Answer

1 O_2 has a higher M_r than H_2. Higher entropy value as the particles are heavier.
2 Oxygen is an element.
3 a) $\Delta H^{\ominus} = 4(+90) + 6(-242) - 4(-46)$
$= -908\,kJ\,mol^{-1}$
b) $\Delta S^{\ominus} = \Sigma S^{\ominus}_{products} - \Sigma S^{\ominus}_{reactants}$
$= (4 \times 211 + 6 \times 189) - (4 \times 193 + 5 \times 205) = +181\,J\,K^{-1}\,mol^{-1}$.
c) Exothermic reaction and entropy increases so reaction is feasible at all temperatures.

TIP
This type of Hess's Law calculation should be revised from Year 1 of the course (AS).

Factors affecting feasibility of a reaction

The feasibility of a reaction depends on whether a reaction is endothermic or exothermic and also whether there is an increase in entropy or a decrease in entropy. The table below shows how these factors affect ΔG and the feasibility of a reaction.

$\Delta H^{\ominus}$	$\Delta S^{\ominus}$	$\Delta G^{\ominus}$	Feasibility
Negative	Negative	May be positive or negative	Feasible below certain temperatures
Negative	Positive	Always negative	Feasible at any temperature
Positive	Negative	Always positive	Not feasible at any temperature
Positive	Positive	May be positive or negative	Feasible above certain temperatures

TIP
This kind of qualitative understanding of the relationship between $\Delta H^\ominus$, $\Delta S^\ominus$, T and $\Delta G^\ominus$ is important as well as being able to carry out the calculations. A reaction which is exothermic and shows an increase in entropy will be feasible at all temperatures.

EXAMPLE 9

The equation for the combustion of ethanol is given below:

$$C_2H_5OH(l) + 3O_2(g) \rightarrow 2CO_2(g) + 3H_2O(g)$$

Write an expression for $\Delta G^\ominus$.

Use this expression to explain why this reaction is feasible at all temperatures.

Answer

$$\Delta G^\ominus = \Delta H^\ominus - T\Delta S^\ominus$$

1 mol of liquid and 3 mol of gas become 5 mol of gas so entropy increases.

It is a combustion reaction which is exothermic.

As $\Delta H^\ominus$ is negative and $\Delta S^\ominus$ is positive, $\Delta G^\ominus$ is negative at all temperatures.

So the reaction is feasible at all temperatures.

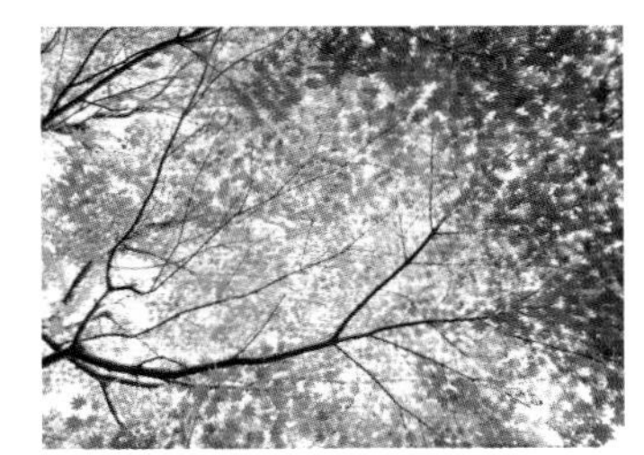

Figure 2.4 The value for $\Delta G^\ominus$ is positive for photosynthesis. The reaction is not feasible until sunlight is present.

Physical changes using $\Delta G^\ominus$

- During a change of state $\Delta G^\ominus = 0$ (zero) kJ mol^{-1}.
- $\Delta G^\ominus$ is positive for temperatures below the melting point as it is not feasible.
- $\Delta G^\ominus$ is negative for temperatures above the melting point as it is feasible.
- The same is true of $\Delta G^\ominus$ for boiling around the boiling point.
- When $\Delta G^\ominus = 0$, the system is at equilibrium and this is what is happening when a change of state occurs.
- If $\Delta G^\ominus$ is given, then other quantities can be calculated by rearranging the expression.

EXAMPLE 10

$\Delta G^\ominus = 0$ kJ mol^{-1} for the melting of ice to water. This physical reaction is endothermic and occurs at 273 K.

$$H_2O(s) \rightarrow H_2O(l)$$

The absolute entropy values for ice and water are given below.

	$H_2O(s)$	$H_2O(l)$
$S^\ominus$/J K^{-1} mol^{-1}	48	70

1. Explain why the melting of water is feasible at 273 K even though this change is endothermic.
2. Calculate the entropy change for the reaction.
3. Calculate the enthalpy change for the reaction.
4. Show that the conversion of ice to water is feasible at 298 K (25 °C) but not feasible at 268 K (−5 °C).

Figure 2.5 Melting ice is an endothermic reaction. It is feasible only at certain temperatures.

Answer

1 Water molecules become more disordered when the solid changes to a liquid so $\Delta S^{\ominus}$ is positive/entropy increases.
$\Delta G^{\ominus}$ is = 0 so $T\Delta S = \Delta H$

2 $\Delta S^{\ominus} = \Sigma S^{\ominus}_{products} - \Sigma S^{\ominus}_{reactants}$
$= 70 - 48 = +22\ J\ K^{-1} mol^{-1}$

3 $\Delta G^{\ominus} = 0\ kJ\ mol^{-1}$
$T = 273\ K$
$\Delta S^{\ominus} = +22\ J\ K^{-1}\ mol^{-1}$
$\Delta G^{\ominus} = \Delta H^{\ominus} - T\Delta S^{\ominus}$
$\Delta H^{\ominus} = \Delta G^{\ominus} + T\Delta S^{\ominus}$
$\Delta H^{\ominus} = 0 + 273(0.022)$
$\Delta H^{\ominus} = +6\ kJ\ mol^{-1}$
This value was given on the first page of this topic.

4 At 298 K
$\Delta G^{\ominus} = \Delta H^{\ominus} - T\Delta S^{\ominus}$
$\Delta G^{\ominus} = +6 - 298(0.022) = -0.556\ kJ\ mol^{-1}$
$\Delta G^{\ominus}$ is less than zero so the conversion of ice to water at 298 K is feasible.
At 268 K
$\Delta G^{\ominus} = \Delta H^{\ominus} - T\Delta S^{\ominus}$
$\Delta G^{\ominus} = +6 - 268(0.022) = +0.104\ kJ\ mol^{-1}$
$\Delta G^{\ominus}$ is greater than zero so the conversion of ice to water at 268 K is not feasible

TIP
$T\Delta S^{\ominus}$ being equal to or a greater numerical value than $\Delta H^{\ominus}$ is important in ensuring that $\Delta G^{\ominus}$ is equal to or less than zero so the reaction is feasible. In a change of state $\Delta G^{\ominus} = 0$.

TIP
The same theory can be applied to any change in state as $\Delta G^{\ominus} = 0$ for a change in state as the change of state is viewed as a reaction at equilibrium.

EXAMPLE 11

The following data is given for the condensation of ammonia.

$NH_3(g) \rightarrow NH_3(l) \qquad \Delta H^{\ominus} = -23.35\ kJ\ mol^{-1}$

The absolute entropy value for $NH_3(g) = 193\ J\ K^{-1}\ mol^{-1}$

The boiling point of ammonia is 240 K.

Calculate the absolute entropy value of $NH_3(l)$.

Answer

For a change of state, $\Delta G^{\ominus} = 0\ kJ\ mol^{-1}$

$\Delta G^{\ominus} = \Delta H^{\ominus} - T\Delta S^{\ominus}$

$0 = -23.35 - 240(\Delta S^{\ominus})$

$\Delta S^{\ominus} = \frac{23.35}{-240} = -0.09729\ kJ\ K^{-1}\ mol^{-1} = -97.29\ J\ K^{-1}\ mol^{-1}$

$\Delta S^{\ominus} = \Sigma S^{\ominus}_{products} - \Sigma S^{\ominus}_{reactants} = S^{\ominus}(NH_3(l)) - 193$
$= -97.29\ J\ K^{-1}\ mol^{-1}$

$S^{\ominus}(NH_3(l)) = -97.29 + 193 = 95.71\ J\ K^{-1}\ mol^{-1}$

TIP
You would expect this change to have a negative entropy change as the liquid is more ordered than the gas.

TIP
Being able to rearrange an expression to calculate another quantity is important. You can check your answers are correct on the way through a calculation by substituting them back into the expression you used to ensure it gives you the expected answer.

EXAMPLE 12

The change of state from liquid to gas for water has an entropy change of +119 J K^{-1} mol^{-1}. The enthalpy change for the physical change is: +44.39 kJ mol^{-1}.

$H_2O(l) \rightarrow H_2O(g)$

Calculate the boiling point of water. Assume that $\Delta G^\ominus = 0$ kJ mol^{-1} for the change of liquid water to a gas.

Answer

$\Delta G^\ominus = \Delta H^\ominus - T\Delta S^\ominus$

$\Delta G^\ominus = 0$

$\Delta H^\ominus - T\Delta S^\ominus = 0$

$+44.39 - T(0.119) = 0$

$44.39 = T(0.119)$

$T = \frac{44.39}{0.119} = 373$ K

The boiling point of water is 100 °C which equals 373 K.

TEST YOURSELF 3

1 Two reactions which are involved in the extraction of iron in the blast furnace are:

A $Fe_2O_3(s) + 3CO(g) \rightarrow 2Fe(s) + 3CO_2(g)$
$\Delta H^\ominus = -25.0$ kJ mol^{-1}
$\Delta S^\ominus = +15.2$ J K^{-1} mol^{-1}

B $Fe_2O_3(s) + 3C(s) \rightarrow 2Fe(s) + 3CO(g)$
$\Delta H^\ominus = +491.0$ kJ mol^{-1}
$\Delta S^\ominus = +542.9$ J K^{-1} mol^{-1}

Which reactions are feasible at

a) 800 K
b) 1800 K

2 For the reaction $Cu(s) + 2H_2O(l) \rightarrow Cu(OH)_2(s) + H_2(g)$

a) Calculate the enthalpy change for the reaction given the following enthalpies of formation.

	$\Delta_f H^\ominus$/kJ mol^{-1}
$H_2O(l)$	−286
$Cu(OH)_2(s)$	−450

b) Calculate $\Delta S^\ominus$ given the following absolute entropy values.

	$S^\ominus$/J K^{-1} mol^{-1}
Cu(s)	33
$H_2O(l)$	70
$Cu(OH)_2(s)$	75
$H_2(g)$	131

c) Calculate the temperature at which $\Delta G^\ominus = 0$.

3 For the reaction $C_2H_4(g) + H_2(g) \rightarrow C_2H_6(g)$

a) Calculate the enthalpy change for the reaction given the following standard enthalpies of combustion (in kJ mol^{-1}) $C_2H_4(g) = -1411$; $H_2(g) = -286$; $C_2H_6(g) = -1560$

b) Calculate $\Delta S^\ominus$ given the following standard entropies.

	$S^\ominus$/J K^{-1} mol^{-1}
$C_2H_4(g)$	220
$H_2(g)$	131
$C_2H_6(g)$	230

c) Calculate the temperature at which the reaction becomes feasible.

4 The change of state of bromine from liquid to gas is represented by the equation:

$Br_2(l) \rightarrow Br_2(g)$ $\Delta H^\ominus = +30$ kJ mol^{-1}

The standard entropy values of bromine are given below:

	$Br_2(l)$	$Br_2(g)$
$S^\ominus$/J K^{-1} mol^{-1}	75.8	166.4

a) Calculate the entropy change for this reaction.
b) Assuming $\Delta G^\ominus = 0$ for a change of state, calculate the boiling point, in K, of bromine.

Graphical calculations

MATHS

$y = mx + c$ is the equation of a straight line on a graph. Any straight line is in this form where m is the gradient of the line and c is the intercept with the y axis. For more detail on this see chapter 16, Maths for chemistry.

- The equation $\Delta G^\ominus = \Delta H^\ominus - T\Delta S^\ominus$ can be plotted as a straight line graph.
- This follows the pattern $y = mx + c$.

  $\Delta G^\ominus = \Delta H^\ominus - T\Delta S^\ominus$
- For the straight line graph $y = mx + c$, m is the gradient and c is in the intercept on the y axis.
- If a graph of $\Delta G^\ominus$ against T is plotted, the gradient is $-\Delta S^\ominus$ and the intercept with the $\Delta G^\ominus$ axis is $\Delta H^\ominus$.

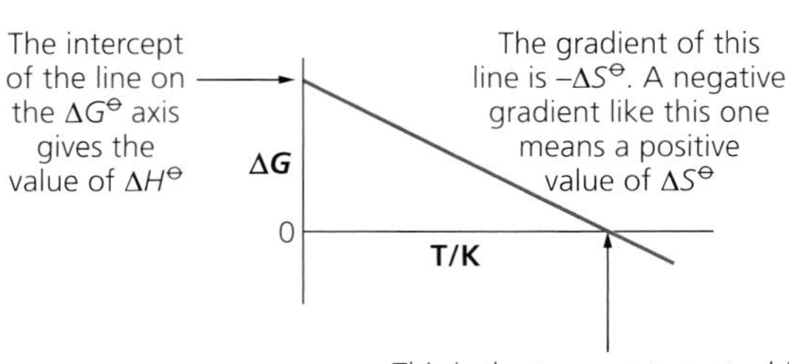

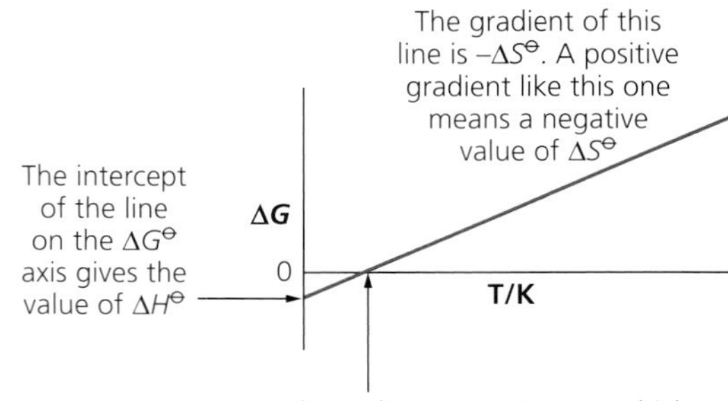

- $\Delta G^\ominus$ and $\Delta H^\ominus$ have the same units of $kJ\,mol^{-1}$.
- When $T = 0$, $\Delta G^\ominus = \Delta H^\ominus$ from the expression $\Delta G^\ominus = \Delta H^\ominus - T\Delta S^\ominus$.
- The gradient of the line has units of $kJ\,K^{-1}\,mol^{-1}$. The gradient can be calculated from the graph and the negative of this value gives $\Delta S^\ominus$.
- $\Delta S^\ominus$ can be calculated in $J\,K^{-1}\,mol^{-1}$ by multiplying the negative gradient by 1000.

EXAMPLE 13

When lead(II) nitrate is heated it decomposes according to the equation

$$2Pb(NO_3)_2(s) \rightarrow 2PbO(s) + 4NO_2(g) + O_2(g)$$

The graph shows how the free energy change ($\Delta G^\ominus$) for this reaction varies with temperature.

Use the graph to find $\Delta H^\ominus$ and $\Delta S^\ominus$.

What is the minimum temperature required for the decomposition of lead(II) nitrate?

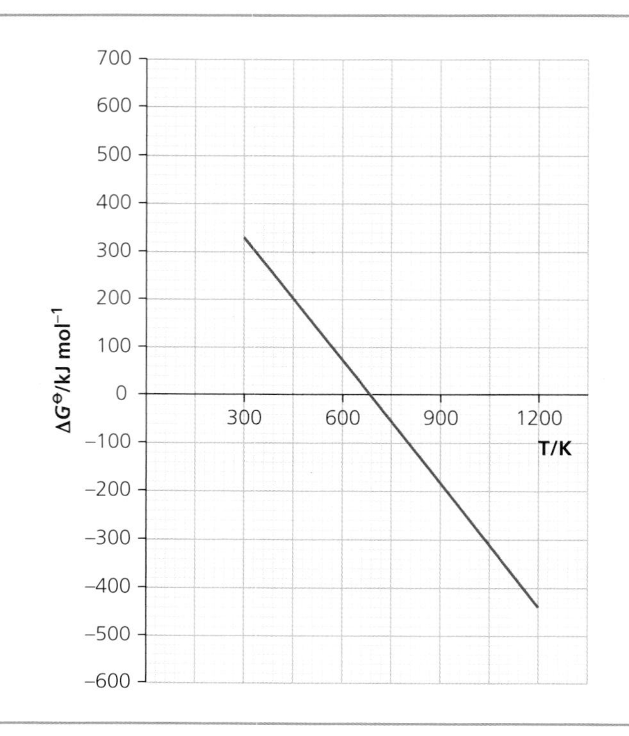

Answer

Extrapolating back to $T = 0$ the line intercepts the $\Delta G^\ominus$ axis at $+600\,kJ\,mol^{-1}$.

From the expression $\Delta G^\ominus = \Delta H^\ominus - T\Delta S^\ominus$, when $T = 0$, $\Delta G^\ominus = \Delta H^\ominus$.

$\Delta H^\ominus = +600\,kJ\,mol^{-1}$.

The gradient is found by dividing rise by run $-\frac{600}{690} = -0.875\,kJ\,mol^{-1}\,K^{-1}$.

The gradient is negative as the line falls from left to right.

The gradient is $-\Delta S^\ominus$ so $\Delta S^\ominus = 0.875\,kJ\,mol^{-1}\,K^{-1}$, which equals $+875\,J\,K^{-1}\,mol^{-1}$.

The temperature at which the line crosses the T axis is the temperature at which $\Delta G^\ominus = 0$. This value is 690 K. The minimum temperature required for the decomposition of lead(II) nitrate is 690–K.

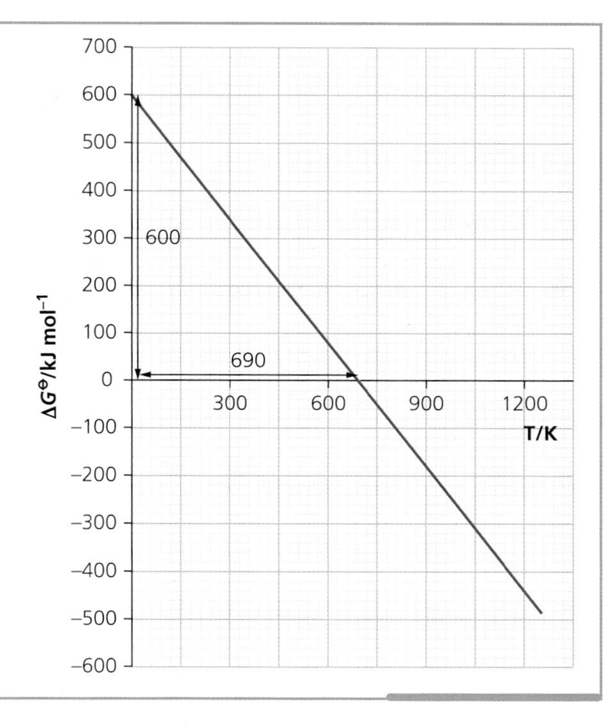

TIP

If you are unsure about how to calculate the gradient of a graph, refer to page 314 in the mathematics chapter of this book.

ACTIVITY

The production of ammonia

Ammonia has a major use in making fertilisers and is used to produce many other chemicals as shown in Figure 2.6.

1 Write a balanced symbol equation for the production of ammonium sulfate fertiliser from ammonia.
2 Nitric acid is manufactured by the oxidation of ammonia, which initially produces nitrogen monoxide and water. The nitrogen monoxide is further oxidised to nitrogen dioxide, which is then reacted with more air and water to produce nitric acid. Write balanced symbol equations for each of the three steps in this manufacture.

In the Haber process, nitrogen reacts with hydrogen to form ammonia according to the equation:

$$N_2(g) + 3H_2(g) \rightleftharpoons 2NH_3(g)$$

The graph shows how the free energy change ($\Delta G^\ominus$) for this reaction varies with temperature.

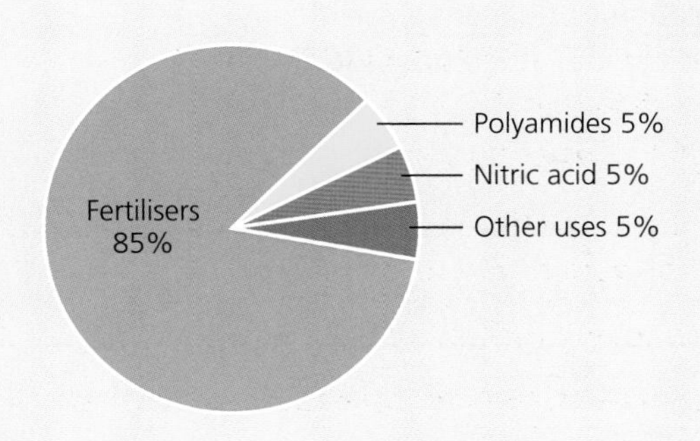

Figure 2.6 The major uses of ammonia.

3 a) Use the graph to determine $\Delta S^\ominus$.

b) Use the graph to determine $\Delta H^\ominus$.

c) Calculate the temperature above which the reaction is no longer feasible.

d) The line is not drawn below a temperature of 240 K because its gradient changes at this point. Suggest what happens to the ammonia at 240 K that causes the slope of the line to change.

e) In terms of the reactants and products and their physical states, account for the sign of the entropy change that you calculated in part (a).

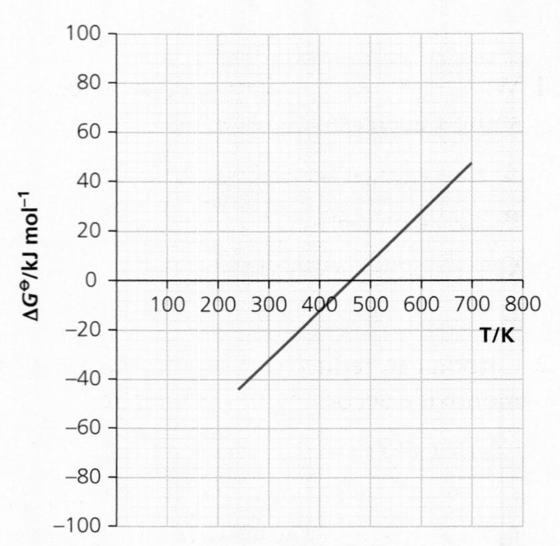

Figure 2.7 Graph of $\Delta G^\ominus$ against T for the production of ammonia.

TEST YOURSELF 4

1 What is the gradient of a $\Delta G^\ominus$ against temperature graph?

2 Explain how the value of $\Delta H^\ominus$ is determined from a $\Delta G^\ominus$ against temperature graph.

3 Consider the following graph of $\Delta G^\ominus$ against T for a reaction.

a) Give an estimate of the value of $\Delta H^\ominus$ for this reaction using the graph.

b) Explain using the graph whether the reaction shows an increase or decrease in entropy.

c) At what temperature does $\Delta G^\ominus = 0$?

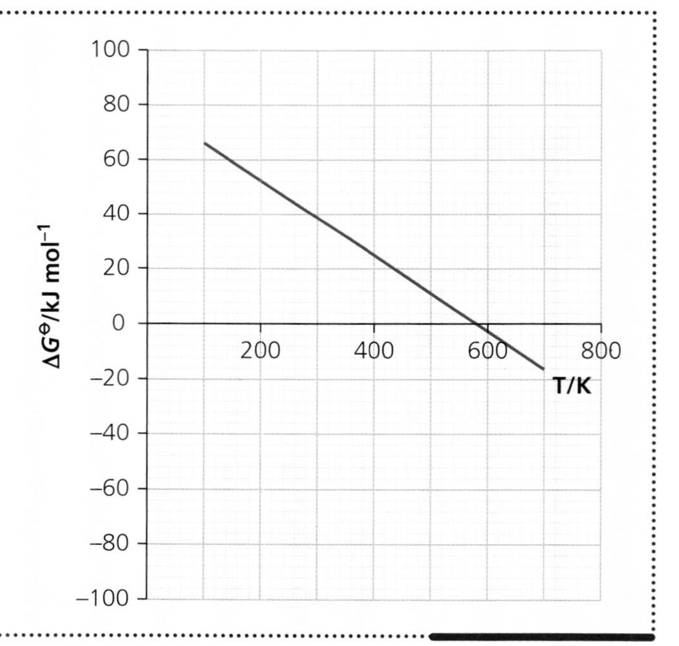

Practice questions

1 Which one of the reactions shown below would show a decrease in entropy?

A $Ca(s) + \frac{1}{2}O_2(g) \rightarrow CaO(s)$

B $Ca(s) + 2HCl(aq) \rightarrow CaCl_2(aq) + H_2(g)$

C $CaCO_3(s) \rightarrow CaO(s) + CO_2(g)$

D $CaCO_3(s) + 2HNO_3(aq) \rightarrow Ca(NO_3)_2(aq) + CO_2(g) + H_2O(l)$ *(1)*

2 Carbon can reduce carbon dioxide to carbon monoxide according to the equation:

$C(s) + CO_2(g) \rightarrow 2CO(g)$ $\Delta H^\ominus = +173\,kJ\,mol^{-1}$

If $\Delta S^\ominus = +175.9\,J\,K^{-1}\,mol^{-1}$, at which of the following temperatures is the reaction feasible?

A 750 K **B** 850 K **C** 950 K **D** 1050 K *(1)*

3 The reaction of copper with steam is represented by the equation:

$Cu(s) + H_2O(g) \rightarrow CuO(s) + H_2(g)$

The table gives the standard entropy values and the standard enthalpies of formation.

	Cu(s)	**$H_2O(g)$**	**CuO(s)**	**$H_2(g)$**
Standard enthalpy of formation ($kJ\,mol^{-1}$)	0	−242	−155	0
Standard entropy value ($J\,K^{-1}\,mol^{-1}$)	33	189	43.5	131

Which one of the following correctly describes the reaction of copper with steam?

A Feasible at any temperature.

B Not feasible at any temperature.

C Feasible at low temperature but not feasible at high temperature.

D Not feasible at low temperature but feasible at high temperature. *(1)*

4 On heating, silver(I) oxide decomposes into silver and oxygen

$2Ag_2O(s) \rightarrow 4Ag(s) + O_2(g)$

$\Delta H^\ominus = +62\,kJ\,mol^{-1}$

$\Delta S^\ominus = +132.8\,J\,K^{-1}\,mol^{-1}$

a) Calculate the free energy change, $\Delta G^\ominus$, for this reaction at:

i) 400 K *(1)*

ii) 600 K and comment on the feasibility of this reaction at this temperature. *(2)*

b) Calculate the temperature at which $\Delta G^\ominus$ has a value of 0 for the decomposition of silver(I) oxide. *(2)*

5 For the reaction:

$CaO(s) + H_2O(l) \rightarrow Ca(OH)_2(s)$
$\Delta H^\ominus = -65\,kJ\,mol^{-1}$

a) Calculate $\Delta S^\ominus$ given the following absolute entropy values (in $J\,K^{-1}\,mol^{-1}$): $CaO(s) = 39.7$; $H_2O(l) = 70$; $Ca(OH)_2(s) = 83.4$. *(2)*

b) Calculate the temperature at which $\Delta G^\ominus = 0$. *(2)*

6 Magnesium carbonate and barium carbonate both decompose on heating according to the equation:

$XCO_3(s) \rightarrow XO(g) + CO_2(g)$

The table below details the values of $\Delta_f H^\ominus$ and absolute entropy.

	$MgCO_3(s)$	**MgO(s)**	**$CO_2(g)$**	**$BaCO_3(s)$**	**BaO(s)**
$\Delta_f H^\ominus$/ $kJ\,mol^{-1}$	−1096	−602	−394	−1216	−554
$S^\ominus$/$J\,K^{-1}\,mol^{-1}$	66	27	214	112	70

a) Calculate $\Delta H^\ominus$ and $\Delta S^\ominus$ for both reactions, and use these values to calculate $\Delta G^\ominus$ at 900 K for both reactions. Which (if either) of the carbonates will decompose on heating to 900 K? *(4)*

b) Calculate the minimum temperature needed to decompose each carbonate based on when $\Delta G^\ominus = 0$. *(4)*

7 When heated strongly, copper(II) sulfate decomposes according to the equation:

$CuSO_4(s) \rightarrow CuO(s) + SO_3(g)$

The graph below shows how the free energy change ($\Delta G^\ominus$) for this reaction varies with temperature.

a) Use the graph to find $\Delta H^\ominus$ and $\Delta S^\ominus$. *(2)*

b) At what temperature does $\Delta G^\ominus = 0$? *(1)*

c) What is the minimum temperature required for the decomposition of copper(II) sulfate? *(1)*

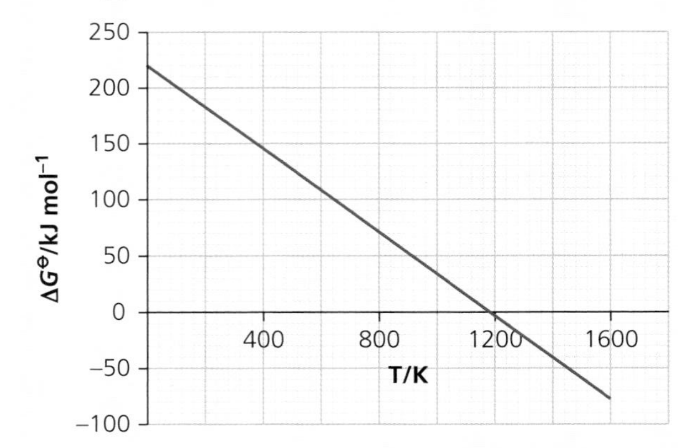

8 Sodium melts at 371 K.

The absolute entropy value of Na(s) is 51.0 J K^{-1} mol^{-1}.

For the change:
Na(s) → Na(l) $\Delta H^\ominus = +2.6$ kJ mol^{-1}.

Calculate the absolute entropy value of Na(l). (3)

9 Aluminium oxide can be reduced using carbon as shown in the equation below.

$$Al_2O_3(s) + 3C(s) \rightarrow 2Al(s) + 3CO(g)$$

The table below gives values for standard enthalpy of formation and absolute entropy for the reactants and products.

	$Al_2O_3(s)$	C(s)	Al(s)	CO(g)
$\Delta_f H^\ominus$/kJ mol^{-1}	−1669	0	0	−111
$S^\ominus$/J K^{-1} mol^{-1}	51	6	28	198

a) Calculate the enthalpy change of the reaction shown above using the enthalpies of formation. *(2)*

b) Explain why the enthalpy of formation of aluminium is 0 (zero). *(1)*

c) Calculate the entropy change for this reaction. *(2)*

d) Calculate the value of $\Delta G^\ominus$ for this reaction at 298 K. State the units. *(2)*

e) Explain whether the reaction is feasible at 298 K. Explain your answer. *(1)*

f) Calculate the temperature at which this reaction becomes feasible. *(2)*

Stretch and challenge

10 The following reaction becomes feasible at temperatures above 5440 K.

$$H_2O(g) \rightarrow H_2(g) + \tfrac{1}{2}O_2(g)$$

The entropies of the species involved are shown in the following table.

	$H_2O(g)$	$H_2(g)$	$O_2(g)$
$S^\ominus$/J K^{-1} mol^{-1}	189	131	205

a) Calculate the entropy change $\Delta S^\ominus$ for this reaction. *(1)*

b) Calculate a value, with units, for the enthalpy change for this reaction at 5440 K. State the units. *(4)*

c) The standard enthalpy of formation of water is −286 kJ mol^{-1}. Using Hess's Law, calculate the enthalpy change for the change $H_2O(l) \rightarrow H_2O(g)$. *(3)*

Rate equations

PRIOR KNOWLEDGE

It is expected that you are familiar with all of the content of the **Kinetics** unit in Year 1 (AS) of the A-level Chemistry course. The following are some key points of Prior Knowledge:

- The rate of reaction is the change in concentration of a reactant or product in a given period of time.
- The rate of a reaction is dependent on the frequency of successful collisions.
- A successful collision is a collision that has the minimum amount of energy for the reacting particles to react.
- The activation energy is the minimum amount of energy that the reacting particles require for a successful collision.
- A catalyst increases the rate of reaction without being used up.
- A catalyst works by providing an alternative reaction pathway (or route) of lower activation energy.
- A Maxwell-Boltzmann distribution is a plot of the number of gaseous molecules against the energy they have at a fixed temperature.
- The energy value of the peak on a Maxwell-Boltzmann distribution is the most probable energy (E_{mp}) of the molecules.
- E_{mp} is only affected by changes in temperature.
- Increasing temperature increases E_{mp} and increases the number of molecules above the activation energy.

TEST YOURSELF ON PRIOR KNOWLEDGE 1

1 State three factors that can affect the rate of a chemical reaction.

2 State the effect, if any, on the most probable energy of decreasing the temperature.

3 State the effect, if any, on the most probable energy of changing the concentration.

4 From the reactions below:

Reaction A: $C_2H_4(g) + Br_2(aq) \longrightarrow C_2H_4Br_2(l)$

Reaction B: $C_3H_6(g) + H_2(g) \longrightarrow C_2H_6(g)$

Reaction C: $C_2H_5OH(l) + 3O_2(g) \longrightarrow 2CO_2(g) + 3H_2O(l)$

a) Which reaction is homogeneous?

b) Explain your answer to part (a).

5 What is meant by the term activation energy?

Conventions

- Kinetics is the study of the rate of a chemical reaction.
- The rate of a reaction can be measured based on how fast the concentration of a reactant is decreasing or how fast the concentration of a product is increasing, i.e. the change in concentration per unit time.
- The units of rate are concentration per unit time, for example $mol\,dm^{-3}\,s^{-1}$ (mol per dm^3 per second).
- Square brackets around a substance, for example [A] or $[H^+]$ or $[CH_3OH]$ or $[O_2]$ all indicate concentration of the substance which is inside the brackets.
- Concentration is always measured in $mol\,dm^{-3}$ (mol per dm^3).
- For example $[H^+]$ = concentration of hydrogen ions measured in $mol\,dm^{-3}$.
- Initial rates of reaction are often used as the initial rate of reaction at t = 0 s (time = zero seconds, i.e. when the reaction has just started) is directly related to the concentrations of the reactants which were used in setting up the experiment. As the experiment progresses the reactants are being used up so it is more difficult to relate the concentrations of reactants to the rate as the reaction proceeds.

TIP
The concentrations of gases and solutions are frequently measured in $mol\,dm^{-3}$ as it expresses the number of moles of a substance in a volume (solution or space) of $1\,dm^3$. This convention is used in equilibrium calculation as well as kinetics.

The rate equation

The *rate equation* is an expression showing how the rate of reaction is linked to the concentrations of the reactants.

- Rate is equal to the *rate constant* (k) multiplied by the concentration of each reactant raised to certain whole number powers (called orders).
- Some reactants may be zero order which means they do not appear in the rate equation as $[X]^0 = 1$.
- The rate equation for general reaction: $A + B \longrightarrow C + D$
  is: rate = $k[A]^m[B]^n$
  where m is the order of reaction with respect to reactant A
  and n is the order of reaction with respect to reactant B
  and k is the rate constant
- The **order of reaction** with respect to a particular reactant is the power to which the concentration of this reactant is raised in the rate equation.
- The overall order of the reaction is the sum of all the orders in the rate equation. In this general example it would be $m + n$.
- The **rate constant** is the proportionality constant which links the rate of reaction to the concentrations in the rate equation.

Wilhelm Ostwald was a German physical chemist and was the first person to introduce the term 'reaction order'. He identified the action of catalysts in lowering activation energy. As a result, Ostwald was awarded the 1909 Nobel Prize in Chemistry. Ostwald is usually credited with inventing the Ostwald process used in the manufacture of nitric acid.

TIP
Make sure that the rate constant is included in a rate equation as it is easily forgotten. Also make sure it is a lower case *k* as a capital *K* is an equilibrium constant.

Effect of changes in concentration on rate

For the reaction: B + C + D → E + F, the rate equation is: rate = k $[B]^2[D]$. This means that the order of reaction with respect to B is 2, the order of reaction with respect to C is zero (as it does not appear in the rate equation) and the order of reaction with respect to D is 1. The overall order of reaction is 3 (2+1).

- Doubling the concentration of B would increase the rate by a factor of 4 as $(\times 2)^2 = \times 4$.
- Doubling the concentration of C would have no effect on the rate as $(\times 2)^0 = \times 1$.
- Doubling the concentration of D would double the rate as $(\times 2)^1 = \times 2$.
- Doubling the concentration of B, C and D would increase the rate by a factor of 8 as $(\times 2)^2(\times 2)^0(\times 2)^1 = \times 8$ or using the overall order $(\times 2)^3 = \times 8$.

MATHS

Any value raised to the power zero is equal to 1. $5^0 = 1$; $[A]^0 = 1$. A power of 1 is not usually written, for example rate = $k[A]$ means that the order of reaction with respect to A is 1.

TIP

Orders with respect to reactants are 0 (zero), 1 or 2. In terms of rate of reaction, the concentration of a zero order reactant has no effect on the rate of the reaction. Third order reactions are not common but examples are used to illustrate the principles.

Units of the rate constant

- Rate has units of $\text{mol dm}^{-3}\text{s}^{-1}$ and concentration has units of mol dm^{-3}.
- The units of the rate constant, k, depend on overall order of reaction and the units can be calculated as shown:
- For a general rate equation = $k[A]^2[B]$ the overall order is 3 (2 + 1). $[B] = [B]^1$ so the order with respect to B is 1.
- rate = k × $(\text{concentration})^3$
- Putting in the units: $\text{mol dm}^{-3}\text{s}^{-1} = k\,(\text{mol dm}^{-3})^3$
- Rearranging to find k: $k = \dfrac{\text{mol dm}^{-3}\text{s}^{-1}}{(\text{mol dm}^{-3})^3} = \dfrac{\text{mol dm}^{-3}\text{s}^{-1}}{\text{mol}^3\text{dm}^{-9}}$
- Treat each term separately: $\dfrac{\text{mol}}{\text{mol}^3} = \text{mol}^{-2}$ and $\dfrac{\text{dm}^{-3}}{\text{dm}^{-9}} = \text{dm}^{-3-(-9)} = \text{dm}^6$.
- Units of the rate constant k = $\text{mol}^{-2}\text{dm}^6\text{s}^{-1}$

The table shows the units of the rate constant for common overall orders.

Overall order of Reaction	Units of rate constant, *k*
1	s^{-1}
2	$\text{mol}^{-1}\text{dm}^3\text{s}^{-1}$
3	$\text{mol}^{-2}\text{dm}^6\text{s}^{-1}$
4	$\text{mol}^{-3}\text{dm}^9\text{s}^{-1}$
5	$\text{mol}^{-4}\text{dm}^{12}\text{s}^{-1}$

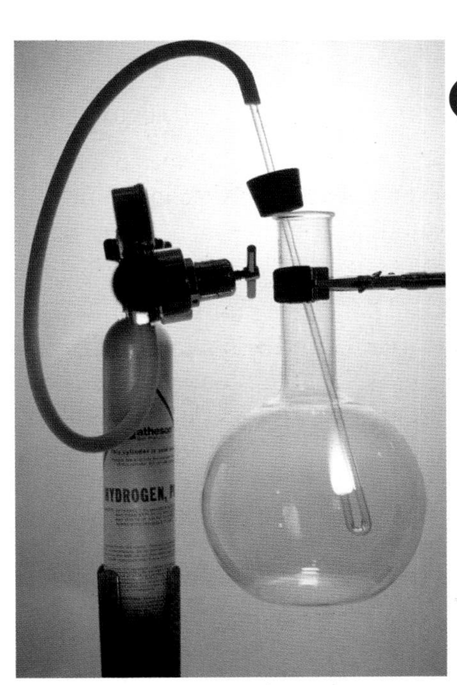

Figure 3.1 The photochemical reaction of hydrogen and chlorine is a zero order reaction.

TEST YOURSELF 2

1 The reaction C + 2D → E + 2F is zero order with respect to C and second order with respect to D. Write a rate equation for the reaction.

2 In the rate equation: rate = $k[A]^2[B]$
  a) What is represented by k?
  b) What is the order of reaction with respect to A?
  c) What is the order of reaction with respect to B?
  d) What is the overall order of reaction?

3 For the reaction P + Q → S + T the rate equation is rate = $k[P]^2$
  a) What is the order of reaction with respect to P?
  b) What is the order of reaction with respect to Q?
  c) What are the units of the rate constant (k)?

4 For the following general rate equation: rate = $k[X]^x[Y]^y$
  a) What is meant by [A]?
  b) What are the units of rate?
  c) What is the overall order of reaction in terms of *x* and *y*?
  d) What are the units of the rate constant if x = 0 and y = 1?
  e) What are the units of the rate constant if x = 1 and y =2?

Deducing and using orders of reaction and the rate constant and its units

Often questions are set with experimental data *at a constant temperature* and you are asked to deduce the order of reaction with respect to one or several reactants or write a rate equation and calculate the value and units of the rate constant.

The key to these types of questions is understanding that if you change the concentration of one reactant, the effect it has on the rate is determined by the order with respect to this reactant.

> **TIP**
>
> '*At a constant temperature*' is very important as temperature affects the rate of reaction so the data are only able to be used to deduce orders of reaction if the series of experiments are carried out at constant temperature.

Using the orders of reaction from a given rate equation

With a given rate equation the orders can be used to work out the effect of the changing concentrations of reactants on the rate.

Also you may be given a table of data with values missing which have to be completed using the orders of reaction from the rate equation.

> **EXAMPLE 1**
>
> For the general reaction A + B + C → D + E, the rate equation is:
> rate = $k[B]^2[C]$
>
> State the effect on the rate of reaction of trebling the concentration of each reactant individually.
>
> State the effect on the rate of reaction of trebling the concentrations of all three reactants.
>
> **Answer**
>
> The order is zero with respect to A, second with respect to B and first with respect to C.
>
> If you treble the concentration of A, this will not change the rate as $(\times 3)^0 = (\times 1)$
>
> Remember anything to the power of zero is equal to 1 so that the concentration of a reactant with order zero has no effect on the rate of the reaction.
>
> If you treble the concentration of B, this will multiply the rate by a factor of nine as $(\times 3)^2 = (\times 9)$
>
> If you treble the concentration of C, this will triple the rate as $(\times 3)^1 = (\times 3)$
>
> If you treble all the reactants, A, B and C, this will multiply the rate by a factor of 27 as $(\times 3)^0(\times 3)^2(\times 3)^1 = (\times 1)(\times 9)(\times 3) = (\times 27)$

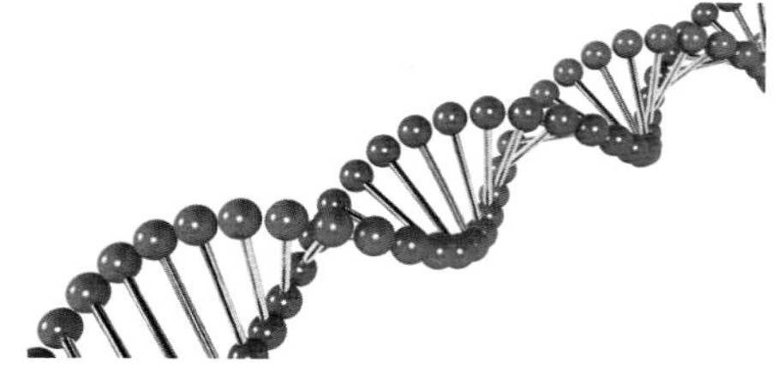

Figure 3.2 Many important biological reactions, such as the formation of double-stranded DNA from two complementary strands, can be described using second order kinetics.

EXAMPLE 2

In the reaction A + B → C + D, at constant temperature, doubling the concentration of A doubles the rate of reaction, whereas when the concentration of B is doubled, this has no effect on the rate.

Determine the orders of reaction with respect to A and B and write a rate equation for the reaction.

Answer

Doubling the concentration of A doubles the rate so order with respect to A is 1.

Doubling the concentration of B has no effect on the rate so the order with respect to B is 0 (zero).

Rate equation is: rate = k[A].

You can write rate = k[A][B]^0, but the zero order reactants are usually left out.

TIP

The concentration of a catalyst may be included in the rate equation and the order with respect to a catalyst can be determined in the same way as for any reactant. Often reactions that are acid catalysed are given and concentrations of [H^+] are given with rate values to allow you to determine the order with respect to H^+ ions.

EXAMPLE 3

The kinetics of the reaction of nitrogen(II) oxide with hydrogen were investigated.

$$2NO(g) + 2H_2(g) \rightarrow 2H_2O(g) + N_2(g)$$

The following data were obtained in a series of experiments about the rate of reaction between NO and H_2 at a constant temperature.

Experiment	Initial concentration of NO/mol dm^{-3}	Initial concentration of H_2/mol dm^{-3}	Initial rate/ mol dm^{-3} s^{-1}
1	2.0	1.0	2.0×10^{-5}
2	6.0	1.0	18.0×10^{-5}
3	6.0	2.0	36.0×10^{-5}

Deduce the order of reaction with respect to NO.

Deduce the order of reaction with respect to H_2.

Write a rate equation for this reaction.

Answer

Each row in the table above corresponds to a different experiment. Write the general rate equation as:

$$\text{rate} = k[NO]^x[H_2]^y$$

If you look at the differences between experiments 1 and 2. [NO] is tripled (×3) but [H_2] does not change (×1) and the rate increased by a factor of 9.

$x = 2$ here as $(\times 3)^2 = (\times 9)$ so the order with respect to NO is 2.

Looking at the differences between the experiments 2 and 3: It is best to choose these as between these rows only the concentration of H_2 is changing.

[H_2] is doubled (×2) but [NO] does not change (×1) and the rate is doubled (×2).

It follows that $y = 1$ here as $(\times 2)^1 = (\times 2)$ so the order with respect to H_2 is 1.

Rate equation is: rate = k[NO]^2[H_2]

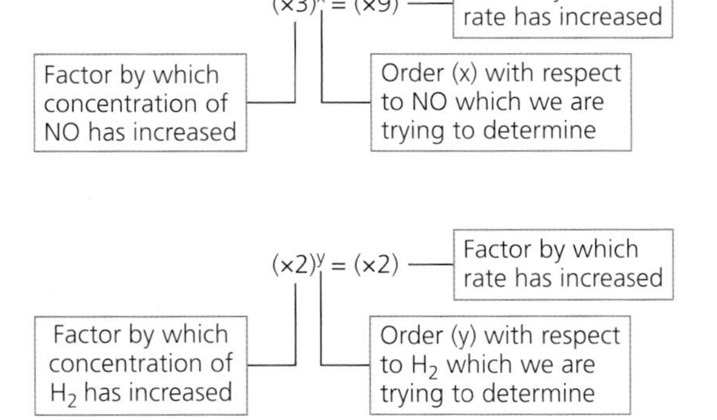

EXAMPLE 4

The data in the table below were obtained in a series of experiments about the rate of reaction between substances P and Q at a constant temperature.

Deduce the orders of reaction with respect to P and Q and write a rate equation for this reaction.

Experiment	Initial concentration of P/mol dm^{-3}	Initial concentration of Q/mol dm^{-3}	Initial rate/mol dm^{-3} s^{-1}
1	0.12	0.22	0.17×10^{-4}
2	0.36	0.22	0.51×10^{-4}
3	0.48	0.44	2.72×10^{-4}

Answer

Write the general rate equation as:

rate = $k[P]^x[Q]^y$

If you look at the differences between experiments 1 and 2. [P] is tripled (×3) but [Q] does not change (×1) and the rate increased by a factor of 3.

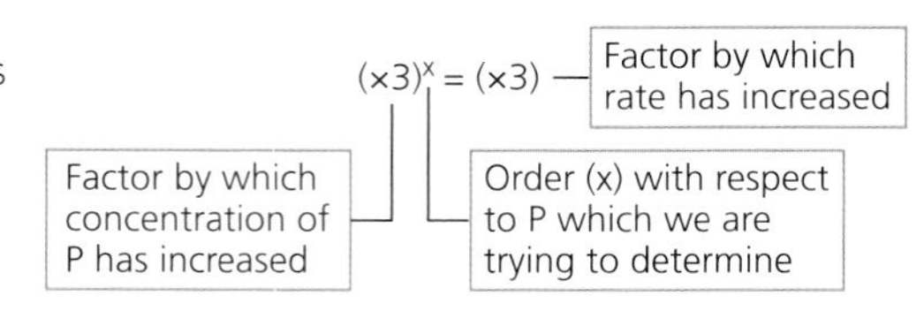

It follows that $x = 1$ here as $(\times 3)^1 = (\times 3)$ so the order with respect to P is 1.

Looking at the differences between experiments 1 and 3 (1 and 3 were chosen as there is a whole number factor increase in the concentrations of both P and Q).

[P] is ×4 and [Q] is doubled (×2). The rate increases by a factor of 16 (×16). We already know the order of reaction with respect to P is 1.

This simplifies to:

$\times 4\ (\times 2)^y = \times 16$

$(\times 2)^y = \dfrac{\times 16}{\times 4} = \times 4$

$(\times 2)^y = \times 4$

$y = 2$ (as $2^2 = 4$)

$(\times 4)^1\ (\times 2)^y = (\times 16)$

Order (y) with respect to Q which we are trying to determine

Factor by which rate has increased

Factor by which concentration of P has increased

Factor by which concentration of Q has increased

Order with respect to P = 1

The order of reaction with respect to Q is 2.

Rate equation is:

rate = $k[P][Q]^2$

EXAMPLE 5

The initial rate of reaction between two gases G and H was measured in a series of experiments at a constant temperature.

The rate equation was determined to be:

rate = $k[G]^2[H]$

Complete the table of data for the reactions between G and H.

Experiment	Initial [G]/ $mol\,dm^{-3}$	Initial [H]/ $mol\,dm^{-3}$	Initial rate/ $mol\,dm^{-3}\,s^{-1}$
1	0.20	0.15	0.32×10^{-3}
2	0.40	0.30	
3	0.60		1.92×10^{-3}
4		0.45	8.64×10^{-3}

Answer

Missing data in experiment 2

Between experiments 1 and 2:

[G] ×2

[H] ×2

rate × x(unknown)

Using rate equation:

rate = $k[G]^2[H]$ rate = $k\,(\times 2)^2(\times 2)$

k is a constant and can be ignored at constant temperature.

rate = ×4×2 = ×8

Missing rate in experiment 2 = $0.32 \times 10^{-3} \times 8 = 2.56 \times 10^{-3}\,mol\,dm^{-3}\,s^{-1}$

Missing data in experiment 3

Between experiments 1 and 3 (chosen as the factor by which [G] increases is a whole number):

[G] ×3

[H] ×a (unknown)

rate changes by a factor of = $\frac{1.92\times10^{-3}}{0.32\times10^{-3}} = 6$

[G] changes by a factor of ×3; rate changes by a factor of ×6

So using rate equation:

rate = $k[G]^2[H] \times 6 = k\,(\times 3)^2(\times a)$

k is a constant can be ignored at constant temperature.

$\times 6 = \times 9 \times a$

Missing concentration in experiment 3 = $0.15 \times \frac{2}{3} = 0.1\,mol\,dm^{-3}$.

Missing data in experiment 4

Between experiments 1 and 4 (chosen as the factor by which [H] increases is a whole number):

[G] ×b (unknown)

[H] ×3

rate changes by a factor of = $\frac{8.64 \times 10^{-3}}{0.32 \times 10^{-3}} = \times 27$

[H] changes by a factor of ×3; rate changes by a factor of ×27

So using rate equation:

rate = $k[G]^2[H] \times 27 = k(\times b)^2(\times 3)$

k is a constant can be ignored at constant temperature.

$\times 27 = \times b^2 \times 3$

$b^2 = \frac{\times 27}{\times 3}$

$b^2 = \times 9$ $b = \times 3$

Missing concentration in experiment 4 = $0.2 \times b = 0.2 \times 3 = 0.6\,mol\,dm^{-3}$

TIP

In these calculations you can generally ignore that × or ÷ in the calculations but these are here to help you to remember to × or ÷ at the end.

Determining the value and units of the rate constant

Often having deduced the orders of reaction for all the reactants (including a catalyst if present) you can be asked to write the rate equation and calculate a value of the rate constant and deduce its units. This combines many of the skills from the previous section.

EXAMPLE 6

The oxidation of nitrogen monoxide is represented by the equation:

$2NO(g) + O_2(g) \rightarrow 2NO_2(g)$

The kinetics of this reaction were studied at a constant temperature and the following results recorded.

Experiment	Initial concentration of NO/ $mol\,dm^{-3}$	Initial concentration of O_2/ $mol\,dm^{-3}$	Initial rate/ $mol\,dm^{-3}\,s^{-1}$
1	4×10^{-3}	1×10^{-3}	6×10^{-4}
2	8×10^{-3}	1×10^{-3}	24×10^{-4}
3	12×10^{-3}	1×10^{-3}	54×10^{-4}
4	8×10^{-3}	2×10^{-3}	48×10^{-4}
5	12×10^{-3}	3×10^{-3}	162×10^{-4}

Using these results, deduce the order of the reaction with respect to nitrogen monoxide and oxygen.

Write a rate equation for the reaction.

Calculate a value for the rate constant at this temperature and deduce its units.

Calculate the rate when the initial concentrations of NO and O_2 are both $5 \times 10\,mol\,dm^{-3}$.

A general rate equation would be:

$rate = k[NO]^x[O_2]^y$

Answers

Experiments 1 and 2:

$[NO] \times 2$

$[O_2] \times 1$

$rate \times 4$

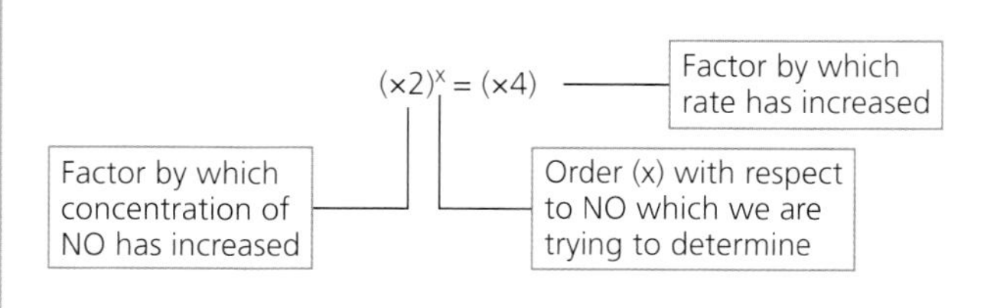

So order with respect to NO is 2 as $(\times 2)^2 = (\times 4)$

Experiments 2 and 4:

$[NO] \times 1$

$[O_2] \times 2$

$rate \times 2$

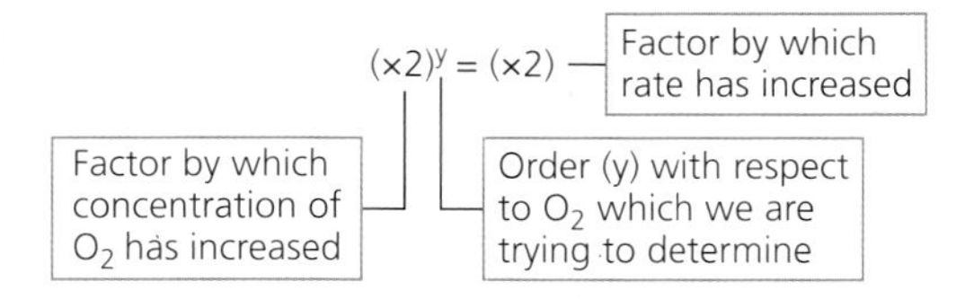

So order with respect to O_2 is 1 as $(\times 2)^1 = (\times 2)$

Rate equation: $rate = k[NO]^2[O_2]$

To calculate a value of the rate constant at this temperature, substitute in any values from the table.

If we use experiment 1:

$[NO] = 4 \times 10^{-3}\,mol\,dm^{-3}$

$[O_2] = 1 \times 10^{-3}\,mol\,dm^{-3}$

$rate = 6 \times 10^{-4}\,mol\,dm^{-3}\,s^{-1}$

The rate equation determined was: $rate = k[NO]^2[O_2]$

So substituting the values from experiment 1 into the equation:

$6 \times 10^{-4} = k(4 \times 10^{-3})^2(1 \times 10^{-3})$

$6 \times 10^{-4} = k(1.6 \times 10^{-8})$

Units of the rate constant were discussed previously. The units of an overall order 3 reaction are $mol^{-2}\,dm^6\,s^{-1}$.

$k = 37\,500\,mol^{-2}\,dm^6\,s^{-1}$.

Using the rate equation and the value of k.

$rate = k[NO]^2[O_2] = 37\,500 \times (5 \times 10^{-3})^2(5 \times 10^{-3}) = 4.6875 \times 10^{-3}\,mol\,dm^{-3}\,s^{-1}$

TIP

In this type of question it may tell you which experiment to use to calculate the rate constant but if it does not, use the first one as this is often the simplest one.

EXAMPLE 7

Gases Y and Z react according to the following equation.

$$2Y(g) + Z(g) \rightarrow S(g) + T(g)$$

The initial rate of reaction was measured in a series of experiments at a constant temperature.

Experiment	Initial [Y]/ $mol\,dm^{-3}$	Initial [Z]/ $mol\,dm^{-3}$	Initial rate/ $mol\,dm^{-3}\,s^{-1}$
1	1.7×10^{-2}	2.4×10^{-2}	7.40×10^{-5}
2	5.1×10^{-2}	2.4×10^{-2}	2.22×10^{-4}
3	8.5×10^{-2}	1.2×10^{-2}	9.25×10^{-5}
4	3.4×10^{-2}	4.8×10^{-2}	5.92×10^{-4}

Deduce the orders of reaction with respect to Y and Z. Write a rate equation for the reaction at this temperature.

Calculate a value for the rate constant (k) at this temperature and deduce its units.

A general rate equation would be:

$$\text{rate} = k[Y]^x[Z]^y$$

Answers

Experiments 1 and 2:

[Y] × 3

[Z] × 1

rate ×3

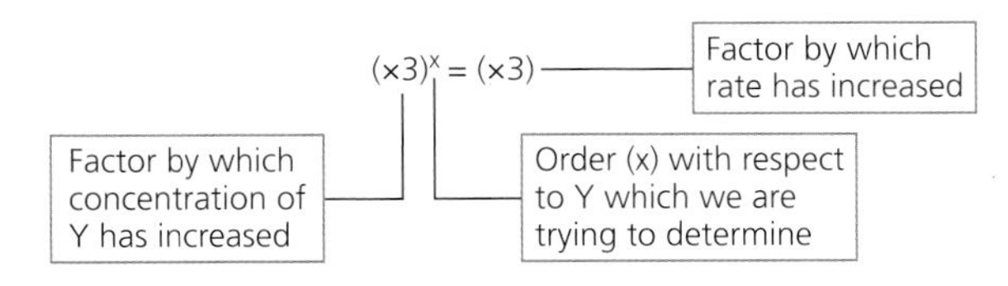

So order with respect to Y is 1 as $(\times 3)^1 = (\times 3)$

There are no two experiments in which the concentration of Y does not change so you can choose any two experiments now that you know the order of reaction with respect to Y.

Experiments 1 and 3:

[Y] × 5

[Z] × 0.5

rate ×1.25

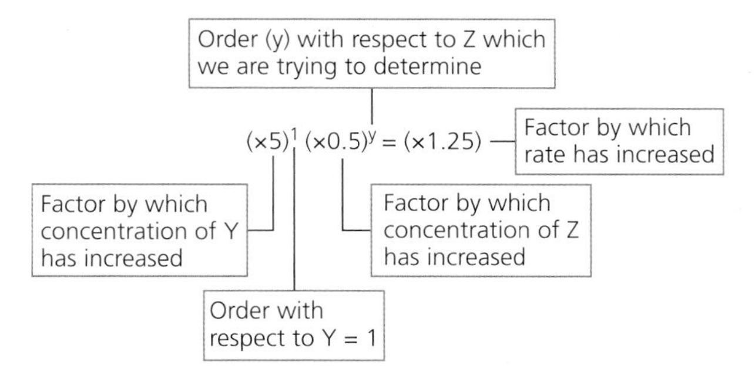

This simplifies to:

$$\times 5\,(\times 0.5)^y = \times 1.25$$

$$(\times 0.5)^y = \frac{\times 1.25}{\times 5} = \times 0.25$$

$$(\times 0.5)^y = \times 0.25$$

$$y = 2 \text{ (as } 0.5^2 = 0.25)$$

The order of reaction with respect to Z is 2.

The rate equation is rate $= k[Y][Z]^2$

TIP

Remember that orders of reaction have to be 0 (zero), 1 or 2.

To calculate a value of the rate constant at this temperature, substitute in any values from the table.

If we use experiment 1:

$$[Y] = 1.7 \times 10^{-2}\,mol\,dm^{-3}$$

$$[Z] = 2.4 \times 10^{-2}\,mol\,dm^{-3}$$

$$\text{rate} = 7.40 \times 10^{-5}\,mol\,dm^{-3}\,s^{-1}$$

The rate equation determined was:

$$\text{rate} = k[Y][Z]^2$$

So substituting the values from experiment 1 into the equation:

$$7.40 \times 10^{-5} = k(1.7 \times 10^{-2})(2.4 \times 10^{-2})^2$$

$$7.40 \times 10^{-5} = k(9.792 \times 10^{-6})$$

Units of the rate constant were discussed previously. The units of an overall order 3 reaction are $mol^{-2}\,dm^6\,s^{-1}$.

$$k = 7.56\,mol^{-2}\,dm^6\,s^{-1}.$$

Effect of temperature on the rate constant

All of the previous work in calculating a value of the rate constant (k) has depended on the measurements being taken at a constant temperature. The rate constant is dependent on temperature. If a constant temperature had not been maintained, the rate would have varied based on changes in concentrations of reactants as well as temperature.

- As temperature increases, the rate constant increases.
- The rate constant increases exponentially as temperature increases.
- A graph of the rate constant (k) against temperature (measured in Kelvin) would be as shown on the left.

The Arrhenius equation

The Arrhenius equation links the rate constant with activation energy and temperature. It is written as

$$k = Ae^{-\frac{E_a}{RT}}$$

where k is the rate constant, A is the Arrhenius constant, e is a mathematical constant (2.71828), E_a is the activation energy, R is the gas constant (8.31 J K^{-1} mol^{-1}) and T is the temperature measured in kelvin (K).

TIP

e^x can be accessed on your calculator as it will be the shift function of natural log (which is written ln). Check you can use e^x; e^1 should be 2.71828 with the 1828 recurring.

In this equation E_a has units of J mol^{-1}. This means that $\frac{E_a}{RT}$ does not have any units as J mol^{-1} is divided by (J K^{-1} mol^{-1} × K). $e^{-\frac{E_a}{RT}}$ will not have any units either, so the Arrhenius constant will have the same units as the rate constant.

EXAMPLE 8

A reaction has an activation energy of 120 kJ mol^{-1} at room temperature (298 K). The gas constant R is 8.31 J K^{-1} mol^{-1}. A = 2.2 × 10^{12} s^{-1}. Calculate a value for this rate constant at this temperature. State its units.

Answer

Initially determine $\frac{E_a}{RT}$. E_a = 120000 J mol^{-1}; R = 8.31 J K^{-1} mol^{-1} and T = 298 K

$$\frac{E_a}{RT} = \frac{120\,000}{8.31 \times 298} = 48.46.$$

$$k = Ae^{-\frac{E_a}{RT}} = 2.2 \times 10^{12} \times e^{-48.46} = 1.98 \times 10^{-9}\ s^{-1}.$$

The units of the rate constant are the same as the units of the Arrhenius constant.

Effect of changes in temperature and presence of a catalyst

Recalculating the value of k at 308 K using the same value for the Arrhenius constant used in Example 8 gives

$$\frac{E_a}{RT} = \frac{120\,000}{831 \times 308} = 46.88.$$

$$k = Ae^{-\frac{E_a}{RT}} = 2.2 \times 1012 \times e^{-46.88} = 9.61 \times 10^{-9}\ s^{-1}.$$

This shows an almost five-fold ($9.61 \times 10^{-9}/1.98 \times 10^{-5} = 4.85$) increase in the rate of reaction with a small increase in temperature.

TIP

Remember that doubling the concentration of a first order reactant will only double the rate of reaction so increasing temperature will have a greater effect on the rate of the reaction than changing the concentration. This was discussed at AS.

However, adding a catalyst has a much more dramatic effect on the rate of reaction. A catalyst for this reaction lowers the activation energy to 100 kJ mol^{-1} at 298 K. The Arrhenius constant is unchanged.

Recalculating the value of k at 298 K:

$$-\frac{E_a}{RT} = \frac{100\,000}{8.31 \times 298} = 40.38.$$

$$k = Ae^{-\frac{E_a}{RT}} = 2.2 \times 10^{12} \times e^{-40.38} = 6.39 \times 10^{-6}\ s^{-1}.$$

A catalyst which lowers the activation energy from 120 kJ mol^{-1} to 100 kJ mol^{-1} increases the rate of reaction by a factor of 3228 ($6.39 \times 10^{-6}/1.98 \times 10^{-9} = 3228.1$). Catalysts have a very significant effect on the rate of a chemical reaction.

Graphical analysis

The Arrhenius equation can be converting using natural logs (represented by ln) to:

$$\ln k = -\frac{E_a}{RT} + \ln A.$$

- This is in the form of $y = mx + c$ so a graph of ln k against 1/T is drawn.
- The gradient of this graph will be $-\frac{E_a}{R}$ and the intercept with the ln k axis will be ln A.
- As the gas constant, R, is given as (8.31 J K^{-1} mol^{-1}), E_a can be calculated from the gradient of the line.
- The intercept is ln A. This can be converted to A using the inverse or shift function on your calculator.

EXAMPLE 9

The rate constant for a reaction was calculated at different temperatures as shown below.

Temperature/K	k/s^{-1}
300	1.75×10^{-3}
310	6.60×10^{-3}
320	2.50×10^{-2}
330	7.60×10^{-2}
340	2.40×10^{-1}

The quantities 1/T and ln k were calculated and presented in the table below.

1/T	ln k
3.33×10^{-3}	-6.35
3.23×10^{-3}	-5.02
3.13×10^{-3}	-3.69
3.03×10^{-3}	-2.58
2.94×10^{-3}	-1.43

Answer

These results are plotted on the graph of ln k against 1/T as shown below:

The intercept on the ln k axis is at 37. This would indicate that ln A = 37 so A = e^{37}. A = 1.17×10^{16} s^{-1}. A has the same units as k given in the table.

The gradient of the line is negative as the line has a negative slope and is −37/0.00285 = −12982.5. This is equal to $-E_a/R$ so $E_a = 12982.5 \times R = 12982.5 \times 8.31 = 107884.6$ J mol^{-1}. This is converted to kJ mol^{-1} by dividing by 1000. So E_a for the reaction at this temperature is 107.9 kJ mol^{-1}.

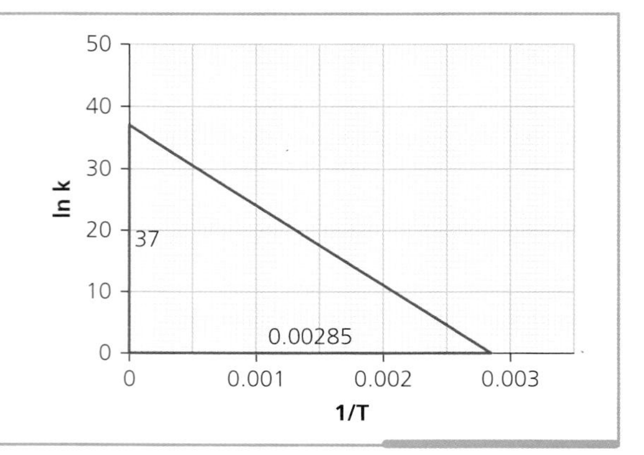

TIP

In these types of question, the equations and any constants required would be given to you with the required units. Remember that E_a should be in J mol^{-1} in the Arrhenius equation and A has the same units as the rate constant.

TEST YOURSELF 3

1 In the rate equation: rate = $k[C][D]^2$

a) State the order of reaction with respect to C and D.

b) State the units of the rate constant (k).

2 A series of experiments were carried out on the rate of reaction between A and B at a constant temperature.
The rate equation for the reaction is rate = $k[A]^2$
The data obtained are given in the table below. Complete the table.

Experiment	Initial [A]/ mol dm^{-3}	Initial [B]/ mol dm^{-3}	Initial rate/ mol dm^{-3} s^{-1}
1	1.5×10^{-2}	2.4×10^{-2}	1.2×10^{-5}
2	3.0×10^{-2}	2.4×10^{-2}	
3		3.6×10^{-2}	4.8×10^{-5}

3 R reacts with S. The kinetics of the reaction were studied at a constant temperature. The data obtained are given in the table below.

Experiment	Initial [R]/ mol dm^{-3}	Initial [S]/ mol dm^{-3}	Initial rate/ mol dm^{-3} s^{-1}
1	0.25	0.15	0.0225
2	0.25	0.30	0.0900
3	0.50	0.30	0.1800
4	0.75	0.45	0.6075

a) Deduce the order of reaction with respect to R and S.

b) Write a rate equation for the reaction.

c) Calculate a value of the rate constant (k) at this constant temperature and deduce its units.

4 What do k, A, R, T and E_a stand for in the expression $k = Ae^{-\frac{E_a}{RT}}$?

Methods of determining rate of reaction

Measuring the rate of a chemical reaction depends on being able to follow the progress of the reaction against time. This involves being able to measure a change in the amount or concentration of a reactant or product during the reaction.

From the equation of the reaction identify a reactant or a product which can be measured. Examples will include:

1 a gaseous product which can be monitored using a *gas syringe* or by *loss in mass*

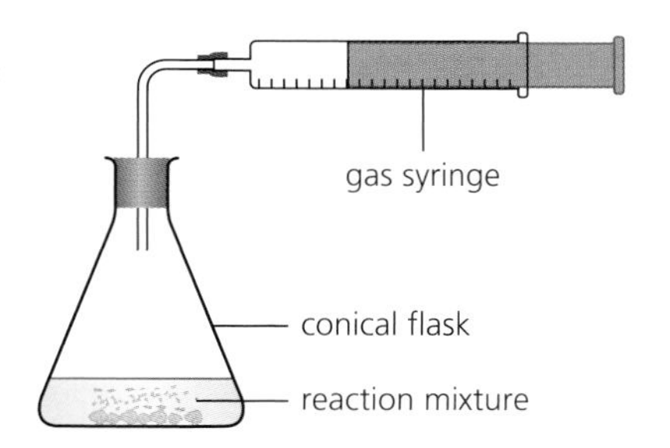

2 a coloured reactant or product which can be monitored using *colorimetry*

3 a titratable reactant or product which can be monitored by *sampling*, *quenching* and *titrating*

4 a directly measurable reactant or product, i.e. H^+ ions or OH^- ions by measuring pH using a pH meter.

Measuring gas volume

- A **gas syringe** is a ground glass syringe which is attached to a sealed reaction vessel and measures the volume of gas produced.
- This is measured against time and a graph of gas volume against time is plotted.
- The gradient of the tangent at $t = 0$ s gives the initial rate of reaction.
- This is the rate of reaction.

Example of gas volume graph

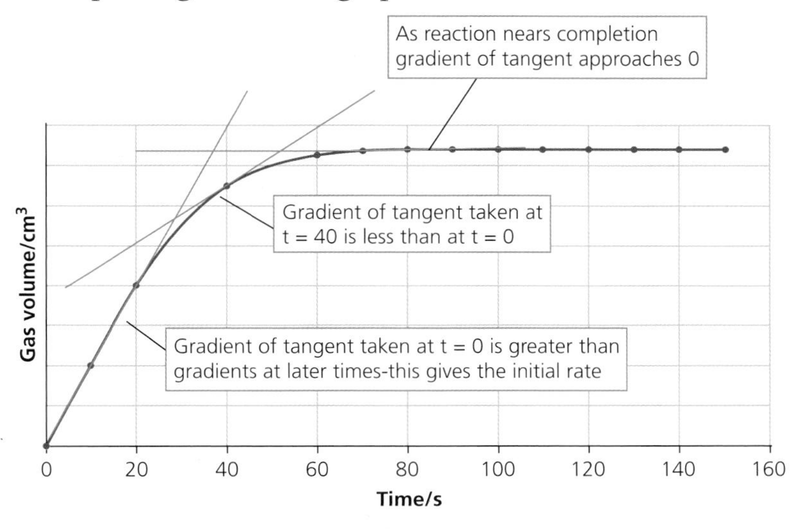

Measuring change in mass

- A reaction in which a gas is produced may also be monitored by measuring the mass over a period of time.
- The curve of the graph decreases again and initial tangent at t = 0 s gives a measure of initial rate of reaction.

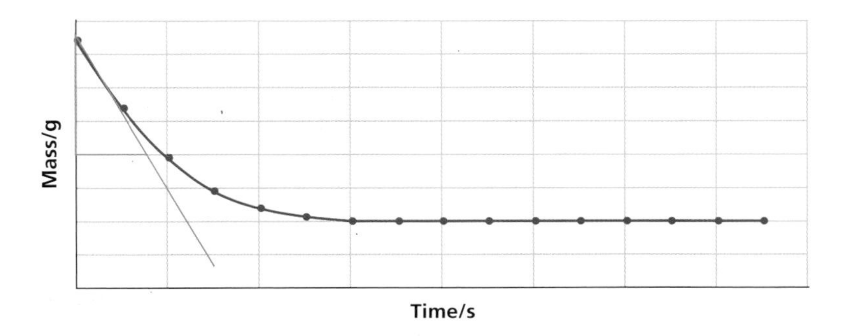

Example of a graph showing change in mass against time

Gas syringe measurements and change in mass measurements are used for **initial rate monitoring**.

> **TIP**
> Mass and gas volume in a specified volume are equivalent to mol $dm^{-3} s^{-1}$ as the gas volume would be divided by a constant (V_m) and the mass lost would be divided by a constant M_r and the reaction would be carried out in a constant volume.

- This means several experiments changing the concentration of the each reactant are carried out and initial rate measurement taken.
- A typical table of results for changes in the concentration of reactant A for the reaction A + B → C would be:

[A]/mol dm^{-3}	Rate/mol dm^{-3} s^{-1}
0.01	0.0025
0.02	0.0050
0.03	0.0075
0.04	0.0100
0.05	0.0125

- The initial rate can be related to the initial concentration and the order determined as shown before or a graph of rate against concentration can be drawn. The shape of this graph gives the order of reaction with respect to the reactant the concentration of which you were changing.
- It could be worked out that the order of reaction with respect for A is 1 as when the concentration doubles the rate doubles.
- A graph of rate against concentration for a first order reaction such as this would look like the graph below.

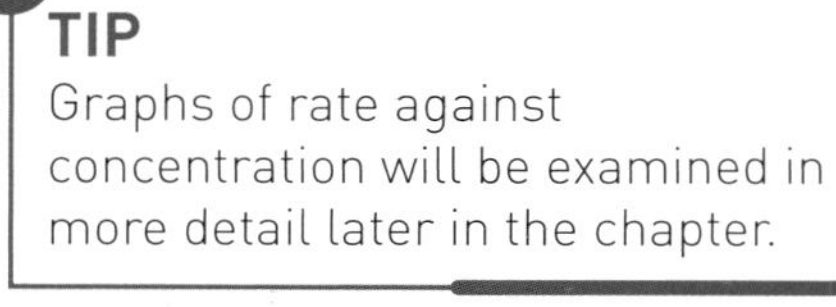

TIP

Graphs of rate against concentration will be examined in more detail later in the chapter.

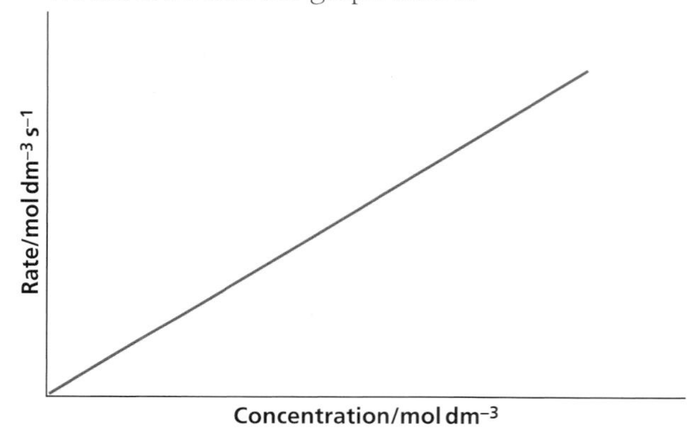

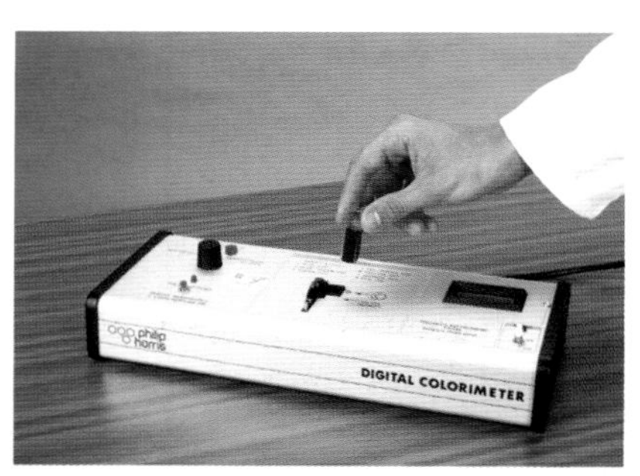

Figure 3.3 A colorimeter

Measuring a coloured reactant or product

A **colorimeter** measures colour intensity of a solution. It must be done with an appropriate coloured filter. For example if you are measuring the intensity of a blue solution, you should use a red filter as the solution is absorbing red light. The amount of red light absorbed relates directly to the concentration. A calibration curve should be set up first with known concentrations of the reactant or product so that you can directly relate the colorimeter reading to the concentration.

A calibration curve

This calibration curve allows absorbance values to be directly converted to concentration. Graphs of concentration against time can then be drawn.

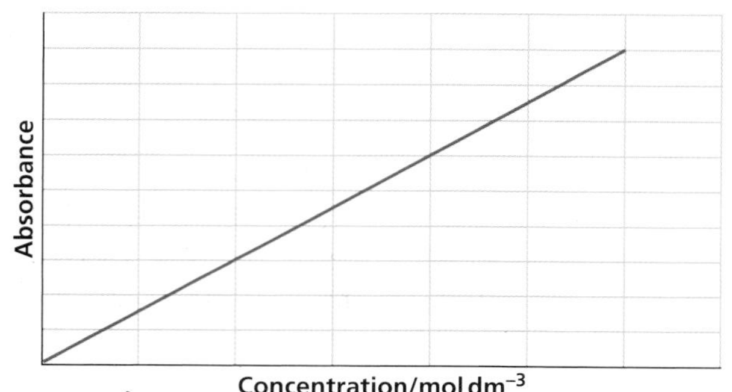

Example of a graph for colorimeter readings for a coloured product

The gradient of the tangent at $t = 0$ s gives the initial rate of reaction.

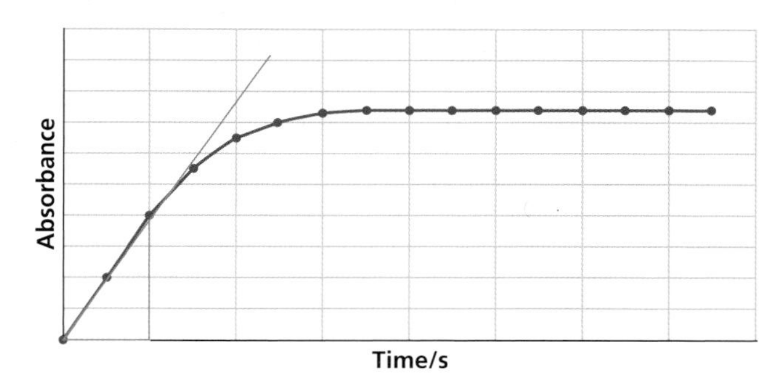

Measurement of a coloured product is used for initial rate monitoring.

- This means several experiments are carried out changing the concentration of each reactant.
- The gradient of the tangents at $t = 0$ s for each concentration give the initial rate of reaction for each of the concentrations used.
- The initial rate can then be related to the initial concentration to determine the order. The order can be determined from a graph of rate against concentration. The shape of this graph gives the order of reaction with respect to the reactant the concentration of which you were changing.

Example of a graph of colorimeter reading for a coloured reactant

Measurement of a coloured reactant can be used for continuous rate monitoring.

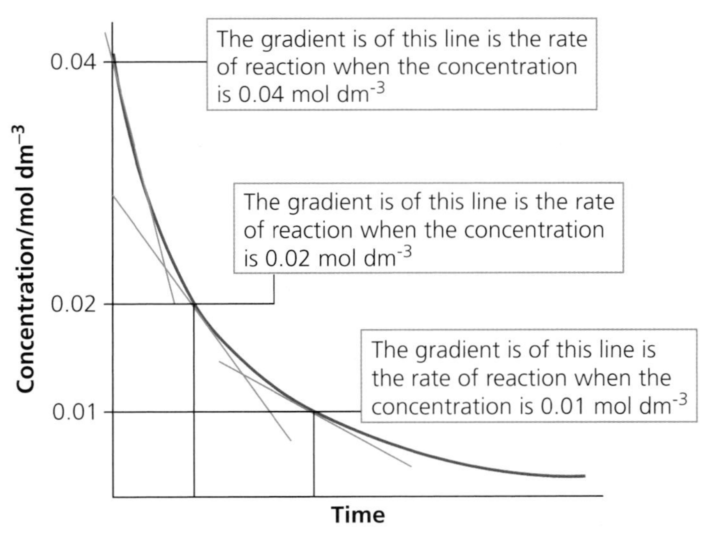

- This means one experiment is carried out and the colorimeter readings are converted to concentration using the calibration curve.
- A graph of concentration against time is drawn and the shape of this curve can give the order with respect to the coloured reactant.
- A gradient of a tangent at any concentration on the graph is a measure of rate at that concentration.
- The rate can then be related to the concentration and the order determined or a graph of rate against concentration can be drawn. The shape of this graph gives the order of reaction with respect to the reactant, the concentration of which you were changing.

Measuring a reactant or product by titration

Sampling, quenching and titrating allows a titratable reactant or product to be measured during the course of the reaction.

A sample is taken at various times, and the reaction is quenched – this means the reaction is stopped. Methods of quenching include rapid cooling, adding a chemical to remove a reactant which is not being monitored or adding a large known volume of water to the sample. The sample may then be titrated to find the concentration of the reactant or product.

From the data, a graph of concentration against time can be drawn.

Measurement of a titratable **product** is used for **initial rate monitoring**.

- This means several experiments changing the concentration of each reactant are carried out and initial rate measurement taken from the graphs of concentration against time.
- The initial rate can then be related to the initial concentration and the order determined as shown before or a graph of rate against concentration can be drawn. The shape of this graph gives the order of reaction with respect to the reactant the concentration of which you were changing.

Measurement of a titratable **reactant** may be used for **continuous rate monitoring**.

- This means one experiment is carried out and a graph of concentration against time is drawn and often the shape of this curve can give the order with respect to the titratable reactant.
- A gradient of a tangent at any concentration on the graph is a measure of rate at that concentration.
- The rate can then be related to the concentration and the order determined as shown before or a graph of rate against concentration can be drawn. The shape of this graph gives the order of reaction with respect to the reactant the concentration of which you were changing.

pH monitoring

$pH = -\log_{10}[H^+]$, so $[H^+] = 10^{(-pH)}$

- pH values can be taken over time and using the above equation they can be converted directly to $[H^+]$
- A graph of $[H^+]$ against time can be drawn.
- If the H^+ ions are a reactant the shape of this curve can give the rate or again gradients of tangents at various $[H^+]$ can be taken which equal rate.
- The rate can then be related to the concentration and the order determined as shown before or a graph of rate against concentration can be drawn. The shape of this graph gives the order of reaction with respect to the reactant the concentration of which you were changing.

TIP

The interconversion between pH and $[H^+]$ will be covered in detail in the **Acids and bases** unit (Chapter 6).

Rate monitoring

The rate of reaction may be measured in various ways but the methods used depend on either initial rate monitoring or continuous rate monitoring.

Initial rate monitoring

Use a reactant or product which is measurable to enable the initial rate of reaction to be measured. For example:

- gas production – measured using a gas syringe
- coloured substance – measured using a colorimeter
- hydrogen ions/hydroxide ions – measured using a pH meter or by quenching and titrating.

Method

1 Pick one reactant for which you will determine the order of reaction.

2 Carry out several experiments *at different concentrations for this reactant* measuring the quantity against time, for example gas volume against time.

3 Plot graphs of the measurable quantity (for example gas volume) against time.

4 Draw a tangent at $t = 0\,s$ and determine the initial gradient of this tangent.

5 The gradient of the tangent at $t = 0\,s$ is the initial rate of the reaction.

6 Repeat for variety of concentrations of this reactant.

7 Plot a graph of the initial rate of reaction against concentration and the shape of the graph gives the order of reaction with respect to the reactant for which the concentration was varied.

8 Repeat for all other reactants to determine order of reaction for each one.

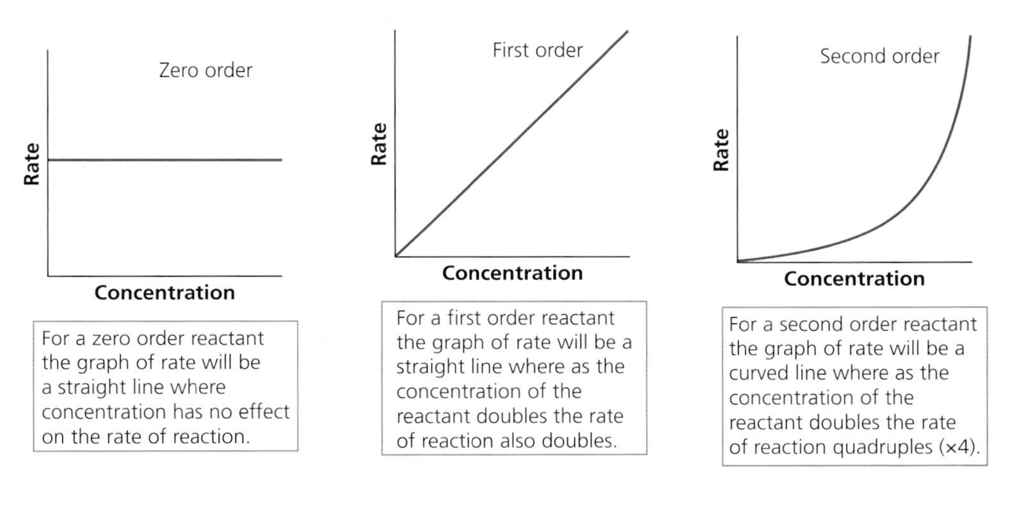

Continuous rate monitoring

If the concentration of a reactant can be determined directly, for example:

- hydrogen ions determined directly from pH measurements or by quenching and titrating
- coloured reactant determined from colorimeter readings using calibration curve.
- a specific reactant may be titrated.

Method

1 Allow the reaction to progress and take readings (colorimeter/pH) or samples at various times.

2 Any samples taken should be quenched to stop the reaction and titrated – quenching can be carried out by rapid cooling/adding large quantities of cold water or by adding a chemical which will remove a reactant and stop the reaction.

3 Plot a graph of concentration against time for this reactant and the shape of the graph gives the order with respect to this reactant.

Here are the three types of graphs, for orders of reaction 0, 1 and 2.

TIP
If you are asked to sketch the graphs, ensure the line or curve starts on the concentration (y) axis.

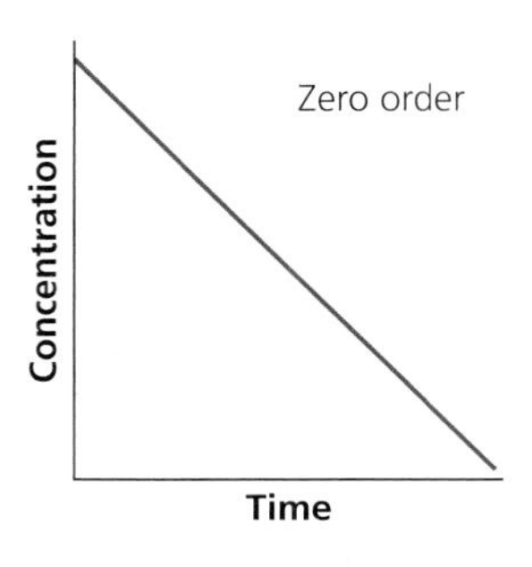

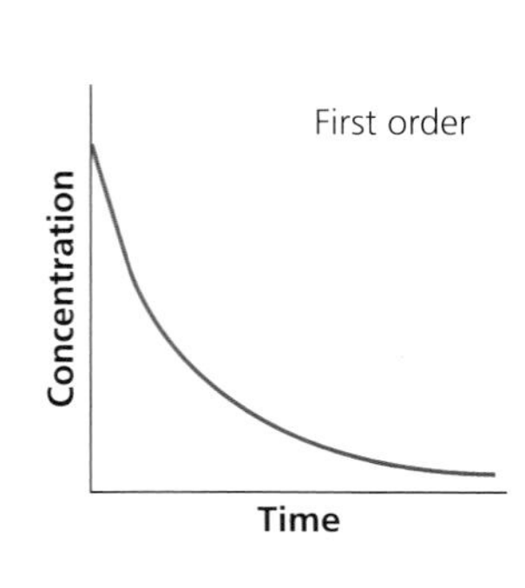

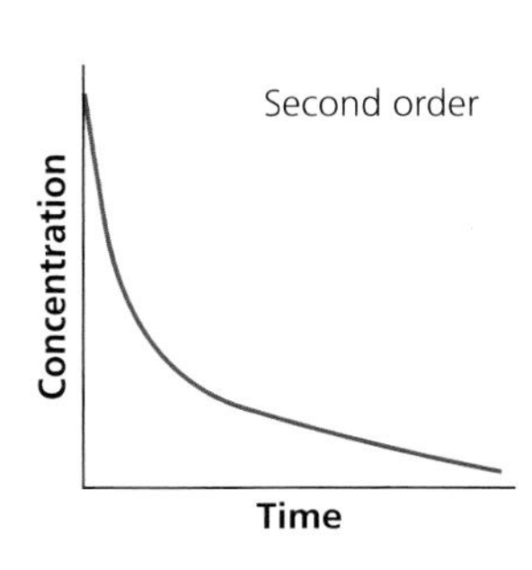

TIP
For reactants which are zero order, the gradient of the concentration against time graph is equal to the rate constant, *k*.

A few rate values are determined at different concentrations and then a rate against concentration graph may be plotted in order to determine a clearer order.

The shape of the rate against concentration graphs can help to determine the order of reaction. These graphs will have the same shapes as shown on page 59 depending on the order of reaction.

Using concentration against time graphs

Gradients of tangents may be taken at various concentrations to determine rate.

For example the graph below shows the concentration of reactant A against time. A tangent to the curve is drawn at a concentration of 0.200 mol dm^{-3}.

At a concentration of A of 0.200 mol dm^{-3} the gradient of the tangent:

$$= \frac{0.240}{3500} = 6.86 \times 10^{-5}\,\text{mol dm}^{-3}\,\text{s}^{-1}.$$

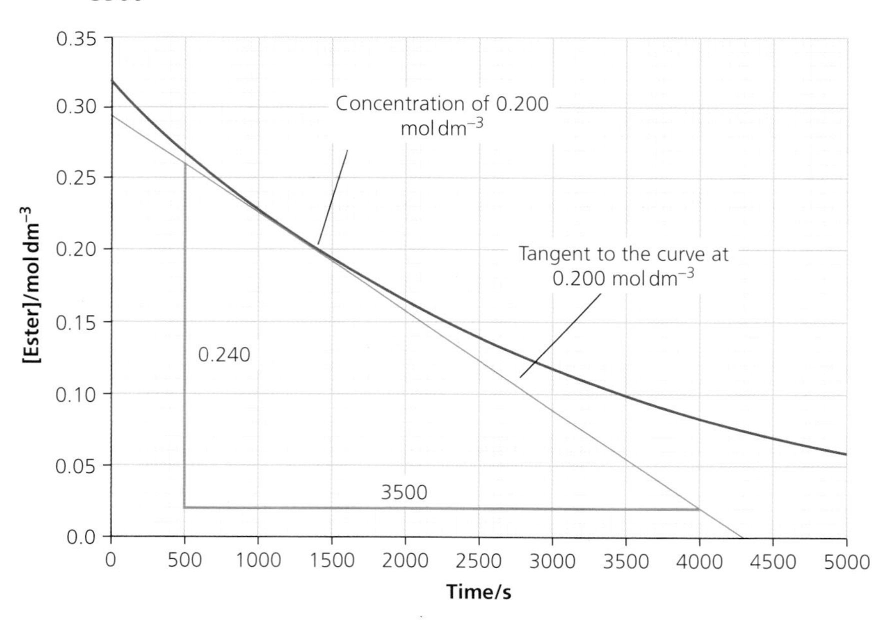

TIP
Any sensible value will be accepted as long as you draw the tangent carefully and use the scale to calculate rise over run. This tangent would have a negative gradient but in terms of rate is taken as the positive value as it is the disappearance of a reactant. You can learn more about tangents and gradients in the maths chapter.

EXAMPLE 10

Methanoic acid undergoes an acid-catalysed reaction with bromine according to the equation:

$$Br_2(aq) + HCOOH(aq) \xrightarrow{H^+(aq)} 2Br^-(aq) + 2H^+(aq) + CO_2(g)$$

Describe an experimental method that could be used to determine the order of reaction with respect to bromine.

The rate equation was found to be:

rate = k$[Br_2]$[HCOOH]

Deduce the units for the rate constant.

On the axes below, sketch the expected shape of the graphs in this reaction.

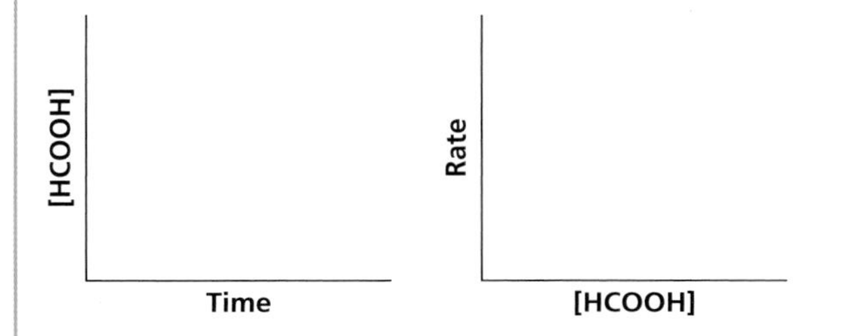

Answers

Measuring absorbance with a colorimeter

$Br_2(aq)$ is brown so use colorimetry with a blue filter:

- measure absorbance against time
- use calibration curve of known concentrations of bromine
- plot bromine concentration against time or rate against bromine concentration and determine order from shape of graph.

Alternative method: measure gas volume:

- measure CO_2 volume using gas syringe against time
- repeat with different bromine concentrations
- plot volume against time and measure tangent at $t = 0$
- plot rate against $[Br_2]$ and determine order from shape of graph.

The answer could be either progressive rate monitoring of the concentration of bromine against time with a colorimeter. A calibration curve may be used to convert from absorbance to concentration. Plotting $[Br_2]$ against time and use the shape of the graph to determine order with respect to Br_2, or take tangents to the $[Br_2]$ against time graph at various concentrations and plot a rate vs $[Br_2]$ graph and use the shape of this graph to determine the order with respect to Br_2.

The second answer involves using initial rate monitoring based on the CO_2 being produced. Several experiments with different initial concentrations of Br_2 were carried out and the volume of carbon dioxide measured using a gas syringe against time. Graphs of gas volume against time were plotted and the gradient of the tangent at $t = 0$ were taken which gave the initial rate. A graph of initial rate against the initial concentrations of Br_2 was plotted and the shape of this graph is used to determine the order with respect to Br_2.

The reaction is second order overall.

$$\text{rate} = k(\text{concentration})^2$$

$$\text{mol dm}^{-3}\text{s}^{-1} = k \times (\text{mol dm}^{-3})^2$$

$$\text{mol dm}^{-3}\text{s}^{-1} = k \times \text{mol}^2\text{dm}^{-6}$$

$$k = \frac{\text{mol dm}^{-3}\text{s}^{-1}}{\text{mol}^2\text{dm}^{-6}} = \text{mol}^{1-2}\text{dm}^{-3-(-6)}\text{s}^{-1} = \text{mol}^{-1}\text{dm}^{3}\text{s}^{-1}$$

The units of the rate constant are $mol^{-1} dm^3 s^{-1}$

From the rate equation, the reaction is first order with respect to HCOOH

A concentration against time graph for a first order reactant should look like this:

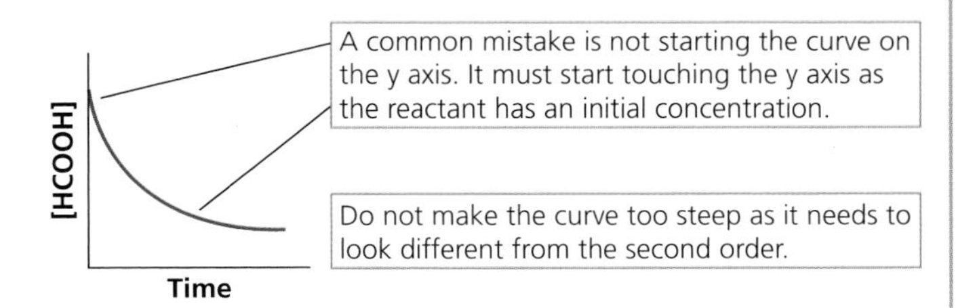

A rate against concentration graph for a first order reactant should look like this:

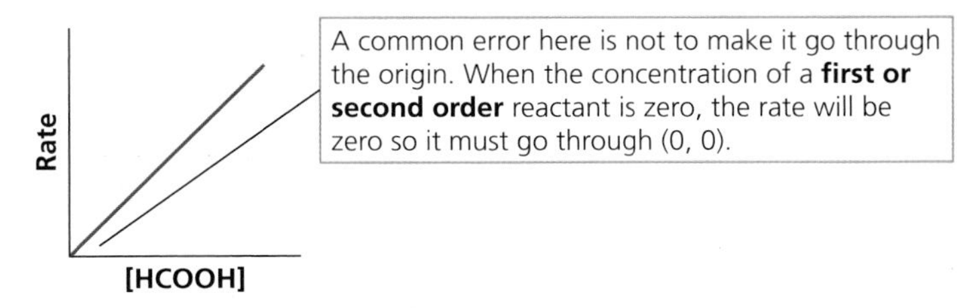

REQUIRED PRACTICAL 7

Investigation of the reaction of iodate(V) ions in acidic solution with sulfate(IV) ions to determine the order of the reaction with respect to hydrogen ions

The iodate(V) ion is an oxidising agent and reacts with sulfate(IV) ions in acidic solution to produce iodine in solution, according to the following equations:

Stage 1: $IO_3^- + 3SO_3^{2-} \rightarrow I^- + 3SO_4^{2-}$

Stage 2: $IO_3^- + 5I^- + 6H^+ \rightarrow 3H_2O + 3I_2$

Overall: $2IO_3^- + 5SO_3^{2-} + 2H^+ \rightarrow I_2 + 5SO_4^{2-} + H_2O$

Iodine is only liberated when acid is added. It is possible to determine the effect of changing the concentration of acid on the initial rate of this reaction by timing how long it takes for iodine to be produced. This is indicated by the formation of a blue-black colour with starch.

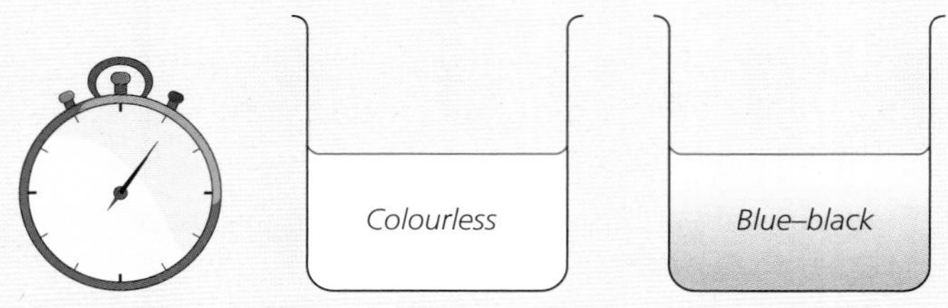

Figure 3.4 When the colourless solutions of acid, iodate(V) and sulfate(IV) are mixed in the presence of starch, there is no visible reaction. After a short time delay, the colour changes suddenly from colourless to blue–black. The colour change is sharp and dependable like a clock – the abrupt colour change occurs in a reproducible time lapse. This reaction is therefore known as an 'iodine clock reaction'.

Method

$15\,cm^3$ of the potassium iodate(V) solution, $85\,cm^3$ of deionised water and $5\,cm^3$ of starch were placed in a beaker. $15\,cm^3$ of sodium sulfate(IV) solution was mixed with $85\,cm^3$ of $0.1\,mol\,dm^{-3}$ sulfuric acid and poured into the beaker. The stop-clock was started simultaneously. The contents were stirred with a glass rod and the time taken (t, seconds) for the blue-black colour to appear was recorded. The experiment was repeated, varying the concentration of sulfuric acid by changing the volume of sulfuric acid and making the total volume of the up to $100\,cm^3$ by the addition of deionised water. The results are presented in the table below.

Table 3.1 Data from the iodine clock experiment

Expt	1	2	3	4	5	Initial
Acid volume/cm^3	25	35	45	55	70	85
Water volume/cm^3	60	50	40	30	15	0
Na_2SO_3 volume/cm^3	15	15	15	15	15	15
t(s)	42	28	20	15	11	9

All of the concentrations were constant except the concentration of acid, hence the rate equation for this reaction simplifies to:

$$\text{rate} = k[H^+]^z$$

where z is the order of reaction with respect to hydrogen ion concentration.

We could write

$$\log(\text{rate}) = z\log[H^+] + \text{constant}$$

The event being measured is fixed, i.e. the first appearance of the blue–black colour, so it is possible for the rate to be expressed as:

$$\text{Rate} = 1/t$$

The total volume of solution in each experiment is constant and so the $[H^+]$ can be represented by the volume of sulfuric acid. Therefore the rate expression becomes:

$$\log(1/t) = z\log(\text{volume of sulfuric acid}) + \text{constant}$$

1 For all the time results in the table
   a) Calculate $1/t$ (to 3 significant figures).
   b) Calculate log $1/t$ (to 3 significant figures).
   c) Calculate log (volume of sulfuric acid).
2 Plot a graph of log $(1/t)$ (y axis) against log V (volume)(x axis). Draw a best-fit straight line through the points.
3 Use your graph to determine the gradient of the straight line that you have drawn. Give your answer to 2 decimal places.
4 The experiment was designed to determine the order of reaction with respect to hydrogen ions in the reaction. What changes would you make to the experiment so that the order of reaction with respect to iodate(V) ions could be determined?
5 a) The reaction is first order with respect to hydrogen ions. In an experiment to determine the order of this reaction a value of 0.963 was obtained. Calculate the percentage error in this result.
   b) The experimental error resulting from the use of the apparatus was determined to be 2.1%. Explain what this means in relation to the practical technique used.

TIP

The graph is of the type $y = mx + c$ and therefore the gradient m will give a value for the order, z. See Chapter 16 for more information about logarithms.

TEST YOURSELF 4

1 For the reaction:
$CaCO_3(s) + 2HCl(aq) \rightarrow CaCl_2(aq) + CO_2(g) + H_2O(l)$
Suggest a method of monitoring the rate of reaction including practical detail.

2 For the reactions:
Reaction A: $2H_2O_2(aq) \rightarrow 2H_2O(l) + O_2(g)$
Reaction B: $CH_3COCH_3(aq) + I_2(aq) \rightarrow CH_3COCH_2I(aq) + H^+(aq) + I^-(aq)$
Reaction C: $2Na_2S_2O_3(aq) + I_2(aq) \rightarrow Na_2S_4O_6(aq) + 2NaI(aq)$

a) Explain how reaction A is monitored using initial rate monitoring using a gas syringe.
b) Explain why reaction B may be monitored in two ways.
c) Explain how reaction C is monitored using initial rate monitoring.

Order linked to mechanism of a reaction

The order of a reactant is clear from the rate equation and this may give a clue as to the details of the mechanism for the reaction.

Mechanisms

The mechanism for a reaction is a series of reactions which shows how the reaction occurs.

For example, the reaction

$$A + 2B \rightarrow AB_2$$

may occur in two stages:

$$A + B \rightarrow AB$$

$$AB + B \rightarrow AB_2$$

Adding equations to determine an overall equation

Taking the steps shown above, if these steps are added together, you will get the overall equation for the reaction.

$$A + B \rightarrow AB$$

$$AB + B \rightarrow AB_2$$

Write down all the species on the left-hand side of both equations and then cancel out anything which appears on both sides of the overall equation.

$$A + B + \cancel{AB} + B \rightarrow \cancel{AB} + AB_2$$

The overall equation is:

$$A + 2B \rightarrow AB_2$$

EXAMPLE 11

In the chlorination of methane the following reactions occur:

$Cl_2 \rightarrow 2Cl\bullet$

$CH_4 + Cl\bullet \rightarrow CH_3\bullet + HCl$

$CH_3\bullet + Cl_2 \rightarrow CH_3Cl + Cl\bullet$

$2Cl\bullet \rightarrow Cl_2$

Write an overall equation for the reaction.

Answer

$Cl_2 + CH_4 + Cl\bullet + CH_3\bullet + Cl_2 + 2Cl\bullet \rightarrow 2Cl\bullet + CH_3\bullet + HCl + CH_3Cl + Cl\bullet + Cl_2$

This simplifies to:

$\cancel{2}Cl_2 + CH_4 + \cancel{3Cl\bullet} + \cancel{CH_3\bullet} \rightarrow \cancel{3Cl\bullet} + \cancel{CH_3\bullet} + HCl + CH_3Cl + \cancel{Cl_2}$

This may be written:

$Cl_2 + CH_4 \rightarrow CH_3Cl + HCl$

This is the overall equation for this reaction.

EXAMPLE 12

For the following two-step process:

$2NO(g) + O_2(g) \rightarrow 2NO_2(g)$

$3NO_2(g) + H_2O(l) \rightarrow 2HNO_3(aq) + NO(g)$

Write an overall equation for this reaction.

Answer

The key to this is the fact that $2NO_2$ is produced in the first reaction but $3NO_2$ is used in the second reaction.

Multiply the first reaction by 3 and the second reaction by 2 before adding them.

$6NO(g) + 3O_2(g) \rightarrow 6NO_2(g)$

$6NO_2(g) + 2H_2O(l) \rightarrow 4HNO_3(aq) + 2NO(g)$

Adding the equations gives:

$\cancel{6}NO(g) + 3O_2(g) + \cancel{6NO_2(g)} + 2H_2O(l) \rightarrow \cancel{6NO_2(g)} + 4HNO_3(aq) + \cancel{2NO(g)}$

The overall equation is:

$4NO(g) + 3O_2(g) + 2H_2O(l) \rightarrow 4HNO_3(aq)$

Rate-determining step

Any of the steps in a mechanism for a reaction may be the rate-determining step. The rate-determining step is the slowest step and it dictates the overall rate of the reaction.

The species (atoms, molecules, ions or molecules) that react in the rate-determining step are the species which are present as concentrations with non-zero orders in the rate equation.

Reactants in the overall equation which have zero order are not involved in the rate-determining step.

Finding the rate-determining step

The rate-determining step may be determined by examining the rate equation and the steps in the mechanism for the reaction.

If there is a reactant in the overall equation which is zero order (i.e. it does not appear in the rate equation), this reactant should not appear in the rate-determining step.

EXAMPLE 13

In the hydrolysis of C_4H_9Br, the following overall reaction occurs:

$C_4H_9Br + OH^- \rightarrow C_4H_9OH + Br^-$

The mechanism for the reaction can be described in two steps:

Step 1: $C_4H_9Br \rightarrow C_4H_9^+ + Br^-$

Step 2: $C_4H_9^+ + OH^- \rightarrow C_4H_9OH$

The rate equation for the reaction is:

rate = $k[C_4H_9Br]$

Deduce the rate-determining step in this two-step process. Explain your answer.

Answer

The rate-determining step is step 1.

Explanation: OH^- ions are zero order so therefore not part of the rate-determining step. Step 1 does not involve OH^- ions.

EXAMPLE 14

For the overall reaction:

$X + 2Y \rightarrow XY_2$

The mechanism for the reaction occurs in two steps:

Step 1: $X + Y \rightarrow XY$

Step 2: $XY + Y \rightarrow XY_2$

The rate equation for this reaction is given as:

rate = $k[X][Y]$

Determine which step in the reaction is the rate-determining step.

Answer

This rate equation shows that the orders of reaction with respect to both X and Y are 1.

The first step in the mechanism shows X and Y reacting in a 1:1 ratio.

This suggests that step 1 is the rate-determining step and both X and Y are present in the rate equation in a 1:1 ratio.

TIP

If the rate equation had been rate = $k[X][Y]^2$ then the second step would be the rate-determining step as it (when combined with the first step) gives the ratio of X:Y as 1:2. The orders of reaction are equal to the number of reacting particles.

TEST YOURSELF 5

1 For the reaction:
$2X + Y \rightarrow Z$
The reaction occurs in a two-step process
Step 1: $2X \rightarrow X_2$
Step 2: $X_2 + Y \rightarrow Z$
The overall rate equation for the reaction is:
rate = $k[X]^2$

a) What is order of reaction with respect to X?
b) What is order of reaction with respect to Y?
c) State and explain which step is the rate-determining step.

2 In the hydrolysis of an ester A (CH_3COOCH_3) with OH^- ions, the order of reaction with respect to the ester, A, is first order and first order with respect to OH^- ions.

a) State the overall order of reaction.
b) Write a rate equation for the hydrolysis of ester A using A to represent the ester.
c) State the units of the rate constant.

3 For the following two-step process
Step 1: $A + B \rightarrow AB$
Step 2: $AB + A \rightarrow A_2B$

a) Write an overall equation for the reaction.
b) The order of reaction with respect to A is 1 and the order of reaction with respect to B is 1.
i) State the overall order of reaction.
ii) Write a rate equation for the reaction.
iii) State and explain which step is the rate-determining step.

Practice questions

1 For the rate equation: rate = $k[A][B]^2$

Which one of the following are the units of the rate constant, k?

A s^{-1} **C** $mol^{-2}\,dm^6\,s^{-1}$

B $mol^{-1}\,dm^3\,s^{-1}$ **D** $mol^{-3}\,dm^9\,s^{-1}$ *(1)*

2 X reacts with Y to form Z according to the overall equation:

$X + 2Y \rightarrow Z$

The mechanism for the reaction is:

$2Y \rightarrow W$ (slow step)

$W + X \rightarrow Z$ (fast step)

Which one of the following is most likely to be the rate equation for the overall reaction?

A rate = k [Y] **C** rate = k [X][Y]

B rate = k $[Y]^2$ **D** rate = k $[X][Y]^2$ (1)

3 The following experimental data were determined for the reaction:

$X + Y \rightarrow Z$

$[X]/mol\,dm^{-3}$	$[Y]/mol\,dm^{-3}$	initial rate of reaction/ $mol\,dm^{-3}\,s^{-1}$
0.1	0.1	0.007
0.2	0.1	0.028
0.2	0.2	0.056

a) Deduce the orders with respect to X and Y. (2)

b) What is the overall order of reaction? *(1)*

c) Write a rate equation for this reaction. *(1)*

d) Calculate a value of the rate constant (k) and deduce its units. (2)

4 In the following series of reactions of A and B at constant temperature the following data were obtained.

Experiment	Initial [A]/ $mol\,dm^{-3}$	Initial [B]/ $mol\,dm^{-3}$	Initial rate/ $mol\,dm^{-3}\,s^{-1}$
1	1.4×10^{-2}	2.0×10^{-2}	3.760×10^{-4}
2	2.8×10^{-2}	2.0×10^{-2}	7.520×10^{-4}
3	2.8×10^{-2}	4.0×10^{-2}	1.504×10^{-3}

a) Deduce the order of reaction with respect to A. *(1)*

b) Deduce the order of reaction with respect to B. *(1)*

c) Write a rate equation for the reaction. *(1)*

d) Calculate a value for the rate constant (k) at constant temperature and deduce its units. (2)

5 P reacts with Q. The kinetics of the reaction were studied at a constant temperature.

The data obtained are given in the table below.

Experiment	Initial [P]/ $mol\,dm^{-3}$	Initial [Q]/ $mol\,dm^{-3}$	Initial rate/ $mol\,dm^{-3}\,s^{-1}$
1	0.10	0.10	0.0020
2	0.20	0.10	0.0040
3	0.30	0.10	0.0060
4	0.40	0.20	0.0160

a) Deduce the order of reaction with respect to P and Q. (2)

b) Write a rate equation for the reaction. *(1)*

c) Calculate a value of the rate constant (k) at this temperature and deduce its units. (2)

6 Two experiments were carried out on the rate of reaction between B and C at a constant temperature.

The rate equation for the reaction is rate = $k[B][C]^2$

The data obtained are given in the table below.

Experiment	Initial [B]/ $mol\,dm^{-3}$	Initial [C]/ $mol\,dm^{-3}$	Initial rate/ $mol\,dm^{-3}\,s^{-1}$
1	2.0×10^{-2}	4.0×10^{-2}	2.7×10^{-7}
2	4.0×10^{-2}	5.0×10^{-2}	To be calculated

a) Calculate a value for the rate constant (k) at this temperature using the results of experiment 1 and deduce its units. (2)

b) Calculate a value for the initial rate in experiment 2. *(1)*

7 a) Write the Arrhenius equation. *(1)*

b) A graph of ln k against 1/T was plotted for a first order reaction and the gradient of the line found to be –15024. The intercept on the ln k axis was 21.4. The gas constant, R, = 8.31 J K^{-1} mol^{-1}.

Using the expression: $\ln k = -E_a/RT + \ln A$

Calculate the value of A and E_a for this reaction.

State the units. *(4)*

c) Given that e = 2.71828, calculate a value for k at 300 K using the value of E_a and A calculated in (b). If you did not calculate a value for E_a or A in (b) use E_a = 130 kJ mol^{-1} and A =2.00 × 1010. State the units. *(3)*

8 The kinetics of the acid-catalysed hydrolysis of an ester were studied and the concentration of the ester plotted against time as shown below.

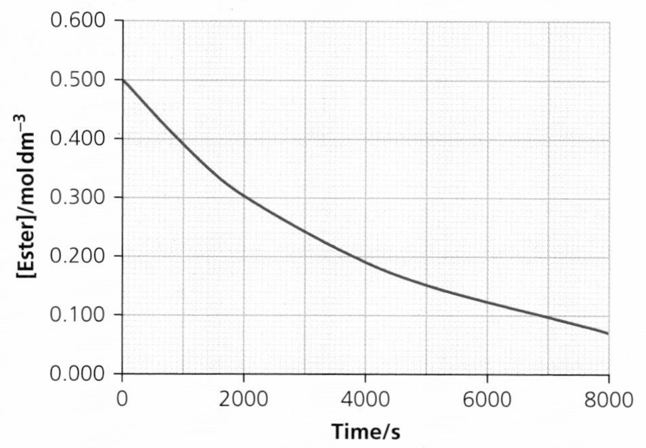

a) Draw a tangent to the curve when the concentration is 0.300 mol dm^{-3}. Use this tangent to determine the rate of reaction at this concentration of the ester. *(3)*

b) Based on the shape of the graph what is the order of reaction with respect to the ester? *(1)*

9 The reaction between propanone and iodine is catalysed by the presence of hydrogen ions, H^+.

$$CH_3COCH_3 + I_2 \rightarrow CH_3COCH_2I + HI$$

a) Suggest a method by which the order of reaction with respect to iodine could be determined using continuous rate monitoring. *(2)*

b) The graph below shows iodine concentration against time.

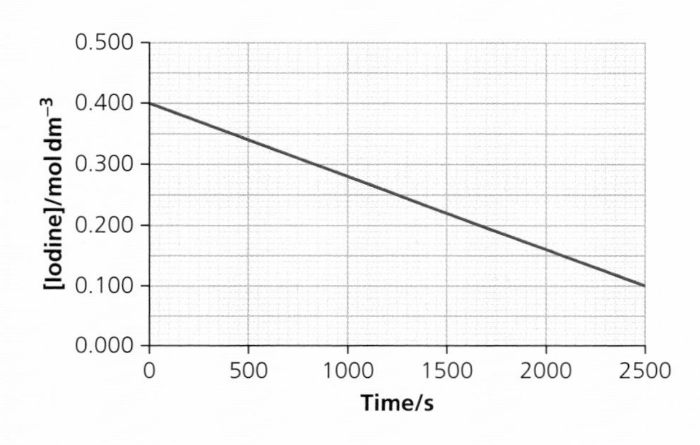

i) What is the order of reaction with respect to iodine? *(1)*

ii) Sketch a graph of rate of reaction against concentration of iodine. *(1)*

c) The rate equation for the reaction is

rate = $k[CH_3COCH_3][H^+]$

i) How does this rate equation support your answer to (b)(i). *(1)*

ii) The table below gives some initial rate data for this reaction. Use this data to calculate a value for the rate constant and state its units. *(2)*

Initial [H+]/ mol dm^{-3}	$[CH_3COCH_3]$/ mol dm^{-3}	Initial $[I_2]$/ mol dm^{-3}	Initial rate/ mol dm^{-3} s^{-1}
0.50	8.00	0.04	1.20×10^{-4}

Stretch and challenge

10 For the reaction D + E → F + G

The rate equation is rate = $k[D][E]^2$

The table below gives details of a series of experiments for this reaction.

Experiment	Initial concentration of D/mol dm^{-3}	Initial concentration of E/mol dm^{-3}	Initial rate/ mol dm^{-3} s^{-1}
1	0.05	0.10	2.25×10^{-3}
2	0.05	0.20	
3	0.10		4.50×10^{-3}
4		0.20	2.70×10^{-2}
5	0.10		8.82×10^{-2}
6			5.76×10^{-4}

a) Complete the missing data in the table for experiments 2 to 5. *(4)*

b) Determine a value for the rate constant using the results of experiment 1 and state its units. *(2)*

c) In experiment 6, the initial concentrations of D and E were in the ratio 1:4. The initial rate of reaction was determined to be 5.76×10^{-4} mol $dm^{-3}s^{-1}$. Using the value of the rate constant determined in (b) and the rate of reaction, determine the initial concentrations of D and E and fill these values into the table.

If you were unable to calculate a value for the rate constant in (b) use 10.0. This is not the correct value. *(4)*

Equilibrium constant K_p for homogeneous systems

PRIOR KNOWLEDGE

It is expected that you are familiar with all of the content of the **Chemical equilibria**, **Le Chatelier's principle and K_c** units in Year 1 of the A-level Chemistry course. The following are some key points of Prior Knowledge:

- A 'dynamic equilibrium' is an equilibrium where both the forward and reverse reactions are occurring at the same rate and the concentrations of the reactants and product remain constant.
- A homogeneous equilibrium is one in which all the reactants and products are in the same state.
- Le Chatelier's principle states that if a factor is changed which affects a system in equilibrium, the position of equilibrium will move in a direction so as to oppose the change.
- The position of equilibrium is often affected by changes in temperature, changes in concentrations of reactants or products, changes in pressure.
- For an equilibrium reaction where the enthalpy change is positive (endothermic), an increase in temperature moves the position of equilibrium from left to right.
- For an equilibrium reaction where the enthalpy change is negative (exothermic), an increase in temperature moves the position of equilibrium from right to left.
- For an equilibrium reaction, an increase in the total pressure moves the position of equilibrium to the side with the fewer moles of gas.
- For an equilibrium reaction, an increase in the concentration of a reactant will move the position of equilibrium from left to right.
- For an equilibrium reaction, an increase in the concentration of a product will move the position of equilibrium from right to left.
- A catalyst allows equilibrium to be achieved more rapidly but it has no effect on the position of equilibrium.
- The equilibrium constant, K_c, for the reaction $N_2(g) + 3H_2(g) \rightleftharpoons 2NH_3(g)$ is given by the expression:

  $$K_c = \frac{[NH_3]^2}{[N_2][H_2]^3}$$

  and the units of this K_c would be $mol^{-2}\,dm^6$.
- The value of K_c is only affected by a change in temperature.

TEST YOURSELF ON PRIOR KNOWLEDGE 1

1 What is meant by *dynamic equilibrium*?

2 State Le Chatelier's principle.

3 For the equilibrium:

$A(g) + 2B(g) \rightleftharpoons C(g) \quad \Delta H = -52\ kJ\ mol^{-1}$

a) Explain how the yield of C would change if the temperature were increased.

b) Explain how the yield of C would change if the total pressure were increased.

c) Explain why a catalyst does not affect the position of equilibrium.

d) Explain why this equilibrium is described as a homogeneous equilibrium.

4 For the gaseous homogeneous equilibrium given below:

$CH_4(g) + 2H_2O(g) \rightleftharpoons CO_2(g) + 4H_2(g)$

a) Write an expression for the equilibrium constant K_c for the reaction

b) 1.00 mol of CH_4 is mixed with 2.00 mol of $H_2O(g)$ in a container of volume 5.00 dm^3. At equilibrium there are 2.20 mol of hydrogen present in the equilibrium mixture. Calculate a value of the equilibrium constant, K_c, and state its units. Give your answer to 3 significant figures.

Mole fractions

The **mole fraction** of a component, A, in a gaseous mixture is denoted x_A and is calculated by dividing the amount, in moles, of A by the total amount of moles of gas in the mixture:

For

$$A(g) + B(g) \rightleftharpoons C(g)$$

$$x_A = \frac{n_A}{n_A + n_B + n_C}$$

where n_A, n_B and n_C are the amounts, in moles, of A, B and C present in the mixture.

Mole fractions can only have values in the range 0 to 1. The sum of the mole fractions of all of the components in a mixture will be equal to 1.

EXAMPLE 1

A mixture contains 1.42 g of nitrogen and 2.41 g of oxygen. Calculate the mole fraction of the gases in this mixture.

Answer

Moles of $N_2 = \frac{1.42}{28.0} = 0.0507$ mol

Moles of $O_2 = \frac{2.41}{32.0} = 0.0753$ mol

Total moles of gas = 0.0507 + 0.0753 = 0.126 mol

Mole fraction of $N_2 = \frac{0.0507}{0.126} = 0.402$

Mole fraction of $O_2 = \frac{0.0753}{0.126} = 0.598$

TIP

The total of the mole fractions should add up to 1 mol. For Example 1: 0.402 + 0.598 = 1.00

Partial pressure

The partial pressure of a component, A, in a mixture of gases is the contribution which that gas makes to the total pressure of the gas mixture. It is also the pressure if the same amount of A were the only substance present in a container of the same size. The partial pressure of A is denoted by p_A or $p(A)$ or pA or P_A.

The partial pressure of a component can be calculated by multiplying the total pressure of the mixture by the mole fraction of that component.

For example the partial pressure of A, $p(A)$, in a mixture is given by the expression:

$$p(A) = x_A \times P$$

where x_A is the mole fraction of A and P is the total pressure.

Although partial pressures and concentrations are not the same thing, they are both proportional to the amount of the substance present.

EXAMPLE 2

A mixture contains 1.00 g of ammonia and 1.00 g of nitrogen. The total pressure is 5.00×10^5 Pa. Calculate the partial pressure of each gas in the mixture.

Answer

Moles of $NH_3 = \frac{1.00}{17.0} = 0.0588$ mol

Moles of $N_2 = \frac{1.00}{28.0} = 0.0357$ mol

Total moles of gas = 0.588 + 0.0357 = 0.0945 mol

Mole fraction of $NH_3 = \frac{0.0588}{0.0945} = 0.622$

Mole fraction of $N_2 = \frac{0.0357}{0.0945} = 0.378$

Checking the calculation, the sum of the mole fractions is (0.622+0.378) = 1.

Partial pressure of NH_3 = mole fraction × total pressure

$= 0.622 \times 5.00 \times 10^5 = 3.11 \times 10^5$ Pa

Partial pressure of N_2 = mole fraction × total pressure

$= 0.378 \times 5.00 \times 10^5 = 1.89 \times 10^5$ Pa

Again a check that the sum of the partial pressures is equal to the total pressure $(3.11 \times 10^5 + 1.89 \times 10^5) = 5.00 \times 10^5$ Pa.

All given values are to 3 significant figures in this calculation so the answers should also be given to 3 significant figures.

TEST YOURSELF 2

1. A mixture contains 2.00 g of oxygen and 2.00 g of nitrogen. Calculate the mole fraction of each gas in the mixture.
2. Two gases were mixed and the total pressure was 100 kPa. One of the gases had a partial pressure of 40.0 kPa. Calculate the partial pressure of the other gas.
3. In the equilibrium, $N_2(g) + 3H_2(g) \rightleftharpoons 2NH_3(g)$, the equilibrium mixture of nitrogen, hydrogen and ammonia at 350 °C contains 7.00 g of nitrogen, 4.50 g of hydrogen and 36.5 g of ammonia.
   a) Calculate the mole fraction of each substance in the equilibrium mixture.
   b) Calculate the partial pressure of each gas in the mixture if the total pressure is 100 kPa.

K_p

For the general case of a gaseous homogeneous equilibrium:

$$aA(g) + bB(g) \rightleftharpoons cC(g) + dD(g)$$

the equilibrium law states that

$$K_p = \frac{p(C)^c p(D)^d}{p(A)^a p(B)^b}$$

where $p(A)^a$ is the partial pressure of A raised to the power of a, etc.

$$N_2(g) + 3H_2(g) \rightleftharpoons 2NH_3(g)$$

$$K_p = \frac{p(NH_3)^2}{p(N_2)p(H_2)^3}$$

Partial pressures in Pa

$$\text{units of } K_p = \frac{(Pa)^2}{(Pa)\,(Pa)^3} = \frac{(Pa)^2}{(Pa)^4} = \frac{1}{(Pa)^2} = Pa^{-2}$$

Calculation of K_p

Calculations to find K_p are similar to those to find K_c. The only difference is that after finding the number of moles at equilibrium, instead of using concentration $= \frac{\text{moles}}{\text{volume}}$, partial pressures are used where

partial pressure = mole fraction × total pressure.

MATHS

The unit most commonly used for pressure is Pascal (Pa). kPa (10^3 Pa) and MPa (10^6 Pa) may also be used. If the total pressure is given in kPa, then the partial pressure will also be in kPa. 100 kPa is the most frequently used value.

EXAMPLE 3

For the equilibrium

$$2NO(g) + O_2(g) \rightleftharpoons 2NO_2(g)$$

10.7 g of nitrogen(II) oxide and 7.33 g of oxygen were mixed and at equilibrium 13.4 g of nitrogen(IV) oxide were formed at a total pressure of 100 kPa and a temperature of 400 K.

1 Calculate the partial pressures of the three gases present in the equilibrium mixture.
2 Calculate a value for K_p at 400 K.

Answers

1 Initial moles of NO $= \frac{10.7}{30.0} = 0.357$ mol

Initial moles of $O_2 = \frac{7.33}{32.0} = 0.229$ mol

Equilibrium moles of $NO_2 = \frac{13.4}{46.0} = 0.291$ mol

	2NO	+	O_2	⇌	$2NO_2$
Initial moles	0.357		0.229		0.00
Reacting moles	−0.291		−0.146		+0.291
Equilibrium moles	0.066		0.083		0.291

Initial amounts in moles (given in the example):

0.357 mol of NO is mixed with 0.229 mol of O_2

Amount in moles of one substance at equilibrium (given in the example):

0.291 mol of NO_2 is present in the equilibrium mixture

Reacting moles of this substance (calculated from given information)

0.291 mol of NO_2 has formed (+0.291)

Reacting moles of other substances (calculated from balancing numbers in the equation):

0.291 mol of NO reacts with 0.146 mol of O_2 and formed 0.291 mol of NO_2

Equilibrium moles of other substances (the initial moles minus the reacting moles):

0.357 − 0.291 = 0.066 mol of NO remaining;
0.229 − 0.146 = 0.083 mol of O_2 remaining.

The total equilibrium moles present is = 0.066 + 0.083 + 0.291 = 0.440 mol

	2NO	O_2	$2NO_2$	
Initial moles	0.357	0.229	0.00	
Reacting moles	−0.291	−0.146	+0.291	
Equilibrium moles	0.066	0.083	0.291	Total equilibrium moles = 0.440
Mole fraction	$\frac{0.066}{0.440} = 0.150$	$\frac{0.083}{0.440} = 0.189$	$\frac{0.291}{0.440} = 0.661$	Check total of mole fractions = 1 0.150 + 0.189 + 0.661 = 1.00
Partial pressure	0.150 × 100 = 15.0 kPa	0.189 × 100 = 18.9 kPa	0.661 × 100 = 66.1 kPa	Check total of partial pressures = 100 kPa 15.0 + 18.9 + 66.1 = 100 kPa

2 $K_p = \frac{p(NO_2)^2}{p(NO)^2\, p(O_2)} = \frac{(66.1)^2}{(15.0)^2(18.9)} = 1.03\ \text{kPa}^{-1}$

TIP

Check the colour code system used at AS for K_c calculations in the Equilibrium, Le Chatelier's Principle and K_c chapters.

EXAMPLE 4

Hydrogen reacts with bromine according to the equilibrium:

$H_2(g) + Br_2(g) \rightleftharpoons 2HBr(g)$

1.00 mol of hydrogen was mixed with 1.00 mol of bromine and the mixture allowed to attain equilibrium. 0.824 mol of hydrogen bromide were present in the equilibrium mixture. Calculate a value for K_p.

TIP
P is used for the total pressure as the total pressure is not given in the question. P will cancel out in the K_p expression.

Answer

	H_2	+	Br_2	$\rightleftharpoons$	2HBr	
Initial moles	1.00		1.00		0.00	
Reacting moles	−0.412		−0.412		+0.824	
Equilibrium moles	0.588		0.588		0.824	Total equilibrium moles = 0.588 + 0.588 + 0.824 = 2.00
Mole fraction	$\frac{0.588}{2.00} = 0.294$		$\frac{0.588}{2.00} = 0.294$		$\frac{0.824}{2.00} = 0.412$	Check total of mole fractions = 1 0.294 + 0.294 + 0.412 = 100
Partial pressures	0.294 P		0.294 P		0.412 P	0.294 P + 0.294 P + 0.412 P = P

$$K_p = \frac{p(HBr)^2}{p(H_2)\,p(Br_2)} = \frac{(0.412\,P)^2}{(0.294\,P)(0.294\,P)} = 1.96$$

Note that using the equilibrium moles will give the same value, i.e.

$$K_p = \frac{p(HBr)^2}{p(H_2)p(Br_2)} = \frac{(0.824)^2}{(0.588)(0.588)} = 1.96.$$

It is advisable to use the partial pressures as this is correct for the K_p expression. If a total pressure had been given then use this to calculate the partial pressures.

TIP
K_p has no units because the pressures cancel out in the expression.

TIP
Practise carrying out the calculation shown above on your calculator as it is easy to make mistakes with the bottom line (denominator) of the fraction.

Using K_p

Some calculations may give a value of K_p and you may be expected to calculate the equilibrium moles of one component or the mole fraction or partial pressure.

TIP

Using 2x moles of HI avoids $\frac{x}{2}$ moles of H_2 and I_2 but the calculation would work either way.

EXAMPLE 5

For the following gaseous homogeneous equilibrium:

$$2HI(s) \rightleftharpoons H_2(g) + I_2(g)$$

2.00 mol of hydrogen iodide was allowed to reach equilibrium at 500 K. K_p for this equilibrium at 500 K is 0.360. Calculate the partial pressure of hydrogen in the equilibrium mixture if the total pressure is 100 kPa.

Answer

As the number of moles which reacts is not known, 2*x* moles of HI react to form *x* moles of H_2 and *x* moles of I_2.

	2HI	⇌	H_2	+	I_2	
Initial moles	2.00		0		0	
Reacting moles	−2x		+x		+x	
Equilibrium moles	2.00 − 2x		x		x	Total equilibrium moles = 2 − 2x + x + x = 2
Mole fraction	$\frac{2.00-2x}{2}$		$\frac{x}{2}$		$\frac{x}{2}$	Check total of mole fractions = 1
Partial pressure	$\frac{2.00-2x}{2} \times 100$		$\frac{x}{2} \times 100$		$\frac{x}{2} \times 100$	$\frac{2.00-2x}{2} \times 100 + \frac{x}{2} \times 100 + \frac{x}{2} \times 100 = \frac{200}{2} = 100$

$$K_p = \frac{p(H_2)\,p(I_2)}{p(HI)^2} = \frac{\left(\frac{100x}{2}\right)^2}{\left(\frac{200-200x}{2}\right)^2} = \frac{x^2}{(2-2x)^2} = 0.360$$

$$\frac{x^2}{(2-2x)^2} = \left(\frac{x}{2-2x}\right)^2 = 0.360$$

$$= \left(\frac{x}{2-2x}\right) = 0.600 \text{ as } \sqrt{0.360} = 0.600$$

$$x = 0.600\,(2-2x)$$

$$x = 1.20 - 1.20x$$

$$2.20x = 1.20$$

$$x = \frac{1.20}{2.20} = 0.545$$

The partial pressure of hydrogen in the equilibrium mixture is given by $\frac{x}{2} \times 100$ so the partial pressure is $\frac{0.545}{2} \times 100 = 27.3$ kPa.

TIP

In a calculation where K_p is unitless, the mole fractions or the equilibrium moles may be used in the K_p expression and will give the same answer. However, as the total pressure is given in this question it is best practice to use it.

Calculations of this sort are more complex and should be worked through carefully. You can check the value of *x* you obtain by substituting it into either the partial pressures or the mole fractions. The sum of the mole fractions should add up to 1 and the sum of the partial pressures should add up to the total pressure in this case 100 kPa.

Also in an equilibrium where 1 mol of two products are formed, the equilibrium moles, mole fractions and partial pressures of these two products are the same throughout for any equilibria at any temperature.

Calculating total pressure

The total pressure in a K_p calculation may be represented by the letter P. P is used in the K_p expression and the expression may be solved to find P.

EXAMPLE 6

At equilibrium, 0.200 mol of nitrogen, 0.400 mol of hydrogen and 0.400 mol of ammonia are present during the Born-Haber process:

$$N_2(g) + 3H_2(g) \rightleftharpoons 2NH_3(g)$$

Calculate a value for the total pressure on the system at 390 K if K_p has a value of 0.172 K Pa^{-2}.

	N_2	+	$3H_2$	$\rightleftharpoons$	$2NH_3$	
Equilibrium moles	0.200		0.400		0.400	Total equilibrium moles = 0.200 + 0.400 + 0.400 = 1.00
Mole fraction	$\frac{0.200}{1.00} = 0.200$		$\frac{0.400}{1.00} = 0.400$		$\frac{0.400}{1.00} = 0.400$	Check total of mole fractions = 1 0.200 + 0.400 + 0.400 = 1.00
Partial pressure	0.200 P		0.400 P		0.400 P	

$$K_p = \frac{p(NH_3)^2}{p(N_2)p(H_2)^3} = \frac{(0.400P)^2}{(0.200P)(0.400P)^3} = 0.172$$

Cancelling $(0.400P)^2$

$$\frac{(0.400P)^2}{(0.200P)(0.400P)^3} = \frac{1}{(0.200P)(0.400P)} = 0.172$$

$$1 = 0.172(0.200P)(0.400P)$$

$1 = 0.01376P^2$ as $0.172 \times 0.200 \times 0.400 = 0.01376$ and $P \times P = P^2$

$$P^2 = \frac{1}{(0.01376)}$$

$$P^2 = 72.674$$

$$P = \sqrt{72.674} = 8.52\,kPa$$

Again this can be checked by filling the total pressure (8.52 kPa) into the equilibrium expression and if you get the correct value for K_p, the value of P is correct.

TIP
All initial values were given to 3 significant figures. This would mean that the final answers should also be given to 3 significant figures.

Changes in temperature and other factors

- K_p is constant at a particular temperature.
- The equilibrium constant, K_p is affected by changes in temperature, but it is not affected by changes in pressure, the presence of a catalyst or changes in concentration or amount of substances present in the mixture.
- This is the same for all equilibrium constants which are constant at a particular temperature and are not affected by changes in any other quantities.

Figure 4.1 The equilibrium between NO_2 and N_2O_4 is shown in cold and hot water baths. A change in temperature shifts the equilibrium between the two species. When more NO_2 is produced, the gas inside the tube becomes darker. The K_p value will change with temperature.

- For the equilibrium:
  $N_2(g) + 3H_2(g) \rightleftharpoons 2NH_3(g)$ $\Delta H = -92\,kJ\,mol^{-1}$
  Increasing the temperature will favour the reverse reaction and the position of equilibrium will move to the left. This will decrease the partial pressure of NH_3 and the value of K_p will decrease.
- Increasing the pressure will favour the forward reaction and the position of equilibrium will move to the right. This will have no effect on the value of K_p as long as the temperature remains constant.
- Adding more hydrogen to the mixture will increase the partial pressure of the hydrogen and it will move the position of equilibrium to the right. It will have no effect on the value of K_p as long as the temperature remains constant.

TEST YOURSELF 3

1 1.00 moles of PCl_5 are allowed to reach equilibrium.

$PCl_5(g) \rightleftharpoons PCl_3(g) + Cl_2(g)$

a) At equilibrium there are 0.400 mol of Cl_2 present at 400 K. Calculate a value for the equilibrium constant, K_p, when the total pressure on the system is 100 kPa. State the units of K_p (if any). Give your answer to 3 significant figures.

b) State the effect, if any, of an increase in pressure on the value of K_p.

2 In the following equilibrium $CO_2(g) + H_2(g) \rightleftharpoons CO(g) + H_2O(g)$ at a temperature of 373 K and a pressure of 700 kPa, the amounts of each gas present at equilibrium were 0.210 mol of CO, 0.210 mol of H_2O, 1.04 mol of CO_2 and 1.04 mol of H_2. Calculate a value for K_p and state its units (if any). Give your answer to 3 significant figures.

3 In the equilibrium $N_2O_4(g) \rightleftharpoons 2NO_2(g)$, the value of K_p is 0.389 MPa at 350 K. The partial pressure of NO_2 at equilibrium was found to be 0.0700 MPa. Calculate the partial pressure of N_2O_4 at equilibrium and the total pressure on the system. Give your answer to 3 significant figures.

Practice questions

1 Which one of the following could be units of K_p for the reaction shown below?

$2NH_3(g) + 3O_2(g) \rightleftharpoons N_2(g) + 3H_2O(g)$

A KPa **C** kPa^2

B kPa^{-1} **D** kPa^{-2} *(1)*

2 Which one of the following is correct when pressure is increased on the following equilibrium at constant temperature?

$2NO(g) + O_2(g) \rightleftharpoons 2NO_2(g)$ *(1)*

	Value of K_p	Partial pressure of NO(g)
A	increase	decrease
B	increase	no change
C	no change	increase
D	no change	decrease

3 For the equilibrium below, calculate a value fo K_p if the partial pressure of the gases present a equilibrium are: O_2 = 100 kPa, SO_2 = 240 kPa and SO_3 = 500 kPa. Give your answer to 3 significant figures.

$2SO_2(g) + O_2(g) \rightleftharpoons 2SO_3(g)$ (

4 2.00 mol of A was mixed with 2.00 moles of B and the mixture was allowed to reach equilibrium at 500 °C. The equilibrium mixture was found to contain 1.00 mole of A. Calculate a value for K_p.

$$2A(g) + B(g) \rightleftharpoons 3C(g)$$ *(3)*

5 Analysis of the equilibrium system, $N_2(g) + 3H_2(g) \rightleftharpoons 2NH_3(g)$ showed 25.1 g of NH_3, 12.8 g of H_2 and 59.6 g of N_2. Calculate a value for K_p if the total pressure is 1.32×10^5 Pa. Give your answer to 3 significant figures. *(4)*

6 When hydrogen iodide is heated in a sealed container, it reaches equilibrium according to the equation:

$$2HI(g) \rightleftharpoons H_2(g) + I_2(g)$$

K_p for this reaction at 700 K is 0.0185 and partial pressure of the hydrogen iodide in the equilibrium mixture is 1.80 MPa. Calculate the partial pressures of the hydrogen and iodine in the equilibrium mixture. Give your answer to 3 significant figures. *(4)*

7 In the equilibrium $CO(g) + H_2O(g) \rightleftharpoons CO_2(g) + H_2(g)$, 1.00 mol of carbon monoxide is mixed with 1.00 mol of water vapour at 200 °C and 400 kPa pressure. The equilibrium constant, K_p, at 200 °C is 0.0625. Calculate the number of moles of carbon monoxide present at equilibrium. *(4)*

8 At 107 °C, the reaction $CO(g) + 2H_2(g) \rightleftharpoons CH_3OH(g)$ reaches equilibrium under a pressure of 1.59 MPa with 0.122 mol of CO, 0.298 mol of H_2 and 0.500 mol of CH_3OH present in a vessel of volume 1.04 dm^3.

Calculate K_c and K_p and state the units of both. Give both answers to 3 significant figures. *(6)*

Stretch and challenge

9 For the reaction, $H_2(g) + I_2(g) \rightleftharpoons 2HI(g)$, 2.00 mol of hydrogen and 2.00 mol of iodine are heated to 700 K in a sealed vessel until equilibrium is attained at a pressure of 100 kPa. The value of K_p at 700 K for this reaction is 0.106.

a) i) Write an expression for the equilibrium constant, K_p. *(1)*

ii) Calculate the number of moles of each gas present at equilibrium. *(5)*

iii) Determine the percentage of hydrogen which reacted. *(1)*

b) Using the same initial amount in moles, the equilibrium was re-established at 100 kPa pressure and 400 K. The value of K_p at 400 K is 0.0111. Calculate the number of moles of hydrogen remaining at equilibrium and hence the percentage of hydrogen which reacted. *(4)*

5 Electrode potentials and cells

PRIOR KNOWLEDGE

- Oxidation state is the relative state of oxidation or reduction of an element.
- Elements have an oxidation state of 0 (zero).
- In compounds, Group 1 elements have an oxidation state of +1.
- In compounds, Group 2 elements have an oxidation state of +2.
- In compounds, aluminium has an oxidation state of +3.
- In a simple binary compound, the more electronegative element has the negative oxidation state.
- Oxygen has an oxidation state of -2 in all compounds except peroxides where its oxidation state is -1 and in the compound OF_2 where its oxidation state is +2.
- Hydrogen has an oxidation state of +1 in all compounds except hydrides where its oxidation state is -1.
- In compounds, d block and p block elements have variable oxidation state.
- Oxidation is the loss of electrons or increase in oxidation state.
- Equations showing oxidation have $+e^-$ on the right-hand side.
- Reduction is the gain of electrons or the decrease in oxidation state.
- Equations showing reduction have $+e^-$ on the left have side.
- The number of moles of electrons is equal to the change in oxidation state.

TEST YOURSELF ON PRIOR KNOWLEDGE 1

1 Give the oxidation state of nitrogen in the following compounds and ions:

a) NO **b)** NH_3 **c)** NO_3^-
d) N_2H_4 **e)** N^{3-}

2 State the full name of the following compounds:

a) Na_2SO_4 **b)** NH_4NO_3 **c)** N_2O
d) K_2CrO_4

3 For the following reactions:

Reaction A: $Mg \longrightarrow Mg^{2+} + 2e^-$
Reaction B: $Cl_2 + H_2O \longrightarrow HCl + HOCl$
Reaction C: $8HI + H_2SO_4 \longrightarrow 4I_2 + 4H_2O + H_2S$
Reaction D: $NaCl + H_2SO_4 \longrightarrow NaHSO_4 + HI$
Reaction E: $NO_3^- + 4H^+ + 3e^- \longrightarrow NO + 2H_2O$

a) Which reaction shows only a reduction reaction?
b) Which reaction is not a redox reaction?
c) Which reaction shows only an oxidation reaction?
d) Which reaction shows the largest changes in oxidation state?

Redox reactions involve electron transfer. An electrochemical cell uses electron transfer reactions to produce electrical energy. Most batteries contain an electrochemical cell and as such, they are found in cameras, laptops, phones and even hybrid cars. Figure 5.1 shows a lemon being used as a cell which provides electrical current sufficient to power a clock.

Figure 5.1 The lemon is being used as an electrolytic cell, the electrolyte being citric acid. Four electrodes – two copper and two zinc – are placed in the lemon and connected together. Zinc atoms on the electrode are oxidised, losing two electrons per atom and dissolving into solution. The electrons pass through the wires to the copper electrode where they combine with hydrogen ions from the citric acid to liberate hydrogen gas. The movement of electrons between electrodes forms the current.

Redox equilibria

When a metal in dipped into a solution containing its simple ions, an equilibrium is established between the metal ions and the metal atoms. This type of arrangement is called a half cell.

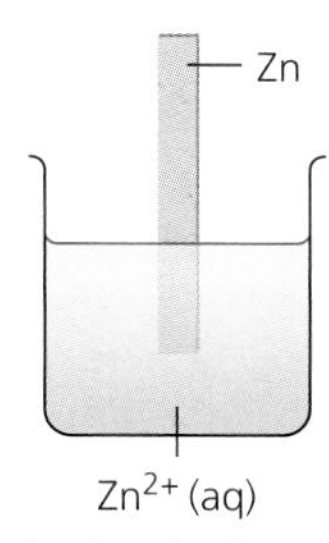

The zinc ions in the solution are in equilibrium with the zinc atoms in the metal.

$Zn^{2+}(aq) + 2e^- \rightleftharpoons Zn(s)$

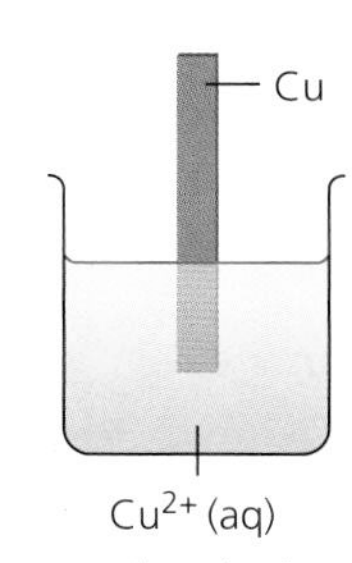

The copper ions in the solution are in equilibrium with the copper atoms in the metal.

$Cu^{2+}(aq) + 2e^- \rightleftharpoons Cu(s)$

It is impossible to measure the equilibrium in one half cell unless another half cell is connected to it. Two half cells joined together create a cell.

An example of a cell is shown below.

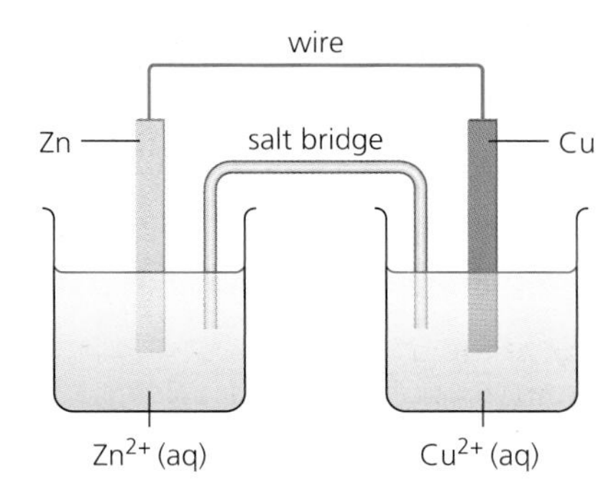

TIP

In any half cell, the metal is the electrode but often the entire half cell is referred to as an electrode as in the standard hydrogen electrode.

The two metals in the half cells are connected externally using a conducting wire and the solutions are connected using a salt bridge. A salt bridge can be one of two things:

1 a piece of filter paper soaked in a solution of potassium chloride or potassium nitrate

2 potassium chloride dissolved in agar gel and set in a U-tube.

The salt bridge has mobile ions that complete the circuit. Potassium chloride/potassium nitrate are used to ensure that there is no precipitation as chlorides and nitrates are usually soluble.

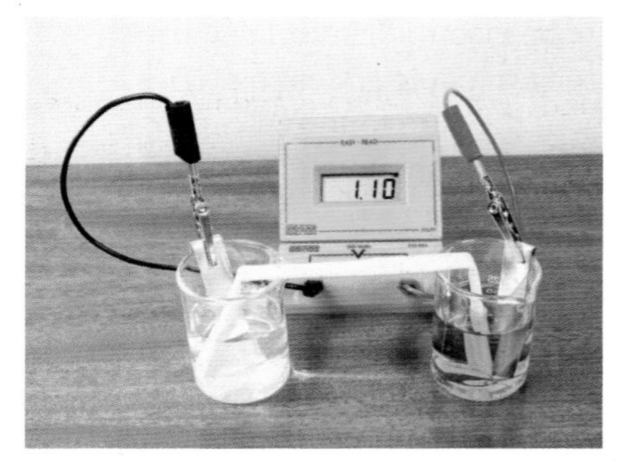

Figure 5.2 The photograph shows a zinc copper cell. On the right, copper is dipped in a copper sulfate solution, and on the left zinc is dipped in zinc sulfate. The zinc has a lower affinity for electrons than copper, and zinc from the zinc strip forms ions and electrons flow along the electrical wires and pull copper ions out of solution to form solid copper on the copper strip. The salt bridge is the white filter paper soaked in potassium nitrate solution.

TIP

Solid arrows rather than equilibrium arrows are often used in these electrode half equations.

TIP

NEGATOX is a good way of remembering that the NEGactive electrode is where OXidation occurs.

However, care should be taken with potassium chloride solution as chloride ions can react with some metal ions such as Cu^{2+} to form complexes such as $[CuCl_4]^{2-}$. Potassium nitrate is better to use for the salt bridge for solutions in which the metal ion might form a complex with the chloride ions.

In the complete cell above, an electrochemical reaction occurs that is based on the equilibrium for each electrode. If the circuit is connected for a while and the mass of the metals measured before and after, the zinc would have lost mass and the copper would have gained mass.

Therefore the reactions which are occurring are:

$$Zn \rightarrow Zn^{2+} + 2e^-$$

$$Cu^{2+} + 2e^- \rightarrow Cu$$

An electrode where oxidation occurs is called the negative electrode whereas an electrode where reduction occurs is called the positive electrode.

- In this cell, the zinc forms the negative electrode and the copper forms the positive electrode.
- The electrons flow externally in the circuit from the zinc side to the copper side. The zinc atoms lose electrons and the electrons flow from the zinc metal to the copper metal and the copper ions in the solution gain electrons to form copper.
- Copper forms on the copper electrode and increases the mass and some of the zinc metal forms zinc ions and this decreases the mass of the zinc electrode.
- This circuit allows electrons to flow.
- This shows us that the zinc is more likely to form its ions than the copper.
- In order to measure this, a **high-resistance voltmeter** is connected in the external circuit to measure the potential difference without allowing any electrons to flow in the circuit. This maintains the concentration of the ions in solution by not allowing any current to flow yet can still measure the potential difference of the electrons trying to flow.

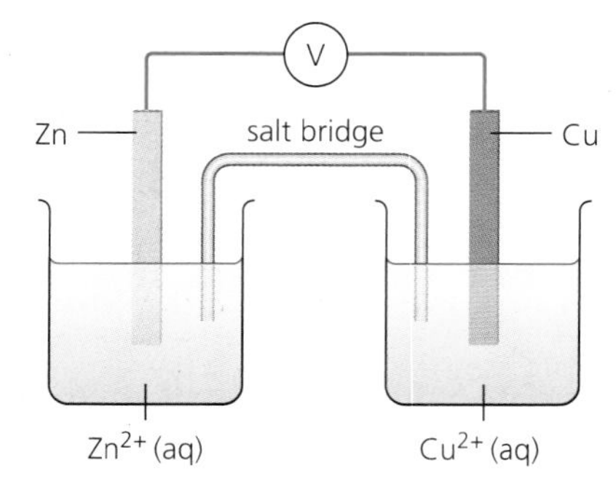

The voltage measured in this circuit is +1.10 V. This indicates the potential difference between the zinc half cell (or electrode) on the left and the copper half cell (or electrode) on the right. This does not give a measure of the potential difference in one half cell.

temperature = 298 K
hydrogen gas at 100 kPa pressure
platinum wire
platinum foil covered in porous platinum
[H⁺] = 1.00 mol dm⁻³

Figure 5.3 Standard hydrogen electrode.

Electrode potentials

To measure the potential difference for a single half cell, a standard must be used to which all other potential differences can be compared. This standard is the **standard hydrogen electrode** or standard hydrogen half cell.

The standard hydrogen electrode is an electrode consisting of hydrogen gas in contact with hydrogen ions, H^+, on a platinum surface.

The standard conditions for this half cell (electrode) apply to all other half cells to ensure that they can be compared. All solutions should have an ion concentration of 1.00 mol dm^{-3}. All gases should be under 100 kPa pressure and the whole cell should be at 298 K.

Measuring standard electrode potentials

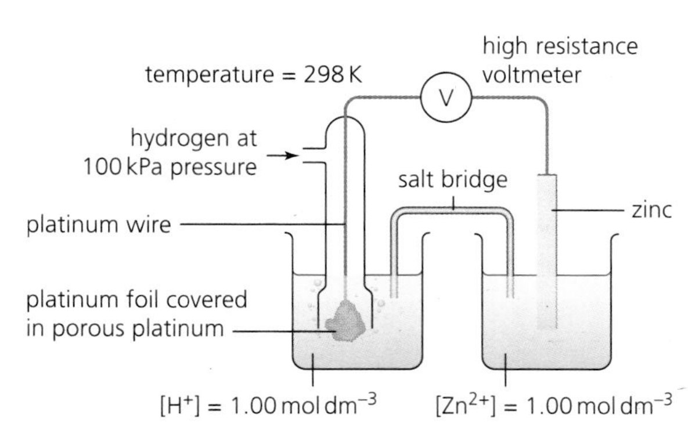

A high-resistance voltmeter should be connected to the standard hydrogen electrode (half cell) and the other electrode (half cell) connected to the voltmeter. A salt bridge will connect the two solutions.

$[H^+]$ = concentration of hydrogen ions; $[Zn^{2+}]$ = concentration of Zn^{2+} ions. Both are measured in mol dm^{-3}.

When the standard hydrogen electrode is used, by convention, it must be the negative electrode (anode) – i.e. an oxidation reaction must occur. This oxidation reaction is:

$H_2(g) \rightarrow 2H^+(aq) + 2e^-$

This means that, by convention, a reduction occurs at the other half cell (electrode) connected to the standard hydrogen electrode. All standard electrode potentials are written as reductions and the value is given for the reduction reaction.

This means that the standard electrode potential for the zinc electrode is −0.76 V. Electrons are trying to flow from the zinc electrode to the hydrogen electrode and giving a negative electrode potential.

The **standard electrode potential** is defined as: the electrode potential of a standard electrode with ion concentration of 1.00 mol dm^{-3} at 298 K connected to a standard hydrogen electrode (1.00 mol dm^{-3} H^+ ions, 100 kPa H_2 gas at 298 K) using a high-resistance voltmeter and a salt bridge.

TIP
Platinum is often used in electrochemical cells as it is a good conductor of electricity and it is inert. Porous platinum gives a larger surface area for reaction.

TIP
The standard electrode potential can also be called a standard redox potential or standard reduction potential. It is represented by the symbol $E^{\ominus}$.

Electrodes containing two ions

If the standard electrode potential of a reduction between two ions is to be measured, it is set up as follows:

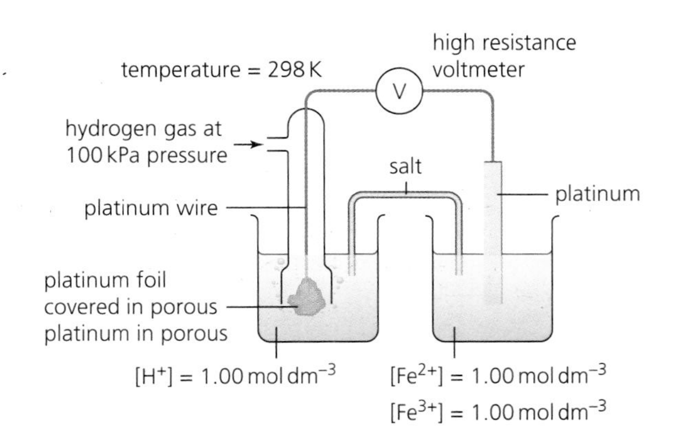

For the right-hand electrode, the solution contains two different iron ions, Fe^{2+} and Fe^{3+}, both at a concentration of $1\,mol\,dm^{-3}$. The equilibrium is established between them by using a platinum electrode.

Again the standard hydrogen electrode should, by convention, have an oxidation reaction occurring. In reality this time the oxidation reaction does occur at the standard hydrogen electrode. The value on the high-resistance voltmeter is the actual standard electrode potential for the Fe^{3+} to Fe^{2+} reduction reaction.

This means that the value for the reduction reaction on the right-hand side is:

$$Fe^{3+}(aq) + e^- \rightarrow Fe^{2+}(aq) \qquad E^\ominus = +0.77\,V$$

In this case the value shown on the voltmeter in a practical setting is the same as the actual standard electrode potential for the reduction reaction in the right-hand cell.

TEST YOURSELF 2

1. Give two reasons why platinum is used as the contact in a standard hydrogen electrode.
2. State the conditions required for the standard hydrogen electrode.
3. Draw a labelled diagram of the apparatus that could be connected to a standard hydrogen electrode in order to measure the standard electrode potential of the $Cu^{2+}(aq) + 2e^- \rightarrow Cu(s)$ electrode.
   In your diagram, show how this electrode is connected to the standard hydrogen electrode and to a voltmeter. Do **not** draw the standard hydrogen electrode.
   State the conditions under which this cell should be operated to measure the standard electrode potential.

Cell conventions and electromotive force (EMF)

Convention means a lot in drawing a complete cell and in calculating standard electrode potentials.

When connecting any two electrodes and drawing a diagram of the overall cell, the oxidation electrode should be drawn on the left and the reduction electrode on the right. All cells drawn previously have been drawn like this.

Obviously for the two electrodes, the oxidation and reduction reactions must be determined before the cell is drawn. The standard electrode potential with the more negative value will form the oxidation electrode.

The more negative electrode potential is the left-hand (oxidation) side of the cell.

EXAMPLE 1

A cell is made from a $Zn^{2+}|Zn$ electrode and a $Cu^{2+}|Cu$ electrode. The values of the standard electrode potentials are:

$$Zn^{2+}(aq) + 2e^- \rightarrow Zn(s) \quad E^\ominus = -0.76\,V$$

$$Cu^{2+}(aq) + 2e^- \rightarrow Cu(s) \quad E^\ominus = +0.34\,V$$

Calculate the EMF of this cell.

Answer

The more negative standard electrode potential is the zinc one, which means it will form the left-hand electrode and the copper one will form the right-hand electrode. Oxidation will occur at the zinc electrode and reduction will occur at the copper electrode.

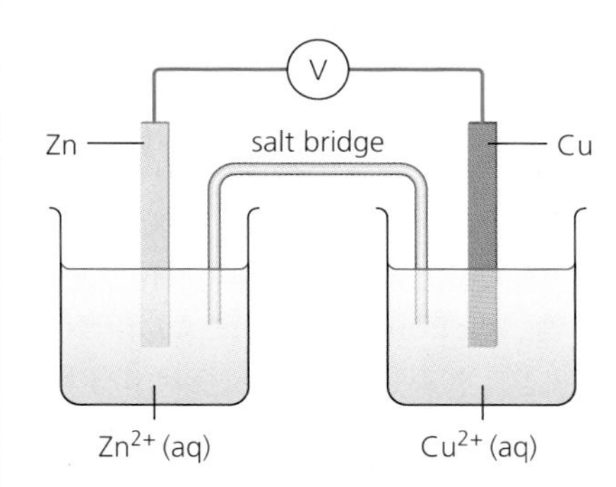

The overall voltage measured on a cell like this is called the **electromotive force** (EMF). For this electrochemical reaction, the zinc being oxidised has a potential of +0.76 V and the copper being reduced has a potential of +0.34 V so the EMF of this cell is +1.10 V.

$Zn \leftrightharpoons Zn^{2+} + 2e^-$	+0.76 V
$Cu^{2+} + 2e^- \leftrightharpoons Cu$	+0.34 V
$Zn + Cu^{2+} \leftrightharpoons Zn^{2+} + Cu$	+1.10 V
This is the overall ionic equation for the complete cell (without electrons)	This is the EMF of the cell

In real terms this means that when zinc is placed in a solution containing copper(II) ions, the zinc will displace the copper(II) ions to form zinc ions and copper. Zinc is more reactive than copper.

Using an ammeter in a cell

The high-resistance voltmeter is only used to measure the potential of a cell under standard conditions. Often an ammeter can be used to show the flow of current in the cell and in particular to show the direction of electron flow in the cell.

Cells not under standard conditions

If an electrochemical cell is created with two copper electrodes, the standard electrode potential for $Cu + 2e^- \rightarrow Cu$ would appear to be the same on both sides, which would give an overall EMF of 0.00 V. There should be no flow of current under standard conditions.

However this only applies if all conditions are standard. In the cell shown below the concentration of $CuSO_4(aq)$ on the left-hand side is greater than the concentration of the same solution on the right-hand side.

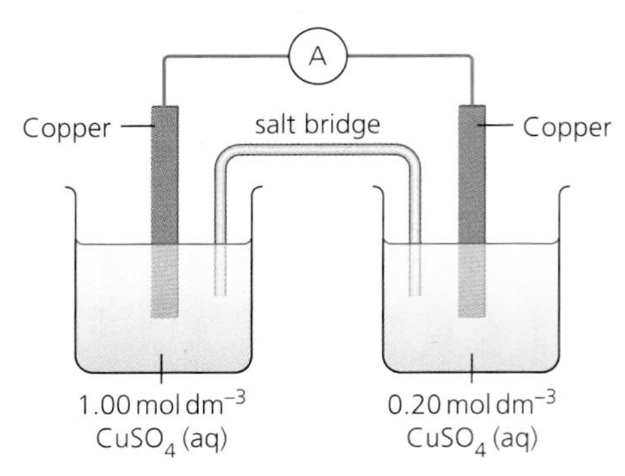

- The ammeter allows current to flow.
- As the concentration of Cu^{2+} ions is greater on the left-hand side, $Cu^{2+} + 2e^- \rightarrow Cu$ occurs in preference on the left-hand side.
- The left-hand electrode in this diagram is the positive electrode.
- The oxidation $Cu \rightarrow Cu^{2+} + 2e^-$ occurs at the right-hand electrode (anode) to provide electrons for the reduction on the left-hand side.
- Electrons flow through the ammeter from the right-hand side to the left-hand side.
- In this cell the current would eventually fall to zero once the concentrations of the solutions were equal.

Redox reactions and feasibility of reactions

Redox potentials are always given as reduction reactions (always with gain of electrons) and the feasibility of the reaction is given by the standard electrode potential. The more positive the electrode potential the more feasible is the reduction reaction.

Redox reactions consist of two parts.

1 an oxidation (which is a standard reduction reaction reversed) and

2 a reduction.

The total of the standard electrode potentials gives the overall feasibility of the redox reaction. This value is called the EMF (electromotive force) of the reaction. If the EMF is positive, then the redox reaction is feasible and if it is negative then it is not feasible.

The table below gives the standard electrode potentials of some of the most commonly used half equations.

strong oxidising agents ↑	Half reaction	$E^{\ominus}$/V	strong reducing agents ↓
	$F_2(g) + 2e^- \rightarrow 2F^-(aq)$	+2.87	
	$Cl_2(g) + 2e^- \rightarrow 2Cl(aq)$	+1.36	
	$Br_2(l) + 2e^- \rightarrow 2Br^-(aq)$	+1.07	
	$Fe^{3+}(aq) + e^- \rightarrow Fe^{2+}(aq)$	+0.77	
	$I_2(s) + 2e^- \rightarrow 2I^-(aq)$	+0.54	
	$Cu^{2+}(aq) + 2e^- \rightarrow Cu(s)$	+0.34	
	$Sn^{4+}(aq) + 2e^- \rightarrow Sn^{2+}(aq)$	+0.14	
	$2H^+(aq) + 2e^- \rightarrow H_2(g)$	0.00	
	$Pb^{2+}(aq) + 2e^- \rightarrow Pb(s)$	−0.13	
	$Fe^{2+}(aq) + 2e^- \rightarrow Fe(s)$	−0.44	
	$Zn^{2+}(aq) + 2e^- \rightarrow Zn(s)$	−0.76	
	$K^+(aq) + e^- \rightarrow K(s)$	−2.93	
	$Li^+(aq) + e^- \rightarrow Li(s)$	−3.05	

Fluorine most easily undergoes reduction to fluoride ions hence fluorine is the strongest oxidising agent (or oxidant).

Lithium ions, Li^+, are most resistant to reduction. As the reverse reaction (oxidation) has a high positive value this means that lithium metal is the most easily oxidised and so is the strongest reducing agent (reductant).

The overall EMF of a redox reaction can be calculated using the standard electrode potentials.

EXAMPLE 2

In the reaction between chlorine and potassium bromide solution, the equation for the reaction is:

$2KBr + Cl_2 \rightarrow 2KCl + Br_2$

The potassium ion, K^+, is the spectator ion and can be disregarded as it does not take part in the redox reaction, being the same at the beginning as at the end. With this ion removed, the equation is now an ionic equation:

$2Br^- + Cl_2 \rightarrow 2Cl^- + Br_2$

Note that this consists of two half equations:

$2Br^- \rightarrow Br_2 + 2e^-$ Loss of electrons = oxidation

and

$Cl_2 + 2e^- \rightarrow 2Cl^-$ Gain of electrons = reduction

The standard electrode potentials for these reactions are given below:

$E^{\ominus}$ value for $Cl_2 + 2e^- \rightarrow 2Cl^-$ +1.36 V

$E^{\ominus}$ value for $Br_2 + 2e^- \rightarrow 2Br^-$ +1.07 V

However these are both reductions and the second must be reversed to form an oxidation step. The EMF is the total of the standard electrode potentials when the sign of one has been changed to make it an oxidation reaction.

EMF for the reaction = +1.36 + (−1.07) = +0.29 V

This would indicate that the redox reaction:

$2Br^- + Cl_2 \rightarrow 2Cl^- + Br_2$

is feasible and does occur.

However in the following reaction:

$2Cl^- + Br_2 \rightarrow 2Br^- + Cl_2$

The half equations and standard electrode potentials for these reactions are given below:

$E^{\ominus}$ value for $Cl_2 + 2e^- \rightarrow 2Cl^-$ +1.36 V

$E^{\ominus}$ value for $Br_2 + 2e^- \rightarrow 2Br^-$ +1.07 V

The top half equation must be reversed to make it an oxidation.

EMF of the redox reaction is = 1.07 − (+1.36) = −0.29 V

This would indicate that the redox reaction:

$2Cl^- + Br_2 \rightarrow 2Br^- + Cl_2$

is not feasible and does not occur.

Conventional cell representation (cell notation)

It would be cumbersome to have to draw a full cell diagram every time to represent the electrodes.

A simple representation of a cell can be used. The rules for writing conventional cell representations are:

1. The oxidation electrode is placed on the left with the components in order of being oxidised, e.g. Zn(s) first then Zn^{2+}(aq).
2. The reduction electrode is written on the right with the component in order of being reduced, e.g. Cu^{2+}(aq) then Cu(s).
3. A double vertical line (||) is used between the two electrodes to represent the salt bridge.
4. Single vertical lines are used between components in an electrode to represent a phase boundary (difference in physical state), e.g. Cu^{2+}(aq)|Cu(s).
5. If Pt is used as a contact in an electrode for gas/ions in solution or two ions in solution it should be placed at the extreme left of the left-hand electrode and the extreme right of the right-hand electrode. Phase boundary lines may also be used to separate it from other components of the electrode.
6. Commas are used to separate components in an electrode which are in the same phase, e.g. Fe^{3+}(aq), Fe^{2+} (aq).

The cell notation for the zinc copper cell is shown below.

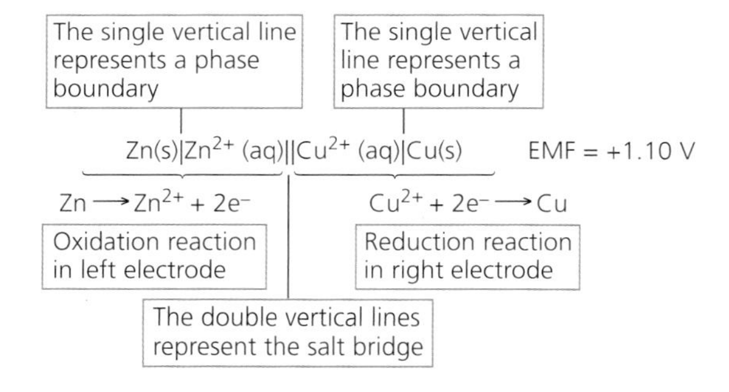

The single vertical lines are important between the components of an electrode. As the zinc is a solid and the Zn^{2+} ions are aqueous, there should be a vertical phase boundary line between them.

The EMF of this type of arrangement can be determined from the standard electrode potentials using an expression which is:

$$\text{EMF} = E^{\ominus}_{\text{RHS}} - E^{\ominus}_{\text{LHS}} \quad \text{(RHS = right-hand side; LHS = left-hand side)}$$

This means that the standard electrode potential of the right-hand electrode minus the standard electrode potential of the left-hand electrode is equal to the EMF.

The minus takes into account that there is an oxidation occurring at the left-hand electrode.

It can also be written:

$$\text{EMF} = E^{\ominus}_{\text{(reduction reaction)}} - E^{\ominus}_{\text{(oxidation reaction)}}$$

TIP

However it is best often to work out which electrode is undergoing oxidation and change the sign of the standard electrode potential and add it to the standard electrode potential for the electrode undergoing reduction. This will also give the EMF of the cell.

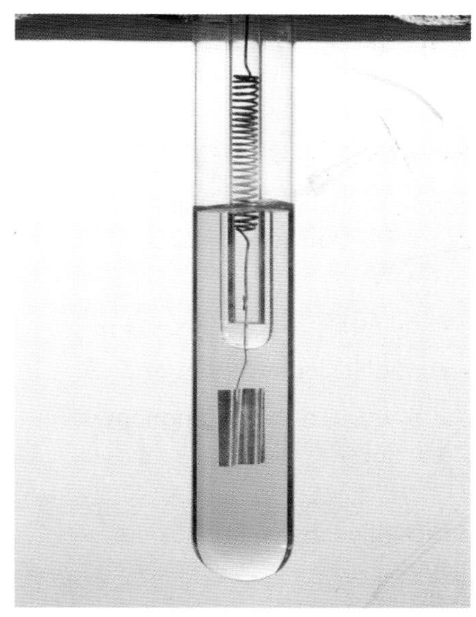

Figure 5.4 A platinum electrode (square, lower centre) in a solution of iron(II) and iron(III) ions. The iron(III) ions are being reduced to iron(II) ions, gaining an electron in the process. The electron movements are powered by the oxidation of hydrogen in the smaller tube seen inside the larger tube. Hydrogen gas is adsorbed onto the platinum coil and loses two electrons to form two hydrogen ions in the surrounding water. A platinum-hydrogen electrode is the standard used to measure redox potentials of ions. In this case the iron(III) to iron(II) redox potential is +0.77 V.

If a standard hydrogen electrode is used in the cell, it should be on the left as it should always be the oxidation electrode (by convention), for example:

$$Pt|H_2(g)|H^+(aq)||Zn^{2+}(aq)|Zn(s) \qquad EMF = -0.76\,V$$

The platinum contact in the electrode must be included. For the left-hand electrode, Pt is placed before the component and again a phase boundary (|) is in place. The phase boundaries are also in place between the components of the electrodes.

The same is true of a solution containing two ions such as the $Fe^{3+}|Fe^{2+}$ electrode, for example:

$$Pt|H_2(g)|H^+(aq)||Fe^{3+}(aq), Fe^{2+}(aq)|Pt \qquad EMF = +0.77\,V$$

In this example both electrodes have a platinum contact. The standard hydrogen electrode is again always on the left-hand side.

The Pt is placed at the end of the reduction side (right-hand side) and again a phase boundary (|) is in place.

As both Fe^{3+} and Fe^{2+} are in the same phase on the right, a phase boundary line is not used but a comma instead.

A comma should be used between components of an electrode that are in the same phase.

Constructing a normal cell from two electrodes:

$$Zn^{2+}(aq) + 2e^- \rightarrow Zn(s) \qquad E^\ominus = -0.76\,V$$

$$Mg^{2+}(aq) + 2e^- \rightarrow Mg(s) \qquad E^\ominus = -2.37\,V$$

The more negative standard electrode potential is for magnesium so this becomes the left-hand electrode.

$$Mg(s)|Mg^{2+}(aq)||Zn^{2+}(aq)|Zn(s)$$

The overall EMF is:

$$E_{RHS} - E_{LHS} = -0.76 - (-2.37) = +1.61\,V$$

or

$$Mg \rightarrow Mg^{2+} + 2e^- \qquad +2.37\,V$$

$$Zn^{2+} + 2e^- \rightarrow Zn \qquad -0.76\,V$$

$$Mg + Zn^{2+} \rightarrow Mg^{2+} + Zn \qquad EMF = +1.61\,V$$

In real terms, this means that when magnesium is placed in a solution containing zinc ions, the magnesium will displace the zinc ions to form magnesium ions and zinc. Magnesium is more reactive than zinc.

This allows us to create a **reactivity series**: magnesium is more reactive than zinc, which is more reactive than copper.

Remember: a positive EMF value for a cell indicates that a reaction given by the overall cell equation is feasible. A negative EMF value for a cell indicates that a reaction given by the overall cell equation is not feasible.

TIP

If a reaction is determined to be feasible as it has a positive EMF, this does not mean that it will happen. The activation energy may be so high as to prevent it reacting at any appreciable rate under standard conditions.

EXAMPLE 3

Which one of the following species will oxidise iron(II) to iron(III) in aqueous solution.

$Fe^{3+}(aq) + e^- \rightarrow Fe^{2+}(aq) \quad E^\ominus = +0.77\,V$

A iodine $I_2 + 2e^- \rightarrow 2I^-$ +0.54 V
B chlorine $Cl_2 + 2e^- \rightarrow 2Cl^-$ +1.36 V
C copper(II) ions $Cu^{2+} + 2e^- \rightarrow Cu$ +0.34 V
D nickel(II) ions $Ni^{2+} + 2e^- \rightarrow Ni$ −0.25 V

Answer

For this question, the iron(II) ions are to be oxidised to iron(III) ions so this must be the left-hand electrode. The EMF will then be calculated for each of the substances in A to D.

For A: $Fe^{2+}|Fe^{3+}||I_2|2I^-$
EMF = +0.54 − (+0.77) = −0.23 V

This is not feasible – iodine will not oxidise iron(II) ions to iron(III) ions

For B: $Fe^{2+}|Fe^{3+}||Cl_2|2Cl^-$
EMF = +1.36 − (+0.77) = +0.59 V

This is feasible – chlorine will oxidise iron(II) ions to iron(III) ions

For C: $Fe^{2+}|Fe^{3+}||Cu^{2+}|Cu$
EMF = +0.34 − (+0.77) = −0.43 V

This is not feasible – copper (II) ions will not oxidise iron(II) ions to iron(III) ions

For D: $Fe^{2+}|Fe^{3+}||Ni^{2+}|Ni$
EMF = −0.25 − (+0.77) = −1.02 V

This is not feasible – nickel(II) ions will not oxidise iron(II) ions to iron(III) ions.

Hence chlorine will oxidise iron(II) ions to iron(III) ions.

EXAMPLE 4

The table below shows some standard electrode potential data.

Electrode half equation	$E^\ominus$/V
$Mg^{2+}(aq) + 2e^- \rightarrow Mg(s)$	−2.37
$Al^{3+}(aq) + 3e^- \rightarrow Al(s)$	−1.66
$Zn^{2+}(aq) + 2e^- \rightarrow Zn(s)$	−0.76
$Fe^{2+}(aq) + 2e^- \rightarrow Fe(s)$	−0.44
$V^{2+}(aq) + 2e^- \rightarrow V(s)$	−0.26
$Ni^{2+}(aq) + 2e^- \rightarrow Ni(s)$	−0.25
$2H^+(aq) + 2e^- \rightarrow H_2(g)$	0.00
$Cu^{2+}(aq) + 2e^- \rightarrow Cu(s)$	+ 0.34

1 Which species in the table above is the strongest reducing agent? Explain your answer.
2 a) Use the data in the table to deduce the equation for the overall cell reaction of a cell which has an EMF of +0.50 V.
b) Give the conventional cell representation for this cell.
c) Identify the negative electrode.
3 A conventional cell representation is shown below:
$Cr(s)|Cr^{3+}(aq)||Cu^{2+}(aq)|Cu(s)$
The EMF of this cell is +1.08 V. Use the data in the table above to calculate a value for the standard electrode potential of the chromium electrode.

Answer

1 The strongest reducing agent is most easily oxidised.
$Mg \rightarrow Mg^{2+} + 2e^-$ would have the most positive value.
Mg is the strongest reducing agent
2 a) $Zn \rightarrow Zn^{2+} + 2e^-$
$V^{2+} + 2e^- \rightarrow V$
overall equation: $Zn + V^{2+} \rightarrow Zn^{2+} + V$
The zinc and vanadium redox potentials show a difference of 0.50 between the values, −0.76 V and −0.26 V. When the zinc is reversed it gives +0.76 V and −0.26 V is added to give +0.50 V.
b) $Zn|Zn^{2+}||V^{2+}|V$
c) The negative electrode is where an oxidation occurs. The negative electrode is the zinc electrode.
3 $Cu^{2+}|Cu$ +0.34 V
$Cr|Cr^{3+}$ + x V
= EMF = +1.08 V
x = 1.08 − 0.34 = +0.74 V
This is for the oxidation reaction ($Cr \rightarrow Cr^{3+} + 3e^-$)
Standard electrode potential for $Cr^{3+} + 3e^- \rightarrow Cr$ = −0.74 V

EXAMPLE 5

The table below shows some standard electrode potential data.

Reaction	Electrode half equation	$E^{\ominus}/V$
1	$O_2(g) + 4H^+(aq) + 4e^- \rightarrow 2H_2O(l)$	+1.23
2	$O_2(g) + 2H^+(aq) + 2e^- \rightarrow H_2O_2(aq)$	+0.68
3	$O_2(g) + 2H_2O(l) + 4e^- \rightarrow 4OH^-(aq)$	+0.40
4	$Br_2(l) + 2e^- \rightarrow 2Br^-(aq)$	+1.07
5	$Ag^+(aq) + e^- \rightarrow Ag(s)$	+0.80

1 A solution of hydrogen bromide reacts with oxygen gas.
Determine the overall equation for the reaction which occurs and calculate the EMF of this reaction using reactions 1 and 4 above.
2 Determine the overall reaction when a solution containing silver(I) ions are added to hydrogen peroxide solution. Calculate the EMF of this reaction.
3 a) Using the standard electrode potentials, explain why silver reacts with bromine.
b) Identify the negative electrode.

Answers

1 Reaction 1: $O_2 + 4H^+ + 4e^- \rightarrow 2H_2O$ $E^{\ominus} = +1.23$ V
Reaction 4: $Br_2 + 2e^- \rightarrow 2Br^-$ $E^{\ominus} = +1.07$ V
Reverse reaction 4 and ×2
$O_2 + 4H^+ + 4e^- \rightarrow 2H_2O$
$4Br^- \rightarrow 2Br_2 + 4e^-$
Overall equation is:
$4H^+ + 4Br^- + O_2 \rightarrow 2H_2O + 2Br_2$
The equation may also be written: $4HBr + O_2 \rightarrow 2H_2O + 2Br_2$
EMF = +1.23 − 1.07 = +0.16 V
2 $O_2 + 2H^+ + 2e^- \rightarrow H_2O_2$ $E^{\ominus} = +0.68$ V
$Ag^+ + e^- \rightarrow Ag$ $E^{\ominus} = +0.80$ V
Reverse first equation and second equation ×2
$H_2O_2 \rightarrow O_2 + 2H^+ + 2e^-$
$2Ag^+ + 2e^- \rightarrow 2Ag$
Overall equation is:
$H_2O_2 + 2Ag^+ \rightarrow O_2 + 2Ag + 2H^+$
EMF = +0.80 − 0.68 = +0.12 V
3 a) $E^{\ominus}(Br_2|Br^-) > E^{\ominus}(Ag^+|Ag)$
b) Negative electrode is where oxidation occurs. Ag is oxidised to Ag^+/silver electrode is the negative electrode

TEST YOURSELF 3

1 A cell is set up using the following two electrodes:
$Fe^{3+}(aq) + e^- \rightarrow Fe^{2+} = (aq)$ $E^{\ominus} = +0.77$ V
$Zn^{2+}(aq) + 2e^- \rightarrow Zn(s)$ $E^{\ominus} = -0.76$ V
a) Calculate the EMF of the overall cell.
b) Which electrode forms the positive electrode?
c) Write an overall equation for the cell reaction.
d) Write the conventional representation for the cell using platinum contacts for the $Fe^{3+}|Fe^{2+}$ electrode.

2 A conventional representation of a cell is shown below:
$Fe(s)|Fe^{2+}(aq)||Cu^{2+}(aq)|Cu(s)$
a) Write an equation to represent the oxidation reaction in this cell.
b) Write an equation to represent the reduction reaction in this cell.
c) Write an overall equation for the cell reaction.
d) The EMF of this cell is +0.78 V. The half equation $Fe^{2+}(aq) + 2e^- \rightarrow Fe(s)$ has a standard electrode potential of −0.44 V. Determine the standard electrode potential of the copper electrode.

3 The table below shows some redox half equations with their standard electrode potentials.

Half equation	$E^{\ominus}/V$
$Zn^{2+}(aq) + 2e^- \rightarrow Zn(s)$	−0.76
$O_2(g) + 4H^+(aq) + 4e^- \rightarrow 2H_2O(l)$	+1.23
$Cl_2(g) + 2e^- \rightarrow 2Cl^-(aq)$	+1.36
$F_2(g) + 2e^- \rightarrow 2F^-(aq)$	+2.87

a) Name the strongest reducing agent from the table above.
b) In terms of electrons, state what happens to a reducing agent in a redox reaction.
c) Fluorine reacts with water, oxidising the oxygen in water to oxygen.
i) Use data from the table to explain why fluorine reacts with water.
ii) Write an equation for the reaction which occurs.
d) Zinc reacts with chlorine.
i) Write a conventional cell representation for the reaction which occurs.
ii) Calculate the EMF of this cell.

REQUIRED PRACTICAL 8

Does the concentration of silver ions in a solution affect the electrode potential of a cell?

A silver chloride electrode is a type of reference electrode commonly used in electrochemical measurements. For example, it is usually the internal reference electrode in pH meters. Silver chloride electrodes are also used in electrocardiography (ECG).

In a laboratory experiment a silver cell was set up as shown in the diagram below.

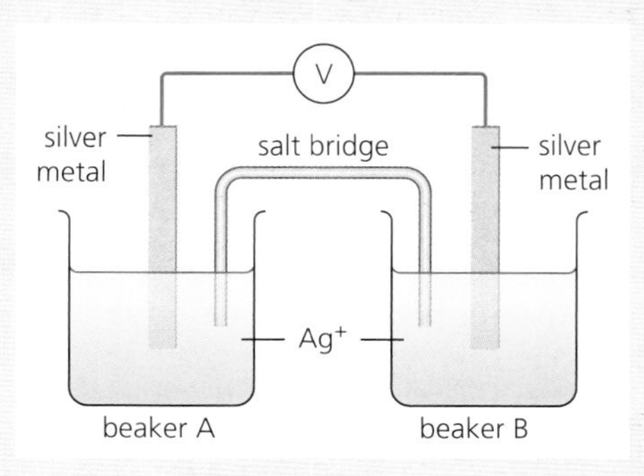

The concentration of silver ions in the solution in beaker B was changed and the electrode potential recorded in the table below.

$[Ag^+]$ in beaker A /mol dm^{-3}	$[Ag^+]$ in beaker B /mol dm^{-3}	$Log[Ag^+]_B$	Electrode potential/V
0.1	0.00001		−0.200
0.1	0.0001		−0.150
0.1	0.001		−0.100
0.1	0.01		−0.050
0.1	0.1		0.000

1 Why is a salt bridge necessary in this experiment?
2 Suggest how the salt bridge may be made.
3 Explain why there is a zero voltage if the solutions in both beakers are identical.
4 Why is a high-resistance voltmeter used in this experiment?
5 Complete the table by calculating values for $Log[Ag^+]_B$ and then plot a graph of Log $[Ag^+]_B$ (x axis) against electrode potential (y axis). Why is a log scale useful in this case?
6 How could you use this experiment to determine the silver ion concentration of an unknown solution?

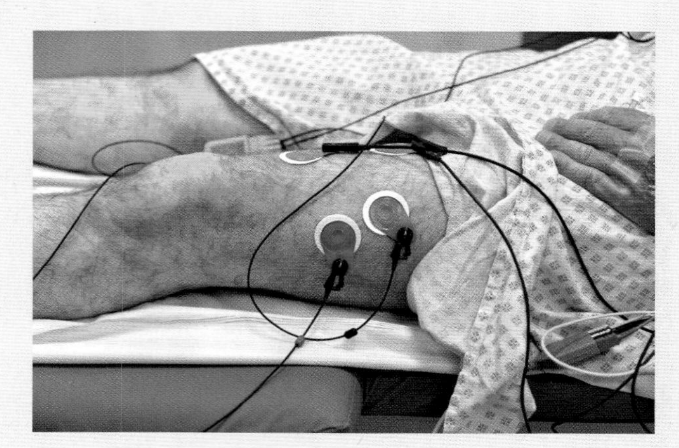

Figure 5.5 Electrocardiogram (ECG) electrodes placed on a patient's leg. The electrodes measure the electrical activity of the heart and are also placed on the chest and arms. The electrodes are silver/silver chloride sensors made by coating a thin layer of silver on plastic, and the outer layer is converted to silver chloride.

TIP

Further information on plotting a log graph is given on page 320 in the mathematics for chemistry chapter.

Commercial applications of electrochemical cells

Electrochemical cells can be used as commercial source of electrical energy. They are commonly called cells or batteries. The electromotive force of the electrochemical cell is the voltage of the cell or battery. The two electrodes combine to produce the voltage. Some cells are non-rechargeable and others are rechargeable.

Primary cells are often referred to as single-use or disposable. The materials in the cell cannot be regenerated by recharging. Some examples include alkaline cells which are commonly used in torches. Secondary cells are able to be recharged and the original reagents in the cell regenerated by recharging. These include the lead-acid batteries used in cars and the rechargeable lithium ion and nickel metal hydride cells used in many electronic devices such as mobile phones.

Non-rechargeable cells

Common examples of non-rechargeable cells include alkaline batteries and dry cells. These are called primary cells which can only be used once and then discarded.

The diagram shows a non-rechargeable cell.

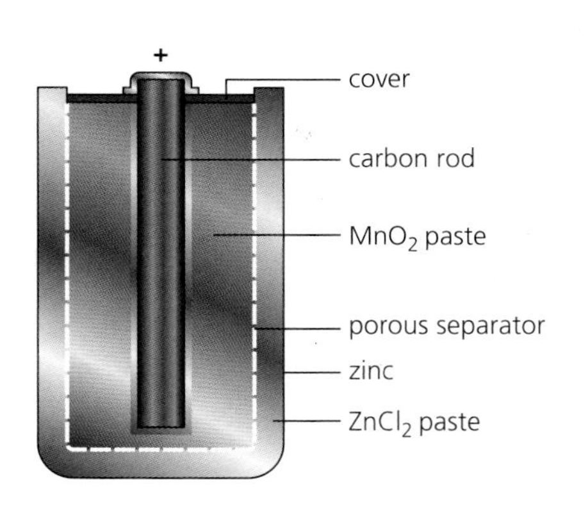

A typical non-rechargeable cell

The porous separator allows ions to pass through it. The carbon rod is made of graphite as it conducts electricity and allows the movement of the electrons through it. The zinc is used as a container for the cell but the cell leaks after being used for a long time as the zinc is used up. The zinc becomes compromised and the contents may leak out of the cell.

At the negative electrode, oxidation occurs:

$$Zn(s) \rightarrow Zn^{2+}(aq) + 2e^-$$

At the positive electrode, reduction occurs:

$$MnO_2(s) + H_2O(l) + e^- \rightarrow MnO(OH)(s) + OH^-(aq)$$

The standard electrode potentials for the reduction reactions are given as:

$$MnO_2(s) + H_2O(l) + e^- \rightarrow MnO\ (OH)(s) + OH^-(aq) \qquad E^{\ominus} = +0.74\,V$$

$$Zn^{2+}(aq) + 2e^- \rightarrow Zn(s) \qquad E^{\ominus} = -0.76\,V$$

The overall EMF of this cell is: +0.74 + 0.76 = 1.50 V.

This is the voltage provided by this cell.

The oxidation state of the zinc in the oxidation reaction changes from 0 in Zn to +2 in Zn^{2+}. The manganese is reduced from +4 in MnO_2 to +3 in MnO(OH).

The overall reaction when the cell discharges is:

$$Zn(s) + 2MnO_2(s) + 2H_2O(l) \rightarrow Zn^{2+}(aq) + 2MnO(OH)_2(s) + 2OH^-(aq)$$

TIP

Try working out this overall equation from the oxidation and reduction reactions occurring in the cell.

EXAMPLE 6

The zinc-silver(I) oxide cell is non-rechargeable. It may be used as a source of electrical energy.

Using the following standard electrode potentials:

$Zn^{2+}(aq) + 2e^- \rightarrow Zn(s)$ $\quad E^\ominus = -0.76\,V$

$Ag_2O(s) + 2H^+(aq) + 2e^- \rightarrow 2Ag(s) + H_2O(l)$ $\quad E^\ominus = +0.34\,V$

1 Write the cell notation of the cell used to represent the zinc-silver(I) oxide cell.
2 Which half equation forms the negative electrode of this cell?
3 Calculate the EMF of this cell.
4 Write an overall equation occurring when this cell is discharging.

Answers

1 $Zn \mid Zn^{2+} \parallel Ag_2O, Ag$
2 The negative electrode is where oxidation occurs. The negative electrode is the zinc electrode as zinc is oxidised.
3 EMF = + 0.34 + 0.76 = 1.10 V
4 $Zn(s) \rightarrow Zn^{2+}(aq) + 2e^-$
$Ag_2O(s) + 2H^+(aq) + 2e^- \rightarrow 2Ag(s) + H_2O(l)$
Combining these:
$Zn(s) + Ag_2O(s) + 2H^+(aq) \rightarrow Zn^{2+}(aq) + 2Ag(s) + H_2O(l)$

TIP

As Ag_2O and Ag are in the same state, they should be separated by a comma rather than a phase line (|). If state symbols are required, the answer would be $Zn(s)|Zn^{2+}(aq)||Ag_2O(s), Ag(g)$.

TIP

Non-rechargeable cells cannot be recharged as the reaction that occurs is not reversible.

Rechargeable cells

Rechargeable cells (often called secondary cells) can be used many times. There are many types of these such as the nickel-cadmium cell and the lithium ion cell.

Nickel-cadmium cell

In this cell the standard electrode potentials are:

$Cd(OH)_2(s) + 2e^- \rightarrow Cd(s) + 2OH^-(aq)$ $\quad E^\ominus = -0.88\,V$

$NiO(OH)(s) + H_2O(l) + e^- \rightarrow Ni(OH)_2(s) + OH^-(aq)$ $\quad E^\ominus = +0.52\,V$

For the cell to work, the Ni is reduced from +3 in $NiO(OH)_2$ to +2 in $Ni(OH)_2$ and the Cd is oxidised from 0 in Cd to +2 in $Cd(OH)_2$. This gives a positive EMF for this cell. The EMF is +0.52 + 0.88 = +1.40 V.

At the negative electrode, oxidation occurs:

$$Cd(s) + 2OH^-(aq) \rightarrow Cd(OH)_2(s) + 2e^-$$

At the positive electrode, reduction occurs:

$$NiO(OH)(s) + H_2O(l) + e^- \rightarrow Ni(OH)_2(s) + OH^-(aq)$$

TIP
The reaction that recharges a rechargeable cell is the reverse of the reaction which discharges it.

The overall reaction occurring when the cell is discharged is:

$$Cd(s) + 2NiO(OH)(s) + 2H_2O(l) \rightarrow Cd(OH)_2(s) + 2Ni(OH)_2(s)$$

When the cell is recharged the reaction above is reversed to regenerate the reagents:

$$Cd(OH)_2(s) + 2Ni(OH)_2(s) \rightarrow Cd(s) + 2NiO(OH)(s) + 2H_2O(l)$$

Rechargeable cells are more environmentally advantageous as they can be reused and prevent waste. Supplies of the metal and other reagents are not depleted as quickly and less energy is used to extract metals.

Lead-acid cell

The lead-acid cell is the cell used in cars and other vehicles. It is recharged as the vehicle moves. Sulfuric acid is used to provide the acid in these types of battery.

The half equations for the reaction are:

$$PbO_2(s) + 3H^+(aq) + HSO_4^-(aq) + 2e^- \rightarrow PbSO_4(s) + 2H_2O(l) \qquad E^\ominus = +1.69\,V$$

$$PbSO_4(s) + H^+(aq) + 2e^- \rightarrow Pb(s) + HSO_4^-(aq) \qquad E^\ominus = -0.46\,V$$

At the negative electrode, oxidation occurs:

$$Pb(s) + HSO_4^-(aq) \rightarrow PbSO_4(s) + H^+(aq) + 2e^-$$

At the positive electrode, reduction occurs:

$$PbO_2(s) + 3H^+(aq) + HSO_4^-(aq) + 2e^- \rightarrow PbSO_4(s) + 2H_2O(l)$$

The overall reaction occurring when the cell is discharged is:

$$PbO_2(s) + 2H^+(aq) + 2HSO_4^-(aq) + Pb(s) \rightarrow 2PbSO_4(s) + 2H_2O(l)$$

When the cell is recharged the reaction above is reversed to regenerate the reagents:

$$2PbSO_4(s) + 2H_2O(l) \rightarrow PbO_2(s) + 2H^+(aq) + 2HSO_4^-(aq) + Pb(s)$$

At the negative electrode, Pb is oxidised from 0 in Pb to +2 in $PbSO_4$.

At the positive electrode, Pb is reduced from +4 in PbO_2 to +2 in $PbSO_4$.

The formation of lead(II) sulfate can be a problem if a lead-acid cell is discharged for long periods of time. The insoluble lead(II) sulfate build up in the cell and the cell cannot be recharged.

Lithium ion cell

Rechargeable lithium ion cells are rechargeable and often used to provide electrical energy for cameras, laptops, tablets and mobile phones.

The positive electrode (cathode) in the cell is represented by the half equation:

$$Li^+ + CoO_2 + e^- \rightarrow Li^+[CoO_2]^- \qquad \text{reduction}$$

The negative electrode (anode) is represented by the half equation:

$$Li \rightarrow Li^+ + e^- \qquad \text{oxidation}$$

TIP
Remember to use NEGATOX to identify the **ne**gative electrode as the electrode at which an **ox**idation occurs.

The overall cell may be written as:

$Li|Li^+||Li^+, CoO_2|LiCoO_2|Pt$

The overall reaction is found by combining the half equations:

$Li + CoO_2 \rightarrow Li^+[CoO_2]^-$

At the negative electrode, lithium is being oxidised from 0 in Li to +1 in Li^+.

At the positive electrode, cobalt is being reduced from +4 oxidation state in CoO_2 to +3 oxidation state in $LiCoO_2$.

The reactants in this electrochemical cell are absorbed into graphite powder which acts as a support medium. The ions are able to react in this support medium without the need for a solvent such as water. Water could not be used as it reacts with lithium.

Fuel cells

- A fuel cell is an electrical cell which converts the chemical energy of a redox reaction into electrical energy. The cell will continue to function as long as the fuel and oxygen are supplied to it.
- The most common fuel cells use hydrogen or ethanol as the fuel though hydrocarbons may also be used.
- Oxidation occurs at the anode and the electrons released travel through a wire and into the external circuit.
- The central electrolyte allows ions and molecules to move through it but not electrons. The ions react with another substance and this is the reduction reaction which takes electrons from the external circuit.
- The oxidation and reduction reactions are both catalysed.

Hydrogen fuel cell

The hydrogen fuel call can operate in acidic or alkaline conditions. In the acidic hydrogen fuel cell this oxidation is the reverse of the following standard electrode potential:

$2H^+(aq) + 2e^- \rightarrow H_2(g) \qquad E^{\ominus} = 0.00\,V$

The oxidation reaction at the negative electrode (anode) is:

$H_2(g) \rightarrow 2H^+(aq) + 2e^-$

The reduction reaction at the positive electrode (cathode) is:

$O_2(g) + 4H^+(aq) + 4e^- \rightarrow 2H_2O(l) \qquad E^{\ominus} = +1.23\,V$

The overall reaction is:

$2H_2(g) + O_2(g) \rightarrow 2H_2O(l) \qquad EMF = +1.23\,V.$

The cell notation for a hydrogen fuel cell operating in acidic conditions is:

$Pt|H_2(g)|H^+(aq)||O_2(g)|H^+(aq), H_2O(l)|Pt$

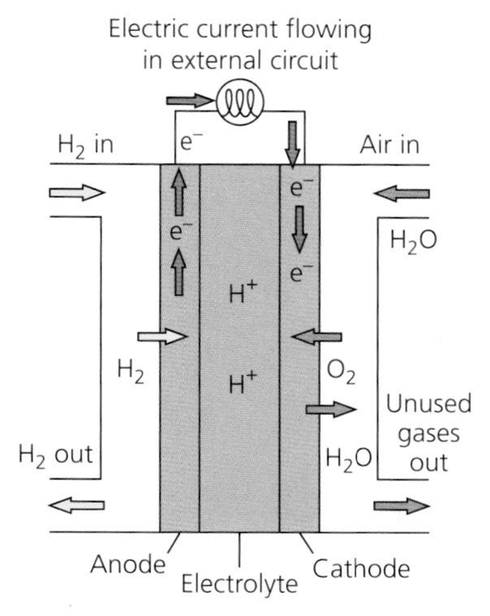

Hydrogen fuel cell in acidic conditions

TIP

The overall reaction is the same so the EMF is the same for both the acidic hydrogen fuel cell and the alkaline hydrogen fuel cell.

The diagram on the left shows these processes:

In the alkaline hydrogen fuel cell, the oxidation reaction at the anode is the reverse of the following standard electrode potential:

$$2H_2O(l) + 2e^- \rightarrow H_2(g) + 2OH^-(aq) \quad E^\ominus = -0.83\,V$$

The oxidation reaction at the negative electrode (anode) is:

$$H_2(g) + 2OH^-(aq) \rightarrow 2H_2O(l) + 2e^-$$

The reduction reaction at the positive electrode (cathode) is:

$$O_2(g) + 2H_2O(l) + 4e^- \rightarrow 4OH^-(aq) \quad E^\ominus = +0.40\,V$$

The overall reaction is:

$$2H_2(g) + O_2(g) \rightarrow 2H_2O(l) \quad EMF = +1.23\,V.$$

The conventional cell representation for a hydrogen fuel cell operating in alkaline conditions is:

$$Pt|H_2(g)|OH^-(aq), H_2O(l)||O_2(g)|H_2O(l), OH^-(aq)|Pt$$

Commercial alkaline hydrogen fuel cells use platinum electrodes in contact with concentrated aqueous hydrogen hydroxide. Porous platinum is used to give a larger surface area.

Fuel cells do not have to be recharged as the fuel is supplied continuously to the cell so the voltage output does not change. However, fuel cells use fuels that may have been produced using an energy source that is not carbon neutral. Ethanol fuel cells can use ethanol produced from fermentation of crops, which are regarded as carbon neutral as the CO_2 produced in the fuel cell is captured again during photosynthesis to generate the carbohydrates used to make the ethanol.

Ethanol fuel cell

Ethanol oxygen fuel cells are used an as alternative to the hydrogen fuel cell.

Ethanol is oxidised in the fuel cell to carbon dioxide and water. The overall reaction is:

$$C_2H_5OH(l) + 3O_2(g) \rightarrow 2CO_2(g) + 3H_2O(l)$$

The oxidation reaction occurring at the negative electrode (anode) reaction is:

$$C_2H_5OH(l) + 3H_2O(l) \rightarrow 2CO_2(g) + 12H^+(aq) + 12e^-$$

The reduction reaction occurring at the positive electrode (cathode) is:

$$12H^+(aq) + 3O_2(g) + 12e^- \rightarrow 6H_2O(l)$$

EXAMPLE 7

The EMF of the ethanol fuel cell is +1.00 V. The standard electrode potential of the following reaction is:

$$2CO_2(g) + 12H^+(aq) + 12e^- \rightarrow C_2H_5OH(l) + 3H_2O(l) \qquad E^\ominus = +0.23\,V$$

Calculate the standard electrode potential of the following reaction:

$$4H^+(aq) + O_2(g) + 4e^- \rightarrow 2H_2O(l)$$

Answer

The oxidation reaction is the reverse of the first reaction:

$$C_2H_5OH(l) + 3H_2O(l) \rightarrow 2CO_2(g) + 12H^+(aq) + 12e^-$$

The reduction reaction is:

$$12H^+(aq) + 3O_2(g) + 12e^- \rightarrow 6H_2O(l)$$

The EMF is the total of the oxidation and reduction potentials.

Reduction potential = to be determined = x

Oxidation potential = −0.23 V

EMF = 1.00 V = x + (−0.23)

x = +1.23 V

Standard electrode potential is +1.23 V

EXAMPLE 8

The EMF of the methanol fuel cell is +1.21 V. The standard electrode potential of the following reaction occurring at the positive electrode is:

$$4H^+(aq) + O_2(g) + 4e^- \rightarrow 2H_2O(l) \qquad E^\ominus = +1.23\,V$$

1 Write an overall equation for the reaction occurring in the fuel cell.
2 Write an equation for the reaction occurring at the negative electrode.
3 Calculate the standard electrode potential for the reaction occurring at the negative electrode.

Answer

1 The overall reaction is the oxidation of methanol:
   $CH_3OH + 1\frac{1}{2}O_2 \rightarrow CO_2 + 2H_2O$
2 The basic equation is:
   $CH_3OH + H_2O \rightarrow CO_2 + H^+ + e^-$
   It is balanced by balancing the atoms and the electrons balance the charge.
   $CH_3OH + H_2O \rightarrow CO_2 + 6H^+ + 6e^-$
3 The EMF is the total of the oxidation and reduction potentials.
   Electrode potential to be determined = x
   Reduction potential = +1.23 V
   EMF = 1.21 V = 1.23 − x
   x = + 0.02 V
   Standard electrode potential as a reduction = +0.02 V

TEST YOURSELF 4

1 The reactions which occur in a lithium ion cell are:
Reaction A: $Li^+ + CoO_2 + e^- \rightarrow Li^+[CoO_2]^-$
Reaction B: $Li \rightarrow Li^+ + e^-$
  a) Write an overall equation for the reaction occurring in the lithium ion cell.
  b) Which reaction (A or B) occurs at the electrode which is the negative electrode in this cell? Explain your answer.

2 Explain why the voltage produced by a hydrogen fuel cell remains constant whereas that of a lead-acid cell decreases eventually after continued use.

3 The conventional representation of the hydrogen fuel cell operating in acidic conditions is:
$Pt|H_2(g)|H^+(aq)||O_2(g)|H^+(aq), H_2O(l)|Pt$
  a) Write an equation for the oxidation reaction which is occurring.
  b) Write an equation for the reduction reaction which is occurring.
  c) Write an overall equation for the reaction in the fuel cell.

Practice questions

1 Which one of the following statements about the standard hydrogen electrode is incorrect?

A the hydrogen gas is at a pressure of 100 kPa

B the value of $E^\ominus$ is 1.00 V

C the hydrogen ion concentration in solution is 1.00 mol dm^{-3}

D the temperature is 298 K *(1)*

2 Using the standard electrode potentials below, choose the reducing agent capable of reducing vanadium from the +5 to the +3 oxidation state but not to the +2 state.

	$E^\ominus$/V
$Zn^{2+}(aq) + 2e^- \rightarrow Zn(s)$	−0.76
$Fe^{2+}(aq) + 2e^- \rightarrow Fe(s)$	−0.44
$V^{3+}(aq) + e^- \rightarrow V^{2+}(aq)$	−0.26
$SO_4^{2-}(aq) + 4H^+(aq) + 2e^- \rightarrow 2H_2O(l) + SO_2(g)$	+0.17
$VO^{2+}(aq) + 2H^+(aq) + e^- \rightarrow V^{3+}(aq) + H_2O(l)$	+ 0.32
$I_2(aq) + 2e^- \rightarrow 2I^-(aq)$	+0.54
$VO_2^+(aq) + 2H^+(aq) + e^- \rightarrow VO^{2+}(aq) + H_2O(l)$	+1.00

A iodide ions **B** iron

C sulfur dioxide **D** zinc *(1)*

3 Use the following data to predict which of the reactions listed below will proceed as written.

	$E^\ominus$/V
$Cr^{3+} + 3e^- \rightarrow Cr$	−0.74
$Fe^{2+} + 2e^- \rightarrow Fe$	−0.44
$Fe^{3+} + 2e^- \rightarrow Fe^{2+}$	+0.77
$Cr_2O_7^{2-} + 14H^+ + 6e^- \rightarrow 2Cr^{3+} + 7H_2O$	+1.33
$BrO_3^- + 6H^+ + 5e^- \rightarrow \frac{1}{2}Br_2 + 3H_2O$	+1.52

A $2Cr^{3+} + 7H_2O + 3Fe^{2+} \rightarrow Cr_2O_7^{2-} + 14H^+ + 3Fe$

B $\frac{1}{2}Br_2 + 3H_2O + 5Fe^{3+} \rightarrow 5Fe^{2+} + BrO_3^- + 6H^+ + 5e^-$

C $Cr_2O_7^{2-} + 14H^+ + 2Cr \rightarrow 4Cr^{3+} + 7H_2O$

D $Br_2 + 6H_2O + 5Fe^{2+} \rightarrow 2BrO_3^- + 12H^+ + 10e^- + 5Fe$ *(1)*

4 $V^{3+}(aq) + e^- \rightarrow V^{2+}(aq)$ $E^\ominus = -0.26$ V

a) Draw a labelled diagram of the apparatus which could be connected to a standard hydrogen electrode in order to measure the standard electrode potential of the V^{3+}/V^{2+} electrode.

b) In your diagram, show how this electrode is connected to the standard hydrogen electrode and to a voltmeter. Do not draw the standard hydrogen electrode.

c) State the conditions under which this cell should be operated to measure the standard electrode potential. (5)

5 For the conventional representation of a cell shown below:

$$Pt|Fe^{2+}(aq), Fe^{3+}(aq)||MnO_4^-(aq), Mn^{2+}(aq)|Pt$$

a) State which electrode is the negative electrode and explain your answer. (2)

b) Write the half equation representing the oxidation and reduction reactions in this cell and combine these to write an overall equation for the cell reaction. (3)

c) The EMF of this cell is +0.74 V. The standard electrode potential of the manganate(VII)/manganese(II) cell is +1.51 V. Calculate a value for the standard electrode potential of the Fe^{3+}/Fe^{2+} electrode. (2)

6 The half equations for two electrodes used to make an electrochemical cell are shown below:

$$ClO_3^-(aq) + 6H^+(aq) + 6e^- \rightarrow Cl^-(aq) + 3H_2O(l) \quad E^\ominus = +1.45\,V$$

$$VO_2^+(aq) + 2H^+(aq) + e^- \rightarrow VO^{2+}(aq) + H_2O(l) \quad E^\ominus = +1.00\,V$$

a) Write the conventional cell representation for the cell using platinum contacts. (2)

b) Write an overall equation for the cell reaction. (1)

c) Identify the oxidising agent and the reducing agent in this reaction. (2)

d) Calculate the EMF of the cell. (1)

e) Identify the negative electrode. (1)

7 The table below shows some standard electrode potentials:

Electrode half equation	$E^\ominus$/V
$Au^+(aq) + e^- \rightarrow Au(s)$	+1.68
$Cl_2(g) + 2e^- \rightarrow 2Cl^-(aq)$	+1.36
$O_2(g) + 4H^+(aq) + 4e^- \rightarrow 2H_2O(l)$	+1.23
$Ag^+(aq) + e^- \rightarrow Ag(s)$	+0.80
$Fe^{3+}(aq) + e^- \rightarrow Fe^{2+}(aq)$	+0.77
$Cu^{2+}(aq) + 2e^- \rightarrow Cu(s)$	+0.34
$Fe^{2+}(aq) + 2e^- \rightarrow Fe(s)$	−0.44
$Zn^{2+}(aq) + 2e^- \rightarrow Zn(s)$	−0.76

a) Au^+ ions react with water.

i) Use data from the table to explain why Au^+ ions react with water. (1)

ii) Write an equation for the reaction which would occur between Au^+ ions and water. (1)

iii) Write a conventional cell representation of this cell. (2)

b) Silver nitrate solution is used to test for the presence of halide ions.

i) Explain why a redox reaction does not occur when silver(I) ions are mixed with a solution containing chloride ions. (1)

ii) What is observed when a solution containing silver(I) ions is added to a solution containing chloride ions? (1)

iii) Write an ionic equation for the reaction between silver(I) ions and chloride ions.

(c) Predict the products of the following reactions and write equation(s) to represent the reactions which occur.

i) Zinc metal is placed in a solution containing Fe^{3+} ions. (3)

ii) Iron metal is placed in a solution containing Ag^+ ions. (3)

iii) Copper metal is placed in silver(I) nitrate solution. (3)

8 The diagram below shows an electrochemical cell:

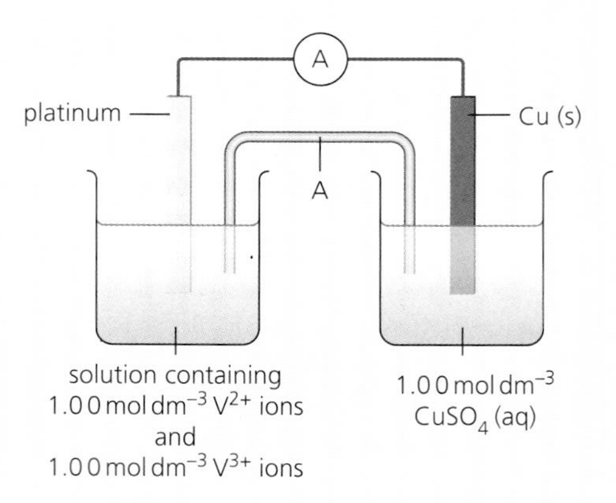

The standard electrode potentials are:

$V^{3+}(aq) + e^- \rightarrow V^{2+}(aq)$ $E^\ominus = -0.26\,V$

$Cu^{2+}(aq) + 2e^- \rightarrow Cu(s)$ $E^\ominus = +0.34\,V$

a) i) Write a conventional cell representation for this cell. *(2)*

ii) What is represented by A? *(1)*

iii) Explain why electrons flow from left to right in this cell. *(2)*

iv) Calculate the EMF of this cell. *(1)*

v) Write an overall equation for the reaction occurring in this cell. *(1)*

b) The copper electrode is replaced by a platinum contact in a solution containing VO_2^+ ions and VO^{2+} ions both of concentration 1.00 mol dm^{-3}.

The EMF of the cell using a high-resistance voltmeter is determined to be +1.26 V.

i) Determine a value for the standard electrode potential for the $VO_2^+|VO^{2+}$ electrode. *(2)*

ii) Write an overall equation for the reaction which occurs in the cell. *(2)*

9 The conventional cell representation for a hydrogen fuel cell operating in alkaline conditions is:

$Pt|H_2(g)|OH^-(aq), H_2O(l)||O_2(g)|H_2O(l), OH^-(aq)|Pt$

a) Write an equation for the oxidation reaction occurring in the fuel cell. *(1)*

b) Write an equation for the reduction reaction occurring in the fuel cell. *(1)*

c) Write an overall equation for the reaction occurring in the fuel cell. *(1)*

d) The EMF of the hydrogen fuel cell is +1.23 V. The standard electrode potential for the positive electrode is given below:

$O_2(g) + 2H_2O(l) + 4e^- \rightarrow 4OH^-(aq)$ $E^\ominus = +0.40\,V$

i) Calculate a value for the standard electrode potential for the other electrode. *(2)*

ii) Explain why fuel cells maintain a constant voltage. *(2)*

Stretch and challenge

10 A table of standard electrode potentials is given below:

Half electrode equation	$E^\ominus$/V
$Zn^{2+}(aq) + 2e^- \rightarrow Zn(s)$	−0.76
$Fe^{2+}(aq) + 2e^- \rightarrow Fe(s)$	−0.44
$Sn^{2+}(aq) + 2e^- \rightarrow Sn(s)$	−0.14
$Fe^{3+}(aq) + e^- \rightarrow Fe^{2+}(aq)$	+0.77
$Cr_2O_7^{2-}(aq) + 14H^+(aq) + 6e^- \rightarrow 2Cr^{3+}(aq) + 7H_2O(l)$	+1.33

a) i) Using the table above, write the formula for the strongest oxidising agent. *(1)*

ii) Describe, without the use of a diagram, how the standard electrode potential of the dichromate(VI)/chromium(III) electrode could be determined. *(4)* SHE

iii) Write the equation for the cell reaction when the dichromate(VI)/chromium(III) and $Fe^{3+}(aq)/Fe^{2+}(aq)$ half cells are combined. *(2)*

b) Tin plate is used to prevent rusting of steel cans. Rusting is promoted if the tin is scratched and the steel is exposed. An electrochemical cell is set up between the iron and tin.

i) Calculate the EMF of this cell. *(1)*

ii) Write conventional cell representation for this cell. *(2)*

c) i) Explain using the standard electrode potentials how galvanising (coating in zinc) could prevent iron from rusting even if the coating is scratched. *(2)*

ii) Write an overall equation for the reaction which occurs when the zinc coating on a piece of galvanised iron is scratched. *(1)*

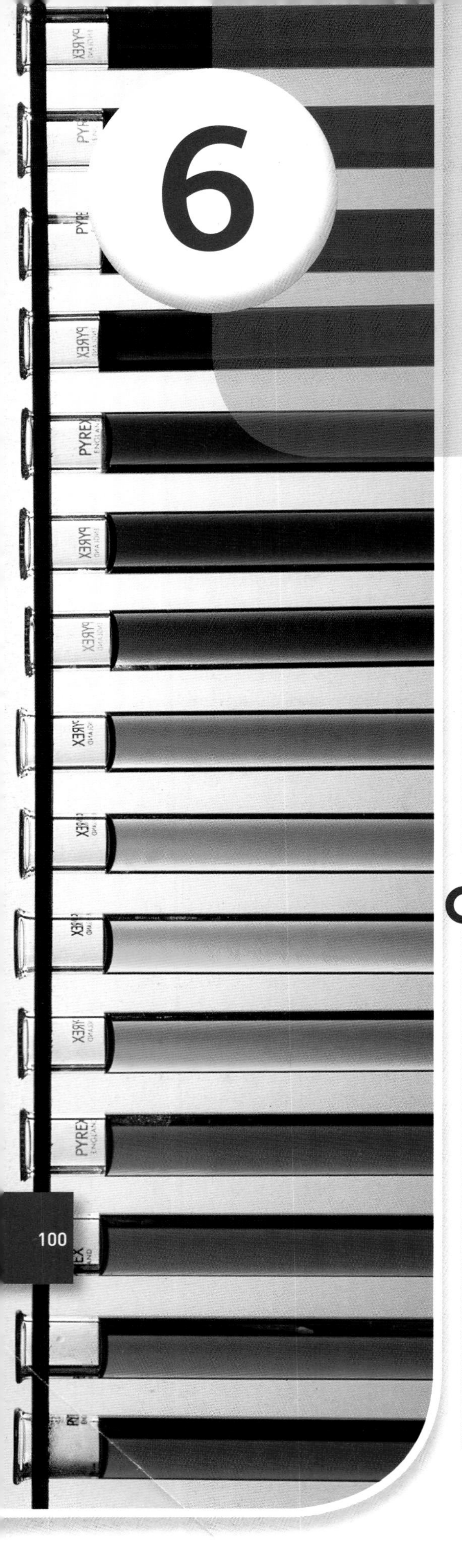

Acids and bases

PRIOR KNOWLEDGE

- Acids are compounds that dissolve in water to produce hydrogen ions.
- Hydrogen ions are H^+.
- Hydrochloric acid (HCl), sulfuric acid (H_2SO_4) and nitric acid (HNO_3) are strong acids.
- Ethanoic acid (CH_3COOH) is a weak acid.
- The pH scale gives a numeric value that measures the strength of an acid or base (alkali); the scale at a simple level goes from 0 to 14.
- A pH value of 7 is neutral.
- pH values <7 are acidic; 0–2 are pH values for strong acids; pH values >2 and <7 are for weak acids.
- A base is a substance that reacts with an acid producing a salt and water.
- Alkalis are soluble bases.
- Alkalis dissolve in water to produce hydroxide ions.
- Hydroxide ions are OH^-
- The ionic equation for neutralisation is $H+(aq) + OH^-(aq) \rightarrow H_2O(l)$
- In general, pH values >7 are alkaline; 12–14 are pH values for strong alkalis; pH values >7 and <12 are for weak alkalis.
- Bases (and alkalis) react with acids forming salts.
- Sodium hydroxide and potassium hydroxide solutions are strong alkalis.
- Sulfuric acid forms salts called sulfates; hydrochloric acid forms salts called chlorides; nitric acid forms salts called nitrates; ethanoic acid forms salts called ethanoates.

TEST YOURSELF ON PRIOR KNOWLEDGE 1

1 From the following acids:
H_2SO_4 HCl HNO_3 CH_3COOH
a) Name each acid.
b) Which one of the acids is a weak acid?

2 From the following bases:
$NaOH$ CuO KOH MgO
a) Which two bases are also alkalis?
b) Name the salt formed when CuO reacts with H_2SO_4.
c) Name the salt produced when MgO reacts with HCl.

3 Solution A has a pH of 9
Solution B has a pH of 1
Solution C has a pH of 14
Solution D has a pH of 7
Solution E has a pH of 5
a) Which solution is neutral?
b) Which solution is a weak acid?
c) Which solution is a strong alkali?
d) Which solution is a weak alkali?
e) Which solution is a strong acid?

Many of the reactions studied in chemistry involve those of acids and bases. Acids are important in many industrial processes from the manufacture of fertilisers, dyes, explosives, pharmaceutical drugs to paint and pigments. Bases are used in household cleaning products, soap manufacture, in the refining of crude oil and the manufacture of explosives and fertilisers.

Brønsted-Lowry theory of acids and bases

The Brønsted-Lowry-Lowry definition of acids and bases depends on protons. A hydrogen ion, H^+, is a proton so the term hydrogen ion and proton are interchangeable.

- A **Brønsted-Lowry acid** is defined as a *proton donor*.
- A **Brønsted-Lowry base** is defined as a *proton acceptor*.

Strong and weak acids and bases

Acids (and bases) may be classified as strong or weak depending on the degree to which they are dissociated in solution.

Strong acids and *strong bases* are completely dissociated in aqueous solution.

Weak acids and *weak bases* are slightly dissociated in aqueous solution.

Strong acids

The strong acids most commonly used are hydrochloric acid (HCl), sulfuric acid (H_2SO_4) and nitric acid (HNO_3).

Strong acids are usually covalent molecules that dissolve in water and dissociate (or ionise) completely.

Acidic hydrogen atoms in a molecule are the hydrogen atoms which will produce $H^+(aq)$ or protons when the acid dissolves in water.

The acidic hydrogen atoms are often bonded to electronegative atoms such as oxygen, fluorine and chlorine. In oxy-acids like sulfuric acid and nitric acid, the acidic hydrogen atoms are bonded to oxygen atoms.

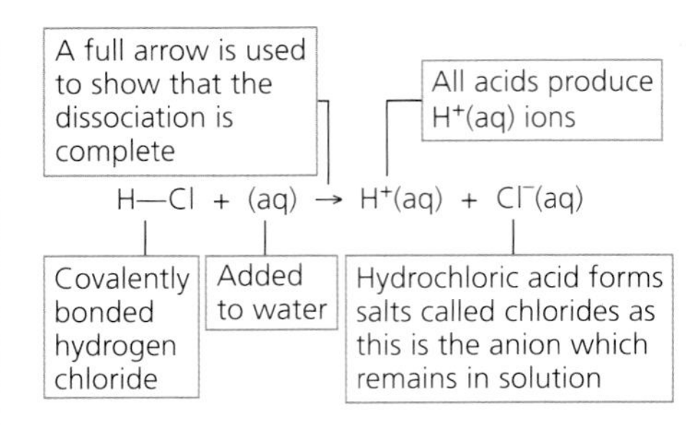

Hydrogen chloride gas dissolves in water to form hydrochloric acid.

Normally the '+(aq)' is left out of the equation.

The equation on the left can be simplified to:

$$HCl \rightarrow H^+ + Cl^-$$

All of the HCl molecules in solution dissociate into ions. This makes hydrochloric acid a strong acid as it dissociates completely.

Hydrochloric acid is described as a **strong monoprotic acid** as 1 mol of hydrogen chloride produces 1 mol of $H^+(aq)$.

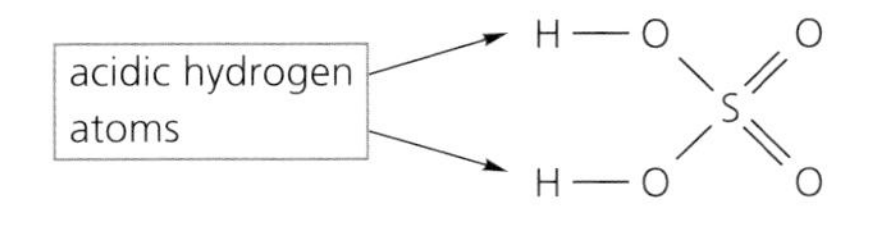

Sulfuric acid is H_2SO_4. It has the structure:

$$H_2SO_4 \rightarrow 2H^+ + SO_4^{2-}$$

Sulfuric acid dissolves in water and dissociates completely (or ionises completely).

This makes sulfuric acid a strong acid as it ionises (or dissociates) completely.

Sulfuric acid is described as a **strong diprotic acid** as 1 mol of sulfuric acid produces 2 mol of hydrogen ions (protons).

Weak acids

The most common weak acid is ethanoic acid (CH_3COOH). Most carboxylic acids are weak acids.

Weak acids are usually covalent molecules which dissolve in water and dissociate (or ionise) slightly.

Acidic hydrogen atoms in a molecule are the hydrogen atoms which will produce $H^+(aq)$ or protons when the acid dissolves in water.

The acidic hydrogen atoms are often bonded to electronegative atoms such as oxygen, fluorine and chlorine. In ethanoic acid the acidic hydrogen atom is the one bonded to the oxygen atom. The other hydrogen atoms bonded to carbon are not acidic hydrogen atoms.

The structure of ethanoic acid is:

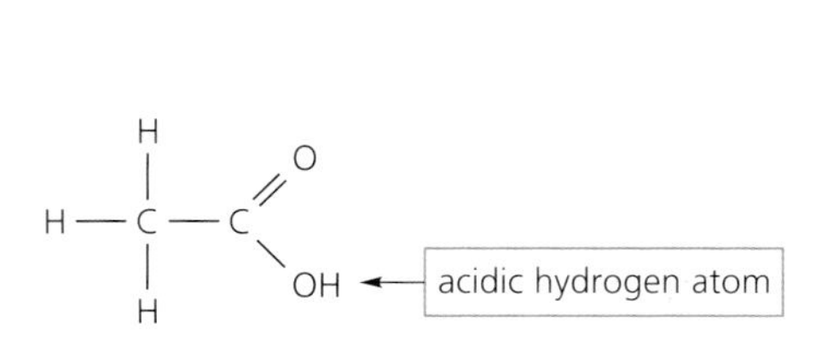

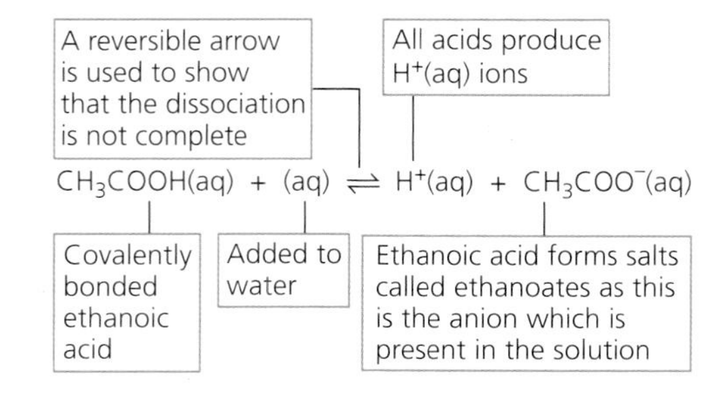

Normally the '+(aq)' is left out of the equation.

The above equation can be simplified to:

$$CH_3COOH \rightleftharpoons H^+ + CH_3COO^-$$

Only some of the CH_3COOH molecules in solution dissociate into ions. This makes ethanoic acid a weak acid as it dissociates slightly.

Ethanoic acid is described as a weak monoprotic acid as 1 mol of ethanoic acid could produce 1 mol of $H^+(aq)$ if it dissociated completely.

Bases and alkalis

Any substance that reacts with an acid and accepts a proton from the acid is classified as a base.

An alkali is any substance which produces hydroxide ions, OH^- (aq), when they are in aqueous solution.

All alkalis are bases but not all bases are alkalis.

Some substance can act as a base but they are not soluble in water and so do not produce hydroxide ions, OH^- (aq) in aqueous solution.

For example copper(II) oxide acts as a base when it reacts with acid. The acid donates protons to the oxide ion to form water.

$$O^{2-} + 2H^+ \rightarrow H_2O$$

In this way the oxide ion, O^{2-}, in the copper(II) oxide can act as a base as it accepts protons. However, copper(II) oxide does not dissolve in water and so does not produce hydroxide ions, $OH^-(aq)$, in aqueous solution. Copper(II) oxide is a base but it is not an alkali.

Sodium hydroxide acts as a base when the hydroxide ion reacts with acid and accepts a proton to produce water:

$$H^+(aq) + OH^-aq) \rightarrow H_2O(l)$$

Sodium hydroxide also dissolves in water and produces aqueous hydroxide ions, $OH^-(aq)$. Sodium hydroxide is also an alkali.

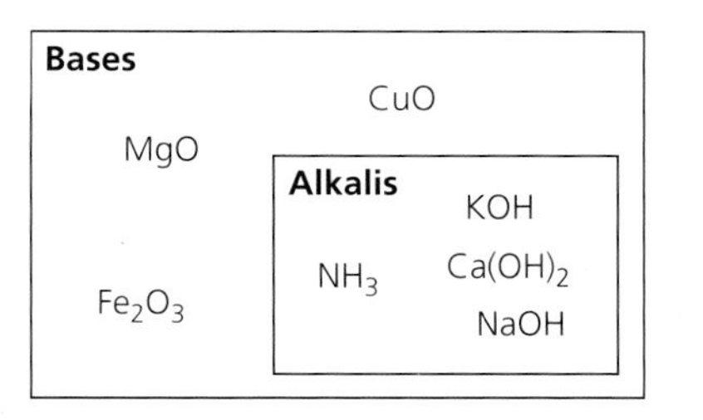

Figure 6.1 This diagram shows the distinction between the terms base and alkali. As all alkalis are bases, we will use the term base for all substances discussed in this unit.

Strong bases

A strong base is completely dissociated in aqueous solution.

The strong bases most commonly used are sodium hydroxide and potassium hydroxide.

Strong bases are usually ionic compounds which dissolve in water and dissociate (or ionise) completely.

Strong bases produce hydroxide ions $OH^-(aq)$ if the base dissolves in water.

A full arrow is used to show that the dissociation is complete

When a base dissolves in water it produces hydroxide ions, $OH^-(aq)$

$$NaOH(s) + (aq) \rightarrow Na^+(aq) + OH^-(aq)$$

Sodium hydroxide

Added to water

Sodium hydroxide forms sodium salts as this is the cation which is present in the solution

Normally the '+(aq)' is left out of the equation.

The above equation can be simplified to:

$$NaOH \rightarrow Na^+ + OH^-$$

All of the NaOH added to water dissociates into ions. This makes sodium hydroxide a strong base as it dissociates completely.

Weak bases

A weak base is slightly dissociated in aqueous solution.

The weak base most commonly used is ammonia.

Weak bases are often covalent substances which dissolve in water and dissociate (or ionise) slightly.

Weak bases produce hydroxide ions $OH^-(aq)$ if the base dissolves in water.

Figure 6.2 Have you ever found hair blocking the plug hole and water pipes? Pouring sodium hydroxide drain cleaner down the drain will help. This strong base will hydrolyse the proteins in hair and help unblock clogged pipes.

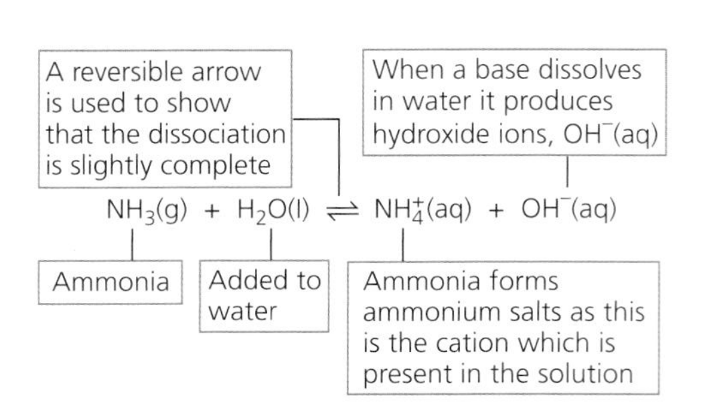

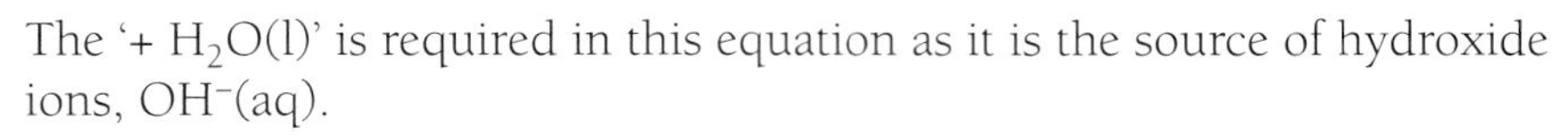

The '+ $H_2O(l)$' is required in this equation as it is the source of hydroxide ions, $OH^-(aq)$.

The state symbols may be left out of the equation to simplify it:

$$NH_3 + H_2O \rightleftharpoons NH_4^+ + OH^-$$

Only some of the NH_3 added to water dissociates into ions. This makes ammonia a weak base as it dissociates (or ionises) slightly.

Equations for acid dissociation

The equations for acid dissociation may be written including water.

The simplified equation for the dissociation of hydrochloric acid is given as:

$$HCl \rightarrow H^+ + Cl^-$$

This can be rewritten including water as:

$$HCl + H_2O \rightarrow H_3O^+ + Cl^-$$

H_3O^+ is the hydronium ion or hydroxonium ion or oxonium ion. It is the ion formed when acids react with water. All three names are used hydronium would be the most common.

This better represents the acid–base reaction as it shows the acid donating a proton to the water and water accepting a proton and acting as a base.

However, for the majority of calculations involving acids, the simplified equation is used.

For ethanoic acid the simplified equation is given as:

$$CH_3COOH \rightleftharpoons CH_3COO^- + H^+$$

and the equation for the reaction with water is:

$$CH_3COOH + H_2O \rightleftharpoons CH_3COO^- + H_3O^+$$

Again the simplified equation is perfectly suitable for calculations involving weak acids like ethanoic acid.

Identifying Brønsted-Lowry acids and bases

The English chemist Thomas Lowry and the Danish chemist Johannes Brønsted both independently proposed the same definitions for acids and bases in 1923. This is usually referred to as the Brønsted-Lowry theory of acids and bases.

It is important to be able to identify Brønsted-Lowry acids and bases in a reaction.

The Brønsted-Lowry acid is the proton donor and the Brønsted-Lowry base the proton acceptor in the reaction.

This best applies to reversible reactions in which one molecule or ion donates a proton and one molecule or ion accepts a proton.

EXAMPLE 1

In the following reaction:

$$NH_3 + H_2O \leftrightharpoons NH_4^+ + OH^-$$

NH_3 accepts a proton to become NH_4^+.

NH_3 acts as a Brønsted-Lowry base.

H_2O donates a proton.

H_2O acts as a Brønsted-Lowry acid.

TIP

Water can act as both a Brønsted-Lowry acid and base. Other species can do this as well such as the hydrogenphosphate ion, HPO_4^{2-}, dihydrogenphosphate ion, $H_2PO_4^-$ and the hydrogencarbonate ion, HCO_3^-.

EXAMPLE 2

In the following reaction:

$$CH_3COOH + H_2O \rightleftharpoons CH_3COO^- + H_3O^+$$

H_2O accepts a proton to become H_3O^+.

H_2O acts as a Brønsted-Lowry base.

CH_3COOH donates a proton.

CH_3COOH acts as a Brønsted-Lowry acid.

pH

pH (always written with a small p and a capital H) is a logarithmic scale which gives a measure of the H^+ concentration, in mol dm^{-3}, in a solution.

The H in pH relates to the hydrogen.

Neutral solutions have a pH value of 7.00 at 25°C.

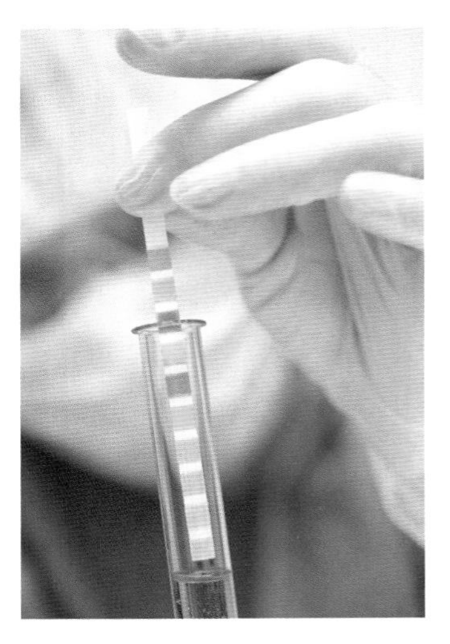

Figure 6.3 Dipsticks as shown in the photo above are used by doctors to measure the pH of urine. Urine pH normally ranges between 2.5 and 8. High acidity can indicate diabetes and high alkalinity can indicate urinary tract infections or kidney stones.

TIP

$\log_{10}$ is the standard log button on your calculator. Shift log will access the antilog function which may appear as 10^x. Try a few calculations to make sure you can convert from $[H^+]$ to pH and from pH to $[H^+]$. A $[H^+]$ of 0.0100 mol dm^{-3} will give a pH of 2.00. A pH of 4.50 will give a $[H^+]$ of 3.16×10^{-5} mol dm^{-3}.

Acidic solutions have a pH of less than 7.00 and alkaline solutions have a pH of greater than 7.00.

An alkali is a soluble base and when bases dissociate in solution they are called alkalis.

Calculating pH from $[H^+]$

The following expression is used to calculate pH from the concentration of hydrogen ions.

$$pH = -\log_{10}[H^+]$$

- In this expression, $[H^+]$ represents the concentration of H^+ ions in solution measured in mol dm^{-3}.
- $\log_{10}$ is often written log so $pH = -\log[H^+]$ is common. The square brackets are essential to indicate the concentration of the hydrogen ions.

To calculate the hydrogen ion concentration from the pH, reverse the calculation:

$$[H^+] = 10^{(-pH)}$$

The H in pH relates to the hydrogen ion concentration but the origin of the p is disputed. The term was first used by Sørenson in the Carlsberg Laboratory in 1909 and was supposed to mean 'power' of hydrogen which would relate to the logarithmic scale. However it has also been called 'potential of hydrogen'. For our purposes, pH means $-\log_{10}[H^+]$.

pH of strong acids

The diagram below shows the links between concentration of the acid, [acid], concentration of hydrogen ions, $[H^+]$, and pH for a strong acid.

Monoprotic acids like hydrochloric acid and nitric acid have a proticity of 1, whereas diprotic acid like sulfuric acid have a proticity of 2.

[Acid] → (× proticity of acid) → $[H^+]$ → ($pH = -\log_{10}[H^+]$) → pH

pH → ($[H^+] = 10^{(-pH)}$) → $[H^+]$ → (÷ proticity of acid) → [Acid]

EXAMPLE 3

Calculate the pH of 0.0500 mol dm^{-3} hydrochloric acid.

Answer

For strong monobasic acids, the proticity is 1 so

$[H^+] = 1 \times [acid]$

$HCl \rightarrow H^+ + Cl^-$

If $[HCl] = 0.0500$ mol dm^{-3}, then $[H^+] = 0.0500$ mol dm^{-3}

$pH = -\log_{10}[H^+] = -\log(0.05)$
$= 1.30$ (to 2 decimal places)

EXAMPLE 4

Calculate the pH of 1.00 mol dm^{-3} sulfuric acid.

Answer

For strong diprotic acids, the proticity is 2 so $[H^+] = 2 \times [acid]$

$H_2SO_4 \rightarrow 2H^+ + SO_4^{2-}$

If $[H_2SO_4] = 1.00$ mol dm^{-3} then $[H^+] = 2.00$ mol dm^{-3}

$pH = -\log[H^+] = -\log(2.00)$
$= -0.30$ (to 2 decimal places)

TIP

pH values are normally given between 0 and 14, but values below 0 are possible for very high $[H^+]$. Values above 14 are also possible for very high hydroxide ion concentrations $[OH^-]$. pH values are almost always quoted to 2 decimal places.

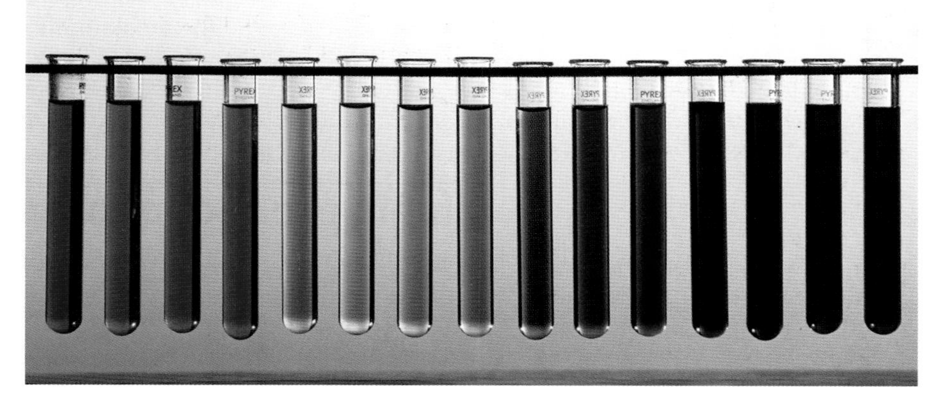

Figure 6.4 When universal indicator is added to a solution, it changes colour depending on the solution's pH. The tubes contain solutions of pH 0 to 14 from left to right.

Determining concentration of an acid from its pH

If you are given the pH of a strong acid you can calculate the concentration of the hydrogen ions and so the concentration of the acid.

EXAMPLE 5

Determine the concentration, in mol dm^{-3}, of nitric acid which has a pH of 0.71. Give your answer to 3 significant figures.

Answer

$[H^+] = 10^{(-pH)} = 10^{(-0.71)} = 0.1950$ mol dm^{-3} (to 4 significant figures)

Nitric acid (HNO_3) is a strong monobasic acid so the $[acid] = [H^+]$

The concentration of nitric acid = 0.195 mol dm^{-3} (to 3 significant figures).

EXAMPLE 6

Determine the concentration of sulfuric acid which has a pH of 1.00.

Answer

$[H^+] = 10^{(-pH)} = 10^{(-1.00)} = 0.100\ mol\ dm^{-3}$

Sulfuric acid (H_2SO_4) is a strong diprotic acid so the [acid]

$= \frac{[H^+]}{2}$

$= \frac{0.100}{2} = 0.0500\ mol\ dm^{-3}$

The concentration of this sulfuric acid = $0.0500\ mol\ dm^{-3}$.

TEST YOURSELF 2

1 Write an expression for pH.

2 Calculate the pH of the following acidic solutions. Give your answer to all questions to 2 decimal places.

a) $0.0240\ mol\ dm^{-3}$ hydrochloric acid

b) $0.0170\ mol\ dm^{-3}$ nitric acid

c) $0.0140\ mol\ dm^{-3}$ sulfuric acid

3 Calculate the concentration of the acid solution which gives a pH of the following values. Give your answers to 3 significant figures.

a) hydrochloric acid of pH 1.80

b) nitric acid of pH 0.50

c) sulfuric acid of pH −0.10

4 Calculate the concentration, in $mol\ dm^{-3}$, of a solution of hydrochloric acid which has the same pH as a solution of $0.185\ mol\ dm^{-3}$ sulfuric acid.

Ionic product of pure water

Water slightly dissociates into hydrogen ions and hydroxide ions according to the equilibrium.

$$H_2O \rightleftharpoons H^+ + OH^-$$

K_w is the ionic product of water and $K_w = [H^+][OH^-]$.

The units of K_w are always $mol^2\ dm^{-6}$.

At 25 °C, $K_w = 1.00 \times 10^{-14}\ mol^2\ dm^{-6}$.

Calculating the pH of pure water

In pure water $[H^+] = [OH^-]$ so $K_w = [H^+]^2$.

$$[H^+] = \sqrt{K_w} \text{ and } pH = -\log_{10}[H^+]$$

The pH of pure water can be calculated from the value of K_w.

EXAMPLE 7

Calculate the pH of water at 25 °C when $K_w = 1.00 \times 10^{-14}\ mol^2\ dm^{-6}$.

Answer

At 25 °C, $K_w = 1.00 \times 10^{-14}\ mol^2\ dm^{-6}$.

$K_w = [H^+][OH^-]$

In water $[H^+] = [OH^-]$ so $K_w = [H^+]^2$

$[H^+]^2 = 1.00 \times 10^{-14}\ mol^2\ dm^{-6}$

$[H^+] = \sqrt{100 \times 10^{-14}} = 1.00 \times 10^{-7}\ mol\ dm^{-3}$.

$pH = -\log_{10}[H^+] = -\log_{10}(1.00 \times 10^{-7}) = 7.00$

The pH of pure water at 25 °C is 7.00.

EXAMPLE 8

At 40 °C, $K_w = 2.92 \times 10^{-14}\ \text{mol}^2\ \text{dm}^{-6}$.

Calculate the pH of water at 40 °C. Give your answer to 2 decimal places.

Answer

At 40 °C $K_w = 2.92 \times 10^{-14} = [H^+]^2$

$[H^+] = \sqrt{2.92 \times 10^{-14}} = 1.709 \times 10^{-7}\ \text{mol dm}^{-3}$ (to 4 significant figures)

$\text{pH} = -\log_{10}[H^+] = -\log_{10}(1.709 \times 10^{-7}) = 6.7673$ (to 4 decimal places)

pH = 6.77 to 2 decimal places

We have become accustomed to think that the pH of pure water is 7, but this is only true at 25 °C. As temperature increases above 25 °C, the pH of water drops below 7. The increase in K_w as temperature increases also indicates that the dissociation of water into H^+ and OH^- ions is endothermic as the equilibrium is moving from left to right (more H^+) as temperature increases.

Mineral water gets its name from the dissolved mineral rocks that it contains. Mineral water from limestone rocks contains calcium hydrogen carbonate and as a result the pH is around 8.00.

TIP

Water at 40 °C is not acidic (even though it has a pH of 6.77) as the concentration of $[H^+] = [OH^-]$.

Calculating K_w from the pH of pure water

The pH of pure water at a certain temperature can be used to calculate $[H^+]$. In pure water $[H^+] = [OH^-]$ and so $K_w = [H^+]^2$.

EXAMPLE 9

The pH of pure water at 60 °C is 6.52. Calculate the value of K_w at 60 °C. Give your answer to 3 significant figures.

Answer

$[H^+] = 10^{(-\text{pH})} = 10^{(-6.52)} = 3.020 \times 10^{-7}\ \text{mol dm}^{-3}$ (to 4 significant figures)

$K_w = [H^+]^2 = (3.020 \times 10^{-7})^2 = 9.12 \times 10^{-14}\ \text{mol}^2\ \text{dm}^{-6}$

pH of strong bases

When a soluble base dissolves in water, hydroxide ions are present in the solution.

The concentration of the hydroxide ions, $[OH^-]$ can be converted to the $[H^+]$ using K_w as at 25 °C $K_w = 1.00 \times 10^{-14}\ \text{mol}^2\ \text{dm}^{-6}$.

In pure water $[H^+] = [OH^-]$ but this is not true in acid or alkaline solutions.

K_w and $[OH^-]$ can be used to calculate $[H^+]$, which is then used to determine pH.

The diagram below shows the links between concentration of the strong bases, [base], concentration of hydroxide ions, $[OH^-]$, concentration of hydrogen ions, $[H^+]$, and pH for a strong base.

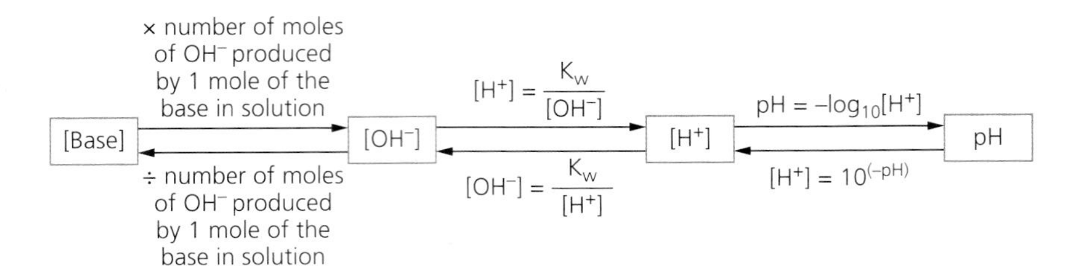

Calculating pH of a strong bases

For strong bases, the concentration of OH^- is equal to the concentration of the base × the number of hydroxide ions produced in solution.

EXAMPLE 10

Calculate the pH of a 0.250 mol dm^{-3} solution of sodium hydroxide ($K_w = 1.00 \times 10^{-14}$ mol^2 dm^{-6}). Give your answer to 2 decimal places.

Answer

Sodium hydroxide is a strong base and so dissociates fully in solution.

$NaOH \rightarrow Na^+ + OH^-$

Also 1 mol of sodium hydroxide contains 1 mol of OH^- ions.

[NaOH] = 0.250 mol dm^{-3} so [OH^-] = 0.250 mol dm^{-3}

$[H^+] = \frac{K_w}{[OH^-]} = \frac{1.00 \times 10^{-14}}{0.250}$ so $[H^+] = 4.00 \times 10^{-14}$ mol dm^{-3}

$pH = -\log_{10}(4.00 \times 10^{-14}) = 13.3979$ (to 4 decimal places)

pH = 13.40 (to 2 decimal places)

The majority of strong bases (alkalis) you will encounter will have 1 mol of OH^- ions per mole of the base. However, sometimes a question is set on a Group 2 hydroxide where you are asked to assume that the base is strong and so the number of moles of OH^- per mole of base for $Ca(OH)_2$ is 2.

Sometimes the pH calculation may be at temperature other than 25 °C and the K_w value at this temperature is given. The method is exactly the same using the K_w value.

EXAMPLE 11

At 50 °C, $K_w = 5.48 \times 10^{-14}$ mol^2 dm^{-6}. Calculate the pH of 0.175 mol dm^{-3} potassium hydroxide solution at 50 °C.

Answer

Potassium hydroxide is a strong base and so dissociates fully in solution.

$KOH \rightarrow K^+ + OH^-$

Also 1 mol of potassium hydroxide contains 1 mol of OH^- ions.

[KOH] = 0.175 mol dm^{-3} so [OH^-] = 0.175 mol dm^{-3}

$[H^+] = \frac{K_w}{[OH^-]} = \frac{5.48 \times 10^{-14}}{0.175}$ = so $[H^+] = 3.131 \times 10^{-13}$ mol dm^{-3}

$pH = -\log_{10}(3.131 \times 10^{-13}) = 12.5043$ (to 4 decimal places)

pH = 12.50 (to 2 decimal places)

Calculating the concentration of a base from its pH

The concentration of hydrogen ions can be calculated from pH.

The $[OH^-]$ can be calculated from $[H^+]$ using K_w.

The concentration of the base can be calculated from the concentration of hydroxide ions.

TIP

Concentration may be in units of $g\,dm^{-3}$ or $mg\,dm^{-3}$ rather than $mol\,dm^{-3}$. To convert between $mol\,dm^{-3}$ and $g\,dm^{-3}$, simply multiply by the M_r. To convert between $mol\,dm^{-3}$ and $mg\,dm^{-3}$, multiply by the M_r of the base and multiply by 1000 to convert mass in g to mass in mg.

EXAMPLE 12

Calculate the concentration of potassium hydroxide solution in $mol\,dm^{-3}$ with a pH of 13.70 at 25 °C. $K_w = 1.00 \times 10^{-14}\ mol^2\,dm^{-6}$ at 25 °C. Give your answer to 2 decimal places.

Answer

$[H^+] = 10^{(-pH)} = 10^{(-13.70)} = 1.995 \times 10^{-14}\ mol\,dm^{-3}$ (to 4 significant figures)

$$[OH^-] = \frac{K_w}{[H^+]} = \frac{1.00 \times 10^{-14}}{1.995 \times 10^{-14}}$$

$= 0.5013\ mol\,dm^{-3}$ (to 4 significant figures)

Potassium hydroxide (KOH) has 1 mol of OH^- ions per mole of base so the concentration of KOH, $[KOH] = 0.5013\ mol\,dm^{-3}$.

Concentration of KOH solution = 0501 $mol\,dm^{-3}$ (to 3 significant figures)

Calculating a value for K_w

In some calculations the pH of a solution of a base is given with its concentration at a particular temperature.

$[H^+]$ may be calculated from the pH, $[OH^-]$ may be calculated from the concentration.

EXAMPLE 13

At 40 °C, a 0.270 $mol\,dm^{-3}$ solution of sodium hydroxide has a pH of 12.98. Calculate the value for K_w at 40 °C. Give your answer to 3 significant figures.

Answer

Using pH to calculate $[H^+]$

$[H^+] = 10^{(-12.98)} = 1.047 \times 10^{-13}\ mol\,dm^{-3}$ (to 4 significant figures)

As NaOH contains 1 mol of OH^- per mole of NaOH

$[OH^-] = [NaOH] = 0.270\ mol\,dm^{-3}$

$K_w = [H^+][OH^-] = (1.047 \times 10^{-13}) \times 0.270 = 2.83 \times 10^{-14}\ mol^2\,dm^{-6}$

pK_w, pH and pOH

pK_w may be used in place of K_w to carry out calculations involving strong bases.

$$pK_w = -\log_{10}K_w$$

$$pK_w = 14.00 \text{ at } 25\,°C.$$

TIP

Remember $pK_w = -\log_{10}K_w$ is used, but $-\log K_w$ is equally acceptable without the log base number.

$$pOH = -\log_{10}[OH^-]$$

$$pH + pOH = pK_w$$

This is an alternative way to use K_w values. The pOH is calculated from the hydroxide ions concentration. The total of pH and pOH is equal to pK_w.

EXAMPLE 14

Calculate the pH of a 2.00 g dm^{-3} solution of sodium hydroxide, NaOH, given that $pK_w = 14.00$ at 25 °C. Give your answer to 2 decimal places.

Answer

M_r of NaOH = 40.0

Concentration of NaOH solution $= \frac{2.00}{40.0} = 0.0500\ \text{mol dm}^{-3}$

$[OH^-] = 0.050\ \text{mol dm}^{-3}$ as NaOH contains 1 mol of OH^- per mole of NaOH

$pOH = -\log[OH^-] = -\log(0.0500) = 1.3010$ (to 4 decimal places)

$pK_w = 14.00 = pOH + pH$

$pH = 14.00 - 1.3010 = 12.699$

The pH of a 2.00 g dm^{-3} solution of sodium hydroxide is 12.70 (to 2 decimal places).

TEST YOURSELF 3

1 Write an expression for the ionic product of water, K_w.

2 At 15 °C, $K_w = 4.65 \times 10^{-13}\ \text{mol}^2\ \text{dm}^{-6}$.

a) Calculate the pH of pure water at 15 °C. Give your answer to 2 decimal places.

b) Calculate the pH of a solution of 0.170 mol dm^{-3} potassium hydroxide at 15 °C. Give your answers to 2 decimal places.

3 The pH of water at 45 °C is 6.70. Calculate the value of K_w at 45 °C. Give your answer to 3 significant figures.

4 The pH of a solution of sodium hydroxide is 13.68 at 25 °C. K_w at 25 °C is $1.00 \times 10^{-14}\ \text{mol}^2\ \text{dm}^{-6}$. Calculate the concentration of the solution of sodium hydroxide at 25 °C. Give your answer to 3 significant figures.

pH of weak acids

Weak acids are slightly dissociated in solution. This is represented using a reversible arrow ($\rightleftharpoons$).

K_a

The equilibrium constant for the acid dissociation is represented by K_a.

For a general acid dissociation

$$HA \rightleftharpoons H^+ + A^-$$

$$K_a = \frac{[A^-][H^+]}{[HA]}$$

Figure 6.5 The oceans have absorbed about half of the carbon dioxide produced by burning fossil fuels since 1800. As carbon dioxide in the ocean increases, ocean pH decreases. This is called ocean acidification. With ocean acidification, corals cannot absorb the calcium carbonate they need to maintain their skeletons and the stony skeletons that support corals and reefs will dissolve.

TIP

It is important to be able to write K_a expressions for weak acids. It is always the [anion] and [H⁺] on the top and [undissociated acid] on the bottom.

For example:

$$CH_3COOH \rightleftharpoons H^+ + CH_3COO^-$$

$$K_a = \frac{[CH_3COO^-]\,[H^+]}{[CH_3COOH]}$$

K_a always has units of mol dm^{-3}.

HA is the undissociated acid. Remember that the undissociated acid does not cause it to be acidic. It is the concentration of H^+ that causes acidity when the acid dissociates. HA is not acidic until it dissociates.

The value of K_a gives a measure of the strength of the acid. A higher K_a value indicates a 'stronger' weak acid. For example ethanoic acid has a K_a value of 1.74×10^{-5} mol dm^{-3} whereas hydrocyanic acid has a K_a of 4.90×10^{-10} mol dm^{-3}.

For a weak acid, $[H^+]$ is calculated from the concentration of the acid and the K_a value.

For a weak acid of concentration 1.00 mol dm^{-3}, the following equilibrium applies:

	HA	⇌	H^+	+	A^-
Initial Concentration/mol dm^{-3}:	1.00		0		0
Equilibrium Concentration/mol dm^{-3}:	$1.00 - x$		x		x

$$K_a = \frac{[A^-][H^+]}{[HA]} = \frac{x^2}{100 - x}$$

But as this is a weak acid, the concentration of the acid at equilibrium $(1 - x)$ will be approximately equal to the initial concentration of the weak acid (1 mol dm^{-3}) as x is very small compared to the initial concentration.

$$K_a = \frac{x^2}{1.00} \quad \text{so in this example } K_a = [H^+]^2$$

In general

$$K_a = \frac{[H^+]^2}{[\text{weak acid}]} \qquad \text{so } [H^+]^2 = K_a \times [\text{weak acid}]$$

and

$$[H^+] = \sqrt{K_a \times [\text{weak acid}]}$$

pH is then calculated using $pH = -\log_{10}[H^+]$.

To determine the concentration of a weak acid from its pH and K_a value

$[H^+]$ is determined from pH in the usual way and then the concentration of the weak acid is calculated using $\frac{[H^+]^2}{K_a}$.

The overall process is described in the diagram below. [weak acid] = the concentration of the weak acid.

[Weak acid] → $[H^+] = \sqrt{K_a \times [\text{weak acid}]}$ → $[H^+]$ → $pH = -\log_{10}[H^+]$ → pH

pH → $[H^+] = 10^{(-pH)}$ → $[H^+]$ → $[\text{Weak acid}] = \frac{[H^+]^2}{K_a}$ → [Weak acid]

pK_a

The pK_a of a weak acid may be given in place of its K_a. pK_a = $-\log_{10}K_a$.

To convert between a pK_a value and K_a, use the following: $K_a = 10^{-pK_a}$.

The higher the pK_a value, the weaker the acid.

A lower pK_a indicates a less weak acid.

Calculating the pH of a weak acid

TIP

The use of unusual units in these types of calculations is common. Remember that there are 1000 (10^3) mg in 1 g and 1 000 000 (10^6) μg in 1 g. The same applies to volume units: 1000 (10^3) ml in 1 litre and 1 000 000 (10^6) μl in 1 litre. 1 cm^3 = 1 ml and 1 dm^3 = 1 litre. These questions can be awkward as mistakes can be made when converting between units but as long as you know the units, it makes it easier.

EXAMPLE 15

The K_a for ethanoic acid (CH_3COOH) is 1.74×10^{-5} mol dm^{-3}. Calculate the pH of a 0.105 mol dm^{-3} solution of ethanoic acid. Give the answer to 2 decimal places.

Answer

$[H^+] = \sqrt{K_a \times [\text{weak acid}]} = \sqrt{1.74 \times 10^{-5} \times 0.105} = \sqrt{1.872 \times 10^{-6}}$

$= 1.352 \times 10^{-3}$ mol dm^{-3} (to 4 significant figures)

$pH = -\log_{10}[H^+] = -\log_{10}(1.352 \times 10^{-3}) = 2.8691$ (to 4 decimal places)

pH = 2.87 (to 2 decimal places)

Calculating the concentration of a weak acid from its pH

EXAMPLE 16

Determine the concentration, in mg dm^{-3}, of a solution of propanoic acid (CH_3CH_2COOH) with a pK_a value of 2.89 with a pH of 2.50.

Answer

$K_a = 10^{(-pKa)} = 10^{(-2.89)} = 1.288 \times 10^{-3}$ mol dm^{-3} (to 4 significant figures)

$[H^+] = 10^{(-pH)} = 10^{(-2.50)} = 3.162 \times 10^{-3}$ mol dm^{-3}

$[\text{weak acid}] = \frac{[H^+]^2}{K_a} = \frac{(3.162 \times 10^{-3})^2}{1.288 \times 10^{-3}}$

[weak acid] = 7.76×10^{-3} mol dm^{-3} (to 3 significant figures)

M_r of propanoic acid (CH_3CH_2COOH) = 74.0

[weak acid] = $7.76 \times 10^{-3} \times 74.0 = 0.574$ g dm^{-3}

[weak acid] = $0.574 \times 1000 = 574$ mg dm^{-3}

Calculating the K_a from the pH of a weak acid and its concentration

For a weak acid, HA, $K_a = \frac{[H^+]^2}{[HA]}$

If the pH of a weak acid is known, $[H^+]$ can be calculated using $10^{(-pH)}$.

A value for K_a can be calculated from the concentration of the weak acid, [HA] and the concentration of hydrogen ions, $[H^+]$.

EXAMPLE 17

A weak acid, HA, of concentration 0.0120 mol dm^{-3} has a pH of 4.10. Calculate a value for the acid dissociation constant, K_a, for the weak acid.

Answer

$[H^+] = 10^{(-pH)} = 10^{(-4.10)} = 7.943 \times 10^{-5}$ mol dm^{-3}

[HA] = 0.0120 mol dm^{-3}

$$K_a = \frac{[H^+]^2}{[HA]}$$

$$K_a = \frac{(7.943 \times 10^{-5})^2}{0.0120} = \frac{6.309 \times 10^{-9}}{0.0120} = 5.26 \times 10^{-7} \text{ mol dm}^{-3}$$

TEST YOURSELF 4

1 Write an equation for the acid dissociation of the weak acid ethanoic acid, CH_3COOH.
2 Write an expression for the acid dissociation constant, K_a, for ethanoic acid (CH_3COOH).
3 Calculate the pH of a 0.270 mol dm^{-3} solution of ethanoic acid. K_a for ethanoic acid at 25 °C is 1.74×10^{-5} mol dm^{-3}. Give your answer to 2 decimal places.
4 Calculate the concentration of the solution of a monoprotic weak acid when it has a pH of 2.80 and a pK_a of 3.10. Give your answer to 3 significant figures.
5 Calculate the pH of a 0.0540 mol dm^{-3} solution of a weak acid, HX, which has an acid dissociation constant, K_a, = 1.54×10^{-4} mol dm^{-3}. Give your answer to 2 decimal places.

Dilutions and neutralisations

When an acid or base is neutralised or diluted, the concentrations of the ions in solution change. It is important to be able to calculate the new concentration of H^+ or OH^- and then determine pH of the new solution.

The total volume must be taken into account to determine the new concentration in mol dm^{-3}.

Dilutions

When a solution of an acid or a base is diluted, the concentration of the ions in the solution changes. As the concentration of the ions in solution changes, this changes the pH.

The new concentration of a solution may be calculated by dividing the amount of solute, in moles, in the solution by the new total volume and multiplying by 1000.

$$\text{New concentration of solution} = \frac{\text{amount in moles of solute}}{\text{new total volume of solution}} \times 1000$$

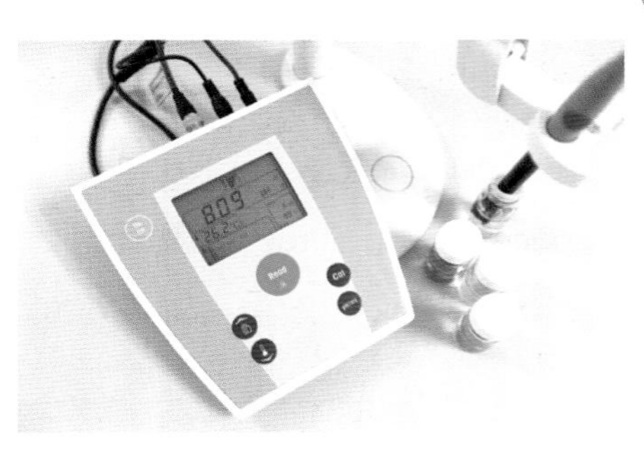

Figure 6.6 The pH of a solution can be accurately measured using a digital pH meter. The pH of this solution is 8.09.

EXAMPLE 18

$20.0\,cm^3$ of a $0.152\,mol\,dm^{-3}$ solution of hydrochloric acid is placed in a $250\,cm^3$ volumetric flask and the volume made up to $250\,cm^3$ using deionised water. Determine the pH of the resulting solution.

Answer

Moles of HCl added $= \frac{20.0 \times 0.152}{1000} = 3.04 \times 10^{-3}\,mol$

New volume $= 250\,cm^3$

New concentration of solution $= \frac{3.04 \times 10^{-3}}{250} \times 1000 = 0.01216\,mol\,dm^{-3}$

As HCl is a strong monoprotic acid, $[H^+] = 0.01216\,mol\,dm^{-3}$

$pH = -\log(0.01216) = 1.92$ (to 2 decimal places)

EXAMPLE 19

$10.0\,cm^3$ of a $1.54\,mol\,dm^{-3}$ solution of potassium hydroxide is diluted to $2.00\,dm^3$ in a volumetric flask. Determine the pH of the resulting solution at 25 °C if $pK_w = 14.0$ at 25 °C. Give your answer to 2 decimal places.

Answer

Moles of KOH added $= \frac{10.0 \times 1.54}{1000} = 0.0154\,mol$

New volume $= 2.00\,dm^3$ $(2000\,cm^3)$

Concentration of new solution $= \frac{0.0154}{2000} \times 1000 = 0.00770\,mol\,dm^{-3}$

As 1 mol of KOH contains 1 mol of OH^-, $[OH^-] = 0.00770\,mol\,dm^{-3}$

$pOH = -\log[OH^-]$ $\qquad pOH = -\log(0.00770) = 2.114$ (to 4 decimal places)

$pK_w = 14 = pH + pOH$

$pH = 14.0 - 2.114 = 11.886$

$pH = 11.89$ (to 2 decimal places)

EXAMPLE 20

Calculate the pH of the solution formed when $15.0\,cm^3$ of $0.114\,mol\,dm^{-3}$ sulfuric acid are added to $485\,cm^3$ of deionised water. Give your answer to 2 decimal places.

Answer

Moles of $H_2SO_4 = \frac{15.0 \times 0.114}{1000}$

$= 1.71 \times 10^{-3}\,mol$

New volume $= 15 + 485 = 500\,cm^3$

Concentration of new solution $= \frac{1.71 \times 10^{-3}}{500} = \times 1000 = 3.42 \times 10^{-3}\,mol\,dm^{-3}$

H_2SO_4 is a diprotic acid so $[H^+] = 3.42 \times 10^{-3} \times 2 = 6.48 \times 10^{-3}\,mol\,dm^{-3}$

$pH = -\log_{10}[H^+] = -\log_{10}(6.48 \times 10^{-3}) = 2.1649$ (to 4 decimal places)

$pH = 2.16$

TIP

In this example the total volume of the solution is not given but it can be calculated from the two volumes $(15.0 + 485 = 500\,cm^3)$.

Calculating the new concentration using dilution factors

You may also determine the dilution factor to calculate the new concentration of acid or base. In example 18 the dilution factor is 12.5. The volume of solution initially was 20.0 cm^3 and the final volume was 250 cm^3.

Dilution factor = $\frac{\text{total volume after dilution}}{\text{initial volume added}}$ as long as the units of volume are the same.

The diluted solution is 12.5 times less concentrated than the original solution

Concentration of the diluted solution = $\frac{0.152}{12.5}$ = 0.01216 mol dm^{-3}.

The dilution factor in Example 19 is 200 $\left(\frac{2000}{10}\right)$, so the diluted solution of KOH is 200 times more dilute than the original solution.

Concentration of the diluted solution = $\frac{1.54}{200}$ = 0.00770 mol dm^{-3}

The remainder of the calculation is carried out as per the examples.

Calculating the volume in a dilution

Some calculations on dilutions may focus on determining the new total volume of the solution or the volume of water added when carrying out the dilution.

The pH of the solution would be given and $[H^+]$ can be calculated from this value.

The concentration of the new solution can be determined from $[H^+]$ and the dilution factor or new total volume of the solution calculated as shown.

TIP

You can reverse the calculation to check that you get 2.10 for the pH of the new solution formed. As the volume was given to 3 significant figures the pH may not be exactly 2.10 but it should be very close. In this case using 169 cm^3 gives a pH of 2.099 or 2.10 to 2 decimal places.

TIP

The volume of water added may be calculated by subtracting the initial volume of solution used from the final volume calculated, in this example 169.21 – 32.0 = 137.21 cm^3.

EXAMPLE 21

32.0 cm^3 of 0.0420 mol dm^{-3} hydrochloric acid were diluted. The pH of the resulting solution was 2.10. Calculate the total volume of the solution formed. Give your answer to 3 significant figures and state the units.

Answer

In this example the pH is first used to calculate $[H^+]$.

As hydrochloric acid is a monoprotic acid, $[HCl] = [H^+]$

$[H^+] = 10^{(-pH)} = 10^{(-2.10)} = 7.943 \times 10^{-3}$ mol dm^{-3} (to 4 significant figures)

The new concentration of HCl after dilution = 7.943×10^{-3} mol dm^{-3}

Initial moles of HCl added = $\frac{32.0 \times 0.0420}{1000} = 1.344 \times 10^{-3}$ mol

New concentration of HCl after dilution = $7.943 \times 10^{-3} = \frac{1.344 \times 10^{-3}}{V} \times 1000$

where V represents the new total volume of solution in cm^3.

Rearranging the expression to calculate V,

$V = \frac{1.344 \times 10^{-3}}{7.943 \times 10^{-3}} \times 1000 = 169$ cm^3 (to 3 significant figures)

Neutralisations

In a neutralisation question, a volume of a certain concentration of a strong acid is added to a volume of a certain concentration of a strong base.

Either the base or the acid may be in excess.

It is the amount, in moles of the acid or base in excess together with the new total volume of the solution which is used to calculate the new concentration of either the acid or the base.

MATHS

Rearranging an expression to change the subject is an important skill. If you are unsure that you have done it correctly, substitute in simple values such as different whole numbers to see what the subject equals before rearrangement. Then use this with other values to see if you obtain the value for the new subject when you use your rearranged expression.

MATHS

The units of volume are important as there are several that could be used. It is a good idea to always put appropriate units after a number, even mol after an amount, in moles.

EXAMPLE 22

25.0 cm^3 of 0.214 mol dm^{-3} sodium hydroxide solution is added to 25.0 cm^3 of 0.258 mol dm^{-3} hydrochloric acid. Calculate the pH of the resulting solution. Give your answer to 2 decimal places.

Answer

This style of question is all about the number of moles of a reactant which are left over.

$NaOH + HCl \rightarrow NaCl + H_2O$

NaOH and HCl react in a 1:1 ratio.

Initial moles of NaOH $= \frac{25.0 \times 0.214}{1000} = 0.00535$ mol

Initial moles of HCl $= \frac{25.0 \times 0.258}{1000} = 0.00645$ mol

The equation for the neutralisation is:

$NaOH + HCl \rightarrow NaCl + H_2O$

0.00535 mol of NaOH reacts with 0.00535 mol of HCl so (0.00645 − 0.00535) = 0.00110 mol of HCl remaining.

New total volume of solution
= 25.0 + 25.0 = 50 cm^3

New concentration of reactant in excess

$$= \frac{\text{moles of reactant in excess}}{\text{new total volume of solution}} \times 1000$$

New concentration of reactant (HCl) in excess

$$= \frac{0.00110}{50.0} \times 1000 = 0.0220 \text{ mol dm}^{-3}$$

As HCl is strong monobasic acid, $[H^+] = 0.0220$ mol dm^{-3}

$pH = -\log_{10}[H^+] = -\log_{10}(0.0220) = 1.66$ (to 2 decimal places)

TIP

In the reaction the ratio of the acid to base is 1 : 1, but if sulfuric acid is used (it is assumed to be completely dissociated) then the ratio of a base to sulfuric acid is often 2 : 1.

EXAMPLE 23

Close to the end point of a titration 24.9 cm^3 of 1.00 mol dm^{-3} potassium hydroxide solution have been added to 25.0 cm^3 of 0.500 mol dm^{-3} sulfuric acid. Calculate the pH of the solution formed. Assume that the sulfuric acid is fully dissociated. Give your answer to 2 decimal places.

Answer

Initial moles of $H_2SO_4 = \frac{25.0 \times 0.500}{1000} = 0.0125$ mol

Initial moles of KOH $= \frac{24.9 \times 1.00}{1000} = 0.0249$ mol

The equation for this neutralisation is:

$$2KOH + H_2SO_4 \rightarrow K_2SO_4 + 2H_2O$$

From the equation, 2 mol of KOH reacts with 1 mol of H_2SO_4.

0.0249 mol of KOH reacts with 0.01245 mol of H_2SO_4. This leaves $(0.0125 - 0.01245) = 5 \times 10^{-5}$ mol of H_2SO_4.

If you are unsure above which reactant is used up (limiting reactant) and which is left over (reactant in excess), follow the method below:

Using the equation for the reaction, write the amount, in moles, you have been told below the equation in a line called 'moles you have'.

	$2KOH$	$+ H_2SO_4$	$\rightarrow$	$K_2SO_4 + 2H_2O$
moles you have:	*0.0249*	*0.0125*		

Using either one of the moles below the reactants, calculate the moles of the other reactant which is needed to react with that number of moles of the first reactant. The ratio in the equation is used here. In this example the moles of KOH is divided by 2 as 2KOH reacts with H_2SO_4.

	$2KOH$	$+ H_2SO_4$	$\rightarrow$	$K_2SO_4 + 2H_2O$
moles you have:	*0.0249*	*0.0125*		
moles you need:	$\div 2$ →	*0.01245*		

For H_2SO_4, you have more moles than needed so the H_2SO_4 is in excess and the KOH is the limiting reactant (it is all used up).

You can now calculate the amount, in moles, of H_2SO_4 left over $(0.0125 - 0.01245) = 5 \times 10^{-5}$ mol.

H_2SO_4 is the reactant in excess.

Moles of H_2SO_4 in excess $= 5 \times 10^{-5}$ mol

New total volume of solution $= 25.0 + 24.9 = 49.9$ cm^3

New concentration of reactant in excess $= \frac{\text{moles of reactant in excess}}{\text{new total volume of solution}} \times 1000$

New concentration of reactant (H_2SO_4) in excess $= \frac{5 \times 10^{-5}}{49.9} \times 1000$

$= 1.002 \times 10^{-3}$ mol dm^{-3} (to 4 significant figures)

$[H^+] = 2 \times [H_2SO_4]$ so $[H^+] = 2 \times 1.002 \times 10^{-3} = 2.004 \times 10^{-3}$ mol dm^{-3}

$pH = -\log_{10}[H^+] = -\log_{10}(2.004 \times 10^{-3}) = 2.6981$ (to 4 decimal places)

pH = 2.70 (to 2 decimal places)

TIP

A weak acid and a strong base can be used in a neutralisation question as long as the strong base is in excess. If the weak acid is in excess, the solution formed is a buffer. Buffers will be examined later in this topic.

EXAMPLE 24

Calculate the pH of the solution formed when 20.0 cm^3 of 0.150 mol dm^{-3} of aqueous sodium hydroxide are added to 10.0 cm^3 of 0.180 mol dm^{-3} ethanoic acid at 25 °C. At 25 °C, K_w has a value of 1.00×10^{-14} $mol^2 dm^{-6}$. Give your answer to 2 decimal places.

TIP

As K_w is given in this neutralisation question, it is likely that the base (sodium hydroxide) is in excess and so the pH of the solution will be greater than 7.

Answer

Moles of NaOH = $\frac{20.0 \times 0.150}{1000}$ = 0.00300 mol

Moles of CH_3COOH = $\frac{10.0 \times 0.180}{1000}$ = 0.00180 mol

The equation for the neutralisation is:

$$NaOH + CH_3COOH \rightarrow CH_3COONa + H_2O$$

1 mol of NaOH reacts with 1 mol of CH_3COOH

0.00180 mol of CH_3COOH reacts with 0.00180 mol of NaOH so (0.003 – 0.0018) = 0.00120 mol of NaOH remaining.

New total volume of solution = 20.0 + 10.0 = 30.0 cm^3

New concentration of reactant in excess = $\frac{\text{moles of reactant in excess}}{\text{new total volume of solution}} \times 1000$

New concentration of reactant (NaOH) in excess = $\frac{0.00120}{30.0} \times 1000$ = 0.0400 mol dm^{-3}

As 1 mol of NaOH contains 1 mol of OH^-, $[OH^-]$ = 0.0400 mol dm^{-3}

$K_w = [H^+][OH^-]$

$[H^+] = \frac{K_w}{[OH^-]} = \frac{1.00 \times 10^{-14}}{0.0400} = 2.50 \times 10^{-13}$ mol dm^{-3}

pH = $-\log_{10}[H^+] = -\log_{10}(2.50 \times 10^{-13})$ = 12.6021 (to 4 decimal places)

pH = 12.60 (to 2 decimal places)

TEST YOURSELF 5

1 Calculate the pH of the solution formed when 25.0 cm^3 of 0.350 mol dm^{-3} hydrochloric acid is added to 175.0 cm^3 of water. Give your answer to 2 decimal places.

2 Calculate the pH of the solution formed when 20.0 cm^3 of 0.120 mol dm^{-3} aqueous sodium hydroxide solution is added to 150 cm^3 of water at 25 °C. $K_w = 1.00 \times 10^{-14}$ $mol^2 dm^{-6}$ at 25 °C. Give your answer to 2 decimal places.

3 Calculate the pH of the solution formed when 20.0 cm^3 of 0.250 mol dm^{-3} aqueous hydrochloric acid is added to 15.0 cm^3 of 0.300 mol dm^{-3} aqueous sodium hydroxide. Give your answer to 2 decimal places.

4 Calculate the pH of the solution formed when 17.5 cm^3 of 0.100 mol dm^{-3} aqueous sulfuric acid is added 12.5 cm^3 of 0.120 mol dm^{-3} aqueous potassium hydroxide. Give your answer to 2 decimal places.

Titration curves

A titration curve is a graph of pH against volume of the base or acid added. To plot a titration curve, carry out an acid–base titration, adding the base in 5 cm^3 portions, and in smaller portions near the end point. Record the pH after each addition.

A typical curve shows the initial pH of the acid or base and the point when neutralisation occurs.

The shape of the curve shows the type of titration.

The inflection in a curve is where the curve goes vertically showing a rapid change in pH.

The inflection in the curve occurs at the equivalence point in a titration.

Most titration curves show a base added to an acid but the reverse is also possible.

Equivalence points

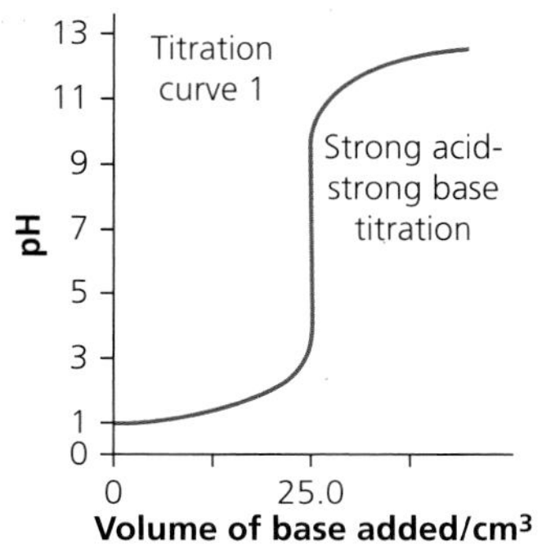

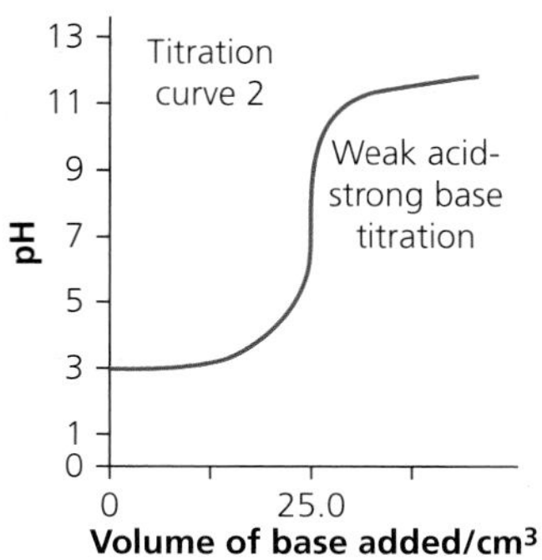

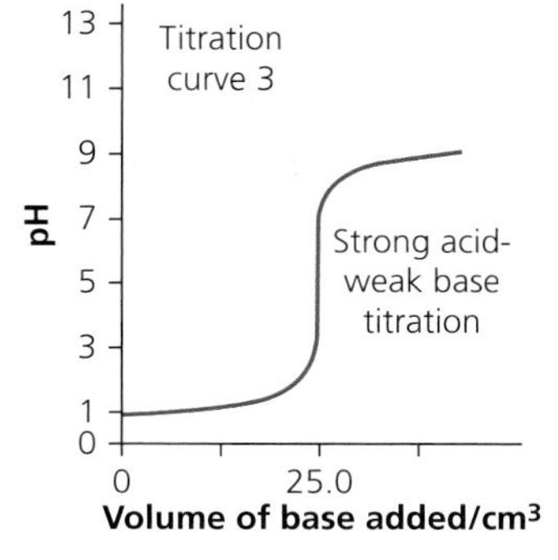

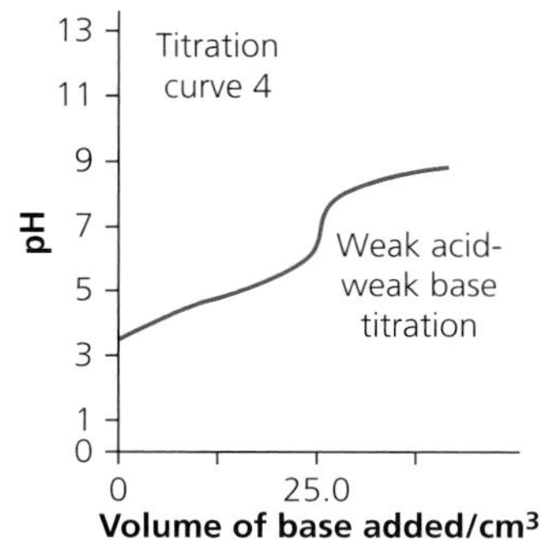

There are four different of titration curves and the equivalence points (points where there is a rapid change in pH and show a vertical inflection in the curve) are different.

For titration curve 1 the inflection occurs between pH 3 and pH 10. This is indicative of a strong acid–strong base titration. The initial pH of the acid is 1 which would suggest a strong acid.

For titration curve 2 the inflection occurs between pH 6 and 10. This is indicative of a weak acid–strong base titration. The initial pH of the acid is almost 3 which would suggest a weak acid.

For titration curve 3 the inflection occurs between pH 3 and 8. This is indicative of a strong acid–weak base titration. The initial pH of the acid is 1 which would suggest a strong acid.

For titration curve 4 there is no major inflection. This is indicative of a weak acid-weak base titration. The initial pH of the acid is between 3 and 4 which would suggest a weak acid.

Understanding titration curves

When examining or choosing a titration curve, the following four features should be considered.

Base added to acid or acid added to base.

The initial pH is the main clue to this if the horizontal axis is not labelled. If the titration curve has an acidic pH below 7 at $0\,cm^3$, the base is being added to the acid. If the initial pH is greater than 7, the acid is being added to the base.

Initial pH

An initial calculation for the pH of the strong or weak acid – this is the starting pH for the curve where the volume of base added is equal to $0\,cm^3$.

Shape of the curve

The names of the acid and base used – this will determine the shape of the titration curve including the length of the inflection at the equivalence point. The information on the shapes of the curves is important as the shape of the curve depends on the type (strong or weak) acid or base used in the titration.

TIP

A sketch of a titration curve is just that – a rough sketch of the shape – so when analysing curves use the (rough) guidelines as to where the inflection (vertical section) occurs, i.e. between approximate pH values.

Volume at which the equivalence point occurs

The concentrations of the acid and base used – the volume at which the equivalence point occurs can be calculated from the volume of base required to neutralise the acid.

Key features of a titration curve

In a titration, $0.200\,mol\,dm^{-3}$ sodium hydroxide solution is added to $25.0\,cm^3$ of $0.100\,mol\,dm^{-3}$ hydrochloric acid.

Considering the four features can help to identify a titration curve from a selection of curves.

Acid added to base or base added to acid

In this titration you are told that the base (sodium hydroxide) is added to the acid so the titration curve will start at an acidic pH.

Initial pH

HCl is a monobasic acid so $[H^+] = 0.100\,mol\,dm^{-3}$

Initial pH $= -\log_{10}(0.100) = 1.00$

Shape of the curve

Strong acid (hydrochloric acid)–strong base (sodium hydroxide) titration so curve shaped like curve 1.

The equivalence point occurs with a change in pH between 3 and 10 approximately.

Volume at which equivalence point occurs

Amount, in moles, of HCl present, $n = \frac{v \times c}{1000} = \frac{25.0 \times 0.100}{1000} = 0.00250\,mol$

$$NaOH + HCl \rightarrow NaCl + H_2O$$

1 mol of NaOH reacts with 1 mol of HCl

0.0025 mol HCl reacts with 0.00250 mol NaOH

Volume of NaOH $= \frac{n \times 1000}{c} = \frac{0.00250 \times 1000}{0.200} = 12.5\,cm^3$

TIP

These types of calculations were first encountered in **Amount of Substance** in AS Chemistry. You should refresh your knowledge of this section to ensure you can carry out these calculations.

TIP

The titration curve may be for a volume of acid added to a volume of base. This is less common. Work out the initial pH of the strong base – this is the initial pH; the volume required for neutralisation is determined in the same way; the shape of curve is the same but reflected, starting at a pH above 7 and coming down.

The titration curve for this neutralisation looks like this, taking into account the four features of the curve.

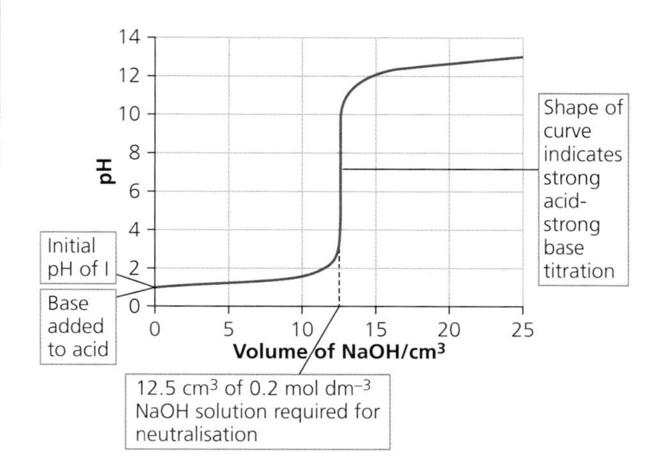

EXAMPLE 25

In a titration, 0.500 mol dm^{-3} aqueous ammonia is added to 25.0 cm^3 of 0.250 mol dm^{-3} aqueous sulfuric acid. Identify the following key features of the titration curve.

Acid added to base or base added to acid

In this titration you are told that the base (aqueous ammonia) is added to sulfuric acid so the titration curve will start at an acidic pH.

Initial pH

H_2SO_4 is a diprotic acid so $[H^+]$ = 0.500 mol dm^{-3}

Initial pH = $-\log_{10}(0.500)$ = 0.3010 (to 4 significant figures)

= 0.30 (2 decimal places)

Shape of the curve

Strong acid (hydrochloric acid)–weak base (ammonia) titration so curve shaped like curve 3.

The equivalence point occurs with a change in pH between 3 and 8 approximately.

Volume at which equivalence point occurs

Amount, in moles, of H_2SO_4 present,

$$n = \frac{V \times c}{1000} = \frac{25.0 \times 0.250}{1000} = 0.00625\ \text{mol}$$

$$2NH_3 + H_2SO_4 \rightarrow (NH_4)_2SO_4$$

2 mol of NH_3 reacts with 1 mol of H_2SO_4

0.00625 mol H_2SO_4 reacts with 0.0125 mol NH_3

$$\text{Volume of } NH_3 = \frac{n \times 1000}{c} = \frac{0.0125 \times 1000}{0.500} = 25.0\ \text{cm}^3$$

The titration curve for this neutralisation looks like this:

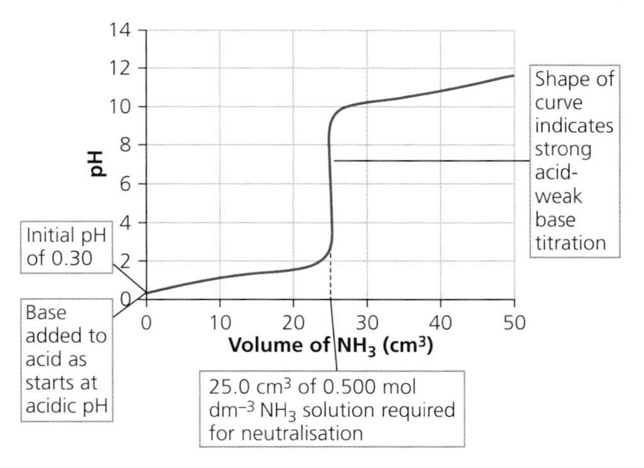

Indicators for titrations

The choice of an indicator for an acid–base titration depends on the pH range in which the indicator changes colour.

The colour change pH range of the indicator must be in the inflection range of the titration curve (the pH range of the equivalence point).

The indicator must change colour with one drop of the substance being added so the rapid change in pH must be in the range of the colour change of the indicator.

The **end point** in a titration is when the indicator changes colour which occurs at the equivalence point when there is a rapid change in pH. This changes the colour of the indicator.

Some indicators and their colour change pH ranges

Indicator	pH range of colour change	Colour in lower pH range	Colour in upper pH range
Thymol blue	1.20–2.80	Red	Yellow
Methyl orange	3.10–4.40	Red	Yellow
Methyl red	4.40–6.20	Red	Yellow
Bromothymol blue	6.00–7.60	Yellow	Blue
Phenolphthalein	8.30–10.00	Colourless	Pink

- For a strong acid–strong base titration the inflection occurs between pH 3 and 10 so any of the indicators except thymol blue would work.
- For a strong acid–weak base titration, the inflection occurs between pH 3 and 8 so any indicator in this range would work (i.e. methyl orange, methyl red and bromothymol blue).
- For a weak acid–strong base titration, the inflection occurs between pH 6 and 10 so any indicator in this range would work (i.e. bromothymol blue or phenolphthalein).
- Methyl orange and phenolphthalein are the main indicators of choice. Both may be used for strong acid–strong base titrations. Methyl orange is used for strong acid–weak base titrations. Phenolphthalein is used for weak acid–strong base titrations. A pH meter must be used for weak acid–weak base titrations as there is no inflection point.

TIP

The colour change observed in the titration is based on the first point of the analysis of the titration curves. Phenolphthalein is colourless in acid and neutral solution but pink in alkaline solution so a rapid change from 3 to 10 would change the colour of phenolphthalein from colourless to pink. If the pH changed from 10 to 3 or 10 or 6 the colour of phenolphthalein would change from pink to colourless.

TEST YOURSELF 6

1 Calculate the volume of 0.250 mol dm^{-3} sodium hydroxide solution required to react with 20.0 cm^3 of 0.150 mol dm^{-3} sulfuric acid.

2 From the list below:

sodium hydroxide	ethanoic acid	ammonia
hydrochloric acid	nitric acid	potassium hydroxide

a) Name one strong acid.
b) Name one weak base.
c) Name one weak acid.

3 The pH colour change of phenolphthalein is between pH 8.3 and 10.0. Explain why phenolphthalein cannot be used for a titration between hydrochloric acid and ammonia.

REQUIRED PRACTICAL 9

Investigating the change in pH when methanoic acid reacts with sodium hydroxide

As early as the 15th century, naturalists were aware that ant hills gave off an acidic vapour. In 1671, the English naturalist John Ray isolated the active ingredient. To do this he collected and distilled a large number of dead ants, and the acid he discovered later became known as formic acid from the Latin word for ant, *formica*. Its proper IUPAC name is now methanoic acid.

Woodland ants are the largest native ant species of the UK and are known to be aggressively territorial. They squirt methanoic acid from their abdomens as a form of protection when provoked. The acid has a strong, penetrating odour and is used to ward off hungry birds such as woodpeckers and jays.

1 Methanoic acid is a weak acid. In an experiment, a calibrated pH meter was used to measure the pH of methanoic acid solution. At 20 °C the pH of a 0.100 mol dm^3 solution was 2.37.

a) Explain why a pH meter should be calibrated before use.

b) Write an equation for the dissociation of methanoic acid and explain what is meant by weak acid.

c) Write an expression for the equilibrium constant, K_a, for the dissociation of methanoic acid in aqueous solution.

d) Use your answer from (c) to calculate the value of K_a for this dissociation at 20 °C. Give your answer to the appropriate precision. Show your working.

2 A student used aqueous sodium hydroxide to determine the titration curve for the reaction of methanoic acid and sodium hydroxide. 25.0 cm^3 of 1.50×10^{-2} mol dm^{-3} methanoic acid was placed in a conical flask at 25 °C. The sodium hydroxide was added in 2 cm^3 portions and the pH of the reaction mixture was measured using a pH meter.

a) Write a balanced symbol equation for the reaction between HCOOH and NaOH.

b) Why was the reaction mixture stirred with a glass rod after the addition of each 2 cm^3 portion of sodium hydroxide?

c) The pH curve for this titration is shown below. Calculate the value of the concentration in mol dm^{-3} of the aqueous sodium hydroxide.

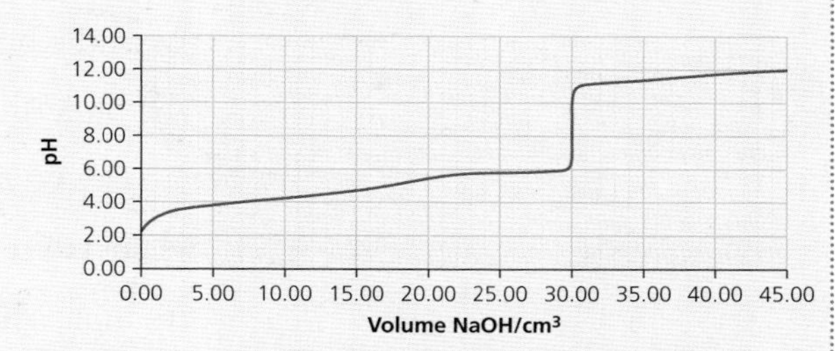

Figure 6.7 A titration curve for the titration of methanoic acid with NaOH

d) The pH ranges in which the colour changes for three acid–base indicators are shown below. Explain which of the three indicators is suitable for this titration.

Indicator	pH range
Metacresol purple	7.40–9.00
2,4,6-trinitrotoluene	11.50–13.00
Ethyl orange	2.4–4.8

TIP

For a weak acid–strong base titration the pH at half neutralisation (pH at half the volume where the equivalence point occurs) is equal to the pK_a. This can be used as a method of determining K_a of a weak acid from a titration curve. At this point of half neutralisation, the acid concentration is equal to the salt concentration so they cancel out and pH = pK_a.

Buffers

Buffers are solutions that can resist changes in pH when small quantities of acid or base are added. An interesting application of buffers is their use in skin creams. Anti-ageing creams are buffered at pH values lower than 5.5. The low pH irritates the skin, causing it to swell hiding wrinkles.

There are two main types of buffer:

1 an acidic buffer

2 a basic buffer.

Acidic buffers

An acidic buffer is a solution formed from a weak acid and its salt.

The buffering action is represented by the general equation:

$$HA \rightleftharpoons H^+ + A^-$$

where HA is the undissociated acid, A^- is the anion and H^+ are hydrogen ions.

Ethanoic acid and sodium ethanoate may be mixed to form an acidic buffer.

Alternatively, the salt may be formed indirectly from the reaction of sodium hydroxide with an excess of ethanoic acid. This leaves a solution containing the excess ethanoic acid and sodium ethanoate formed from the reaction with ethanoic acid.

The equation for this buffering action is:

$$CH_3COOH \rightleftharpoons CH_3COO^- + H^+$$

Explanation of buffering action

When small quantities of acid or base are added, the pH of a buffer remains almost constant.

You can be asked to explain how an acidic buffer resists the change in pH.

Addition of dilute acid

The following points are used to explain how a buffer maintains an almost constant pH when a small amount of a dilute acid is added:

- When a small amount of an acid is added, extra hydrogen ions, H^+, are added.
- The position of equilibrium in the equilibrium $HA \rightleftharpoons H^+ + A^-$ moves to the left to remove the added H^+ ions.
- This keeps the pH almost constant.

If the weak acid and its salt are named then use the correct anion and weak acid. For example, if the buffer is made from ethanoic acid and ethanoate ions, the equation should read $CH_3COOH \rightleftharpoons CH_3COO^- + H^+$ and the position of equilibrium will move to the left to remove the added H^+ ions.

Addition of base

The following points are used to explain how a buffer maintains an almost constant pH when a small amount of a base is added:

- When a small amount of a base is added, extra hydroxide ions, OH^-, are added.
- The hydroxide ions react with the hydrogen ions in the buffer, removing some H^+ ions.
- The position of equilibrium in the equilibrium $HA \rightleftharpoons H^+ + A^-$ moves to the right to replace the H^+ which reacted with the OH^- ions.
- This keeps the pH almost constant

TIP
The equation for the reaction of hydroxide ions, OH^-, with a buffer may be represented by

$$HA + OH^- \rightarrow A^- + H_2O$$

to show how the buffer removed the hydroxide ions which were added.

TIP
Organic bases such as amines can also act as buffers in solution. The basic nature of amines is examined in Chapter 13. No calculation work on the pH of basic buffers is expected.

Again if the weak acid and its salt are named then use the correct weak acid and anion. For example if the buffer is made from hydrocyanic acid and cyanide ions, the equation should read $HCN \rightleftharpoons H^+ + CN^-$ to explain the buffering action. When OH^- is added, they react with H^+ ions in the buffer and the position of equilibrium moves to the right to replace the H^+ that were removed.

Basic buffers

A basic buffer is a solution formed from a weak base and its salt. The most common example involves ammonia in solution with an ammonium salt.

The buffering action of a basic buffer can be explained using the equation:

$$NH_3 + H^+ \rightleftharpoons NH_4^+$$

When small amounts of acid are added, the added hydrogen ions, H^+, are removed as they react with the ammonia in the solution to form ammonium ions. The position of equilibrium moves from left to right.

When small amounts of an alkali are added, the hydroxide ions, OH^-, react with some of the hydrogen ions in the buffer to form water:

$$H^+ + OH^- \rightarrow H_2O.$$

The position of equilibrium moves from right to left to replace the H^+ ions.

Preparation of an acidic buffer

There are several methods which may be used to prepare a buffer. All of the methods result in a solution containing a weak acid and the anion of its salt.

1 Addition of a solid salt of the weak acid to a solution of the weak acid.

2 Addition of a solution of the salt of a weak acid to a solution of the weak acid.

3 Addition of sodium hydroxide solution to an excess of the weak acid.

Method 3 is the most often used.

Calculating pH of an acidic buffer

To determine the pH of an acidic buffer you must determine the amount, in moles, of the anion A^- (this equals the amount, in moles, of the salt) and the amount in moles of the weak acid, HA, from the information given in the method used to prepare the buffer.

In the K_a expression for a weak acid, $K_a = \frac{[H^+]\,[A^-]}{[HA]}$

The $[H^+]$ is determined from the K_a expression and pH is then calculated in the usual way. As both HA and A^- are in the same volume of solution, the amount in moles of HA and A^- may be used in the calculation instead of concentration. It will be shown that the calculations using moles and concentrations give the same pH values for the buffers.

Method 1: Addition of a solid salt of the weak acid to a solution of the weak acid

In this method the concentration of the acid remains unchanged as there is no other solution or water added to dilute it.

The amount in moles of the anion is given or in other examples it may be determined from the mass of the solid salt added.

TIP

If a mass of the salt is given, simply divided the mass, in g, by the M_r of the salt to determine the amount in moles of salt added. The rest of the calculation follows the method shown.

EXAMPLE 26

A buffer solution was prepared by dissolving 0.012 mol of sodium ethanoate in 100 cm^3 of 0.0520 mol dm^{-3} ethanoic acid. The dissociation constant, K_a, for ethanoic acid has the value 1.74×10^{-5} mol dm^{-3} at 25 °C.

Calculate the pH of this buffer solution. Give your answer to 2 decimal places.

Answer

Moles of CH_3COO^- = 0.012

Moles of $CH_3COOH = \frac{100 \times 0.0520}{1000} = 5.20 \times 10^{-3}$

$$K_a = \frac{[H^+][CH_3COO^-]}{[CH_3COOH]} = 1.74 \times 10^{-5} \text{ mol dm}^{-3}$$

$$[H^+] = \frac{K_a \times [CH_3COOH]}{[CH_3COO^-]} = \frac{1.74 \times 10^{-5} \times 5.20 \times 10^{-3}}{0.012}$$

$$= 7.54 \times 10^{-6} \text{ mol dm}^{-3}$$

$pH = -\log_{10}[H^+] = -\log_{10}(7.54 \times 10^{-6}) = 5.1226$ (to 4 decimal places)

pH = 5.12 (to 2 decimal places)

Using concentrations

To show how this works using concentrations:

Concentration of weak acid $[CH_3COOH]$ = 0.052 mol dm^{-3}

The concentration of the anion, in mol dm^{-3}, is determined by dividing the amount (in moles) of the salt, CH_3COONa, by the volume of the solution and multiplying by 1000 to convert to moles per 1 dm^3 (1000 cm^3).

Concentration of the salt: $[CH_3COONa] = [CH_3COO^-]$

$$[CH_3COO^-] = \frac{0.012}{100} \times 1000 = 0.12 \text{ mol dm}^{-3}$$

Concentration of anion $[CH_3COO^-]$ = 0.12 mol dm^{-3}

$$K_a = \frac{[H^+][CH_3COO^-]}{[CH_3COOH]} = 1.74 \times 10^{-5} \text{ mol dm}^{-3}$$

$$[H^+] = \frac{K_a \times [CH_3COOH]}{[CH_3COO^-]} = \frac{1.74 \times 10^{-5} \times 0.0520}{0.12}$$

$$= 7.54 \times 10^{-6} \text{ mol dm}^{-3}$$

$pH = -\log_{10}[H^+] = -\log_{10}(7.54 \times 10^{-6}) = 5.1226$ (to 4 decimal places)

pH = 5.12 (to 2 decimal places)

TIP

All these calculations will give the same answer if concentrations of HA and A^- are determined and used in the calculation of pH but the amount in moles method is shorter and fewer steps means fewer potential errors in calculations.

Method 2: Addition of a solution of the salt of a weak acid to a solution of the weak acid

In this method a certain volume of the solution of the salt (most often a solution of the sodium salt) is added to a certain volume of a solution of the weak acid. Both solutions dilute each other so the concentrations of both the salt and the weak acid change.

EXAMPLE 27

A buffer is prepared by mixing 50.0 cm^3 of a 0.0417 $mol\,dm^{-3}$ solution of the salt of a weak acid, NaX, with 150 cm^3 of a 0.0204 $mol\,dm^{-3}$ solution of the weak acid, HX. The pK_a for HX is 5.12 at 25 °C.

Calculate the pH of this buffer solution. Give your answer to 2 decimal places

Answer

Moles of NaX $= \frac{50.0 \times 0.0417}{1000} = 2.085 \times 10^{-3}$ mol

Moles of HX $= \frac{50.0 \times 0.0204}{1000} = 3.06 \times 10^{-3}$ mol

The pK_a is converted into a K_a using $K_a = 10^{-pKa}$.

$K_a = 10^{(-5.12)} = 7.583 \times 10^{-6}\,mol\,dm^{-3}$ (to 4 significant figures)

$K_a = \frac{[H^+][X^-]}{[HX]} = 7.583 \times 10^{-6}\,mol\,dm^{-3}$

$[H^+] = \frac{K_a \times [HX]}{[X^-]} = \frac{7.583 \times 10^{-6} \times 3.06 \times 10^{-3}}{2.085 \times 10^{-3}}$

$[H^+] = 1.113 \times 10^{-5}\,mol\,dm^{-3}$ (to 4 significant figures)

$pH = -\log_{10}[H^+] = -\log_{10}(1.113 \times 10^{-5}) = 4.9535$ (to 4 decimal places)

pH = 4.95 (to 2 decimal places)

Using concentrations

The new total volume of the solution is 200 cm^3 so concentrations in $mol\,dm^{-3}$ may be determined using the amount of solute, in moles, in 200 cm^3 converted to 1 dm^3 (1000 cm^3) by dividing the amount, in moles, by 200 and multiplying by 1000 (this is the same as multiplying by 5).

Concentration of NaX: $[NaX] = [X^-] = 2.085 \times 10^{-3} \times 5 = 0.010425\,mol\,dm^{-3}$

Concentration of HX: $[HX] = 3.06 \times 10^{-3} \times 5 = 0.0153\,mol\,dm^{-3}$

The pK_a is converted into a K_a using $K_a = 10^{-pKa}$

$K_a = 10^{(-5.12)} = 7.583 \times 10^{-6}\,mol\,dm^{-3}$ (to 4 significant figures)

$K_a = \frac{[H^+][X^-]}{[HX]} = 7.583 \times 10^{-6}\,mol\,dm^{-3}$

$[H^+] = \frac{K_a \times [HX]}{[X^-]} = \frac{7.583 \times 10^{-6} \times 0.0153}{0.010425}$

$[H^+] = 1.113 \times 10^{-5}\,mol\,dm^{-3}$ (to 4 significant figures)

$pH = -\log_{10}[H^+] = -\log_{10}(1.113 \times 10^{-5}) = 4.9535$ (to 4 decimal places)

pH = 4.95 (to 2 decimal places)

The same value for pH is obtained as the operation of ×5 on both the amounts of the acid, HX, and the anion, X^-, would cancel out.

Henderson-Hasselbalch equation

The Henderson-Hasselbalch equation may also be used to calculate the pH of a buffer.

$pH = pK_a + \log\left(\frac{[H^-]}{[HX]}\right)$ where $[X^-]$ is the concentration of the anion and [HX] is the concentration of the acid. Again, moles of X^- and HX may be used in the calculation.

$pK_a = 5.12$, the amount in moles of HX present is 3.06×10^{-3} and the amount of X^- present is 2.085×10^{-3} mol.

$pH = 5.12 + \log\left(\frac{2.085 \times 10^{-3}}{3.06 \times 10^{-3}}\right) = 4.95$

Using concentrations

For this example, the $pK_a = 5.12$, [HA] was calculated to be 0.0153 $mol\,dm^{-3}$ and $[A^-]$ was calculated to be 0.010425 $mol\,dm^{-3}$

$pH = 5.12 + \log\left(\frac{0.010425}{0.0153}\right) = 4.95$

TIP

Practise these styles of calculation on your calculator if you are using the Henderson-Hasselbalch equation to calculate the pH of an acidic buffer.

Method 3: Addition of sodium hydroxide or potassium hydroxide solution to an excess of the weak acid

In this buffer preparation, the sodium hydroxide (or potassium hydroxide) reacts with the weak acid to form the salt in situ.

The weak acid must be in excess so no sodium hydroxide remains and the resulting solution contains the weak acid and its salt.

TIP

Be careful with examples involving a weak acid reacting with a strong base such as ethanoic acid with sodium hydroxide. If the sodium hydroxide is in excess, it is a neutralisation example and NOT a buffer example. Look for K_w being given as you will have to work out the pH of a base. Buffer examples will usually use the word buffer.

EXAMPLE 28

$100\,cm^3$ of $0.0515\,mol\,dm^{-3}$ nitrous acid (HNO_2) was mixed with $50.0\,cm^3$ of $0.0428\,mol\,dm^{-3}$ sodium hydroxide solution. The acid dissociation constant, K_a, for nitrous acid is $7.20 \times 10^{-4}\,mol\,dm^{-3}$ at 25 °C. Calculate the pH of this buffer solution. Give your answer to 2 decimal places.

Answer

Again both solutions dilute each other and the sodium hydroxide reacts with some of the nitrous acid to form the salt, sodium nitrite ($NaNO_2$).

Moles of nitrous acid added (HNO_2) $= \frac{100 \times 0.0515}{1000} = 5.15 \times 10^{-3}\,mol$

Moles of sodium hydroxide added (NaOH) $= \frac{50.0 \times 0.0428}{1000} = 2.14 \times 10^{-3}\,mol$

The equation for the reaction between nitrous acid and sodium hydroxide solution is given below

$$HNO_2 + NaOH \rightarrow NaNO_2 + H_2O$$

2.14×10^{-3} mol of NaOH will react with 2.14×10^{-3} mol of HNO_2 (leaving $5.15 \times 10^{-3} - 2.14 \times 10^{-3} = 3.01 \times 10^{-3}$ mol of HNO_2) and forming 2.14×10^{-3} mol of $NaNO_2$.

Moles of HNO_2 in $150\,cm^3 = 3.01 \times 10^{-3}\,mol$

Moles of $NaNO_2$ in $150\,cm^3 = 2.14 \times 10^{-3}\,mol$

All of the sodium hydroxide has been used up leaving only some moles of the weak acid (HNO_2) and some moles of the salt ($NaNO_2$) in solution. The water formed is in the solution.

$$K_a = \frac{[H^+][NO_2^-]}{[HNO_2]} = 7.20 \times 10^{-4}\,mol\,dm^{-3}$$

$$[H^+] = \frac{K_a \times [HNO_2]}{[NO_2^-]} = \frac{7.20 \times 10^{-4} \times 3.01 \times 10^{-3}}{2.14 \times 10^{-3}}$$

$[H^+] = 1.0127 \times 10^{-3}\,mol\,dm^{-3}$

$pH = -\log_{10}[H^+] = -\log_{10}(1.0127 \times 10^{-3}) = 2.9945$ (to 4 decimal places).

pH = 2.99 (to 2 decimal places)

Using concentrations

Moles of HNO_2 in $150\,cm^3 = 3.01 \times 10^{-3}\,mol$

Moles of $NaNO_2$ in $150\,cm^3 = 2.14 \times 10^{-3}\,mol$

The total volume is $150\,cm^3$ and this may be used to calculate the concentrations of the weak acid and its salt in the solution.

$$[HNO_2] = \frac{3.01 \times 10^{-3}}{150} \times 1000 = 0.02007\,mol\,dm^{-3}$$

$$[NaNO_2] = [NO_2^-] \frac{2.14 \times 10^{-3}}{150} \times 1000 =$$

$0.01427\,mol\,dm^{-3}$ (to 4 significant figures)

$$K_a = \frac{[H^+][NO_2^-]}{[HNO_2]} = 7.20 \times 10^{-4}\,mol\,dm^{-3}$$

$$[H^+] = \frac{K_a \times [HNO_2]}{[NO_2^-]} = \frac{7.20 \times 10^{-4} \times 0.02007}{0.01427}$$

$[H^+] = 1.0126 \times 10^{-3}\,mol\,dm^{-3}$ (to 4 significant figures)

$pH = -\log_{10}[H^+] = -\log_{10}(1.0126 \times 10^{-3}) = 2.9945$ (to 4 decimal places).

pH = 2.96 (to 2 decimal places)

Note the very slight difference in the concentration of the H^+ ion using concentration caused by the rounding of the concentration of HNO_2 and $NaNO_2$ but the pH is the same.

Addition of acid or base to a buffer

A buffer by definition maintains an almost constant pH when small amounts of acid or base are added to it.

The pH of a buffer can be calculated after the addition of a small amount of an acid or a base.

Addition of a small amount of an acid

- When an acid is added, the amount in moles of the acid should be calculated if it is not given.
- The position of equilibrium for the equilibrium $HA \rightleftharpoons H^+ + A^-$ moves to the left to remove the added H^+ ions.
- The amount, in moles, of HA increases by the number of moles of acid added.
- The amount, in moles, of A^- decreases by the number of moles of acid added.
- The new amounts in moles of H^+ and A^- and the K_a expression are used to calculate $[H^+]$. Then $pH = -\log_{10}[H^+]$ is used to calculate the pH of the buffer after the addition of the acid.

EXAMPLE 29

A buffer is prepared by adding 0.0122 mol of salt, NaY to 40.0 cm^3 of a 0.210 mol dm^{-3} solution of the weak acid, HY. The acid dissociation constant, K_a, for the weak acid, HY, is 1.71×10^{-5} mol dm^{-3} at 25 °C. Give all pH answers in this question to 2 decimal places.

1 Calculate the pH of the buffer formed.
2 A 2.00×10^{-4} sample of hydrochloric acid was added to this buffer solution. Calculate the pH of the buffer solution after the hydrochloric acid was added.

Answers

1 moles of A^- added = 0.0122

moles of HA = $\frac{40.0 \times 0.210}{1000} = 8.40 \times 10^{-3}$

$K_a = \frac{[H^+][Y^-]}{[HY]} = 1.71 \times 10^{-5}$ mol dm^{-3}

$[H^+] = \frac{K_a \times [HY]}{[Y^-]} = \frac{1.17 \times 10^{-5} \times 8.40 \times 10^{-3}}{0.0122}$

$[H^+] = 1.177 \times 10^{-5}$ mol dm^{-3} (to 4 significant figures)

$pH = -\log_{10}[H^+] = -\log_{10}(1.177 \times 10^{-5}) = 4.9292$ (to 4 decimal places)

pH = 4.93 (to 2 decimal places)

2 Adding 2.00×10^{-4} mol of HCl, adds 2.0×10^{-4} mol of H^+.

The position of equilibrium for the equilibrium $HA \rightleftharpoons H^+ + A^-$ moves to the left to remove the added H^+ ions.

The amount, in moles, of HA increases by 2.0×10^{-4} mol

Initial amount of HY = $\frac{40.0 \times 0.210}{1000} = 8.40 \times 10^{-3}$ mol

New amount of HY = = $8.40 \times 10^{-3} + 2.00 \times 10^{-4} = 8.60 \times 10^{-3}$ mol

The amount, in moles, of A^- decreases by 2.0×10^{-4} mol

Initial amount of Y^- = 0.0122 mol

New amount of Y^- = $0.0122 - 2.00 \times 10^{-4} = 0.0120$ mol

$K_a = \frac{[H^+][Y^-]}{[HY]} = = 1.71 \times 10^{-5}$ mol dm^{-3}

$[H^+] = \frac{K_a \times [HY]}{[Y^-]} = \frac{1.17 \times 10^{-5} \times 8.60 \times 10^{-3}}{0.0120}$

$[H^+] = 1.226 \times 10^{-5}$ mol dm^{-3} (to 4 significant figures)

$pH = -\log_{10}[H^+] = -\log_{10}(1.226 \times 10^{-5}) = 4.9115$ (to 4 decimal places)

pH = 4.91 (to 2 decimal places)

TIP

As can be seen, the pH of buffer remains relatively constant even with the addition of a small amount of acid. This is the basis of the definition of a buffer.

Addition of a small amount of a base

- When a base is added, the amount in moles of the base should be calculated if it is not given.
- The position of equilibrium for the equilibrium $HA \rightleftharpoons H^+ + A^-$ moves to the right to replace the H^+ ions which were removed by the base.
- The amount, in moles, of A^- increases by the number of moles of base added.
- The amount, in moles, of HA decreases by the number of moles of base added.

EXAMPLE 30

The value of K_a for methanoic acid is 1.78×10^{-4} mol dm^{-3} at 25 °C. A buffer solution contains 0.0150 mol of methanoic acid and 0.0120 mol of sodium methanoate in 500 cm^3 of solution at 25 °C. Give all pH values in this question to 2 decimal places.

1 Calculate the pH of this buffer at 25 °C. Give your answer to 2 decimal places.
2 A 5.00 cm^3 sample of 0.120 mol dm^{-3} sodium hydroxide solution is added to the buffer. Calculate the pH of the buffer solution after this addition.

Answer

1 $K_a = \frac{[H^+][HCOO^-]}{[HCOOH]} = 1.78 \times 10^{-4}$ mol dm^{-3}

$[H^+] = \frac{K_a \times [HCOOH]}{[HCOO^-]} = \frac{1.78 \times 10^{-4} \times 0.0150}{0.01427}$

$[H^+] = 2.225 \times 10^{-4}$ mol dm^{-3}

pH = $-\log_{10}[H^+] = -\log_{10}(2.225 \times 10^{-4}) = 3.6527$ (to 4 decimal places)

pH = 3.65 (to 2 decimal places)

2 5.0 cm^3 of 0.12 mol dm^{-3} NaOH added $= \frac{5.00 \times 0.0120}{1000}$
$= 6.00 \times 10^{-4}$ mol

Adding 6.00×10^{-4} mol of NaOH, adds 6.00×10^{-4} mol of OH^-.
6.00×10^{-4} mol H^+ ions in the buffer react with 6.00×10^{-4} mol of OH^- ions
$H^+ + OH^- \rightarrow H_2O$
The position of equilibrium for the equilibrium $HA \rightleftharpoons H^+ + A^-$ moves to the right to replace the H^+ ions that were removed by OH^- ions.

The amount, in moles, of HA decreases by 6.00×10^{-4} mol
Initial amount of HA = 0.0150 mol
New amount of HA = $0.0150 - 6.00 \times 10^{-4}$ = 0.0144 mol
The amount, in moles, of A^- increases by 6.00×10^{-4} mol
Initial amount of A^- = 0.0120 mol
New amount of A^- = $0.0120 + 6.00 \times 10^{-4}$ = 0.0126 mol

$K_a = \frac{[H^+][HCOO^-]}{[HCOOH]} = 1.78 \times 10^{-4}$ mol dm^{-3}

$[H^+] = \frac{K_a \times [HCOOH]}{[HCOO^-]} = \frac{1.78 \times 10^{-4} \times 0.0144}{0.0126}$

$[H^+] = 2.034 \times 10^{-4}$ mol dm^{-3} (to 4 significant figures)

pH = $-\log_{10}[H^+] = -\log_{10}(2.034 \times 10^{-4}) = 3.6916$ (to 4 decimal places)

pH = 3.69 (to 2 decimal places)

TIP

Any of the buffer questions above can be carried out by converting to concentration of the weak acid and its salt but be careful as the total volume in the last calculation is 505 cm^3 as 5.00 cm^3 of the sodium hydroxide solution was added to 500 cm^3 of the buffer.

TEST YOURSELF 7

1 What is a buffer?
2 Explain why the pH of an acidic buffer solution remains almost constant despite the addition of a small amount of hydrochloric acid.
3 The acid dissociation constant, K_a, for ethanoic acid is 1.74×10^{-5} mol dm^{-3} at 25 °C. Calculate the pH of a buffer formed when 0.00170 mol of sodium ethanoate is added to 75.0 cm^3 of 0.0550 mol dm^{-3} ethanoic acid at 25 °C. Give your answer to 2 decimal places.
4 25.0 cm^3 of 0.200 mol dm^{-3} sodium hydroxide solution was added to 50.0 cm^3 of 0.250 mol dm^{-3} methanoic acid. K_a for methanoic acid at 25 °C is 1.78×10^{-4} mol dm^{-3}. Calculate the pH of the buffer solution formed at 25 °C. Give your answer to 2 decimal places.
5 1.00 dm^3 of a buffer solution contains 0.0880 mol of a weak acid HA and 0.0540 mol of the salt of the acid NaA. pK_a for the weak acid HA is 4.52.
   a) Calculate the pH of the buffer at 25 °C. Give your answer to 2 decimal places.
   b) 4.00 cm^3 of 0.150 mol dm^{-3} nitric acid is added to the buffer. Calculate the pH of the buffer after this addition. Give your answer to 2 decimal places.

Practice questions

1 Which of the following has units of $mol\,dm^{-3}$?

A K_a **B** K_w

C pH **D** pK_a *(1)*

2 What is the pH of a solution of $0.0154\,mol\,dm^{-3}$ H_2SO_4?

A 1.00 **B** 1.51

C 1.81 **D** 2.11 *(1)*

3 Write expressions for the following:

a) pH

b) Acid dissociation constant, K_a, for the weak acid HA

c) K_w *(3)*

4 The acid dissociation constant, K_a, for the weak acid HX is $2.45 \times 10^{-4}\,mol\,dm^{-3}$ at 25°C.

Calculate the pH of a $0.215\,mol\,dm^{-3}$ solution of the weak acid HX at 25°C. *(3)*

5 The ionic product of water, K_w, is $5.48 \times 10^{-14}\,mol^2\,dm^{-6}$ at 50°C.

a) Write an expression for K_w. *(1)*

b) Calculate the pH of pure water at 50°C. Give your answer to 2 decimal places. *(2)*

6 a) In the reaction:

$$CH_3NH_2 + H_2O \rightleftharpoons CH_3NH_3^+ + OH^-$$

Explain whether water is acting as a Brønsted-Lowry acid or base in this reaction. *(2)*

b) In the reaction:

$$H_2SO_4 + HNO_3 \rightleftharpoons H_2NO_3^+ + HSO_4^-$$

State which of the reactants acts as a Brønsted-Lowry base in this reaction. *(1)*

7 Calculate the pH of the solutions produced in the following way at 25°C.
$K_w = 1.0 \times 10^{-14}\,mol^2\,dm^{-6}$ at 25°C.

a) $50.0\,cm^3$ of $0.200\,mol\,dm^{-3}$ sodium hydroxide solution are diluted to $250\,cm^3$ with deionised water. Give your answer to 2 decimal places. *(4)*

b) $5.0\,cm^3$ of $0.0520\,mol\,dm^{-3}$ hydrochloric acid are mixed with $20.0\,cm^3$ of deionised water. Give your answer to 2 decimal places. *(3)*

c) $15.0\,cm^3$ of $0.124\,mol\,dm^{-3}$ sulfuric acid are mixed with $35.0\,cm^3$ of deionised water. Give your answer to 2 decimal places. *(3)*

8 Calculate the pH of the solution formed when $25.0\,cm^3$ of $0.100\,mol\,dm^{-3}$ hydrochloric acid are reacted with $10.0\,cm^3$ of $0.150\,mol\,dm^{-3}$ sodium hydroxide solution. Give your answer to 2 decimal places. *(5)*

9 A buffer is prepared by mixing 0.0271 mol of sodium ethanoate with $125\,cm^3$ of $0.325\,mol\,dm^{-3}$ ethanoic acid. The acid dissociation constant for ethanoic acid, K_a, is $1.74 \times 10^{-5}\,mol\,dm^{-3}$ at 25°C.

a) Explain why the pH of a buffer solution remains almost constant despite the addition of a small amount of an acid. *(2)*

b) Calculate the pH of the buffer formed. Give your answer to 2 decimal places. *(4)*

c) $5.00\,cm^3$ of $0.200\,mol\,dm^{-3}$ hydrochloric acid was added to the buffer solution. Calculate the pH of the buffer solution after the hydrochloric acid was added. Give your answer to 2 decimal places. *(6)*

10 $25.0\,cm^3$ of a solution of $0.125\,mol\,dm^{-3}$ hydrochloric acid was placed in a conical flask and sodium hydroxide solution added from a burette. $22.7\,cm^3$ of sodium hydroxide solution was required to reach the end point.

a) Calculate the pH of the $0.125\,mol\,dm^{-3}$ hydrochloric acid used in this titration. *(2)*

b) Write an equation for the reaction between sodium hydroxide and hydrochloric acid. *(1)*

c) Calculate the concentration, in $mol\,dm^{-3}$, of the sodium hydroxide solution used. Give your answer to 3 significant figures. *(3)*

d) Using your answer to (c), calculate the pH of this concentration of sodium hydroxide solution at 25°C. At 25°C, K_w has the value $1.00 \times 10^{-14}\,mol^2\,dm^{-6}$. If you did not get answer to (c), use $0.145\,mol\,dm^{-3}$ as the concentration of the sodium hydroxide solution. This is not the correct answer to (c). Give your answer to 2 decimal places. *(3)*

11 Titration curves 1 to 4 below were obtained for combinations of different solution of acids and bases.

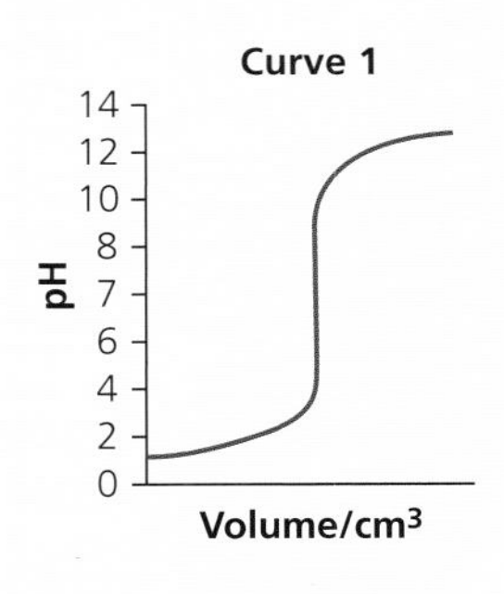

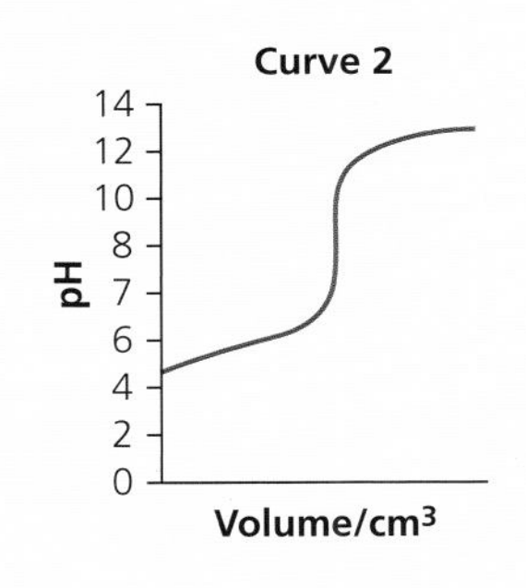

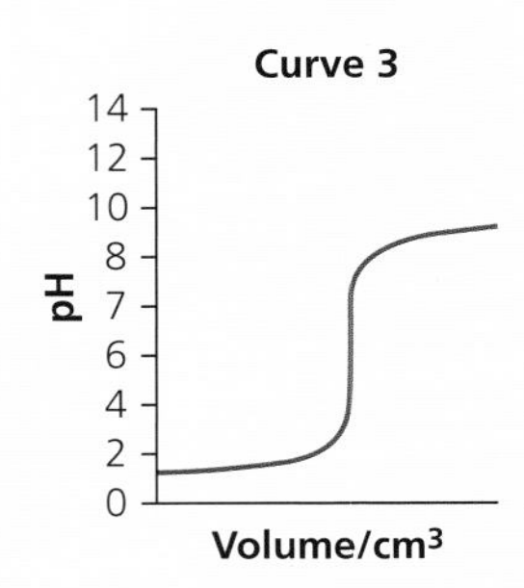

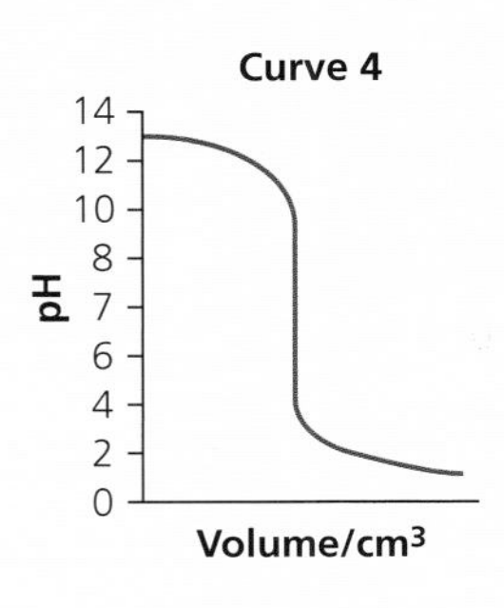

a) From the curves 1, 2, 3 and 4, choose the curve which would be produced by the addition of the following.

i) sodium hydroxide added to ethanoic acid *(1)*

ii) hydrochloric acid added to potassium hydroxide *(1)*

iii) ammonia added to nitric acid *(1)*

b) The following indicators may be used in titrations.

Indicator	pH range of colour change	Colour in acid	Colour in alkali
Thymol blue	1.20–2.80	Red	Blue
Methyl orange	3.10–4.40	Red	Yellow
Methyl red	4.40–6.20	Red	Yellow
Bromothymol blue	6.00–7.60	Yellow	Blue
Phenolphthalein	8.30–10.00	Colourless	Pink

i) Select from the table an indicator which could be used in the titration which produces curve 1 but not in the titration that produces curve 3. *(1)*

ii) Give the colour change at the end point of the titration which produces curve 4 when bromothymol blue is used as the indicator. *(1)*

12 An acidic buffer solution is prepared by mixing $25.0\,cm^3$ of $0.180\,mol\,dm^{-3}$ potassium hydroxide solution with $50.0\,cm^3$ of a $0.200\,mol\,dm^{-3}$ solution of a weak acid HX. K_a for the weak acid HX at 25 °C has the value $2.74 \times 10^{-5}\,mol\,dm^{-3}$.

a) Explain why the pH of a buffer solution remains almost constant despite the addition of a small amount of a base. *(2)*

b) Calculate the pH of the buffer at 25 °C. Give your answer to 2 decimal places. *(6)*

c) 0.00240 mol of sodium hydroxide was added. Calculate the pH of the buffer solution after the addition of the sodium hydroxide. Give your answer to 2 decimal places. *(4)*

Stretch and challenge

13 An acidic buffer is formed by mixing 0.124 mol of ethanoic acid with *x* mol of sodium hydroxide in $1\,dm^3$ of solution. The pH of the buffer formed is 4.81. The K_a for ethanoic acid is $1.74 \times 10^{-5}\,mol\,dm^{-3}$. Calculate a value for *x*. Give your answer to 3 significant figures. *(5)*

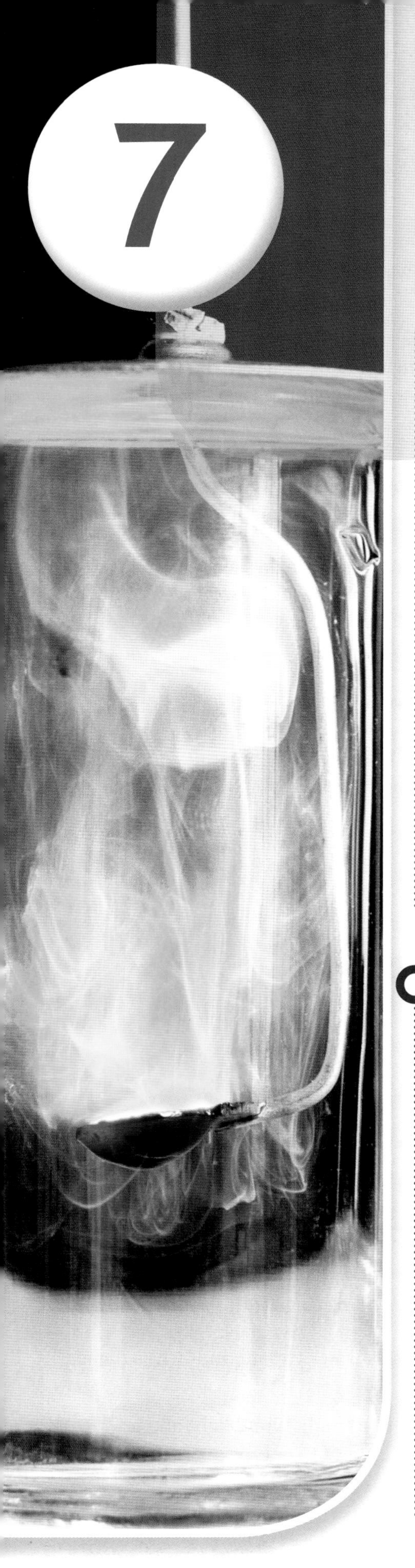

7 Properties of Period 3 elements and their oxides

PRIOR KNOWLEDGE

- Elements in the Periodic Table are grouped into blocks which are the s, p, d and f blocks.
- The position of an element in the Periodic Table is defined by its atomic number (or proton number).
- From sodium to argon, the atomic radius decreases.
- From sodium to argon, first ionisation energy generally increases.
- From sodium to argon, the melting points of the elements increases to silicon and then decreases (apart from a slight rise between phosphorus and sulfur).
- Elements of Group 2 are known as the alkaline earth metals and elements of Group 7 are known as the halogens.
- Group 2 elements show an increase in reactivity down the group whereas Group 7 elements show a decrease in reactivity down the group.
- Group 2 elements react with water forming the metal hydroxide and releasing hydrogen gas (the reaction of magnesium with water is very slow).
- Acidified barium chloride solution is used to test for sulfate ions. A positive test yields a white precipitate of barium sulfate.
- Acidified silver nitrate solution is used to test for halide ions. A white precipitate indicates the presence of chloride ions, a cream precipitate indicates bromide ions and a yellow precipitate indicates iodide ions.

TEST YOURSELF ON PRIOR KNOWLEDGE 1

1 From the following ten elements:

barium	iron	uranium	chlorine	sodium
magnesium	neon	bismuth	silver	vanadium

a) Name the s block elements.
b) Name the p block elements.
c) How many are d block elements?
d) How many are f block elements?

2 Write an equation for the reaction of barium with water.

3 State the colour of the precipitate observed when the following solutions are mixed:
a) Potassium iodide solution and acidified silver nitrate solution
b) Acidified barium chloride solution and magnesium sulfate solution
c) Zinc chloride solution and acidified silver nitrate solution.

4 State the following trends:
a) Melting point across Period 3
b) Reactivity down Group 7
c) Atomic radius across Period 3
d) First ionisation energy down Group 2.

Elements of Period 3

The elements of Period 3 include the metals sodium and magnesium which reacts with water.

Reaction of sodium with water

Sodium metal reacts vigorously with water. A small piece of the metal is cut and added to a trough of water.

Observations: sodium floats on the water, moves about the surface of the water and there is fizzing. The metal melts to form a silvery ball. It eventually disappears and the solution which remains is colourless.

The equation for the reaction is:

$$2Na + 2H_2O \rightarrow 2NaOH + H_2$$

The solution which remains contains sodium ions and hydroxide ions and it has an alkaline pH. The more sodium that is added, the higher the pH. Sodium has a slightly lower density than water and so floats on the water. The reaction is vigorous and very exothermic and this why the sodium melts. The fizzing is caused by the production of hydrogen which also propels the molten bead of sodium across the surface of the water.

Reaction of magnesium with water

Magnesium metal reacts very slowly with water. A piece of magnesium ribbon will produce a small volume of hydrogen when left in contact with water for several weeks. The equation for the reaction is:

$$Mg + 2H_2O(l) \rightarrow Mg(OH)_2 + H_2$$

The solution that remains contains magnesium ions and hydroxide ions and it has a very slightly alkaline pH (just above 7).

Magnesium reacts vigorously with water vapour at temperatures above 100 °C (373 K) in the absence of air. This is often carried out using the apparatus shown in Figure 7.1.

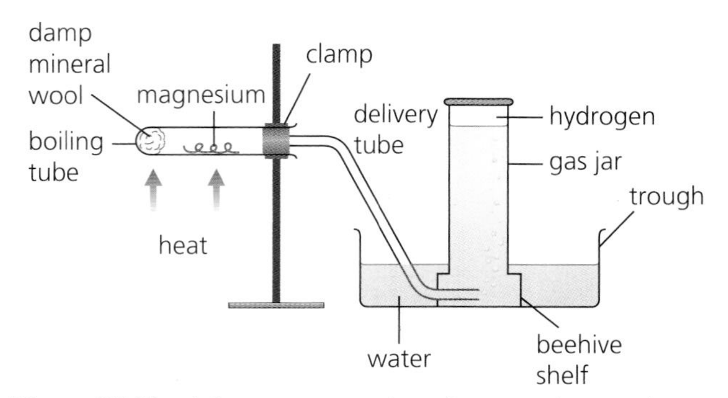

Figure 7.1 The laboratory reaction of magnesium and steam.

Heating the damp mineral wool generates water vapour and drives the air out the boiling tube.

Observations: The magnesium burns with a bright white light and a white solid is formed.

$$Mg + H_2O(g) \rightarrow MgO + H_2$$

The white solid formed is magnesium oxide.

Oxides of elements in Period 3

The oxides of elements in Period 3 will be examined in terms of their formula, type of bonding, structure, acidic, basic or amphoteric nature of the oxide and the reactions of the oxides with water.

Metal oxides

Sodium oxide

- Sodium oxide is an ionic, basic oxide.
- Its formula is Na_2O and its bonding is ionic. Ionic bonding is the electrostatic attraction between the positive and negative ions.
- It is a white solid which has an ionic lattice structure.

> **TIP**
> You should know how to draw a representation of Figure 7.2 with a limited number of ions as discussed in the bonding chapter in the Year 1 AS chemistry book.

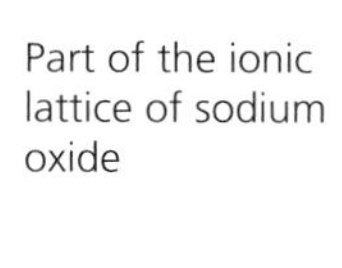

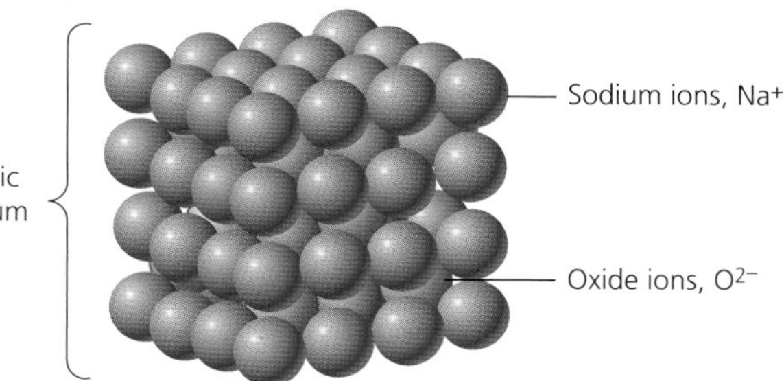

Figure 7.2 The ionic lattice structure of sodium oxide.

> **TIP**
> Lithium oxide melts at 1438 °C or 1711 K. This is due to the smaller Li^+ ion creating greater electrostatic attractions between the ions in lithium oxide.

Sodium oxide has a high melting point (1275 °C or 1548 K) as it consists of an ionic lattice and there are strong forces of attraction between the ions which require a large amount of energy to break.

Formation of sodium oxide

Sodium oxide can be formed from the reaction between sodium and oxygen.

$$4Na + O_2 \rightarrow 2Na_2O$$

A yellow flame is observed and a white solid is formed.

Figure 7.3 Combustion of sodium in a gas jar. This is an exothermic reaction as seen by the bright flame.

Sodium metal reacts spontaneously when exposed to air. Sodium, like the other alkali metals, is stored under oil to prevent it reacting with air. The alkali metals are soft metals which can be cut with a knife exposing a shiny surface. The shiny surface tarnishes (goes dull) rapidly.

During this reaction there will also be some sodium peroxide formed, Na_2O_2 when sodium reacts with oxygen. Sodium peroxide contains the peroxide ion, O_2^{2-}. It is a highly reactive compound.

> **TIP**
> You would not be expected to give the equation for the formation of the peroxide when asked for an equation for sodium reacting with oxygen in air but you should be aware that some forms.

Reactions of sodium oxide

Sodium oxide reacts with water to produce a colourless solution of sodium hydroxide which is alkaline (pH 12–14).

$$Na_2O + H_2O \rightarrow 2NaOH$$

This can be written to show the ions formed:

$$Na_2O + H_2O \rightarrow 2Na^+ + OH^-$$

Sodium oxide is basic so it also reacts with acids producing a salt and water.

$$Na_2O + 2HCl \rightarrow 2NaCl + H_2O$$

$$Na_2O + H_2SO_4 \rightarrow Na_2SO_4 + H_2O$$

Sodium oxide dissolves in water and the oxide ion acts as a base and accepts H^+ ions from the water or the acid:

with water: $O^{2-} + H_2O \rightarrow 2OH^-$

with acids: $O^{2-} + 2H^+ \rightarrow H_2O$

Magnesium oxide

- Magnesium oxide is an ionic, basic oxide.
- Its formula is MgO and its bonding is ionic. Ionic bonding is the electrostatic attraction between the positive and negative ions
- It is a white solid which has an ionic lattice structure.

Figure 7.4 Magnesium oxide is a white solid.

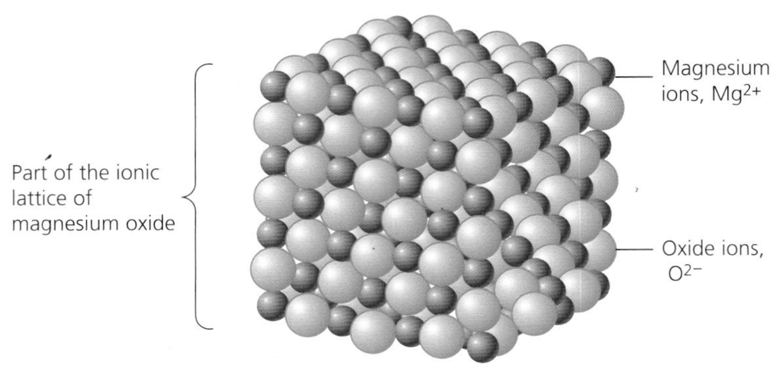

Figure 7.5 Ionic lattice structure of magnesium oxide.

Magnesium oxide has a high melting point (2852 °C or 3125 K) as it consists of an ionic lattice and there are strong forces of attraction between the oppositely charged ions which require a large amount of energy to break.

The melting point of magnesium oxide is substantially higher than that of sodium oxide as the Mg^{2+} ion is smaller and has a higher 2+ charge than the single positive charge on the Na^+ ion. The electrostatic forces of attraction between the Mg^{2+} ions and the O^{2-} ions are stronger than the attraction between the Na^+ ions and O^{2-} ions in sodium oxide.

Formation of magnesium oxide

Magnesium oxide can be formed from the reaction between magnesium and oxygen.

$$2Mg + O_2 \rightarrow 2MgO$$

Magnesium burns in air. A white light is observed and a white solid is formed.

Reactions of magnesium oxide

Magnesium oxide is sparingly soluble in water but some does react to form magnesium hydroxide

$$MgO + H_2O \rightarrow Mg(OH)_2$$

The pH of a solution of magnesium oxide is slightly alkaline, producing a solution of around pH 9 due to some of the magnesium oxide reacting with water. The slightly alkaline pH is due to the fact that magnesium oxide is slightly soluble so only some of the oxide ions dissolve and react with the water producing hydroxide ions. The enthalpy of solution of magnesium oxide is not quoted in data books as it is so sparingly soluble in water.

Magnesium oxide is basic so it reacts with acids producing a salt and water.

$$MgO + 2HCl \rightarrow MgCl_2 + H_2O$$

$$MgO + 2HNO_3 \rightarrow Mg(NO_3)_2 + H_2O$$

The oxide ion accepts protons from the acid to form water:

$$O^{2-} + 2H^+ \rightarrow H_2O$$

Aluminium oxide

- Aluminium oxide is an ionic, amphoteric oxide.
- Its formula is Al_2O_3 and its bonding is ionic.
- Ionic bonding is the electrostatic attraction between the positive and negative ions.
- It is a white solid which has an ionic lattice structure similar to those shown previously for sodium oxide and magnesium oxide.

Aluminium oxide melts at 2072 °C (2345 K). The high melting point is caused by the strong electrostatic attraction between the Al^{3+} and O^{2-} ions.

Formation of aluminium oxide

Aluminium oxide can be formed from the reaction between powdered aluminium and oxygen.

$$4Al + 3O_2 \rightarrow 2Al_2O_3$$

A white solid is formed. Aluminium powder reacts readily but aluminium foil does not react as easily as it has a protective oxide layer that prevents reaction. This layer also protects aluminium from reaction with water so preventing corrosion.

TIP

If the general equations are learnt, then it is simply a matter of including the negative ion for the acid (e.g. Cl^-) to complete the equation, or including the positive metal ion for the hydroxide (e.g. Na^+) to complete the base equation.

Reactions of aluminium oxide

Aluminium oxide does not dissolve in water or react with water. This is due to the strength of the ionic bonds between the oppositely charged, small ions.

A mixture of aluminium oxide and water has a pH of 7 as it does not react with water.

Aluminium oxide is amphoteric as it reacts with both acids and bases:

General equations:

With acid: $Al_2O_3 + 6H^+ \rightarrow 2Al^{3+} + 3H_2O$

With base: $Al_2O_3 + 2OH^- + 3H_2O \rightarrow 2Al(OH)_4^-$

When aluminium oxide reacts with a base, the aluminate ion, $Al(OH)_4^-$, is formed.

TIP

The basic properties of the metallic oxides of Period 3 elements is due to the presence of the oxide ion, O^{2-}, which can act as a base and accept a proton, H^+.

Overall equations:

with hydrochloric acid: $Al_2O_3 + 6HCl \rightarrow 2AlCl_3 + 3H_2O$

with sodium hydroxide: $Al_2O_3 + 2NaOH + 3H_2O \rightarrow 2NaAl(OH)_4$

TEST YOURSELF 2

1 State the structure of sodium oxide.
2 State the type of bonding found in magnesium oxide.
3 Sodium oxide reacts with water.
   a) Write an equation for the reaction showing the ions formed.
   b) State the pH of the solution formed when sodium oxide reacts with water.
4 Write an equation for the formation of aluminium oxide from aluminium and oxygen.

5 Aluminium oxide reacts with both acids and bases.
a) Write an equation for the reaction of aluminium oxide with sulfuric acid.
b) Write an equation for the reaction of aluminium oxide with potassium hydroxide.
c) What term is used for an oxide which reacts with both acids and bases?

Non-metal oxides

Silicon dioxide

- Silicon dioxide is a covalent, acidic oxide.
- Its formula is SiO_2 and its bonding is covalent.
- Silicon dioxide is a white powder and its structure is macromolecular and the bonding within it is covalent.
- The structural term macromolecular can sometimes be called giant covalent.

Macromolecular (giant covalent) structure of SiO_2
The arrangement of the atoms is tetrahedral. Each silicon atom is connected to 4 oxygen atoms by covalent bonds and each oxygen atom is connected to two silicon atoms again by covalent bonds

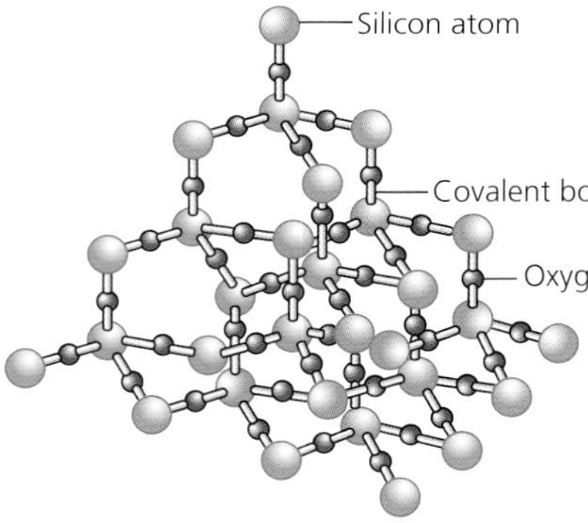

Silicon atom
Covalent bond
Oxygen atom

Figure 7.6 Macromolecular structure of silicon dioxide.

The melting point of silicon dioxide is 1610 °C (1883 K). The high melting point is caused by the macromolecular (giant covalent) structure with many covalent bonds which require a great deal of energy to break them.

Formation of silicon dioxide

Finely divided silicon reacts with oxygen when heated. Silicon dioxide is formed.

$$Si + O_2 \rightarrow SiO_2$$

Reactions of silicon dioxide

Silicon dioxide does not dissolve in or react with water as the water cannot supply enough energy to break the strong covalent bonds in the macromolecular structure. The pH of a mixture of silicon dioxide and water is 7. Silicon dioxide is still referred to as an acidic oxide as it reacts with bases.

Silicon dioxide is an acidic oxide that will react with bases forming silicates. The silicate ion is SiO_3^{2-}. Silicon dioxide reacts with hot concentrated sodium hydroxide sodium forming sodium silicate and water. The reaction is slow at room temperature but alkalis will cause marks on glass containers.

$$SiO_2 + 2NaOH \rightarrow Na_2SiO_3 + H_2O$$

TIP
Remember again that all you have to do is add the metal ion into both sides of the general equation to write an overall equation.

Ionic equation:

$$SiO_2 + 2OH^- \rightarrow SiO_3^{2-} + H_2O$$

Reaction with sodium hydroxide

$$SiO_2 + 2NaOH \rightarrow Na_2SiO_3 + H_2O$$

Na_2SiO_3 is sodium silicate.

Phosphorus(V) oxide

- Phosphorus(V) oxide (phosphorus pentoxide) is a covalent acidic oxide and is a white solid.
- Its molecular formula is P_4O_{10} and its bonding is covalent. The empirical formula is P_2O_5 but each molecule has 4 P atoms and 10 O atoms so it is usually written as P_4O_{10}.
- Its structure is molecular covalent. It consists of simple molecules of P_4O_{10}. The structure can be called simple covalent.

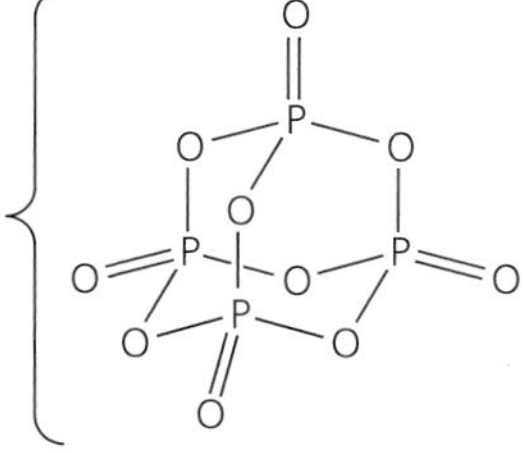

Figure 7.7 Structure of phosphorus(v) oxide.

The melting point of P_4O_{10} is 300 °C (573 K). This melting point is lower than the metallic oxides as P_4O_{10} is molecular covalent so considerably less energy is required to break the weak intermolecular forces of attraction between the P_4O_{10} molecules.

Formation of phosphorus(V) oxide

Phosphorus burns spontaneously in air with a very bright white flame and forms a white smoke which is a mixture of phosphorus(III) oxide, P_4O_6 and phosphorus(V) oxide, P_4O_{10}. In excess oxygen the product is almost all phosphorus(V) oxide.

$$P_4 + 5O_2 \rightarrow P_4O_{10}$$

Phosphorus is normally stored under water to prevent it coming into contact with oxygen in the air.

Figure 7.8 Phosphorus oxide is a white powder. The picture shows the formation of phosphorus oxide from the combustion of phosphorus in a flask. The white solid is seen around the top half of the flask.

Reactions of phosphorus(V) oxide

P_4O_{10} reacts with water producing phosphoric(V) acid, H_3PO_4

$$P_4O_{10} + 6H_2O \rightarrow 4H_3PO_4$$

Phosphorus(V) oxide is an acidic oxide so will react with bases forming phosphate salts. The phosphate ion is PO_4^{3-}.

General equation:

$$12OH^- + P_4O_{10} \rightarrow 4PO_4^{3-} + 6H_2O$$

Overall equation (add the metal ion to both sides of the general equation):

$$12NaOH + P_4O_{10} \rightarrow 4Na_3PO_4 + 6H_2O$$

Basic oxides such as sodium oxide and magnesium oxide also react with P_4O_{10} forming phosphate salts. Na_3PO_4 is sodium phosphate(v) and $Mg_3(PO_4)_2$ is magnesium phosphate(v). They are often simply called sodium phosphate and magnesium phosphate.

$$6Na_2O + P_4O_{10} \rightarrow 4Na_3PO_4$$

$$6MgO + P_4O_{10} \rightarrow 2Mg_3(PO_4)_2$$

Basic oxides also react with phosphoric(V) acid, H_3PO_4. The same phosphate(V) salt is formed but water is also formed.

$$3Na_2O + 2H_3PO_3 \rightarrow 2Na_3PO_4 + 6H_2O$$

$$3MgO + 2H_3PO_4 \rightarrow Mg_3(PO_4)_2 + 3H_2O$$

ACTIVITY

1 Write a balanced symbol equation for the combustion of white phosphorus.

2 What is observed when white phosphorus burns?

White phosphorus matches were replaced in the early 1900s by 'safety matches'. The heads of these matches contained a mixture of phosphorus sesquisulfide P_4S_3 and potassium chlorate(V) $KClO_3$. When the match is struck across a rough surface the heat of friction is sufficient to ignite the phosphorus sesquisulfide; the potassium chlorate(V) decomposes to provide the oxygen needed for combustion.

3 Write a balanced symbol equation for the combustion of phosphorus sesquisulfide to produce phosphorus(V) oxide and sulfur dioxide.

4 Write a balanced symbol equation for the decomposition of potassium chlorate(V) into potassium chloride and oxygen.

5 Using your answers to question 4 and 5, write a single equation to show the overall reaction that occurs between these two substances when a match head ignites.

6 The phosphorus(V) oxide produced when a safety match head ignites is covalently bonded. Magnesium oxide, another Period 3 oxide is an ionically bonded compound. Outline a simple experiment that could be used to demonstrate that magnesium oxide has ionic bonding.

TIP

Some texts describe sulfur trioxide as a colourless solid but this may relate more to the temperature of the environment, given that its melting point is close to room temperature.

Oxides of sulfur

- Sulfur forms two oxides, sulfur(IV) oxide (sulfur dioxide) and sulfur(VI) oxide (sulfur trioxide). Both are covalent acidic oxides.
- Sulfur dioxide has the formula SO_2 and sulfur trioxide has the formula SO_3. The bonding in both is covalent. The structure of both SO_2 and SO_3 is molecular (simple covalent).
- Sulfur dioxide is a colourless gas (melting point −73 °C or 200 K) with a pungent odour. It often appears as misty fumes in most air due to its reaction with the moisture in the air.
- Sulfur trioxide is a colourless liquid (melting point 17 °C or 290 K).

The structure of both sulfur dioxide and sulfur trioxide is described as molecular covalent. The SO_2 molecule is bent and the SO_3 molecule is trigonal planar.

TIP

The bond angle for SO_2 is 119°. The two double bonds and lone pair take up a trigonal planar arrangement and the lone pair reduces the bond angle from 120° to 119°. The bond angle for SO_3 is 120° due to the equal repulsions of the three double bonds. Sulfur dioxide is polar but sulfur trioxide is non-polar because of the symmetry of its shape.

bent shape of SO_2

trigonal planar shape of SO_3

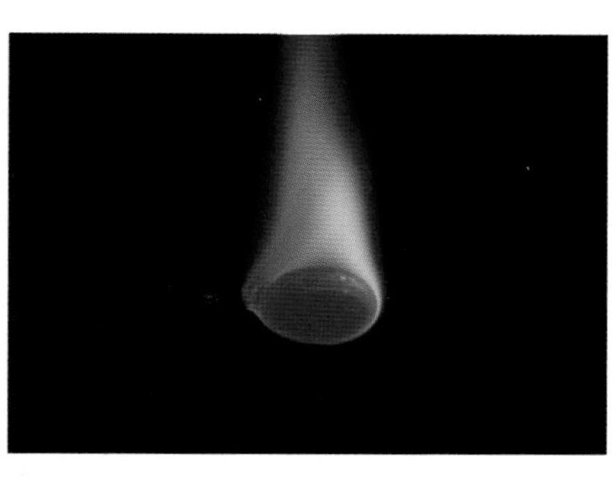

Figure 7.9 Sulfur burning in air, on a combustion spoon. It has an intense blue flame.

Formation of the oxides of sulfur

Sulfur burns in air. The yellow solid sulfur melts to form a red liquid which burns with a blue flame forming misty fumes of a choking and pungent gas.

$$S + O_2 \rightarrow SO_2$$

Sulfur dioxide can be converted to sulfur trioxide on reaction with more oxygen in the presence of a vanadium(V) oxide catalyst under specific conditions.

$$2SO_2 + O_2 \rightarrow 2SO_3$$

Reactions of the oxides of sulfur

Sulfur dioxide reacts with water producing sulfuric(IV) acid, H_2SO_3. Sulfuric(IV) acid is also called sulfurous acid.

$$SO_2 + H_2O \rightarrow H_2SO_3$$

The solution formed when sulfur dioxide reacts with water is weakly acidic. This is due to the reaction being reversible and the position of equilibrium being to the left-hand side.

$$SO_2 + H_2O \rightleftharpoons H^+ + HSO_3^-$$

Sulfur trioxide, SO_3, reacts very vigorously with water producing sulfuric acid, H_2SO_4. Sulfuric acid is also called sulfuric(VI) acid.

$$SO_3 + H_2O \rightarrow H_2SO_4$$

As both sulfur dioxide and sulfur trioxide are acidic oxides, they react with bases. Sulfur dioxide reacts with bases forming sulfate(IV) ions. The sulfate(IV) ion is SO_3^{2-}. Sulfur trioxide reacts with bases forming sulfate(VI) ions. The sulfate(VI) ion is SO_4^{2-}.

The sulfate(IV) ion is also called the sulfite ion and the sulfate(VI) ion is called the sulfate ion.

General equations:

$$SO_2 + 2OH^- \rightarrow SO_3^{2-} + H_2O$$

$$SO_3 + 2OH^- \rightarrow SO_4^{2-} + H_2O$$

Overall equations (adding the metal ion to the left and the right will complete the equation and give you the balanced symbol equation):

$$SO_2 + 2NaOH \rightarrow Na_2SO_3 + H_2O$$

$$SO_3 + 2KOH \rightarrow K_2SO_4 + H_2O$$

P_4O_{10} melts at 300 °C (573 K) whereas SO_2 melts at −73 °C (200 K). SO_3 melts at 17 °C (290 K). All three of these substances are molecular covalent. P_4O_{10} has a higher melting point than the others as it is a larger molecule with more electrons so the van der Waals' forces between the molecules are greater and more energy is required to overcome these forces of attraction between the molecules. SO_3 is a larger molecule than SO_2 so the melting point is higher, again due to larger van der Waals' forces between the molecules of SO_3 compared to SO_2, even though SO_2 is polar and SO_3 is non-polar.

TIP

Examples relating melting point to the structure are common in this section. Make sure you can explain it fully based on either an ionic lattice, a macromolecular (giant covalent) structure or the forces of attraction between molecules for molecular (simple) covalent substances.

Summary table of the reactions of oxides of elements in Period 3 with water

Oxide	Reaction with water	pH of resulting solution
Na_2O	$Na_2O + H_2O \longrightarrow 2Na^+ + OH^-$	14
MgO	$MgO + H_2O \longrightarrow Mg(OH)_2$	9
Al_2O_3	None	–
SiO_2	None	–
P_4O_{10}	$P_4O_{10} + 6H_2O \longrightarrow 4H_3PO_4$	0
SO_2	$SO_2 + H_2O \rightleftharpoons H^+ + HSO_3^-$	3
SO_3	$SO_3 + H_2O \rightarrow H_2SO_4$	0

Trends in the melting points of the Period 3 oxides

The table below gives the melting points of the highest oxidation state oxides of elements of Period 3.

Oxide	Melting point (°C)
Na_2O	1438
MgO	2852
Al_2O_3	2072
SiO_2	1610
P_4O_{10}	300
SO_3	17

For the metal oxides (Na_2O, MgO and Al_2O_3) it is important to remember that, in general, the melting point increases as the charge on the metal ions and as the size of the metal ion decreases. Ionic compounds with smaller ions and higher charge ions have the highest melting points. This explains the increase in melting point from sodium oxide to magnesium oxide. However, the melting point of aluminium oxide is lower than that of magnesium oxide. This is due to the fact that the aluminium ion has such a high charge density that it polarises the electrons in the oxide ion and aluminium oxide has a degree of covalent character. This reduces the melting point.

Silicon dioxide has a macromolecular (giant covalent structure) and so a large amount of energy is required to break the many strong covalent bonds to melt it.

P_4O_{10} and SO_3 are molecular in structure. P_4O_{10} has a higher M_r than SO_3 so it has stronger van der Waals' forces of attraction between the molecules.

Structure of the acids and anions

When acidic oxides react with water they form oxyacids. These oxyacids are covalent compounds which form ions in aqueous solution.

Phosphoric acid

Food-grade phosphoric acid is used to acidify foods and beverages such as various colas. It provides a tangy and sour taste.

- When P_4O_{10} reacts with water, phosphoric(v) acid, H_3PO_4, is formed. The (v) refers to the oxidation state of the phosphorus atom which is +5 in H_3PO_4.

phosphoric(V) acid

- The structure of phosphoric(v) acid is shown on the left.
- Phosphoric(v) acid is often simply called phosphoric acid.
- Phosphorus atoms in H_3PO_4 promote electrons into the 3d sub-level to be able to form five covalent bonds.
- The shape around the phosphorus atom is *tetrahedral (bond angle is 109.5°)* (see structure, left).

The three hydrogen atoms are acidic hydrogen atoms as they are bonded to electronegative oxygen atoms and so can be donated as H^+ ions.

Phosphoric(v) acid is a triprotic acid. It forms three different anions as the H^+ leave the molecule on reaction with bases.

$$NaOH + H_3PO_4 \rightarrow NaH_2PO_4 + H_2O$$

NaH_2PO_4 is sodium dihydrogenphosphate(v). The $H_2PO_4^{-}$ ion is the dihyrogenphosphate(v) ion.

TIP

All of the anions formed from phosphoric acid may have the oxidation state left out after the name of the salt. For example the phosphate(v) ion is often called the phosphate ion and the dihydrogenphosphate(v) ion is often called the dihydrogenphosphate ion.

$$2NaOH + H_3PO_4 \rightarrow Na_2HPO_4 + H_2O$$

Na_2HPO_4 is sodium hydrogenphosphate(v). HPO_4^{-} is the hydrogenphosphate(v) ion.

$$3NaOH + H_3PO_4 \rightarrow Na_3PO_3 + 3H_2O$$

Na_3PO_4 is sodium phosphate(v). PO_4^{3-} is the phosphate(v) ion.

The structures of the three ions are shown below.

All the ions shown above have a tetrahedral shape with a 109.5° bond angle.

dihydrogenphosphate(V) ion, $H_2PO_4^-$

hydrogenphosphate(V) ion, HPO_4^{2-}

phosphate(V) ion, PO_4^{3-}

Sulfuric(IV) acid

- When SO_2 reacts with water, sulfuric(IV) acid, H_2SO_3, is formed. The (IV) refers to the oxidation state of the sulfur atom in H_2SO_3 which is +4.
- The structure of sulfuric(IV) acid is shown on the left.
- Sulfuric(IV) acid is often called sulfurous acid.
- Sulfur atoms in H_2SO_3 promote one electron into the 3d sub-level to be able to form four covalent bonds. The sulfur atom has a lone pair of electrons.
- The shape around the sulfur atom is *pyramidal (bond angle 107.5°)* (see structure, left).

sulfuric(IV) acid

Sulfuric(IV) acid forms two different salts on removal of one or both of the hydrogen atoms as H^+ ions.

$$NaOH + H_2SO_3 \rightarrow NaHSO_3 + H_2O$$

$NaHSO_3$ is sodium hydrogensulfate(IV). The HSO_3^- ion is the hydrogensulfate(IV) ion, which is often called the hydrogensulfite ion.

$$2NaOH + H_2SO_3 \rightarrow Na_2SO_3 + H_2O$$

Na_2SO_3 is sodium sulfate(IV). The SO_3^{2-} ion is the sulfate(IV) ion, which is often called the sulfite ion.

The structures of the two anions are given below.

TIP
The ions shown have a pyramidal shape with a bond angle of 107.5°. This is a similar shape to an ammonia molecule.

hydrogensulfate(IV) ion, HSO_3^- sulfate(IV) ion, SO_3^{2-}

Sulfuric(VI) acid

- When SO_3 reacts with water, sulfuric(VI) acid, H_2SO_4, is formed. The (VI) refers to the oxidation state of the sulfur atom in H_2SO_4 which is +6.
- The structure of sulfuric(VI) acid is shown on the left.
- Sulfuric(VI) acid is often simply called sulfuric acid.
- Sulfur atoms in H_2SO_4 promote two electrons into the 3d sub-level to be able to form six covalent bonds.
- The shape around the sulfur atom is *tetrahedral (bond angle 109.5°)* (see structure, left).

sulfuric(VI) acid

Sulfuric(VI) acid forms two different salts on removal of one or both of the hydrogen atoms as H^+ ions.

$$NaOH + H_2SO_4 \rightarrow NaHSO_4 + H_2O$$

$NaHSO_4$ is sodium hydrogensulfate(VI). The HSO_4^- ion is the hydrogensulfate(VI) ion, which is often called the hydrogensulfate ion.

$$2NaOH + H_2SO_4 \rightarrow Na_2SO_4 + H_2O$$

Na_2SO_4 is sodium sulfate(VI). The SO_4^{2-} ion is the sulfate(VI) ion, which is often called the sulfate ion.

The structures of the two anions are given below.

TIP
Both anions have a tetrahedral shape with a bond angle of 109.5°.

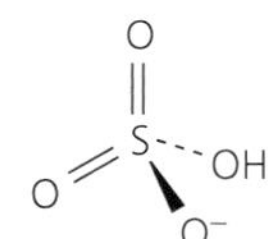
hydrogensulfate(VI) ion, HSO_4^-

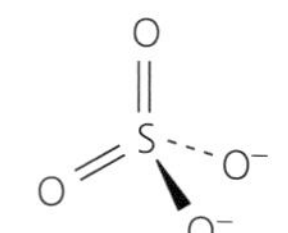
sulfate(VI) ion, SO_4^{2-}

TEST YOURSELF 3

1 Sulfur dioxide and sulfur trioxide both reacts with water.
   a) Write an equation for the reaction of sulfur dioxide with water.
   b) State the pH of the solution formed when sulfur trioxide reacts with water.
   c) Write an equation for the reaction of sulfur dioxide with excess sodium hydroxide.
2 P_4O_{10} is an acidic oxide.
   a) Write an equation for the reaction of P_4O_{10} with water.
   b) State the structure of P_4O_{10}.
3 State what is observed when sulfur burns in air.
4 Write the formula for the following salts of oxyacids.
   a) sodium hydrogensulfate(VI)
   b) magnesium phosphate(V)
   c) potassium hydrogenphosphate(V)
   d) sodium hydrogensulfite
5 Write an equation for the reaction of P_4O_{10} with sodium oxide.

Practice questions

1 Which one of the following oxides reacts with water to form an alkaline solution?

A Al_2O_3 **B** Na_2O

C SiO_2 **D** SO_2 *(1)*

2 What is the formula of iron(III) sulfate(IV)?

A $FeSO_3$ **B** $Fe_2(SO_3)_3$

C $FeSO_4$ **D** $Fe_2(SO_4)_3$ *(1)*

3 Which one of oxides of Period 3 elements reacts with water to form a compound with an element in the +6 oxidation state?

A Aluminium **B** Phosphorus

C Silicon **D** Sulfur *(1)*

4 Which one of the following oxides has a molecular structure?

A Na_2O **B** MgO

C SiO_2 **D** P_4O_{10} *(1)*

5 The melting points of the Period 3 elements are given below.

Element	Sodium	Magnesium	Aluminium	Silicon
Melting point (°C)	98	649	660	1410
State at room temperature and pressure	Solid	Solid	Solid	Solid
Element	**Phosphorus**	**Sulfur**	**Chlorine**	**Argon**
Melting point (°C)	44	114	−101	−189
State at room temperature and pressure	Solid	Solid	Gas	Gas

a) Explain why the melting point of magnesium is greater than the melting point of sodium. *(2)*

b) Explain why silicon has the highest melting point of the Period 3 elements. *(2)*

c) Phosphorus, sulfur and chlorine exist as molecules. Write the formulae of the molecules of each element. *(2)*

d) Explain why the phosphorus and sulfur are solids at room temperature and pressure but chlorine is a gas. *(3)*

6 The melting points of the oxides of the Period 3 are given in the table below.

Oxide	Na_2O	MgO	Al_2O_3	SiO_2	P_4O_{10}	SO_2	SO_3
Melting point (°C)	1275	2852	2072	1610	300	−73	17

a) Explain why the melting points of Na_2O, MgO and Al_2O_3 are high. *(2)*

b) Explain the difference between the melting points of SiO_2 and P_4O_{10}. *(3)*

c) Explain why SO_3 has a higher melting point than SO_2. *(3)*

7 P_4O_{10} is phosphorus(V) oxide.

a) Write an equation for the reaction of P_4O_{10} with water. *(1)*

b) Write an equation for the reaction of P_4O_{10} with magnesium oxide. *(1)*

c) Calculate the mass of P_4O_{10} required to react with 500 cm^3 of water to create a solution of phosphoric acid of concentration 0.15 mol dm^{-3}. *(3)*

8 Sodium oxide and sulfur dioxide have very different melting points.

Sodium oxide melts at 1274 °C whereas sulfur dioxide melts at −73 °C.

a) State the structure of sodium oxide and sulfur dioxide. *(2)*

b) Explain the difference in the melting points of the two substances. *(2)*

c) State the pH of the solution formed if they are dissolved in separate samples of water. *(2)*

d) Write an equation for the reaction of sodium oxide with sulfur dioxide. *(1)*

9 The structures of two oxyacids of elements in Period 3 are shown below:

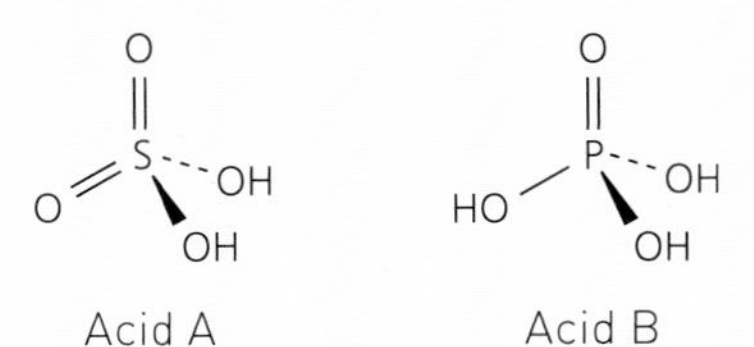

a) Give the oxidation state of sulfur in Acid A. *(1)*

b) Give the oxidation state of phosphorus in Acid B. *(1)*

c) Give the names of both acids. *(2)*

d) Write an equation for the formation of acid A from an oxide of an element in Period 3. *(1)*

e) Write an equation for the reaction of Acid B with calcium oxide. *(1)*

f) Explain why both acids cannot be oxidised further. *(1)*

g) Write the formulae including the charge of two anions formed from Acid A. Name these anions. *(2)*

Stretch and challenge

10 Sodium reacts with water according to the equation:

$$2Na + 2H_2O \rightarrow 2NaOH + H_2$$

Calculate the pH of the solution formed when a sample of 0.0300 g of sodium is added to 1.00 dm^3 of water. ($K_w = 1.00 \times 10^{-14}\,mol^2\,dm^{-6}$). Give your answer to 2 decimal places. *(4)*

8 Transition metals

PRIOR KNOWLEDGE

- Transitions metals are found in the block between Groups 2 and 3 called the d block as their outer shell electrons are located in the d sub-level.
- The 4s sub-level fills before the 3d sub-level.
- The electronic configuration of chromium and copper are different than expected.
- When transition metal atoms form ions they lose their 4s electrons first.
- Coordinate bonds are formed when the shared pair of electrons is donated from one atom.
- The compounds of transition metals are coloured.
- Transition metals form compounds with the transition metal in different oxidation states.
- Oxidation and reduction reactions between these oxidation states are common and there are changes in colour associated with these changes.
- Transition metal compounds can be used in organic chemistry for oxidation reactions.
- Know the shapes of molecules including octahedral, tetrahedral and linear.

TEST YOURSELF ON PRIOR KNOWLEDGE 1

1 Write the electronic configuration of the following transition metals.

a) Sc
b) Cr
c) Ni
d) Cu

2 Write the electronic configuration of the following ions.

a) Ti^{2+}
b) Fe^{2+}
c) Co^{2+}
d) Cu^{2+}

3 Explain how a coordinate bond forms.

4 State the shape of the following molecules:

a) CH_4
b) $BeCl_2$
c) SF_6

General properties of transition elements

The general properties of transition elements are listed below.

- Their atoms or ions have an incomplete d sub-level.
- They form complexes.
- They have variable oxidation states.
- The metals and their compounds show catalytic activity.
- The form coloured ions.

Figure 8.1 Transition metal ion solutions. From left to right these strongly coloured metal solutions are: titanium(II) (Ti^{2+}), vanadium(III) (V^{3+}), vanadium(IV) (VO^{2+}), chromium(III) (Cr^{3+}), dichromate(VI) ($Cr_2O_7^{2-}$), manganese(II) (Mn^{2+}), manganate(VII) (MnO_4^-), iron(III) (Fe^{3+}), cobalt(II) (Co^{2+}), nickel(II) (Ni^{2+}) and copper(II) (Cu^{2+}).

Incomplete d sub-level

- Transition elements are elements in the middle of the Periodic Table that have atoms or ions with an incomplete d sub-level.
- The first transition series runs from scandium to zinc (elements 21 to 30).
- Scandium and zinc are not true transition elements as they do not have variable oxidation states in their compounds and their ions, Zn^{2+} and Sc^{3+} do not have an incomplete d sub-level. Zn^{2+} is [Ar] $3d^{10}$, Sc^{3+} is [Ar].
- The term d block element is a better description if scandium and zinc are to be included. However as these two elements are not mentioned in this part of the specification which specifically addresses Ti to Cu, it is appropriate to use the term transition elements.
- It is the incomplete d sub-level that gives transition metals the other properties listed above.

Electronic configuration

The table below gives the electronic configuration of the atoms of the transition elements from Ti to Cu. Also given are the electronic configurations of the most common simple ions.

Transition element atom	Electronic configuration	Transition element ion	Electronic configuration
Ti	$1s^2 2s^2 2p^6 3s^2 3p^6 3d^2 4s^2$	Ti^{2+}	$1s^2 2s^2 2p^6 3s^2 3p^6 3d^2$
V	$1s^2 2s^2 2p^6 3s^2 3p^6 3d^3 4s^2$	V^{3+}	$1s^2 2s^2 2p^6 3s^2 3p^6 3d^2$
Cr	$1s^2 2s^2 2p^6 3s^2 3p^6$ $\mathbf{3d^5 4s^1}$	Cr^{3+}	$1s^2 2s^2 2p^6 3s^2 3p^6 3d^3$
Mn	$1s^2 2s^2 2p^6 3s^2 3p^6 3d^5 4s^2$	Mn^{2+}	$1s^2 2s^2 2p^6 3s^2 3p^6 3d^5$
Fe	$1s^2 2s^2 2p^6 3s^2 3p^6 3d^6 4s^2$	Fe^{3+}	$1s^2 2s^2 2p^6 3s^2 3p^6 3d^5$
Co	$1s^2 2s^2 2p^6 3s^2 3p^6 3d^7 4s^2$	Co^{2+}	$1s^2 2s^2 2p^6 3s^2 3p^6 3d^7$
Ni	$1s^2 2s^2 2p^6 3s^2 3p^6 3d^8 4s^2$	Ni^{2+}	$1s^2 2s^2 2p^6 3s^2 3p^6 3d^8$
Cu	$1s^2 2s^2 2p^6 3s^2 3p^6$ $\mathbf{3d^{10} 4s^1}$	Cu^{2+}	$1s^2 2s^2 2p^6 3s^2 3p^6 3d^9$

TIP

It is important to remember that the first series of transition metal atoms always lose their 4s electrons first when forming ions.

Metallic nature

Transition metals are all hard and dense. They are good conductors of electricity and heat and possess good mechanical properties.

Their melting and boiling temperatures and standard enthalpies of melting are higher for most transition metal elements than those of s block elements.

All the above properties give a measure of the strength of the metallic bond. With d electrons as well as s electrons available to take part in delocalisation, the metallic bond is strong in transition metals.

Complex formation

Transition metals form complexes.

Ligands have at least one lone pair of electrons that can form a coordinate bond to a metal atom or ion.

A **complex** is a central metal atom or ion surrounded by ligands

A **ligand** is an ion or molecule which forms coordinate bonds with the transition metal atom or ion by donating a pair of electrons.

In transition metal complexes, lone pairs of electrons on the ligand form coordinate (dative covalent) bonds to the central metal atom or ion.

The lone pairs on the ligand are donated into empty orbitals in the transition metal atom or ion.

Hexaaqua cations

When simple ions of transition metals dissolve in water, they form complex ions such as $[Cu(H_2O)_6]^{2+}$, $[Ni(H_2O)_6]^{2+}$ and $[Fe(H_2O)_6]^{3+}$. These are called hexaaqua cations as they have six water ligands coordinately bonded to the metal ion.

Dissolving nickel(II) sulfate in water would produce a green solution containing the $[Ni(H_2O)_6]^{2+}$ complex ion. The sulfate ion would also be present in the solution.

Compounds of metal ions, other than transition metal ions, such as Mg^{2+}, Zn^{2+} and Al^{3+} can also form hexaaqua cations as long as they have available empty orbitals into which the lone pairs of electrons can be donated.

Mg^{2+}, Zn^{2+} and Al^{3+} ions are not transition metal ions but they do form complexes.

TIP

It is important to note that atoms and ions of the first series of transition metals have empty orbitals (some 3d, 4s and 4p) into which the lone pairs of electrons may be placed to form the coordinate bond from the ligand.

Writing complexes

Complexes are written with the metal atom or ion and the ligands inside square brackets. The charge on the complex ion (if there is one) is placed as a superscript outside the square bracket.

EXAMPLE 1

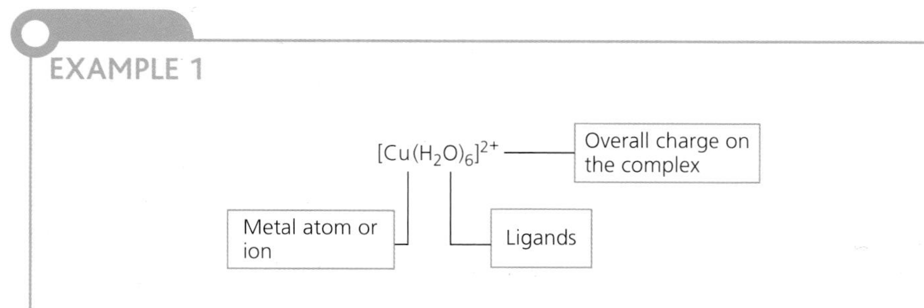

This complex has six water ligands coordinately bonded to the Cu^{2+} ion.

When multiple ligands are present which contain more than one atom, these are placed in a bracket with the number outside the bracket.

Water ligands are neutral (they have no charge).

As the water ligands are neutral, the overall charge of the complex is the same as the charge of the metal ion (2+) or its oxidation number (+2).

EXAMPLE 2

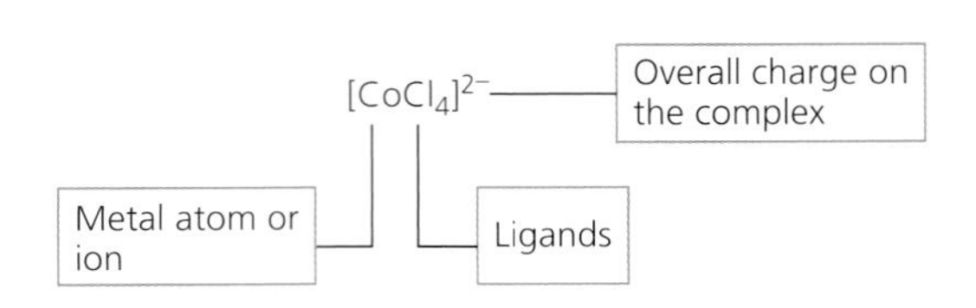

This complex has four chloride ion ligands coordinately bonded to the Co^{2+} ion.

These four ligands do not need a bracket as they do not consist of more than one atom.

Chloride ion ligands have a single negative charge, Cl^-.

The overall charge on the complex is a total of the charge on the metal ion (2+) or its oxidation state (+2) and the total charge of the ligands (4 × 1⁻).

Coordination number

The **coordination number** is the number of coordinate bonds to the metal atom or ion.

Some ligands can form only one coordinate bond to the metal atom or ion.

Some ligands can form more than one coordinate bond to the metal atom or ion at the centre of the complex.

$[Ni(H_2O)_6]^{2+}$ coordination number = 6

$[CuCl_4]^{2-}$ coordination number = 4

$[Fe(H_2O)_6]^{3+}$ coordination number = 6

$[FeEDTA]^-$ coordination number = 6

$[Ag(NH_3)_2]^+$ coordination number = 2

Monodentate ligands

Ligands that form one coordinate bond to a metal atom or ion are called **monodentate ligands**. Examples of monodentate ligands are NH_3, H_2O, Cl^-, OH^-, CN^-.

All the above ligands have at least one lone pair of electrons but even if they have more than one, they are still monodentate as the lone pairs are on the same atom or adjacent atoms. This means that these ligands cannot form more than one coordinate bond to the metal atom or ion as the bonds must be in a certain orientation around the metal ion. This will be discussed when shapes of complexes are considered.

1,2-diaminoethane showing the lone pairs of electrons

ethanedioate ion showing the lone pairs of electrons

Bidentate ligands

Ligands which form two coordinate bonds to a metal atom or ion are called **bidentate ligands**. Examples of a bidentate ligand are 1,2-diaminoethane ($H_2NCH_2CH_2NH_2$) and ethanedioate (oxalate) ions, $C_2O_4^{2-}$.

For a bidentate ligand, the lone pairs of electrons must be on different atoms to allow the ligand to form more than one coordinate bond.

Figure 8.2 Kidney stones, are formed when calcium ethanedioate (calcium oxalate) accumulate on the inner surfaces of the kidney. As they grow they may block the flow of urine out of the kidney, causing extreme pain.

This is important as the coordinate bonds must form in certain orientations around the central ion.

$[Co]^{2+}$

coordinate bonds to the Co^{2+} ion must form in the orientation shown, forming an octahedral shaped complex ion

So even though water, H_2O, has two lone pairs of electrons, both occur on the oxygen atom so it would be impossible for these two lone pairs to form two coordinate bonds to the Co^{2+}, whereas the two lone pairs of 1,2-diaminoethane are separated and so it can acts as a bidentate ligand.

H_2O: Co $^{2+}$

CH_2 CH_2 H_2N H_2N Co $^{2+}$

Even if the coordination number of a complex is 4 or 2, water can only ever act as a monodentate ligand due to the position of the lone pairs of electrons.

Multidentate ligands

Ligands that form many coordinate bonds to a metal atom or ion are called **multidentate ligands**. An example of a multidentate ligand is $EDTA^{4-}$.

EDTA is an abbreviation for **e**thylene**d**iamine**t**etraacetic **a**cid. The structure of EDTA is given below.

```
HOOC–H2C      H   H      CH2–COOH
        \     |   |     /
         N — C — C — N
        /     |   |     \
HOOC–H2C      H   H      CH2–COOH
```

EDTA is usually used in alkaline buffered solution to ensure it is the anion form. All the acid COOH groups lose their hydrogen ions to form the anion $EDTA^{4-}$.

```
⁻:ÖOC–H2C      H   H      CH2–COÖ:⁻
         \     |   |     /
         :N — C — C — N:
         /     |   |     \
⁻:ÖOC–H2C      H   H      CH2–COÖ:⁻
```

The six lone pairs are shown on the diagram. These form coordinate bonds with the central metal ion. This makes EDTA multidentate and hexadentate

Six lone pairs are shown on the diagram. $EDTA^{4-}$ can form six coordinate bonds with the central metal ion and so as well as being described as multidentate, it is also described as **hexadentate**.

Chelates

All multidentate ligands form complexes called **chelates**. The word chelate comes from the Greek word '*chel*', meaning a crab's claw, and refers to the pincer-like manner in which the metal is bound. Multidentate ligands bind more tightly because of the chelate effect. The metal atom or ion is completely surrounded by the chelating ligand such as $EDTA^{4-}$.

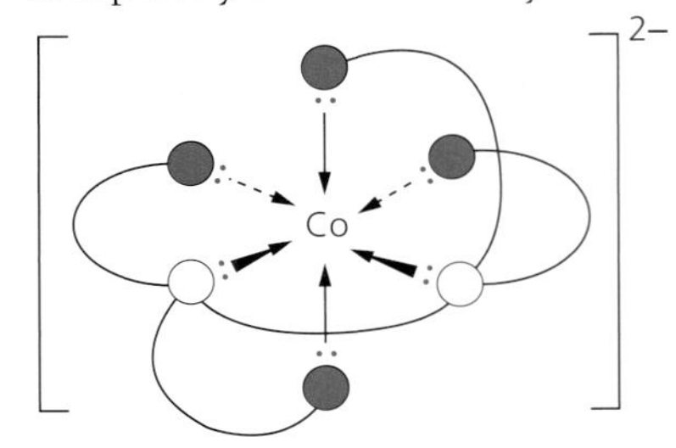

Figure 8.3 A simplified form of the $[CoEDTA]^{2-}$ complex. A chelate as the $EDTA^{4-}$ has surrounded the Co^{2+} ion. Note the overall charge on the complex ion is 2^- as cobalt has an oxidation state of +2 and EDTA has a charge of 4^-. The white circles represent the N atoms of $EDTA^{4-}$ and the green circles represent the O^- of COO^-.

$EDTA^{4-}$ reacts with metal ions in a 1:1 ratio. $EDTA^{4-}$ can be used to titrate metal ions in solution using a suitable indicator such as eriochrome black T for the titration of Mg^{2+} ions and Ca^{2+} ions.

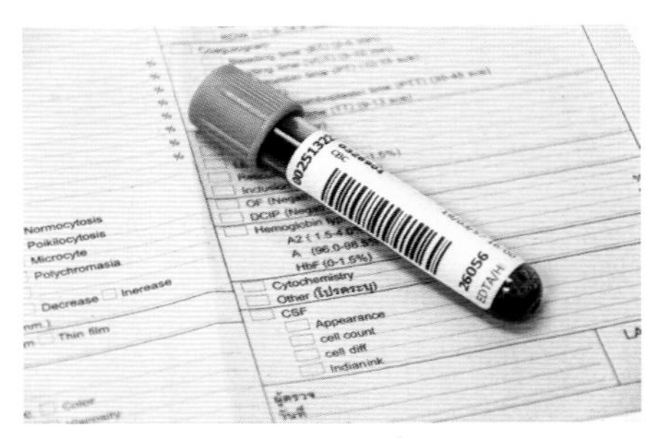

Figure 8.4 Blood collection tubes are pre-sprayed with EDTA on the inside. The $EDTA^{4-}$ removes calcium ions from the blood and so prevents it clotting during tests: $[Ca(H_2O)_6]^{2+} + EDTA^{4-} \rightarrow [CaEDTA]^{2-} + 6H_2O$.

Ligand substitution

A ligand substitution reaction is one in which one ligand which is coordinately bonded to a metal atom or ion in a complex ion is replaced by another ligand.

Ligand substitution with no change in coordination number

In the following examples the coordination number of the complex does not change.

This is common with smaller ligands such as water and ammonia as they have a **similar size** and are both **uncharged.**

It also occurs when bidentate ligands replace monodentate ligand and multidentate ligands replace either monodentate or bidentate ligands.

- Reaction of $[Co(H_2O)_6]^{2+}$ with ammonia.
  Adding ammonia solution to a solution containing $[Co(H_2O)_6]^{2+}$ results in the six water ligands being replaced by six ammonia ligands to give $[Co(NH_3)_6]^{2+}$.
  The equation for this reaction is written as:
  $[Co(H_2O)_6]^{2+} + 6NH_3 \rightleftharpoons [Co(NH_3)_6]^{2+} + 6H_2O$
- Reaction of $[Cu(H_2O)_6]^{2+}$ with ammonia.
  Adding ammonia solution to a solution containing $[Cu(H_2O)_6]^{2+}$ results in four of the six water ligands being replaced by ammonia ligands to give $[Cu(NH_3)_4(H_2O)_2]^{2+}$. The substitution is not complete but the coordination number in both complexes is 6.
  The equation for this reaction is written as:
  $[Cu(H_2O)_6]^{2+} + 4NH_3 \rightleftharpoons [Cu(NH_3)_4(H_2O)_2]^{2+} + 4H_2O$

TIP

The reactions are often written as equilibrium reaction as they can be reversed, but it is also common to see them written with $\rightarrow$ as in many of the reactions the equilibrium is displaced almost completely to the right-hand side.

- Reaction of $[Co(NH_3)_6]^{2+}$ with 1,2-diaminoethane.
  Adding 1,2-diaminoethane to the cobalt-ammonia complex results in the ammonia ligands being replaced by 1,2-diaminoethane.

  $$[Co(NH_3)_6]^{2+} + 3H_2NCH_2CH_2NH_2 \rightleftharpoons [Co(H_2NCH_2CH_2NH_2)_3]^{2+} + 6NH_3$$

  The coordinate number in both complexes is 6. Only three 1,2-diaminoethane ligands are required as it is bidentate and each forms two coordinate bonds to the Co^{2+} ion.
- Reaction of $[Co(H_2NCH_2CH_2NH_2)_3]^{2+}$ with $EDTA^{4-}$
  Adding a solution of $EDTA^{4-}$ to the cobalt-1,2-diaminoethane complex results in another ligand substitution reaction.

  $$[Co(H_2NCH_2CH_2NH_2)_3]^{2+} + EDTA^{4-} \rightleftharpoons [CoEDTA]^{2-} + 3H_2NCH_2CH_2NH_2$$

  Again the coordination number in both complexes is 6. $EDTA^{4-}$ is a multidentate ligand and forms six coordinate bonds to the Co^{2+} ion so only one $EDTA^{4-}$ is needed per Co^{2+} in the complex.

Ligand substitution with a change in the coordination number

Larger charged ligands such as Cl^- may cause a change in the coordination number of the complex when a ligand substitution reaction occurs. Often the coordination number is 4 as only four of these larger ligands will fit around the metal atom or ion. Also the charges on the ligands repel each other.

- Reaction of $[Cu(H_2O)_6]^{2+}$ with concentrated hydrochloric acid.
  Adding concentrated hydrochloric acid to a solution containing $[Cu(H_2O)_6]^{2+}$ results in the water ligands being replaced with Cl^- ligands.
  $[Cu(H_2O)_6]^{2+} + 4Cl^- \rightleftharpoons [CuCl_4]^{2-} + 6H_2O$
  The coordination number of the complex changes from 6 to 4.

TIP
The same reactions occur with $[Co(H_2O)_6]^{2+}$ and $[Fe(H_2O)_6]^{3+}$ with concentrated hydrochloric acid.

Enthalpy and entropy considerations

The enthalpy change and entropy change in a ligand substitution reaction determine whether or not the substitution reaction is feasible, but most often it is the **entropy change** that is used to explain the process.

The **enthalpy change** is a balance of the energy required to break the coordinate bonds between the ligands and the metal atom or ion and the energy released when the bonds are formed between the new ligands and the metal atom or ion.

- Some ligand substitution reactions show no change in the type of coordinate bond or the coordination number of the complex, so the same number and type of coordinate bonds would be formed as are broken. Generally, in these reactions $\Delta H = 0$. For example:

  $$[Co(NH_3)_6]^{2+} + 3H_2NCH_2CH_2NH_2 \rightarrow [Co(H_2NCH_2CH_2NH_2)_3]^{2+} + 6NH_3 \quad \Delta H = 0$$

- There will be change in enthalpy in ligand substitution reactions where there is a change in the coordination number.
- If more coordinate bonds are broken than made, ΔH is positive. If fewer coordinate bonds are broken than made, ΔH is negative. For example:
  $[Cu(H_2O)_6]^{2+} + 4Cl^- \rightarrow [CuCl_4]^{2-} + 6H_2O \qquad \Delta H > 0$
  In this example, six coordinate bonds are broken in $[Cu(H_2O)_6]^{2+}$ and four are made in $[CuCl_4]^{2-}$.

- If fewer coordinate bonds are broken than made, ΔH is negative. For example:
  $[CuCl_4]^{2-} + EDTA^{4-} \rightarrow [CuEDTA]^{2-} + 4Cl^-$ $\Delta H < 0$
  In this example four coordinate bonds are broken in $[CuCl_4]^{2-}$ and six are made in $[CuEDTA]^{2-}$.

The **entropy change** is determined by the number of reacting particles in solution and the number of product particles in solution.

- If the number of particles in solution increases, this would increase the disorder in the solution which would increase the entropy. The entropy change for the reaction would be positive ($\Delta S > 0$).
- The denticity of a ligand has a large effect on the stability of the complexes which it forms.
- A bidentate ligand will form more stable complexes than a monodentate ligand.
- A multidentate ligand, such as the hexadentate ligand $EDTA^{4-}$, will form more stable complexes than a bidentate or monodentate ligand.
- This is often called the **chelate effect** as bidentate and multidentate ligands form chelates with metal ions in solution, and these complexes are more stable than those formed with monodentate ligands.
- Moving from a monodentate ligand to a bidentate or multidentate ligand causes an increase in the number of particles in solution and so an increase in disorder (entropy). The chelate effect can be explained in terms of increasing entropy of the substitution reaction.

TIP

When asked about the chelate effect, always state the total number of particles in solution on the left of the reaction and the right and use this to explain the increase in entropy.

EXAMPLE 3

Explain why the equilibrium is displaced almost completely to the right in the following reaction.

$[Co(NH_3)_6]^{2+} + 3H_2NCH_2CH_2NH_2 \rightleftharpoons [Co(H_2NCH_2CH_2NH_2)_3]^{2+} + 6NH_3$

Answer

Both complexes have a coordination number of 6 so the same type and the same number of coordinate bonds are broken and made. $\Delta H = 0$.

In this example, three molecules of a bidentate ligand replace six molecules of a monodentate ligand. This leads to an increase in the number of particles in the solution, the solution becomes more disordered which leads to higher entropy.

EXAMPLE 4

Explain why the equilibrium is displaced almost completely to the right in the following reaction.

$[Co(H_2NCH_2CH_2NH_2)_3]^{2+} + EDTA^{4-} \rightleftharpoons [CoEDTA]^{2-} + 3H_2NCH_2CH_2NH_2$

Answer

One molecule of a hexadentate ligand replaces three molecules of a bidentate ligand. There is a change in the number of particles in solution from 2 to 4. Again this increases disorder which increases entropy and so the reaction is feasible.

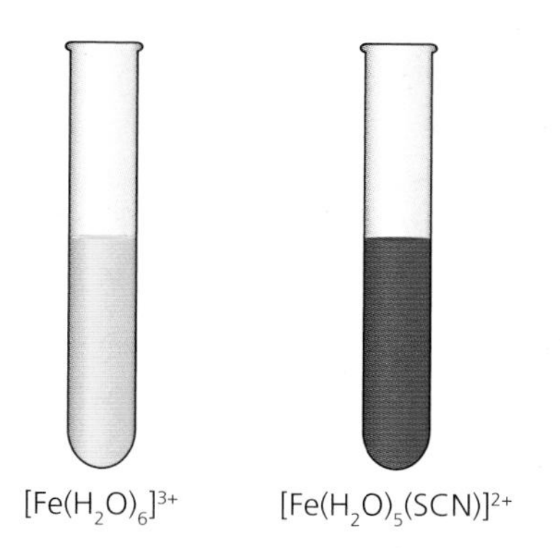

Figure 8.5 A solution of $[Fe(H_2O)_6]^{3+}$ is shown on the left. When thiocyanate ions are added a blood red solution, as shown on the right, is produced. It has formula $[Fe(H_2O)_5SCN]^{2+}$. This ligand substitution reaction is used as a test for iron(III) ions in solution.

Complex ion formation can be detected by elevation of boiling point or depression of freezing point or by a decrease in electrical conductivity of a solution as there are fewer free ions than would be expected.

Charge and oxidation number

The overall charge of a complex ion depends on the oxidation number of the metal and the number and charge of the ligands.

With neutral ligands such as H_2O and NH_3 the oxidation number of the metal is the same as the overall charge of the complex ion.

$[Fe(H_2O)_6]^{3+}$ oxidation number of Fe = +3

$[Co(NH_3)_6]^{2+}$ oxidation number of Co = +2

With anionic ligands, the negative charge of the ligands contributes to the overall charge of the complex ion.

- $[CuCl_4]^{2-}$: as there are four Cl^- ligands; the overall charge is 2– as the oxidation number of the copper is +2.
- $[Fe(H_2O)_5SCN]^{2+}$: the five neutral H_2O ligands have no effect on the overall charge; however, the thiocyano (SCN^-) ligand does. As the overall charge is 2+ and there is one SCN^- ligand, the oxidation number of the Fe = +3
- $[Pt(NH_3)Cl_2]$: the neutral NH_3 ligands have no effect on the overall charge; however, the Cl^- ligands do. As the overall charge of the complex is 0, the oxidation number of the Pt = +2

Haem and haemoglobin

Haem is a complex of an iron(II) ion surrounded by a porphyrin ring. The iron(II) ion can accept 6 pairs of electrons from ligands. The porphyrin ring contains 4 nitrogen atoms which can form 4 coordinate bonded to the iron(II) ion.

The porphyrin ring is a multidentate ligand. An amino acid residue in a protein chain forms the fifth coordinate bond to the iron(II) ion.

Figure 8.6 Haem, which is a complex of Fe^{2+} and a porphyrin ring.

Haemoglobin is a protein structure containing four haem groups bonded to four globular proteins, called α and β units.

Each Fe^{2+}-containing haem complex can accept a pair of electrons and form a coordinate bond from an oxygen molecule. This complex formed is called oxyhaemoglobin. This is how haemoglobin transports oxygen in the blood.

Carbon monoxide also contains a lone pair of electrons on the oxygen atom, and this can form a very stable complex with haemoglobin. Hence when incomplete combustion of fuels occurs and carbon monoxide is produced, people can be poisoned as the stable complex, carboxyhaemoglobin, prevents oxygen being carried in the blood. This is the cause of death in almost all house fires.

TEST YOURSELF 2

1 From the following complexes:
   A $[Cu(H_2O)_6]^{2+}$
   B $[CoCl_4]^{2-}$
   C $[Ni(H_2NCH_2CH_2NH_2)_3]^{2+}$
   D $[Ag(NH_3)_2]^{+}$
   a) In which complex does the transition metal have an oxidation state of +1?
   b) Which complex has a coordination number of 4?
   c) Which complex contains a bidentate ligand?
2 Which ion is present in haemoglobin?
3 In the following ligand substitution reaction:
   $[Co(H_2O)_6]^{2+} + 4Cl^- \rightarrow [CoCl_4]^{2-} + 6H_2O$
   a) State the coordination number of the two complexes in this reaction.
   b) Explain, in terms of entropy, why this ligand substitution reaction occurs.

Shapes of complex ions

The shape of complex ions is based mainly on the coordination number. You should be familiar with working out the shapes of simple compounds using electron pair repulsion theory. Complex ions have bonding pairs of electrons around a central metal atom or ion and these repel each other and create different shapes depending on the ligands and the coordination number of the complex.

TIP
The shapes of complexes can be remembered using LOST. These are the initial letters of the shapes: **L**inear, **O**ctahedral, **S**quare planar, **T**etrahedral.

6-coordinated complex ions

These are complex ions in which there are six coordinate bonds to the metal atom or ion from ligands. These ions have an **octahedral** shape.

Four of the ligands are in one plane, with the fifth one above the plane, and the sixth one below the plane. It does not matter what the ligands are. If you have six coordinate bonds, the shape will be octahedral. The bond angle at the central metal atom or ion in an octahedral complex is 90°.

The diagram shows three examples of octahedral ions.

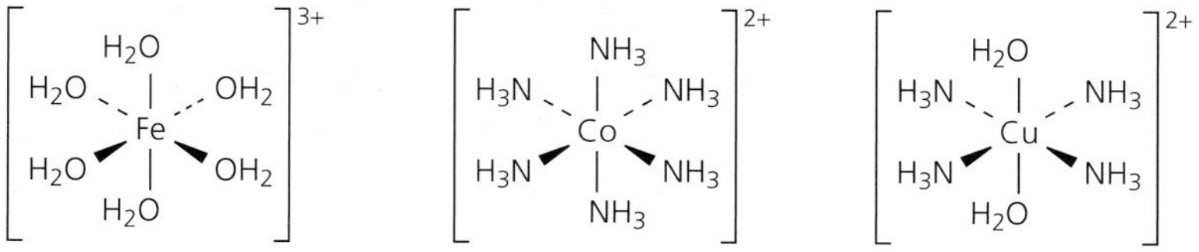

$[Fe(H_2O)_6]^{2+}$ $[Co(NH_3)_6]^{2+}$ $[Cu(NH_3)_4(H_2O)_2]^{2+}$

TIP
Remember that the ligands attached to a wedge-shaped bond are coming out of the screen or paper towards you. Those attached to a dashed bond are behind the plane of the screen or paper. The two ligands attached to the solid bonds are above and below the plane of the rest.

The 6-coordinated copper(II) complexes are actually distorted octahedral complexes as the top and bottom ligands are further away than the others. These ligands are often not replaced when copper complexes undergo ligand replacements reactions.

4-Coordinated complex ions

These are less common, and they can take up one of two different shapes.

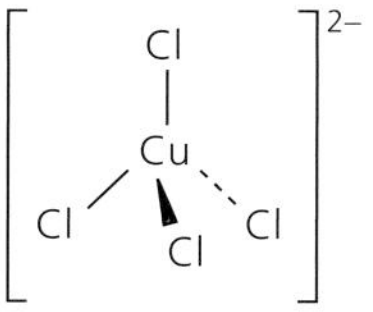

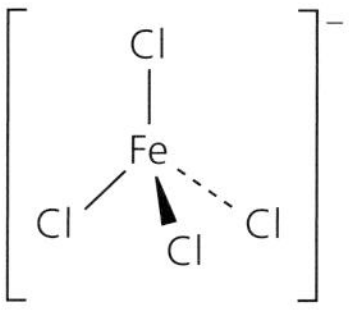

Tetrahedral complex

The $[CuCl_4]^{2-}$, $[CoCl_4]^{2-}$ and $[FeCl_4]^{-}$ complexes are tetrahedral in shape. The bond angle at the central metal atom or ion in 109.5°.

The above two complexes may be formed by adding concentrated hydrochloric acid to $[Cu(H_2O)_6]^{2+}$, $[Co(H_2O)_6]^{2+}$ and $[Fe(H_2O)_6]^{3+}$.

$[CuCl_4]^{2-}$ is yellow/green and $[FeCl_4]^{-}$ is orange.

Square planar complex

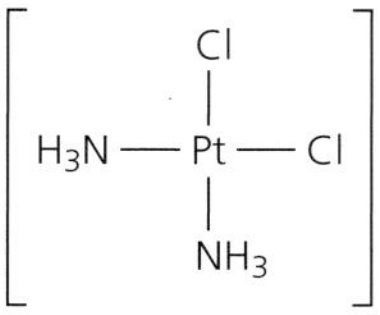

Occasionally a 4-coordinated complex turns out to be square planar. These are fewer in number than the tetrahedral 4-coordinated complexes. Usually a 4-coordinated complex that has no overall charge is square planar, whereas those with a charge tend to be tetrahedral. Also, platinum tends to form square planar complexes. The bond angle in a square planar complex is 90°.

$[Pt(NH_3)_2Cl_2]$

Cisplatin is such a square planar complex ion and is used as an anti-cancer drug.

Cisplatin is a neutral complex, $[Pt(NH_3)_2Cl_2]$. It is neutral because the 2+ charge of the platinum(II) ion is exactly cancelled by the two negative charges supplied by the chloride ions.

TIP
If asked to draw a displayed structure of cisplatin, the coordinate bonds between the N and H atoms in the NH_3 ligand must be shown.

The platinum, the two chlorides and the two nitrogen atoms are all in the same plane.

2-coordinated complex ions

The most common 2-coordinated complex is the one formed between silver ions, Ag^+, and ammonia, $[Ag(NH_3)_2]^+$.

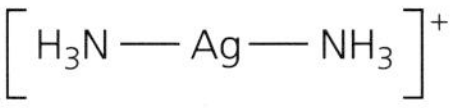

$[H_3N—Ag—NH_3]^+$

It is a linear complex. The bond angle is 180°.

This is the complex used in organic chemistry as Tollens' reagent to distinguish between aldehydes and ketones. It is also the complex formed when the white precipitate of silver chloride redissolves in dilute ammonia solution and the cream precipitate of silver bromide redissolves in concentrated ammonia solution.

Cl
H_3N — Pt — Cl
NH_3

the 'cis' form

Z isomer

NH_3
Cl — Pt — Cl
NH_3

the 'trans' form

E isomer

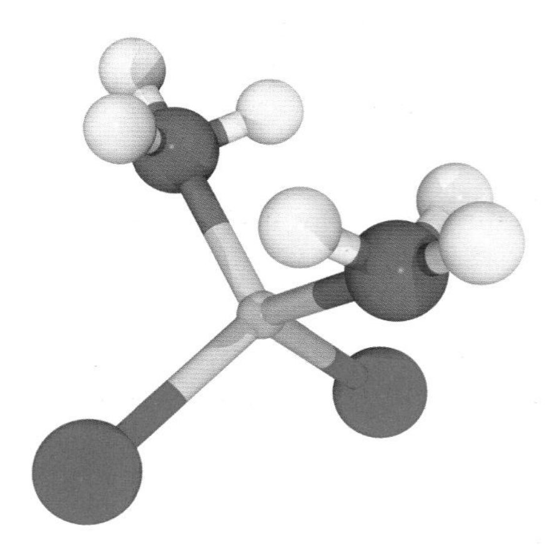

Figure 8.7 The diagram shows a molecular model of cisplatin. Atoms are represented as spheres and are colour-coded: platinum (grey), hydrogen (white), nitrogen (blue) and chlorine (green). Which isomer is shown?

O C O O O H_3N H_3N Pt Cl Cl OH

Figure 8.8 Clinical use of cisplatin is limited as it may develop resistance, earning it the nickname the 'penicillin of cancer'. Anchoring aspirin onto cisplatin could create a cancer treatment capable of overcoming drug resistance in cisplatin resistant cells. This new drug is called Asplatin.

TIP
Trans isomers are sometimes called E isomers and *cis* isomers are sometimes called Z isomers.

Stereoisomerism in complex ions

Some complex ions can show either optical or geometric (E-Z) isomerism.

Tetrahedral complexes can show E-Z isomerism.

Octahedral complexes can show E-Z isomerism with monodentate ligands or optical isomerism with bidentate ligands.

E-Z isomerism in square planar complexes

This occurs in planar complexes like the $Pt(NH_3)_2Cl_2$ as shown on the left. There are two different ways in which the ammonia molecules and chloride ions could arrange themselves around the central platinum ion:

The two structures drawn are isomers because they are non-superimposable. This is the key feature with stereoisomers. There are two forms which are non-superimposable due to the different three-dimensional arrangement of the atoms.

The terms *cis* and *trans* are used in the same way as they are in organic chemistry.

Trans implies 'opposite' – notice that the ammonia molecules are arranged opposite each other in that version, and so are the chloride ions.

Cis implies 'on the same side' – in this instance that just means that the ammonia molecules and the chloride ions are next to each other.

It is better to use the E-Z system of naming these isomers. E (entgegen, meaning across) is equivalent to *trans* and Z (zusammen, meaning together) is equivalent to *cis*. The E-Z system includes many examples which cannot be named using *cis* and *trans*.

Only the Z or *cis* isomer of the complex shows any anti-cancer activity. The E or *trans* form is inactive against cancer. It is thought that cisplatin has its chloride ligands replaced after entering the cell. This occurs because of the low intracellular chloride concentration which favours the replacement of the chloride ions with water.

The aquated complex is able to form links between bases in DNA in the tumour cell. This halts the process of DNA replication in the tumour cells and effectively stops the growth of the tumour. The drug is expensive due to its platinum content and there are side effects such as potential death, organ damage and hair loss.

Cis-trans isomerism in octahedral complexes

For octahedral complexes of the type ML_4A_2 where M represents the metal atom or ion and L and A represent different ligands, there is a *cis* and a *trans* isomer of this complex.

The isomers of $[Co(NH_3)_4Cl_2]^+$ are shown on page 161; the left-hand isomer is the *cis* isomer as the two Cl^- ligands are beside each other, i.e. at 90° to each other.

The right-hand isomer is the *trans* isomer as the two Cl^- ligands are opposite each other, i.e. at 180° to each other.

The charge on the complex is + as the cobalt is in the +3 oxidation state and the two Cl^- ligand bring the overall charge on the complex to +.

cis isomer

trans isomer

Optical isomerism in octahedral complexes

Optical isomers are non-superimposable and their optical activity allows them to rotate the plane of plane polarised light. One isomer rotates the plane to the left and is referred to as an l-isomer and one rotates the plane to the right and is call the d-isomer. The 'l' is from laevorotatory (rotates to the left) and 'd' is from dextrorotatory (rotates to the right).

> **TIP**
> You would not be expected to be able to distinguish between the d and l forms based on a diagram. This would be carried out practically by determining the effect of each on plane polarised light.

Octahedral complexes formed with a bidentate ligand form two optical isomers. It is the connection of the ligands by coordinate bonds within the octahedral shape which gives the different isomers.

A racemic mixture of both isomers shows no overall rotation as there would be equal concentrations of both and the overall rotations would cancel each other out.

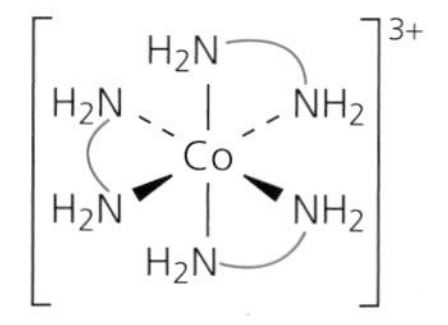

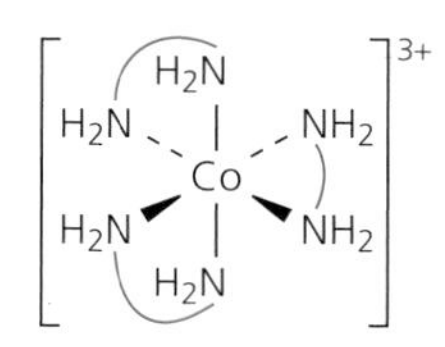

The diagram on the left shows the two optical isomers formed when 1,2-diaminoethane forms a complex with Co^{3+} ions. The ligands are represented by $H_2N—NH_2$. The two optical isomers have a different three-dimensional spatial arrangement of the ligands around the central metal ion.

TEST YOURSELF 3

1 For the following complexes:

A $[CuCl_4]^{2-}$
B $[NiEDTA]^{2-}$
C $[Ag(NH_3)_2]^+$
D $[Co(H_2O)_6]^{3+}$
E $[Co(H_2NCH_2CH_2NH_2)_3]^{2+}$
F $[Pt(NH_3)_2Cl_2]$

a) State the shape of all the complexes.
b) State the coordination number of all the complexes.
c) State the oxidation state of the transition metal in all of the complexes.
d) Which one of the complexes would exhibit *cis–trans* isomerism?
e) Which one of the complexes would exhibit optical isomerism?

Formation of coloured ions

Transition metal ions show distinct colours that can aid in their identification. The main colours of the some common ions are:

- Copper(II) compounds are mainly blue in colour.
- Iron(II) compounds are pale green in colour.
- Cobalt(II) compounds are pink in colour.

In solution these ions become hexaaqua cations, for example $[Cu(H_2O)_6]^{2+}$.

The electromagnetic spectrum

For an examination of coloured compounds, the different colours of visible light needs to be examined. Visible light is part of the electromagnetic spectrum.

The electromagnetic spectrum consists of many bands of electromagnetic radiation which differ in the wavelength (λ), frequency (ν) and energy (E).

The main types of electromagnetic radiation are given in the diagram below.

Visible light is split into seven different colours that have different wavelengths and frequencies.

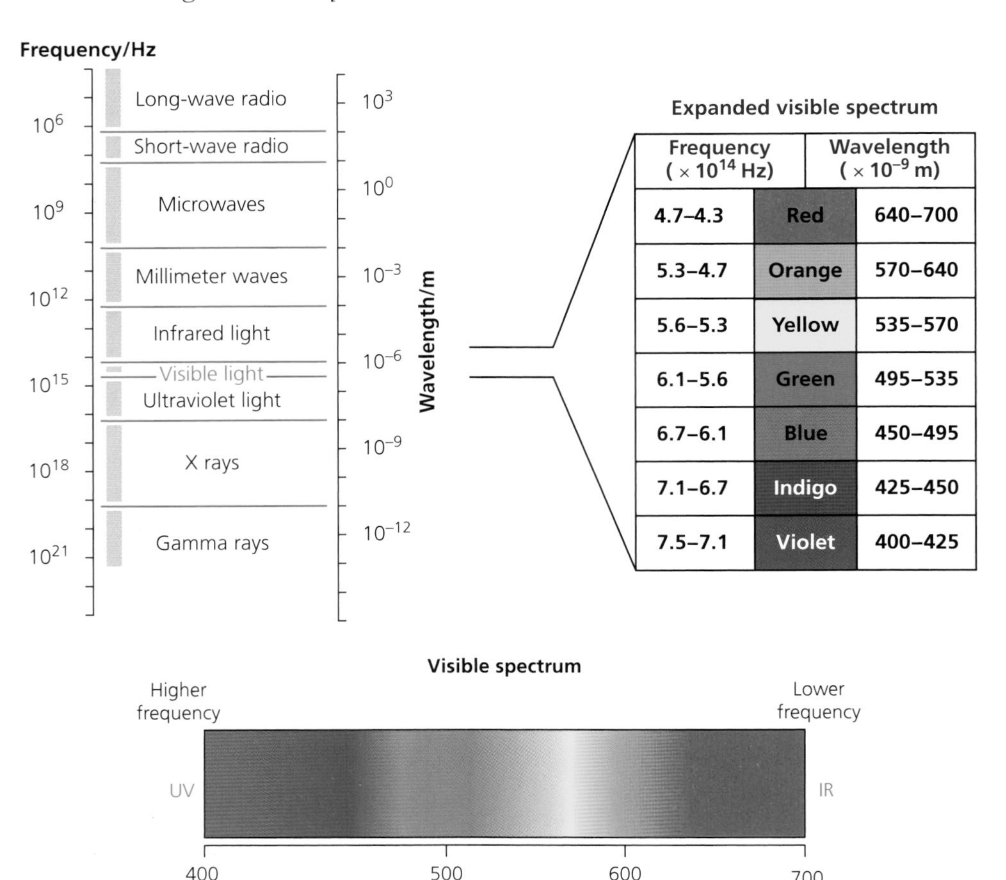

Frequency ($\times 10^{14}$ Hz)		Wavelength ($\times 10^{-9}$ m)
4.7–4.3	Red	640–700
5.3–4.7	Orange	570–640
5.6–5.3	Yellow	535–570
6.1–5.6	Green	495–535
6.7–6.1	Blue	450–495
7.1–6.7	Indigo	425–450
7.5–7.1	Violet	400–425

Colour in transition metal compounds and complexes

The presence of a transition metal ion in a compound or a complex gives colour to the compound and to the complexes they form in solution.

- In these coloured compounds and complexes the d sub-level must be partially filled.
- In complexes the d sub-level is split into two distinct sets of orbitals that have a difference in energy between them (ΔE).

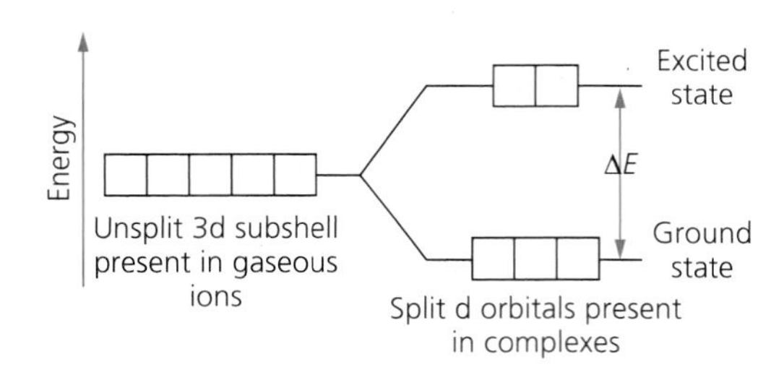

> **TIP**
> This table only serves to act as a guide but in general the colour you see is the colour that is not absorbed.

- Energy in the visible region of the spectrum causes d electron transition between the ground state of the split d orbitals and electrons become excited and are promoted to the higher energy orbitals (excited state).
- Some frequencies of visible light are absorbed and the complementary colour (the colours not absorbed) are observed as they are reflected or transmitted.
- The table below shows the wavelength of visible light in nm, the colour observed when certain colours are absorbed from the visible region of the electromagnetic spectrum.

Wavelength/nm	Colour absorbed	Colour observed
400–430	Violet	Yellow-green
430–460	Blue-violet	Yellow
460–490	Blue	Orange
490–510	Blue-green	Red
510–530	Green	Purple
530–560	Yellow-green	Violet
560–590	Yellow	Blue-violet
590–610	Orange	Blue
610–700	Red	Blue-green

> **TIP**
> Changes in identity of the metal, oxidation state of the metal/charge on the ion, identity of the ligands, coordination number of the complex and shape of the complex all have an effect on ΔE, and change the light absorbed and so the visible colour observed.

- Different metal ions and different charges of the ions (oxidation states) of the same metal and different ligands cause different splitting of the d sub-level, and so different frequencies of visible light are observed.
- The shape of the complex and its coordination number also causes differences in the splitting of the d sub-shell and so the colour of the complexes.

Calculating energy, frequency and wavelength

There is an equation that links the energy difference (ΔE) between the ground state and the excited state of the d electrons with the frequency (and wavelength) of the light absorbed:

$$\Delta E = h\nu$$

where ΔE is the difference in energy measured in joules (J), h is the Planck constant (6.63×10^{-34} J s) and ν is the frequency measured in hertz (Hz) or s^{-1}.

> **TIP**
> ν is written like a v in italics. It is the Greek letter nu and represents frequency. f is also sometimes used for frequency.

The wavelength of the light absorbed may also be used to calculate ΔE or the wavelength may be calculated in metres (m) or nanometres (nm).

The equation relating wavelength to frequency is:

$$c = \nu\lambda$$

where c is the speed of light (usually taken as 3.0×10^8 m s^{-1}), ν is the frequency of light in Hz and λ is the wavelength in m.

The equations may be combined to give:

$$\Delta E = \frac{hc}{\lambda}$$

The diagram on the left shows how the calculations are carried out.

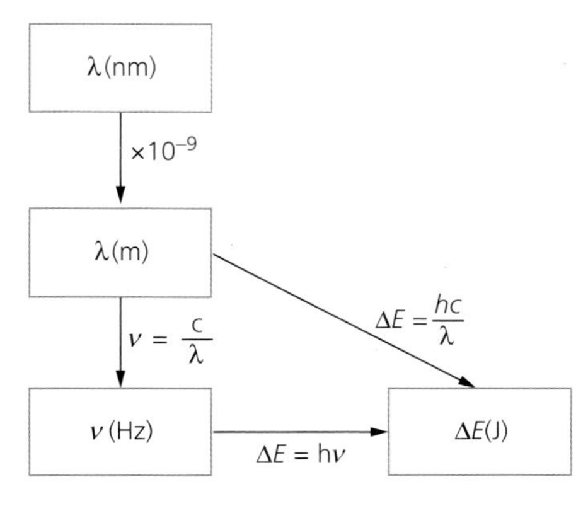

MATHS

The use of units such as nm and m and their interconversion is key in this type of question. 1 nm = 10^{-9} m. Rearranging even a simple expression can allow other quantities in the expression to be calculated. All the calculations in the above diagram can be reversed.

EXAMPLE 5

$[Cu(H_2O)_6]^{2+}$ absorbs red light (700 nm) and the solution appears blue.

1 Calculate a value for the energy, in J, associated with this wavelength of light. The Planck constant, $h = 6.63 \times 10^{-34}$ J s and $c = 3 \times 10^8$ m s^{-1}
2 When a solution containing $[Cu(H_2O)_6]^{2+}$ is mixed with concentrated hydrochloric acid the solution changed from blue to yellow–green and absorbs violet light (400 nm).
   a) Explain whether the energy change, ΔE, between the ground state and excited state of the d electrons is bigger, smaller or the same for a solution which forms a yellow-green solution compared to a blue solution.
   b) State **two** different features of a transition metal complex which causes a change in the value of ΔE

Answers

1 $\lambda = 700$ nm

$\lambda = 700 \times 10^{-9}$ m

$$\Delta E = \frac{hc}{\lambda}$$

$$\Delta E = \frac{hc}{\lambda} = \frac{6.63 \times 10^{-34} \times 3 \times 10^{8}}{700 \times 10^{-9}} = \frac{1.989 \times 10^{-52}}{700 \times 10^{-9}} = 2.84 \times 10^{-19}\text{ J}$$

2 a) Bigger
   b) Identity of the metal; type of ligands which are present; coordination number; oxidation state of the metal or charge on the ion; shape of the complex

EXAMPLE 6

A d electron in a transition metal ion is excited by visible light from its ground state to an excited state. ΔE for this change is 3.81×10^{-19} J.

1 Give the equation which relates the energy change (ΔE) to the frequency of the visible light.
2 Calculate a value for the frequency of the visible light and state its units.
3 Explain why a solution of this transition metal ion is coloured based on the electronic transition.

Answers

1 $\Delta E = h\nu$

2 $$\nu = \frac{\Delta E}{h} = \frac{3.81 \times 10^{-19}}{6.63 \times 10^{-34}} = 5.75 \times 10^{14}\text{ Hz}$$

3 Light is absorbed to excite electrons from ground state to excited state. The complementary colour is the colour observed.

Colorimetry and spectrometry

Colorimetry is the measurement of colour intensity using an instrument called a colorimeter. The intensity of colour in a solution is directly proportional to the concentration of the coloured species.

A colorimeter works by passing a beam of light through the coloured sample and comparing the light that was passed into the sample (incident light) with the light that passes through the sample (transmitted light).

The absorbance is the ratio of the incident light to the absorbed light. Absorbance is also proportional to the path length, which is the actual length of solution through which the light must pass.

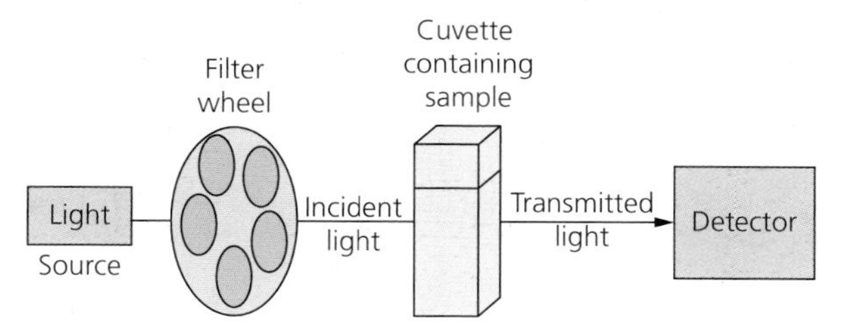

The light from the light source is passed through a filter which changes the light to a specific wavelength (i.e. colour) and the incident light then passes into the sample. The choice of the incident light wavelength (colour) is important. A blue solution is not absorbing blue light so a blue filter would be useless. A red filter would be used to measure the absorbance of a blue solution.

The sample is contained in a cuvette which is a 1 cm × 1 cm glass or plastic vial.

The transmitted light is detected and the difference between the transmitted light and the incident light is the **absorbance**.

The path length is the length through which the solution has to pass in most cases 1 cm.

A simple equation relates absorbance to concentration and path length:

$$A = \varepsilon c l$$

where c is the concentration in mol dm^{-3}, l is path length in cm and ε is the molar absorption coefficient (absorption by 1 mol dm^{-3} solution with a path length of 1 cm). This equation is not required but an understanding of the relationship between absorption and concentration and path length is useful.

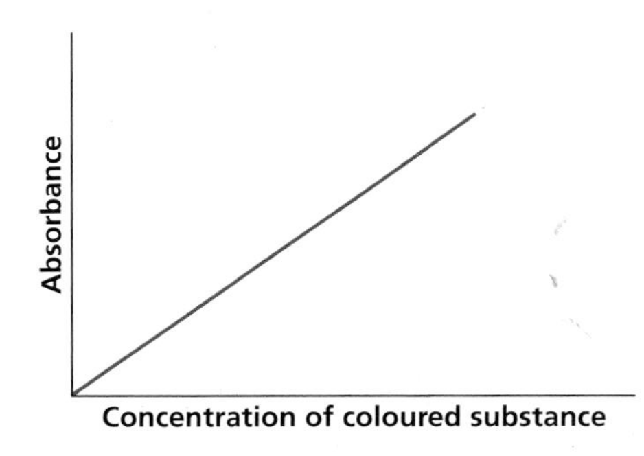

A graph of absorbance against concentration of the coloured species gives a directly proportional line as shown. This type of graph is often used to determine the concentration of an unknown sample of the coloured substance from its absorbance.

Spectroscopy uses a spectrophotometer and it is the same process but the wavelength can be varied over values in the ultraviolet region of the spectrum, as well as in this visible region.

Many molecules and some complexes may appear white/colourless in solution as they do not absorb light in the visible region of the electromagnetic spectrum, but they may have characteristic absorbance in the ultraviolet region of the spectrum. This can be particularly useful in organic chemistry.

The absorbance at a particular wavelength in the ultraviolet region of the spectrum is still directly proportional to the concentration of the substance under examination.

TEST YOURSELF 4

1 A solution of a complex ion absorbs light at a wavelength of 450 nm. $c = 3.0 \times 10^8\,m\,s^{-1}$. Calculate the energy, in J, associated with this wavelength.
2 Suggest why a solution of $[CoCl_4]^{2-}$ is blue.
3 Calculate the wavelength of light, in nm, absorbed by a solution which absorbs energy 3.617×10^{-19} J. $c = 3.0 \times 10^8\,m\,s^{-1}$.

TIP

You should know the test for the anions covered in AS in the chapters on the Halogens and Group 2. You may be asked to identify a transition metal compound that involves identifying the transition metal cation and the anion.

Ions in aqueous solution

Transition metal compounds dissolve in water to form coloured complexes. Other metal compounds also dissolve in water to form complex ions, for example compounds of aluminium. The complexes are called hexaaqua complexes and have the general formula $[M(H_2O)_6]^{2+}$ where M is Fe or Cu or $[M(H_2O)_6]^{3+}$ where M is Al or Fe.

The reaction of these hexaaqua complexes with solutions of sodium hydroxide, ammonia and sodium carbonate can be used to identify them.

Sodium hydroxide solution, NaOH(aq)

Sodium hydroxide solution contains sodium ions and hydroxide ions. The hydroxide ions can cause the precipitation of insoluble metal hydroxides. Sometimes the insoluble metal hydroxide may react with an excess of hydroxide ions and redissolve to form a hydroxo complex.

Ammonia solution, NH_3(aq)

Ammonia, NH_3, reacts with water reversibly to form a solution containing NH_3, NH_4^+ ions and OH^- ions due to the reversible reaction:

$$NH_3 + H_2O \rightleftharpoons NH_4^+ + OH^-$$

The presence of the hydroxide ions causes the precipitation of the insoluble metal hydroxide. Sometimes the insoluble metal hydroxide may react with excess ammonia and redissolve to form an ammine complex.

Sodium carbonate solution, Na_2CO_3(aq)

Sodium carbonate solution contains sodium ions and carbonate ions. For 2+ ions the carbonate ions may react to form an insoluble metal carbonate precipitate. For 3+ ions the carbonate ions react with the solution to form a hydroxide precipitate and carbon dioxide gas is evolved.

Acidity of $[M(H_2O)_6]^{3+}$ solutions

The reaction of these complexes with sodium carbonate is because complexes of the type $[M(H_2O)_6]^{3+}$ form more acidic solutions than complexes of the type $[M(H_2O)_6]^{2+}$. The charge density of a 3+ ion is generally greater than the charge density of the 2+ ion, but it does also depend on the size of the ion. Water ligands bonded to a 3+ ion are polarised and release H^+ ions into the solution.

For example:

$$[Fe(H_2O)_6]^{3+} \rightleftharpoons [Fe(H_2O)_5OH]^{2+} + H^+$$

> **TIP**
> Sodium hydroxide solution or any other solution may be referred to as aqueous sodium hydroxide. Aqueous means dissolved in water and often the two are used but mean the same thing.

This makes the solution acidic and the carbonate ions are broken down to CO_2 and H_2O by the H^+ ions in the solution. H^+ ions continue to be released from the complex until a neutral insoluble complex is formed, i.e. $Fe(OH)_3(H_2O)_3$. The reaction stops when either all the carbonate ions are used up or all of the complex has been converted to the neutral complex.

$[M(H_2O)_6]^{2+}$

For this type of complex, only Fe and Cu are considered.

$[Fe(H_2O)_6]^{2+}$

When iron(II) compounds dissolve in water, they form the hexaaquairon(II) complex, $[Fe(H_2O)_6]^{2+}$. The solution formed is green due to the presence of this complex.

When a solution of sodium hydroxide, ammonia or sodium carbonate is added to a solution containing $[Fe(H_2O)_6]^{2+}$ ions, precipitation reactions occur.

Figure 8.9 Iron(II) hydroxide precipitate can be formed by adding sodium hydroxide solution to a solution containing iron(II) ions.

> **TIP**
> A precipitate is a solid that may be formed on mixing two solutions. Often ppt is used an abbreviation for precipitate.

With sodium hydroxide solution

Observations: A green precipitate of iron(II) hydroxide, $Fe(OH)_2(H_2O)_4$, is formed which slowly changes to a brown solid (caused by the oxidation of iron(II) hydroxide to iron(III) hydroxide by oxygen in the air).

Equation:

$$[Fe(H_2O)_6]^{2+} + 2OH^- \rightarrow \underset{\text{green ppt}}{Fe(OH)_2(H_2O)_4} + 2H_2O$$

The precipitate of iron(II) hydroxide does not redissolve on the addition of more sodium hydroxide solution.

With ammonia solution

Equation:

2NH₃

$$[Fe(H_2O)_6]^{2+} + 2OH^- \rightarrow \underset{\text{green ppt}}{Fe(OH)_2(H_2O)_4} + 2H_2O$$

The precipitate of iron(II) hydroxide does not redissolve on the addition of more ammonia solution.

> **TIP**
> Note that the observations and the chemistry is essentially the same as both solutions contain hydroxide ions.

With sodium carbonate solution
Observations: A green precipitate of iron(II) carbonate, $FeCO_3$, is formed.

Equation:

$$[Fe(H_2O)_6]^{2+} + CO_3^{2-} \rightarrow \underset{\text{green ppt}}{FeCO_3} + 6H_2O$$

> **TIP**
> Make sure all the equations begin with the hexaaqua complex, in this case $[Fe(H_2O)_6]^{2+}$. Some sources will show a simple precipitation reaction as $Fe^{2+} + CO_3^{2-} \rightarrow FeCO_3$ but this is not accepted by AQA as the equation must include the complex.

$[Cu(H_2O)_6]^{2+}$

When copper(II) compounds dissolve in water, they form the hexaaquacopper(II) complex, $[Cu(H_2O)_6]^{2+}$. The solution formed is blue due to the presence of this complex.

When a solution of sodium hydroxide, ammonia or sodium carbonate is added to a solution containing $[Cu(H_2O)_6]^{2+}$ ions, precipitation reactions occur.

With sodium hydroxide solution
Observations: A blue precipitate of copper(II) hydroxide, $Cu(OH)_2(H_2O)_4$, is formed.

Equation:

$$[Cu(H_2O)_6]^{2+} + 2OH^- \rightarrow \underset{\text{blue ppt}}{Cu(OH)_2(H_2O)_4} + 2H_2O$$

The precipitate of copper(II) hydroxide does not redissolve on the addition of more sodium hydroxide solution.

With ammonia solution
Observations: A blue precipitate of copper(II) hydroxide, $Cu(OH)_2(H_2O)_4$, is formed.

Equation:

$$[Cu(H_2O)_6]^{2+} + 2OH^- \rightarrow \underset{\text{blue ppt}}{Cu(OH)_2(H_2O)_4} + 2H_2O$$

The precipitate of copper(II) hydroxide redissolves on the addition of more ammonia solution to form a deep blue solution.

Equation

$$Cu(OH)_2(H_2O)_4 + 4NH_3 \rightarrow \underset{\text{deep blue solution}}{[Cu(NH_3)_4(H_2O)_2]^{2+}} + 2H_2O + 2OH^-$$

> **TIP**
> The deep blue solution may also be described as dark blue or royal blue. It is a very characteristic colour for this reaction and helps identify copper(II) ions.

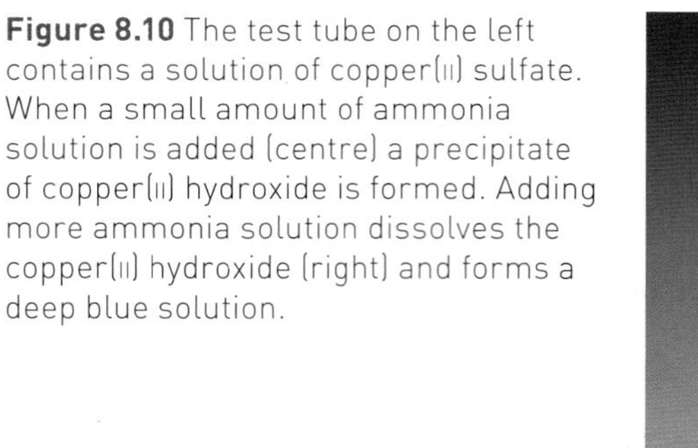

Figure 8.10 The test tube on the left contains a solution of copper(II) sulfate. When a small amount of ammonia solution is added (centre) a precipitate of copper(II) hydroxide is formed. Adding more ammonia solution dissolves the copper(II) hydroxide (right) and forms a deep blue solution.

With sodium carbonate solution
Observations: A green precipitate of copper(II) carbonate, $CuCO_3$, is formed.

Equation:

$$[Cu(H_2O)_6]^{2+} + CO_3^{2-} \rightarrow \underset{\text{green ppt}}{CuCO_3} + 6H_2O$$

$[M(H_2O)_6]^{3+}$

For this type of complex, only Fe and Al are considered.

$[Fe(H_2O)_6]^{3+}$

When iron(III) compounds dissolve in water, they form the hexaaquairon(III) complex, $[Fe(H_2O)_6]^{3+}$. The solution formed is often yellow due to the presence of this complex. The colour of the solution may vary depending on the anion present in the compound. Some solutions of iron(III) compound may appear purple, brown, lilac or violet.

When a solution of sodium hydroxide, ammonia or sodium carbonate is added to a solution containing $[Fe(H_2O)_6]^{3+}$ ions, precipitation reactions occur.

With sodium hydroxide solution
Observations: A brown precipitate of iron(III) hydroxide, $Fe(OH)_3(H_2O)_3$, is formed.

Equation:

$$[Fe(H_2O)_6]^{3+} + 3OH^- \rightarrow \underset{\text{brown ppt}}{Fe(OH)_3(H_2O)_3} + 3H_2O$$

TIP
The brown precipitate of iron(III) hydroxide may also be described as rusty.

The precipitate of iron(III) hydroxide does not redissolve on the addition of more sodium hydroxide solution.

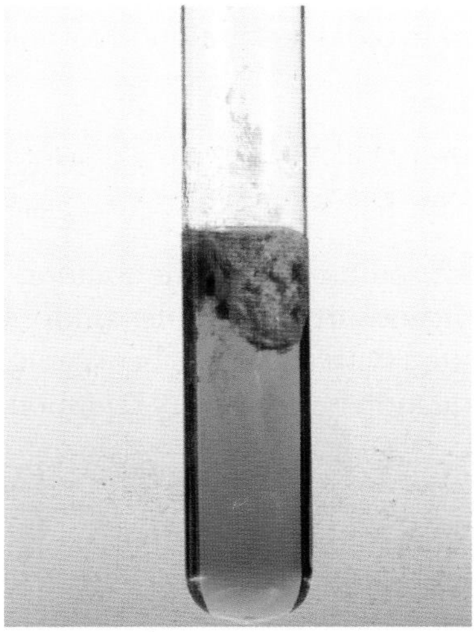

Figure 8.11 Iron(III) hydroxide precipitate can be formed by adding sodium hydroxide solution to a solution containing iron(III) ions.

With ammonia solution
Observations: A brown precipitate of iron(III) hydroxide, $Fe(OH)_3(H_2O)_3$, is formed.

Equation:

$$[Fe(H_2O)_6]^{3+} + 3OH^- \rightarrow \underset{\text{brown ppt}}{Fe(OH)_3(H_2O)_3} + 3H_2O$$

The precipitate of iron(III) hydroxide does not redissolve on the addition of more ammonia solution.

With sodium carbonate solution
Observations: A brown precipitate of iron(III) hydroxide, $Fe(OH)_3(H_2O)_3$, is formed and bubbles of a gas are produced.

Equation:

$$2[Fe(H_2O)_6]^{3+} + 3CO_3^{2-} \rightarrow \underset{\text{brown ppt}}{2Fe(OH)_3(H_2O)_3} + 3CO_2 + 3H_2O$$

TIP
It is important to be aware of the difference between the reactions of solutions containing iron(II) compounds and iron(III) compounds with sodium carbonate solution. Make sure you can describe the observations and write the equations.

$[Al(H_2O)_6]^{3+}$

When aluminium compounds dissolve in water, they form the hexaaquaaluminium(III) complex, $[Al(H_2O)_6]^{3+}$. The solution formed is colourless.

When a solution of sodium hydroxide, ammonia or sodium carbonate is added to a solution containing $[Al(H_2O)_6]^{3+}$ ions, precipitation reactions occur.

With sodium hydroxide solution
Observations: A white precipitate of aluminium hydroxide, $Al(OH)_3(H_2O)_3$, is formed.

Equation:

$$[Al(H_2O)_6]^{3+} + 3OH^- \rightarrow \underset{\text{white ppt}}{Al(OH)_3(H_2O)_3} + 3H_2O$$

The precipitate of aluminium hydroxide redissolves on the addition of more sodium hydroxide solution to form a colourless solution.

Equation:

$$Al(OH)_3(H_2O)_3 + OH^- \rightarrow \underset{\text{colourless solution}}{[Al(OH)_4(H_2O)_2]^-} + H_2O$$

As more sodium hydroxide solution is added more OH^- ligands replace the water ligands in the complex until $[Al(OH)_6]^{3-}$ is formed. The equation can be written with anywhere between 4 and 6 OH^- present in the complex. The equation below is very common:

$$Al(OH)_3(H_2O)_3 + 3OH^- \rightarrow [Al(OH)_6]^{3-} + 3H_2O$$

The solution formed is colourless irrespective of the number of OH^- ligands in the complex.

With ammonia solution
Observations: A white precipitate of aluminium hydroxide, $Al(OH)_3(H_2O)_3$, is formed.

Equation:

$$[Al(H_2O)_6]^{3+} + 3OH^- \rightarrow \underset{\text{white ppt}}{Al(OH)_3(H_2O)_3} + 3H_2O$$

The precipitate of aluminium hydroxide does not redissolve on the addition of more ammonia solution.

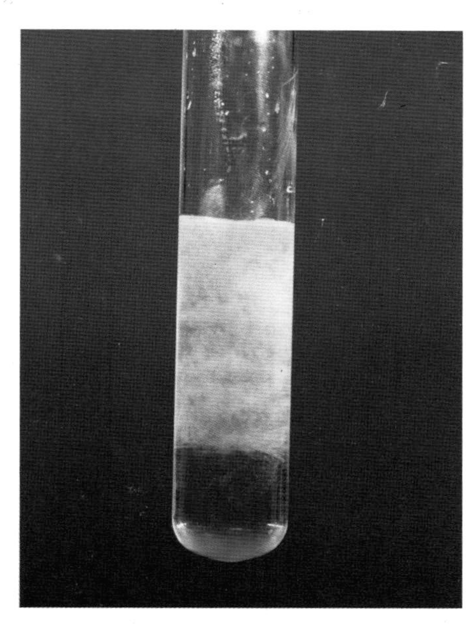

Figure 8.12 Aluminium hydroxide precipitate is formed by adding ammonia solution to a solution containing aluminium ions.

With sodium carbonate solution

Observations: A white precipitate of aluminium hydroxide, $Al(OH)_3(H_2O)_3$, is formed and bubbles of a gas are produced.

Equation:

$$2[Al(H_2O)_6]^{3+} + 3CO_3^{2-} \rightarrow \underset{\text{white ppt}}{2Al(OH)_3(H_2O)_3} + 3CO_2 + 3H_2O$$

The table below summarises the observations and the formulae of the complexes involved.

Hexaaqua complex	$[Fe(H_2O)_6]^{2+}$	$[Cu(H_2O)_6]^{2+}$	$[Fe(H_2O)_6]^{3+}$	$[Al(H_2O)_6]^{3+}$
Colour	Green	Blue	Yellow (or brown or violet or purple or lilac)	Colourless
Reaction with NaOH(aq)	Green ppt	Blue ppt	Brown ppt	White ppt
Complex formed	$Fe(OH)_2(H_2O)_4$	$Cu(OH)_2(H_2O)_4$	$Fe(OH)_3(H_2O)_3$	$Al(OH)_3(H_2O)_3$
Does it redissolve in excess NaOH(aq) and colour of solution formed?	No	No	No	Yes Colourless solution formed
Complex formed				$[Al(OH)_4(H_2O)_2]^-$ or $[Al(OH)_6]^{3-}$
Reaction with NH_3(aq)	Green ppt	Blue ppt	Brown ppt	White ppt
Complex formed	$Fe(OH)_2(H_2O)_4$	$Cu(OH)_2(H_2O)_4$	$Fe(OH)_3(H_2O)_3$	$Al(OH)_3(H_2O)_3$
Does it redissolve in excess NH_3(aq) and colour of solution formed?	No	Yes Deep blue solution formed	No	No
Complex formed		$[Cu(NH_3)_4(H_2O)_2]^{2+}$		
Reaction with Na_2CO_3(aq)	Green ppt	Green ppt	Brown ppt Bubbles of gas released	White ppt Bubbles of gas released
Compound or complex formed	$FeCO_3$	$CuCO_3$	$Fe(OH)_3(H_2O)_3$	$Al(OH)_3(H_2O)_3$

TEST YOURSELF 5

1 Consider the following complexes:
   A $[Cu(H_2O)_6]^{2+}$
   B $[Al(H_2O)_6]^{3+}$
   C $[Fe(H_2O)_6]^{2+}$
   D $[Cu(NH_3)_4(H_2O)_2]^{2+}$
   a) Which complexes are blue?
   b) What is observed when aqueous sodium carbonate is added to complex A?
   c) What is observed when aqueous sodium hydroxide is added to complex B until in excess?
   d) What is observed when aqueous ammonia is added to complex C?

2 Consider the complex $[Al(OH)_4(H_2O)_2]^-$
   a) What is the coordination number of this complex?
   b) What is the oxidation state of aluminium in this complex?
   c) Suggest the shape of this complex.
   d) Write an equation for the formation of this complex from $Al(OH)_3(H_2O)_3$.

3 A sample of iron(III) chloride is dissolved in water.
   a) Write the formula of the iron complex formed when iron(III) chloride dissolves in water.
   b) What is observed when aqueous sodium carbonate is added to the solution containing the complex?
   c) Write an equation for the reaction of aqueous sodium carbonate with this iron complex.

REQUIRED PRACTICAL II

Using test-tube reactions to identify ions in solution

Chemical tests are used by analytical chemists to give an indication of the nature of substances that may be present in a sample. For example, a chemical test may be used to determine the presence of a particular metal ion in a sample of water taken from a polluted river. A forensic scientist may use a simple chemical test, using an alkaline solution, to identify components in a soil sample taken from a crime scene and compare this with a soil sample taken from a suspect's shoes.

To determine the identity of a chemical salt A, the following tests were carried out and the observations recorded.

Test	Observation
1 Describe the appearance of A	Blue crystals
2 Dissolve a spatula measure of A in approximately 50 cm^3 of water	Blue solution
3 Add a few drops of barium chloride solution to 2 cm^3 of the solution of A	White ppt Blue solution remains
4 Add a few drops of silver nitrate solution to 2 cm^3 of the solution of A	Blue solution
5 Add sodium hydroxide solution to 2 cm^3 of the solution of A, drop wise with shaking, until in excess	Blue ppt
6 In a fume cupboard, add concentrated ammonia solution drop wise, until present in excess, to 2 cm^3 of the solution of A	Blue ppt which redissolves on excess to produce a deep blue solution
7 Add a few drops of sodium carbonate solution to cm^3 of the solution of A.	Green ppt

1 Use the evidence in the table to suggest the metal ion present in A.
2 Suggest the formula of the complex ion formed in test 2. Name the shape of this complex.
3 Using the results from tests 3 and 4, determine the anion present in A. State your reasons, giving an ionic equation for any reaction which occurs.
4 State the formula of the complex formed in test 5.
5 Explain using two balanced symbol equations, the reaction occurring in test 6.
6 What is the name of the green ppt formed in test 7?
7 Suggest and explain what would be observed if the salt A was heated gently in a boiling tube.
8 Suggest a name for the salt A.

Practice questions

1 Which of the following ions does not have an electronic configuration of $1s^2\ 2s^2\ 2p^6\ 3s^2\ 3p^6\ 3d^5$?

A Mn^{2+}

B Fe^{3+}

C Cr^{+}

D Cr^{3+} *(1)*

2 In which of the following complexes does the complex have a coordination number of 6 and transition metal have an oxidation state of +2?

A $[CoCl_4]^{2-}$

B $[Fe(H_2O)_6]^{3+}$

C $[Pt(NH_3)_2Cl_2]$

D $[CoEDTA]^{2-}$ *(1)*

3 Which one of the following is correct?

A On adding sodium hydroxide solution to a solution of copper(II) sulfate, a green precipitate is observed.

B On adding sodium carbonate solution to a solution of iron(II) sulfate, a white precipitate is observed.

C On adding ammonia solution to a solution of iron(III) chloride, a brown precipitate is observed.

D On adding sodium carbonate solution to a solution of copper(II) sulfate, a blue precipitate is observed. *(1)*

4 The following tests were carried out to a solution of an unknown compound:

Test 1: To a solution of the compound, sodium carbonate solution was added and a brown precipitate and bubbles of a gas were observed.

Test 2: To a solution of the compound, nitric acid and silver nitrate solution were added and a white precipitate was observed.

Which one of the following is the most likely identity of the compound?

A aluminium carbonate

B copper(II) chloride

C iron(II) bromide

D iron(III) chloride *(1)*

5 $[CuCl_4]^{2-}$ is green–yellow. It absorbs light of wavelength 420 nm. The Planck constant, $h = 6.63 \times 10^{-34}$ J s. The speed of light, $c = 3.0 \times 10^8$ m s^{-1}.

a) Give the equation which relates the energy change ΔE to the Planck constant h and the wavelength of the visible light, λ. *(1)*

b) Calculate a value for ΔE and state its units. *(3)*

c) Explain why the solution is coloured. *(2)*

d) State **three** features of a transition metal complex which can cause a change in the wavelength of light absorbed. *(3)*

6 Some ligands are listed below. All of the ligands are monodentate except the oxalate ligand which is bidentate.

Ligands: thiocyano: SCN^-; cyano: CN^-; aqua: H_2O; hydroxo: OH^-; oxalate: $C_2O_4^{2-}$; ammine: NH_3

a) Explain the meaning of the term bidentate. *(1)*

b) Write the formula of the complex formed between the nickel(II) ion and the ammine ligand where the coordination number is 6. *(1)*

c) Write the formula of the complex formed between the chromium(III) ions and the hydroxo ligand where the coordination number is 6. *(1)*

d) Write the formula of the complex formed between the silver(I) ion and the ammine ligand where the coordination number is 2. *(1)*

e) Suggest the shape of the complexes described in (b), (c) and (d). *(3)*

7 A ligand substitution reaction occurs when a solution containing $EDTA^{4-}$ is added to a solution containing $[Ni(H_2O)_6]^{2+}$ ions.

$$[Ni(H_2O)_6]^{2+} + EDTA^{4-} \rightarrow [NiEDTA]^{2-} + 6H_2O$$

a) $EDTA^{4-}$ is a chelating agent. Explain this term. *(1)*

b) Explain in terms of the chelate effect why this ligand substitution reaction occurs. *(2)*

c) What type of bond is formed between $EDTA^{4-}$ and the nickel ion? *(1)*

d) State the coordination number of the two complexes, $[Ni(H_2O)_6]^{2+}$ and $[NiEDTA]^{2-}$. *(1)*

8 Consider the following reaction scheme. All the complexes are in aqueous solution.

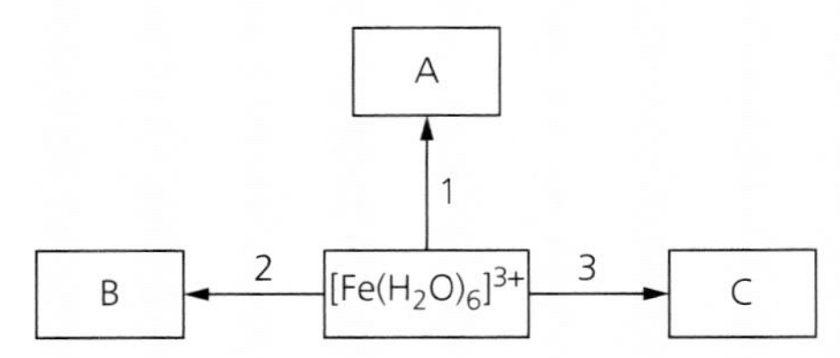

a) In reaction 1, aqueous sodium hydroxide is added to a solution of the complex. Write an equation for the reaction. *(2)*

b) In reaction 2, ammonia solution is added to a solution of the complex. What is observed? *(1)*

c) In reaction 3, a solution of sodium carbonate is added to a solution of the complex.

i) What is observed? *(2)*

ii) Write an equation for the reaction which is occurring. *(2)*

9 Complexes of $[Pt(NH_3)_2Cl_2]$, $[Co(NH_3)_4Cl_2]^+$ and $[Ni(H_2NCH_2CH_2NH_2)_3]^{2+}$ exhibit different forms of stereoisomerism.

a) What is the coordination number and shape of the complex $[Pt(NH_3)_2Cl_2]$? *(2)*

b) State the type of stereoisomerism shown by the three complexes. *(3)*

c) What is a use of an isomer of the $[Pt(NH_3)_2Cl_2]$ complex. *(1)*

d) Explain why the $[Pt(NH_3)_2Cl_2]$ complex has no overall charge. *(2)*

10 Aqueous metal ions can be identified by test tube reactions.

For each of the following describe what you would observe.

Write an equation or equations for any reactions that occur.

a) The addition of aqueous sodium carbonate to solutions containing $[Fe(H_2O)_6]^{2+}$ and another solution containing $[Fe(H_2O)_6]^{3+}$. *(7)*

b) The addition of aqueous ammonia to a solution containing $[Cu(H_2O)_6]^{2+}$. *(4)*

Stretch and challenge

11 Iron rusts when exposed to air and moisture. Rusting occurs in a series of reactions as shown below:

Reaction 1: $Fe(s) \rightarrow Fe^{2+}(aq) + 2e^-$

Reaction 2: $O_2 + 2H_2O + 4e^- \rightarrow 4OH^-$

Reaction 3: $Fe^{2+}(aq) + 2OH^-(aq) \rightarrow Fe(OH)_2(s)$

Reaction 4:
$4Fe(OH)_2(s) + O_2(g) \rightarrow 2Fe_2O_3.H_2O(s) + 2H_2O(l)$

a) Write an overall equation for rusting by combining the equations for reaction 1 to 4 above. *(3)*

b) Which one of the reactions is not a redox reaction? Explain your answer. *(2)*

c) In the process of sacrificial protection of iron, blocks of magnesium are attached to the iron.

$Mg^{2+}(aq) + 2e^- \rightarrow Mg(s) \quad E^\ominus = -2.37\,V$

$Fe^{2+}(aq) + 2e^- \rightarrow Fe(s) \quad E^\ominus = -0.44\,V$

i) Explain why the block of magnesium is described as the negative electrode (anode). *(1)*

ii) Write cell notation for the electrochemical cell in this redox reaction. *(2)*

iii) Calculate the emf of the cell. *(1)*

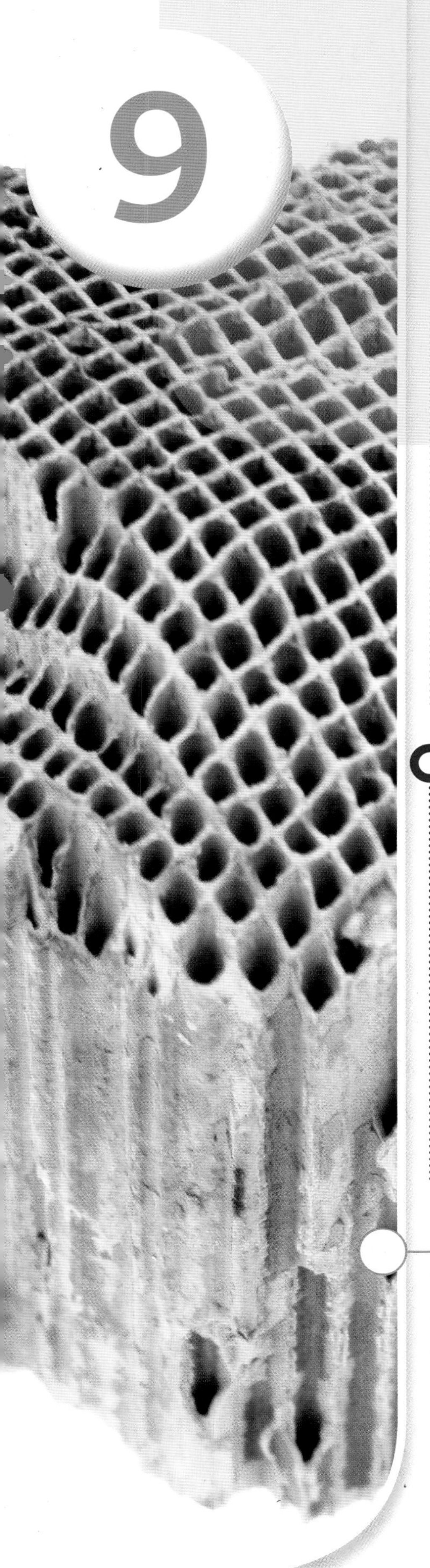

9 Transition metals: Variable oxidation states

PRIOR KNOWLEDGE

- Transition metals form compounds with the transition metal in different oxidation states.
- The oxidation states of transition metals vary from 0 (for the element) to a maximum of +7.
- +2 is a common oxidation state of transition metals.
- Oxidation and reduction reactions between these oxidation states are common and there are changes in colour associated with these changes.
- Transition metal compounds can be used in organic chemistry for oxidation reactions.
- Transitions metals and their compounds may be used as catalysts.
- Catalysts provide an alternative reaction pathway of lower activation energy and so increase the rate of a chemical reaction.

TEST YOURSELF ON PRIOR KNOWLEDGE 1

1 Give the oxidation state of the transition metal in the following compounds and ions.
   a) Fe_2O_3
   b) MnO_4^-
   c) Na_2CrO_4
   d) $K_2Cr_2O_7$
2 Calculate the value of x in the manganate(VI) ion, MnO_x^{2-}.
3 Name the following compounds including the oxidation state of the transition metal.
   a) $Na_2Cr_2O_7$
   b) MnO_2
   c) Co_2O_3
4 Explain how a catalyst increases the rate of a chemical reaction.

Oxidation states of transition metals

Elements of the series of d block elements (Sc → Zn) excluding chromium and copper have the general electronic configuration [Ar] $3d^n$, $4s^2$. The removal of the $4s^2$ electrons creates the ubiquitous 2+ ion.

The oxidation states of the elements vary depending on the number of d electrons available in the atom. The most common oxidation states of the d block elements are as follows:

Sc +3; **Ti** +3, +4; **V** +4, +5; **Cr** +3, +6; **Mn** +2, +4, +7; **Fe** +2, +3; **Co** +2; +3, **Ni** +2; **Cu** +2, **Zn** +2.

The stability of the +2 oxidation state relative to the +3 and higher states increases from left to right across the series. This reflects the increasing difficulty of removing a 3d electron as the nuclear charge increases.

The table below shows all oxidation states of the first series of d block elements. The ones shown in bold are most common oxidation states of the elements.

Element	Sc	Ti	V	Cr	Mn	Fe	Co	Ni	Cu	Zn
Oxidation states					**+7**					
				+6	+6	+6				
			+5	+5	+5	+5				
		+4	**+4**	+4	**+4**	+4	+4	+4		
	+3	**+3**	+3	**+3**	+3	**+3**	**+3**	+3	+3	
	+2	+2	+2	+2	**+2**	**+2**	**+2**	**+2**	**+2**	**+2**
				+1					**+1**	

TIP

Scandium to zinc are the elements in the first series of the d block, but scandium and zinc are d block elements, not transition elements. Transition elements form at least one ion with an incomplete d subshell. Scandium and zinc form ions with empty or incomplete d subshells. The first transition series runs from titanium to copper.

In general, the compounds of the higher oxidation states are found in covalent compounds and molecular ions and they can act as oxidising agents as the transition metal can gain electrons and be reduced to lower oxidation states.

Vanadium chemistry

Vanadium has 4 major oxidation states, +2, +3, +4 and +5. The molecular and simple ions below show vanadium in these oxidation states with the colour in aqueous solution.

Oxidation state	Name of molecular ion	Formula of ion	Colour in aqueous solution
+5	Dioxovanadium(V) ion	VO_2^+	Yellow
	Vanadate(V) ion	VO_3^-	
+4	Oxovanadium(IV) ion	VO^{2+}	Blue
+3	Vanadium(III) ion	V^{3+}	Green
+2	Vanadium(II) ion	V^{2+}	Violet

TIP

You can remember the colours of vanadium using '**Y**ou **B**etter **G**et **V**anadium'. Y in you is yellow, B in better is blue, G in get is green and V in vanadium is violet. Try to create your own rhyme (mnemonic) for this if you cannot remember this one. These are a good way to learn.

An aqueous solution of vanadium in the +5 oxidation state (either vanadate(V) or dioxovanadium(V)) can be reduced from +5 to +2 showing all the oxidation states in between.

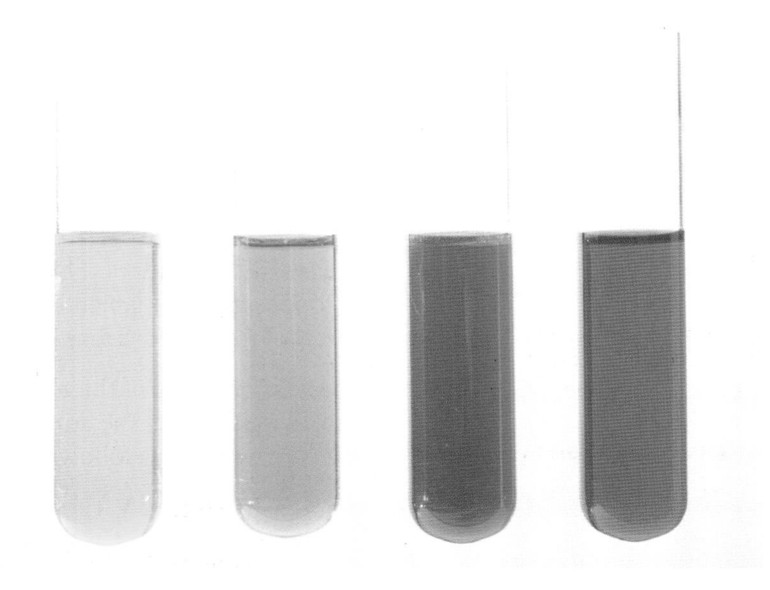

Figure 9.1 Test tubes containing solutions of dissolved vanadium compounds in different oxidation states. Can you identify the oxidation states?

You must be able to write half equations for all these reduction reactions (or oxidation reactions if going from +2 to +5). This is revision of writing half equations from the oxidation, reduction and redox reactions section in Book 1, AS level.

+5 to +4

Step 1: Write the species involved in the reduction.

$$VO_2^+ \rightarrow VO^{2+}$$

Step 2: Check the oxidation numbers and add electrons as appropriate to cause the reduction. (Remember electrons go on the left-hand side of a reduction and they go on the right-hand side of an oxidation.)

$$VO_2^+ + e^- \rightarrow VO^{2+}$$

Step 3: Add H^+ ions to one side and H_2O to the other side to balance the oxygen atoms. (Remember usually H^+ go on the left-hand side in a reduction and H^+ go on the right-hand side in an oxidation.)

$$VO_2^+ + 2H^+ + e^- \rightarrow VO^{2+} + H_2O$$

Step 4: Check the charge balance on each side

Total charge on left = +2 Total charge on right = +2

Equation is balanced both with atoms and charges.

Final equation: $VO_2^+ + 2H^+ + e^- \rightarrow VO^{2+} + H_2O$

+4 to +3

Step 1: Write the species involved in the reduction

$$VO^{2+} \rightarrow V^{3+}$$

Step 2: Check the oxidation numbers and add electrons as appropriate to cause the reduction

$$VO^{2+} + e^- \rightarrow V^{3+}$$

Step 3: Add H^+ ions to one side and H_2O to the other side to balance the oxygen atoms.

$$VO^{2+} + 2H^+ + e^- \rightarrow V^{3+} + H_2O$$

Step 4: Check the charge balance on each side

Total charge on left = +3 Total charge on right = +3

Equation is balanced both with atoms and charges.

Final equation: $VO^{2+} + 2H^+ + e^- \rightarrow V^{3+} + H_2O$

+3 to +2

Step 1: Write the species involved in the reduction

$$V^{3+} \rightarrow V^{2+}$$

Step 2: Check the oxidation numbers and add electrons as appropriate to cause the reduction

$$V^{3+} + e^- \rightarrow V^{2+}$$

All charges balance

Final equation: $V^{3+} + e^- \rightarrow V^{2+}$

TIP

You can revise writing half equations from the Oxidation, reduction and redox equations topic in AS Book 1.

Redox potentials

The reduction potentials for these reactions are given below.

$$VO_2^+ + 2H^+ + e^- \rightarrow VO^{2+} + H_2O \quad +1.00\,V$$

$$VO^{2+} + 2H^+ + e^- \rightarrow V^{3+} + H_2O \quad +0.34\,V$$

$$V^{3+} + e^- \rightarrow V^{2+} \quad -0.26\,V$$

Reduction of vanadium

Vanadium can be reduced from +5 to +2 using a reducing agent like zinc (in the presence of acid).

$$Zn^{2+} + 2e^- \rightarrow Zn \quad -0.76\,V$$

When zinc reduces vanadium, the zinc becomes oxidised so the half equation for the reaction of zinc is:

$$Zn \rightarrow Zn^{2+} + 2e^- \quad +0.76\,V$$

By combining this with each of the reduction half equations for the vanadium reductions, we can see how far zinc can reduce vanadium.

The ionic equation for each overall reaction can be determined by combining the oxidation and reduction half equations to cancel out electrons on both sides.

The electromotive force (EMF) is determined by simply adding the reduction potential and the oxidation potential together. Remember the addition of redox potentials does not depend on the number of moles of each substance reacting – it is simply a total.

The acid is present to provide H^+ ions which are required in the redox reaction:

- A positive EMF means that the reaction is feasible.
- A negative EMF means that the reaction is not feasible.

$$VO_2^+ + 2H^+ + e^- \rightarrow VO^{2+} + H_2O \quad +1.00\,V \text{ (REDUCTION)}$$
$$Zn \rightarrow Zn^{2+} + 2e^- \quad +0.76\,V \text{ (OXIDATION)}$$

EMF =+1.76 V (FEASIBLE)

Ionic equation: $2VO_2^+ + 4H^+ + Zn \rightarrow 2VO_2^+ + 2H_2O + Zn^{2+}$

$$VO^{2+} + 2H^+ + e^- \rightarrow V^{3+} + H_2O \quad +0.34\,V \text{ (REDUCTION)}$$
$$Zn \rightarrow Zn^{2+} + 2e^- \quad +0.76\,V \text{ (OXIDATION)}$$

EMF = +1.10 V (FEASIBLE)

Ionic equation: $2VO^{2+} + 4H^+ + Zn \rightarrow 2V^{3+} + 2H_2O + Zn^{2+}$

$$V^{3+} + e^- \rightarrow V^{2+} \quad -0.26\,V \text{ (REDUCTION)}$$
$$Zn \rightarrow Zn^{2+} + 2e^- \quad +0.76\,V \text{ (OXIDATION)}$$

EMF = +0.50 V (FEASIBLE)

Ionic equation: $2V^{3+} + Zn \rightarrow 2V^{2+} + Zn^{2+}$

Excess zinc, in acid solution, will reduce vanadium from +5 to +2. You would observe a colour change to the solution from yellow to blue to green to violet. You may see an intermediate green colour as the vanadium changes oxidation state from +5 to +4 but this is simply caused by a mixture of the yellow and blue colours of these oxidation states.

Some other substances may partially reduce vanadium. Sulfur dioxide will reduce vanadium from +5 to +3 but not to +2.

$$SO_4^{2-} + 4H^+ + 2e^- \rightarrow SO_2 + 2H_2O \qquad +0.17\,V$$

The sulfur dioxide will be oxidised so the half equation is rewritten as an oxidation reaction and the sign of redox potential is changed.

$$SO_2 + 2H_2O \rightarrow SO_4^{2-} + 4H^+ + 2e^- \qquad -0.17\,V$$

Examining each reaction separately:

$$VO_2^+ + 2H^+ + e^- \rightarrow VO^{2+} + H_2O \qquad +1.00\,V \text{ (REDUCTION)}$$
$$SO_2 + 2H_2O \rightarrow SO_4^{2-} + 4H^+ + 2e^- \qquad -0.17\,V \text{ (OXIDATION)}$$

EMF = +0.83 V (FEASIBLE)

Ionic equation: $2VO_2^+ + SO_2 \rightarrow 2VO^{2+} + SO_4^{2-}$

$$VO^{2+} + 2H^+ + e^- \rightarrow V^{3+} + H_2O \qquad +0.34\,V \text{ (REDUCTION)}$$
$$SO_2 + 2H_2O \rightarrow SO_4^{2-} + 4H^+ + 2e^- \qquad -0.17\,V \text{ (OXIDATION)}$$

EMF = +0.17 V (FEASIBLE)

Ionic equation: $2VO^{2+} + SO_2 \rightarrow 2V^{3+} + SO_4^{2-}$

$$V^{3+} + e^- \rightarrow V^{2+} \qquad -0.26\,V \text{ (REDUCTION)}$$
$$SO_2 + 2H_2O \rightarrow SO_4^{2-} + 4H^+ + 2e^- \qquad -0.17\,V \text{ (OXIDATION)}$$

EMF = −0.43 V (**NOT FEASIBLE**)

Sulfur dioxide can therefore reduce vanadium from +5 to +3. You would observe the solution changing from yellow to blue to green.

Tollens' reagent

Tollens' reagent is an ammoniacal solution of silver nitrate. It contains the complex $[Ag(NH_3)_2]^+$. It can be used to test for the presence of an aldehyde as it indicates the oxidation of an aldehyde. When warmed with an aldehyde, the silver(I) ion in the complex is reduced to silver and a silver mirror coats the inside of the test tube. Ketones cannot be oxidised so the colourless solution of the complex remains colourless when warmed in the presence of a ketone.

$$\underset{\text{Tollens' reagent}}{2[Ag(NH_3)_2]^{2+}} + \underset{\text{aldehyde}}{RCHO} + 3OH^- \rightarrow \underset{\text{silver mirror}}{2Ag} + RCOO^- + 4NH_3 + 2H_2O$$

Silver ions would not normally be reduced by the oxidation of an aldehyde, but when the silver ions complex with ammonia and in alkaline conditions, the redox potential changes so that the oxidation of an aldehyde is enough to make the reduction of silver(I) ions in the complex feasible.

Manganate(VII) ions in acidic and alkaline solution

Manganate(VII) ions, MnO_4^-, are a strong reducing agent in acidic solution but a weaker reducing agent in alkaline solution. pH has an effect on the redox potential of the reduction of transition metal ions from higher to lower oxidation state. In acidic solution, manganate(VII) ions, MnO_4^-, are reduced to Mn^{2+} ions. In alkaline solution, manganate(VII) ions, MnO_4^-, are reduced to manganate(VI) ions, MnO_4^{2-}. In organic chemistry, manganate(VII)

TIP

There is more detail on this in the electrode potentials and electrochemical cells topic and in aldehydes and ketones topic.

ions in acidic solution are such a strong oxidising agent that they will break C–C bonds and completely oxidise the molecule. Manganate(VII) ions in alkaline conditions are a weaker oxidising agent and can oxidise alkenes to the corresponding diol.

TEST YOURSELF 2

1 State the oxidation state and colour of the aqueous solutions of the following compounds of vanadium.

a) VCl_2 **b)** $V_2(SO_4)_3$

c) NH_4VO_3 **d)** VO_2SO_4

2 Write a half equation representing the reduction of the dioxovanadium(V) ion, VO_2^+, to the oxovanadium(IV) ion, VO^{2+}.

3 Consider the half equations below:

$VO^{2+} + 2H^+ + e^- \rightarrow V^{3+} + H_2O$ +0.34 V

$Zn^{2+} + 2e^- \rightarrow Zn$ −0.76 V

a) Write an ionic equation for the reaction between oxovanadium(IV), VO^{2+}, and zinc forming vanadium(III) ions and zinc ions.

b) Calculate the EMF of this reaction.

c) What colour change would be observed in this solution?

Redox titrations

In this section you should revise all the detail of how to carry out a titration from the Amount of Substance section in AS.

There are several types of redox titrations but the most common is the use of a solution of potassium manganate(VII) to determine the amount in moles of a reducing agent present in solution.

Potassium manganate(VII) (or potassium permanganate) has the formula $KMnO_4$ and is a purple solid and forms a purple solution. Manganate(VII) ions are MnO_4^-.

Purple potassium manganate(VII) is easily reduced to pale pink (virtually colourless) manganese(II) ions according to the equation:

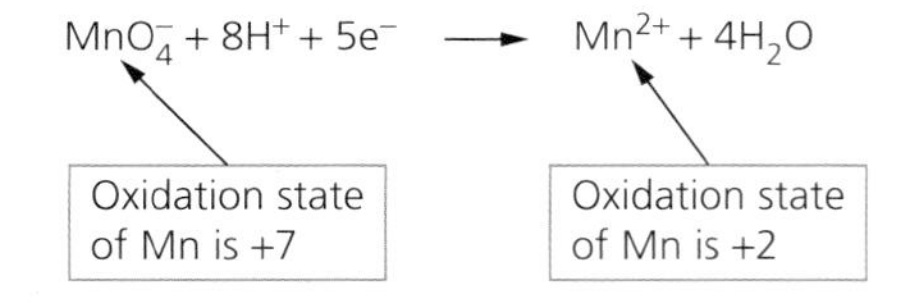

Acidified potassium manganate(VII) solution can be used to determine the concentration of reducing agents in two ways:

Direct titration

1 25.0 cm^3 of a solution of the reducing agent is placed in a conical flask.

2 The mixture is acidified using excess dilute sulfuric acid.

3 The standard solution of potassium manganate(VII) is added from the burette until the solution changes from colourless to pink.

Titration with iron(II) ions in solution produced from reduction of iron(III) ions

1 A known amount (volume and concentration or mass) of a reducing agent is added to a known volume and concentration of a solution containing iron(III) ions which are in excess.

   The reducing agent reduces some of the iron(III) to iron(II)

2 Place a known volume (usually $25.0\,cm^3$) of the reduced solution in a conical flask

3 Acidify the solution with excess dilute sulfuric acid

4 The standard solution of potassium manganate(VII) is added from the burette until the solution changes from colourless to pink

For all these titrations, as each drop of the potassium manganate(VII) solution is added, the purple manganate(VII) is decolourised to colourless manganese(II); when the reducing agent is used up the last drop of manganate(VII) is not decolourised so the last drop makes the solution pink.

Common examples of reducing agents

A reducing agent (or reductant) is an electron donor. It will donate electrons to other species so that they gain electrons and become reduced. A reducing agent becomes oxidised in a redox reaction.

- Iron(II) ions, Fe^{2+}, for example: iron(II) sulfate, ammonium iron(II) sulfate.
- Oxalate ions, $C_2O_4^{2-}$, for example: sodium oxalate, potassium oxalate.

Oxalate ions are also called ethanedioate ions as they are the ion from ethanedioc acid, $(COOH)_2$

Iron(II) ions

Manganate(VII) ions reduction:

$$MnO_4^- + 8H^+ + 5e^- \longrightarrow Mn^{2+} + 4H_2O$$

Iron(II) ions oxidation:

$$Fe^{2+} \longrightarrow Fe^{3+} + e^-$$

Combining equations:

$$MnO_4^- + 8H^+ + 5e^- \longrightarrow Mn^{2+} + 4H_2O$$

$$Fe^{2+} \longrightarrow Fe^{3+} + e^- \text{ (×5 to cancel electrons)}$$

$$MnO_4^- + 8H^+ + 5e^- \longrightarrow Mn^{2+} + 4H_2O$$

$$5Fe^{2+} \longrightarrow 5Fe^{3+} + 5e^-$$

Ionic equation:

$$MnO_4^- + 8H^+ + 5Fe^{2+} \longrightarrow Mn^{2+} + 4H_2O + 5Fe^{3+}$$

The most important part of the above equation is to realise that 1 mol of manganate(VII) ions, MnO_4^-, react with 5 mol of iron(II) ions, Fe^{2+}.

Oxalate ions

Manganate(VII) ions reduction:

$$MnO_4^- + 8H^+ + 5e^- \rightarrow Mn^{2+} + 4H_2O$$

Oxalate ions oxidation:

$$C_2O_4^{2-} \rightarrow 2CO_2 + 2e^-$$

Combining equations:

$$MnO_4^- + 8H^+ + 5e^- \rightarrow Mn^{2+} + 4H_2O \quad (\times 2)$$

$$C_2O_4^{2-} \rightarrow 2CO_2 + 2e^- (\times 5)$$

$$2MnO_4^- + 16H^+ + 10e^- \rightarrow 2Mn^{2+} + 8H_2O$$

$$5C_2O_4^{2-} \rightarrow 10CO_2 + 10e^-$$

Ionic equation:

$$2MnO_4^- + 16H^+ + 5C_2O_4^{2-} \rightarrow 2Mn^{2+} + 10CO_2 + 8H_2O$$

The most important thing here is to realise that 2 mol of manganate(VI) ions, MnO_4^-, reacts with 5 mol of oxalate ions, $C_2O_4^{2-}$.

TIP

These are only two examples, but you can apply the techniques to any titration with manganate(VII) ions. The most important thing is to determine the ratio of the reducing agent to MnO_4^- ions.

Overall ratios

1 mol of MnO_4^-, reacts with 5 mol of Fe^{2+}

2 mol of MnO_4^-, reacts with 5 mol of $C_2O_4^{2-}$

EXAMPLE 1

1.39 g of a sample of hydrated iron(II) sulfate, $FeSO_4.xH_2O$ were dissolved in dilute sulfuric acid and the volume made up using deionised water to 250.0 cm^3 in a volumetric flask. A 25.0 cm^3 sample of this solution was acidified and titrated against 0.005 mol dm^{-3} potassium manganate(VII) solution. The average titre was found to be 20.0 cm^3.

Iron(II) sulfate is usually dissolved in dilute sulfuric acid as it is more soluble in it than in water; often solutions are cloudy when just in water.

1 State the colour change observed at the end point in a manganate(VII) titration.
2 Calculate the value of x in $FeSO_4.xH_2O$.

Answer

Moles of potassium manganate(VII) used = $\frac{20 \times 0.005}{1000}$ = 0.0001 mol

Remember that this is the same as the number of moles of manganate(VII) ions as 1 mol of $KMnO_4$ contains 1 mol of manganate(VII) ions, MnO_4^-.

In the equation, ratio of $MnO_4^- : Fe^{2+} = 1 : 5$

So moles of Fe^{2+} in 25.0 cm^3 in conical flask = $0.0001 \times 5 = 0.0005$ mol

Moles of Fe^{2+} in 250 cm^3 in volumetric flask = $0.0005 \times 10 = 0.005$ mol

Moles of hydrated iron(II) sulfate used = 0.005 mol

Each mole of hydrated iron(II) sulfate, $FeSO_4.xH_2O$ contains 1 mol of Fe^{2+} ions so the number of moles of $FeSO_4.xH_2O$ = moles of Fe^{2+}

TIP

A sketch of a titration is a good starting point so you know the processes that are involved in the preparation of a solution and the titration.

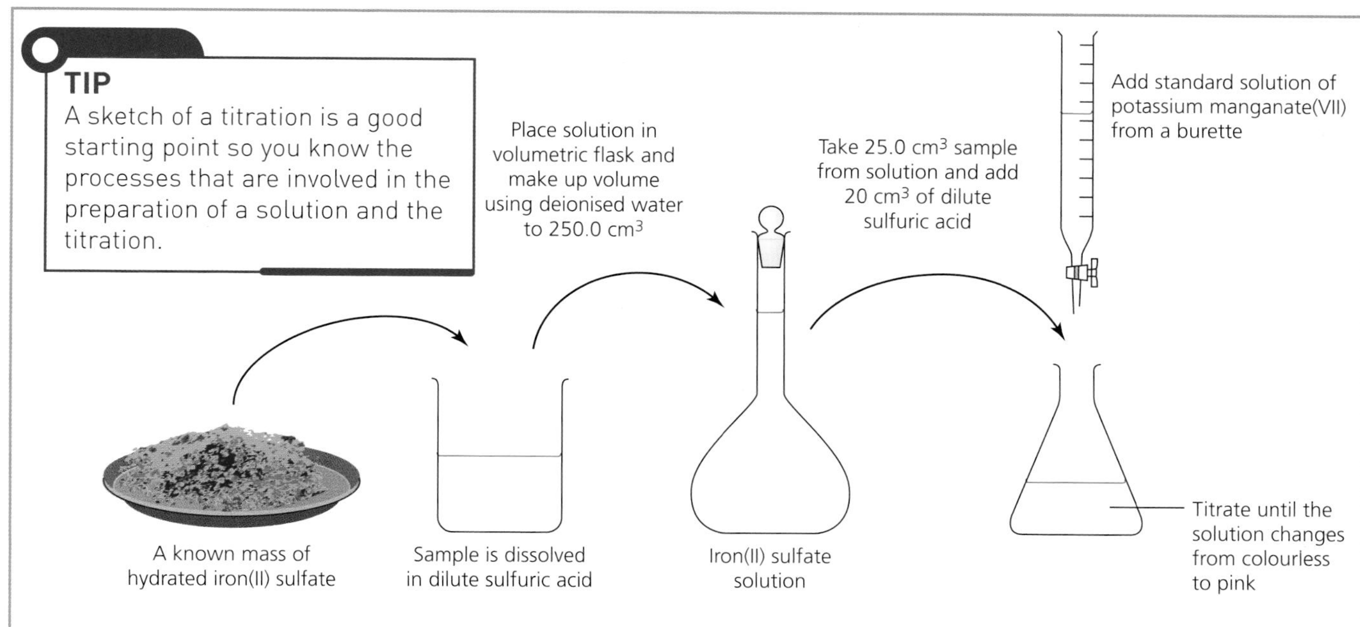

There are two ways of finishing the calculation from here:

Method 1: M_r

M_r of $FeSO_4.xH_2O = \frac{\text{mass}}{\text{moles}} = \frac{1.39}{0.005} = 278$

M_r of $FeSO_4 = 151.9$

Total of the M_r due to $xH_2O = 278 \quad 151.9 = 126.1$

$x = \frac{121.6}{18} = 7$ so the value of x in $FeSO_4.xH_2O = 7$

Method 2: ratios

Mass of anhydrous $FeSO_4$ in sample $= 0.005 \times 151.9$
$= 0.7595\,g$

Mass of water in sample $= 1.39 - 0.7595 = 0.6305\,g$

Moles of water $= \frac{\text{mass}}{M_r} = \frac{0.6305}{18} = 0.03503\,mol$

Ratio of anhydrous $FeSO_4 : H_2O = 0.005 : 0.03503$
$= 1 : 7$ so $x = 7$

EXAMPLE 2

Sulfate(IV) ions (also called sulphite ions) in natural seawater can reduce iron(III) ions to iron(II) ions. A $25.0\,cm^3$ sample of seawater containing sulfate(IV) ions, SO_3^{2-}, is treated with $25.0\,cm^3$ of $0.01\,mol\,dm^{-3}$ iron(III) chloride solution (an excess).

$SO_3^{2-} + H_2O + 2Fe^{3+} \rightarrow SO_4^{2-} + 2Fe^{2+} + 2H^+$

The resulting solution is placed in a conical flask and an excess of dilute sulfuric acid is added. The solution is titrated with $0.001\,mol\,dm^{-3}$ potassium manganate(VII) solution and $14.8\,cm^3$ of this solution was required.

Calculate the concentration of sulfate(IV) ions in the seawater in $mol\,dm^{-3}$.

Answer

moles of MnO_4^- used $= \frac{14.8 \times 0.001}{1000} = 1.48 \times 10^{-5}\,mol$

$MnO_4^- + 8H^+ + 5Fe^{2+} \rightarrow Mn^{2+} + 5Fe^{3+} + 4H_2O$

moles of $Fe^{2+} = 1.48 \times 10^{-5} \times 5 = 7.4 \times 10^{-5}\,mol$

moles of SO_3^{2-} in $25.0\,cm^3 = \frac{7.4 \times 10^{-5}}{2} = 3.7 \times 10^{-4}\,mol$

concentration of SO_3^{2-} in $mol\,dm^{-3} = 3.7 \times 10^{-4} \times 40 =$ $1.48 \times 10^{-3}\,mol\,dm^{-3}$

TIP

At this stage in the A-level course you would be expected to recognise that dividing by 25 and multiplying by 1000 is the same process as multiplying by 40. This is common when converting from an amount, in moles, in $25\,cm^3$ to a concentration in $mol\,dm^{-3}$

EXAMPLE 3

Hydrogen peroxide is used in hair bleach. A sample of hair bleach (10.0 cm^3) was added to a volumetric flask and the volume made up to 250 cm^3 of aqueous solution. A 25.0 cm^3 sample of this solution was acidified and it was found that it reacted completely with 17.25 cm^3 of 0.0274 mol dm^{-3} potassium manganate(VII) solution.

1 a) Write half equations for the reduction of manganate(VII) ions in acidic solution and the oxidation of hydrogen peroxide to oxygen gas.
b) Write an ionic equation for the reaction of acidified manganate(VII) ions with hydrogen peroxide.

2 Calculate the concentration of hydrogen peroxide in the hair bleach in mol dm^{-3}.

Answer

Each half equation is written using H_2O and H^+ to balance the change in oxygen content and then the electrons balance the charge on each side.

1 a) $MnO_4^- + 8H^+ + 5e^- \rightarrow Mn^{2+} + 4H_2O$
$H_2O_2 \rightarrow O_2 + 2H^+ + 2e^-$
b) $2MnO_4^- + 6H^+ + 5H_2O_2 \rightarrow 2Mn^{2+} + 8H_2O + 5O_2$

The ratio of $MnO_4^- : H_2O_2$ in the ionic equation above is 2:5. This ratio is important in calculating the amount, in moles, of hydrogen peroxide which reacted with the manganate(VII) ions.

2 moles of $MnO_4^- = \frac{17.25 \times 0.0274}{1000} = 4.727 \times 10^{-4}$ mol

moles of H_2O_2 in 25 cm^3 = $\frac{4.727 \times 10^{-4}}{2} \times 5$ = 1.182×10^{-3} mol

moles of H_2O_2 in 250 cm^3 = moles of H_2O_2 in original 10 cm^3 = $1.182 \times 10^{-3} \times 10$ = 0.01182 mol

concentration of original solution = $\frac{0.01182}{10} \times 1000$ = 1.182 mol dm^{-3}

TIP

When writing these equations make sure that the charges balance in every equation.

In the next example two ions in the solution can both react with the manganate(VII) ions.

EXAMPLE 4

Iron(II) ethanedioate dihydrate, $FeC_2O_4.2H_2O$, reacts with acidified potassium manganate(VII) solution. Both the iron(II) ion and the ethanedioate ion are oxidised by the manganate(VII) ions.

$5Fe^{2+} + MnO_4^- + 8H^+ \rightarrow 5Fe^{3+} + Mn^{2+} + 4H_2O$

$5C_2O_4^{2-} + 2\,MnO_4^- + 16H^+ \rightarrow 10CO_2 + 2Mn^{2+} + 8H_2O$

A sample of iron(II) ethanedioate was dissolved in excess dilute sulfuric acid and made up to 250 cm^3 of solution. 25.0 cm^3 of this solution decolourised 21.45 cm^3 of a 0.0187 mol dm^{-3} solution of potassium manganate(VII).

1 Calculate the reacting ratio of moles of manganate(VII) ions to moles of iron(II) ethanedioate.

2 Calculate the mass of the sample of iron(II) ethanedioate dihydrate used.

Answer

1 5 mol of FeC_2O_4 in solution produces 5 mol of Fe^{2+} and 5 mol of $C_2O_4^{2-}$. 5 mol of FeC_2O_4 will react with 3 mol of MnO_4^-

2 moles of $MnO_4^- = \frac{21.45 \times 0.0187}{1000} = 4.011 \times 10^{-4}$ mol

moles of FeC_2O_4 in 25 cm^3 = $\frac{4.011 \times 10^{-4}}{3} \times 5$ = 6.685×10^{-4} mol

moles of FeC_2O_4 in 250 cm^3 = $6.685 \times 10^{-4} \times 10$ = 6.685×10^{-3} mol
mass of original sample used = $6.685 \times 10^{-3} \times 179.8$ = 1.202 g

TIP

This can be worked out from the equations as if the equations were combined 3 mol of MnO_4^- would react with 5 mol of Fe^{2+} and $C_2O_4^{2-}$.

TIP

The ratio of ethanedioate ions to manganate(VII) is 5:2. The two ionic equations given in this example are important and should be learnt as it can take time working them out from half equations.

ACTIVITY

Finding the formula of Mohr's salt

Crystalline ammonium iron(II) sulfate is called Mohr's salt after the German chemist Karl Friedrich Mohr, the son of a pharmacist who made many important advances in titration methods in the 19th century, including an improved burette. Before this, chemists used equipment similar to a graduated cylinder.

Mohr's salt is a common laboratory reagent that is used for titrations. Due to the presence of ammonium ions it is less prone to oxidation by air than many other compounds. The oxidation of solutions of iron(II) is very pH dependent, occurring much more readily at high pH.

Mohr's salt has the formula $Fe(NH_4)_2(SO_4)_2.nH_2O$. The symbol *n* represents the number of molecules of water of crystallisation.

To find the value of *n*, 25.0 cm^3 of a solution of ammonium iron(II) sulfate of concentration 31.4 g dm^{-3} were transferred to a conical flask; 10 cm^3 of sulfuric acid were added and the solution then titrated with a solution of potassium manganate(VII) of concentration at 0.0200 mol dm^{-3}. A student obtained the results shown in Table 9.1.

Table 9.1 Results obtained in titration of Mohr's salt solution with potassium manganate(VII)

	Rough	Titration 1	Titration 2	Titration 3
Final volume	21.10	20.05	21.40	20.00
Initial volume	0.10	0.00	1.20	0.05
Titre/cm^3				

1 Why is 10 cm^3 of sulfuric acid added to the conical flask before titration?
2 Why is ethanoic acid not used?
3 Why is ammonium iron(II) sulfate less readily oxidised than iron(II) sulfate solution?
4 Complete the last row in Table 9.1 (Titre/cm^3).
5 Calculate the mean titre and justify your answer.
6 State the colour change at the end point and explain why no indicator is added.
7 The error in the titration value for this experiment is ±0.25 cm^3. Calculate the percentage error in this titration.
8 Write the equation for the reaction of iron(II) ions with acidified permanganate ions (H^+/MnO_4^-).
9 Determine the concentration of Mohr's salt solution and use it to find the molar mass of Mohr's salt and deduce the value of *n*.

TEST YOURSELF 3

1 A solution of iron(II) sulfate was acidified. A 25.0 cm^3 sample of the solution was titrated with 0.0120 mol dm^{-3} potassium manganate(VII) solution. 17.45 cm^3 of the potassium manganate(VII) solution was required for complete reaction.
   a) Write an equation for the reaction between manganate(VII) ions and iron(II) ions.
   b) Calculate the concentration of the iron(II) sulfate solution in mol dm^{-3}. Give your answer to 3 significant figures.
2 A sample of iron(II) ethanedioate, FeC_2O_4, was dissolved in excess sulfuric acid and the volume made up to 250 cm^3 in a volumetric flask. A 25.0 cm^3 portion of this solution was titrated with a 0.00200 mol dm^{-3} solution of potassium permanganate(VII). 24.75 cm^3 were required for complete reaction. Calculate the mass of the sample of iron(II) ethanedioate. Give your answer to 3 significant figures.
3 12.2 g of hydrated iron(II) sulfate, $FeSO_4.xH_2O$, were dissolved in acidic solution and the volume made up to 500 cm^3. A 25.0 cm^3 sample of this solution was titrated against 0.0200 mol dm^{-3} potassium manganate(VII). 21.95 cm^3 of the solution were required. Calculate the value of x in $FeSO_4.xH_2O$.

Catalytic activity

Transition metals and their compounds are important catalysts.

A homogeneous catalyst is one in which the catalyst is in the same physical state or phase as the reactants. Enzymes are homogeneous catalysts as they are in the same physical state or phase as the reactants – all are aqueous.

A heterogeneous catalyst is one in which the catalyst is in a different physical state or phase to the reactants. Iron is the catalyst in the Haber Process:

$$N_2(g) + 3H_2(g) \rightleftharpoons 2NH_3(g)$$

The iron is described as a heterogeneous catalyst as it is in a different state or phase to the reactants – the iron is a solid and the reactants are gases.

Heterogeneous catalysts work by chemisorption. You may be asked to explain the process of chemisorption:

- reactants molecules are adsorbed onto active sites on the surface of the catalyst
- the bonds in the reactants molecules are weakened
- the reactants molecules are held in a more favourable conformation for reaction
- the product molecules are desorbed from the surface of the catalyst.

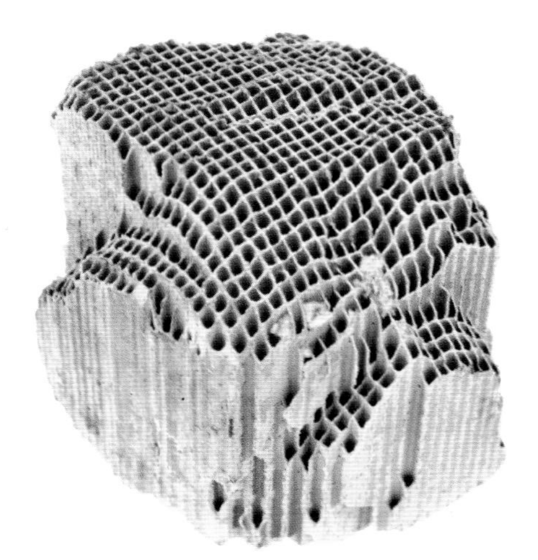

Figure 9.2 The catalyst found in a catalytic converter is usually platinum or rhodium plated onto a ceramic honeycomb structure. The honeycomb structure allows for a larger surface area for the reactant gases to undergo chemisorption on the surface of the catalysts.

Common heterogeneous catalysts

Haber Process:

$N_2(g) + 3H_2(g) \rightleftharpoons 2NH_3(g)$ Iron catalyst

Contact Process:

$2SO_2(g) + O_2(g) \rightleftharpoons 2SO_3(g)$ Vanadium(V) oxide catalyst

Hydrogenation of alkenes:

$C_2H_4(g) + H_2(g) \rightleftharpoons C_2H_6(g)$ Finely divided nickel catalyst

Catalysts in industrial processes

- In any industrial process, the manufacturers want to achieve the maximum yield of product with the minimum of expense.
- Often the use of a high temperature, a high pressure, and a catalyst can increase the yield of their product.
- However, these factors can be expensive and a cost analysis must be carried out to make sure there are significant benefits in terms of profit to justify the expense.
- High pressure is expensive to apply as electrical pumping is needed and valves and other equipment required for high pressures are also more expensive than those which could be used at lower pressures. High pressure is also expensive in terms of the containers needed and it is also dangerous for staff working on the site though many questions ask you not to focus on safety.
- The rate of a chemical reaction increases with temperature, so if a manufacturer uses a lower temperature then the rate is lowered even if more of the product is obtained (but in a much longer time period). A compromise temperature is used which allows enough of the product to be made in as short a time period as possible.

Figure 9.3 Soybeans.

- Development of more effective catalysts can increase the profit of an industrial process. The best catalysts will give the lowest activation energy but there must be a balance between the effectiveness of the catalyst and its cost.

The platinum catalyst used to make hydrogen from the electrolysis of water is a large part of the cost. A new catalyst made from molybdenum and ground soybeans has just been discovered and it is around 1500 times cheaper than platinum. This could make hydrogen a cheaper, clean fuel of the future.

Recognising a catalyst

A series of reactions may be given in which a substance reacts but is reformed in later reactions. This substance is a catalyst as it meets the description of a catalyst and increases the rate of the reaction and is not used up during the reaction.

EXAMPLE 5

In the Contact process for the production of sulfuric acid, the following reactions occur:

$$SO_2 + V_2O_5 \rightarrow SO_3 + V_2O_4$$

The vanadium(V) oxide reacts with the sulfur(IV) oxide to form vanadium(IV) oxide and sulfur(IV) oxide

The second reaction which occurs is:

$$2V_2O_4 + O_2 \rightarrow 2V_2O_5$$

The first step in combining a series of reactions into a single equation is to increase the balancing numbers in one equation if necessary.

In this example there are 2 mol of V_2O_4 and 2 mol of V_2O_5 in the second equation but only 1 mol of both V_2O_4 and V_2O_5 in the first equation. So the first equation should be doubled, giving:

$$2SO_2 + 2V_2O_5 \rightarrow SO_3 + 2V_2O_4$$

$$2V_2O_4 + O_2 \rightarrow 2V_2O_5$$

Write down all substances on the left of both equations then insert an arrow and write out all substances on the right of both equations. Cancel out substances that appear on both sides of the equation.

$$2SO_2 + 2V_2O_5 + 2V_2O_4 + O_2 \rightarrow SO_3 + 2V_2O_4 + 2V_2O_5$$

$2V_2O_4$ and $2V_2O_5$ appear on both sides of the equation so they can be cancelled out.

$$2SO_2 + \cancel{2V_2O_5} + \cancel{2V_2O_4} + O_2 \rightarrow SO_3 + \cancel{2V_2O_4} + \cancel{2V_2O_5}$$

The equation simplifies to:

$$2SO_2 + O_2 \rightarrow 2SO_3$$

V_2O_5 is a catalyst as it takes part in the reaction but is reformed as it does not appear in the overall equation for the reaction. It is not used up.

EXAMPLE 6

Hydrogen peroxide decomposes in the presence of potassium iodide. Making 'elephant toothpaste' is a fun demonstration showing how the catalyst potassium iodide causes the rapid decomposition of hydrogen peroxide, generating a large volume of oxygen, so the oxygen quickly pushes out of the container. The soapy water traps the oxygen, creating bubbles, and turns into foam which looks like giant toothpaste as shown in the photo.

The equations for the decomposition are:

$$H_2O_2 + KI \rightarrow KIO + H_2O$$

$$H_2O_2 + KIO \rightarrow KI + H_2O + O_2$$

In the example the equation will balance out when added together.

The overall added equation is:

$$2H_2O_2 + \cancel{KI} + \cancel{KIO} \rightarrow \cancel{KIO} + \cancel{KI} + 2H_2O + O_2$$

This leaves:

$$2H_2O_2 \rightarrow 2H_2O + O_2$$

Potassium iodide is a catalyst as it takes part in the reaction but is reformed as it does not appear in the overall equation.

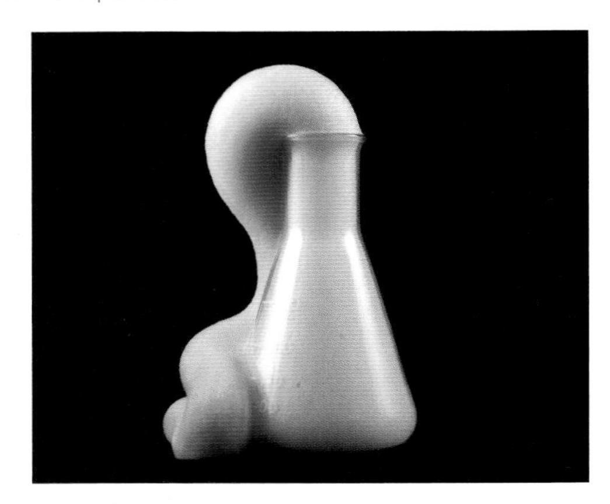

Catalyst poisoning

A heterogeneous catalyst may be poisoned by impurities in the reaction mixture. The heterogeneous catalyst can be coated by impurities that render it passive and inactive. An example of this is the use of lead-containing petrol in cars with a catalytic convertor that would normally take unleaded petrol. The lead in the petrol coats the surface of the catalyst in the convertor and makes it ineffective as a catalyst.

The catalyst is poisoned as the active sites on the surface of the catalyst become blocked and therefore inactive. This has a cost implication as the catalyst has to be replaced. Transition metals and their compounds are often used as heterogeneous catalysts. These catalysts are expensive.

Homogeneous catalysis

The reaction between peroxodisulfate ions, $S_2O_8^{2-}$, and iodide ions, I^- into sulfate(VI) ions and iodine is an example of homogeneous catalysis.

The equation for this reaction is:

$$S_2O_8^{2-} + 2I^- \rightarrow 2SO_4^{2-} + I_2$$

The reaction between $S_2O_8^{2-}$ ions and I^- ions is slow (has a high activation energy) as the two negative ions repel each other.

However, the reaction is catalysed by the either aqueous iron(II) ions, Fe^{2+}, or aqueous iron(III) ions, Fe^{3+}.

The Fe^{2+} ions catalyse the reaction as shown below:

$$2Fe^{2+} + S_2O_8^{2-} \rightarrow 2Fe^{3+} + 2SO_4^{2-}$$

$$2Fe^{3+} + 2I^- \rightarrow 2Fe^{2+} + I_2$$

> **TIP**
> The iron(II) or iron(III) ions are present in solution as are the reactant ions. This is an example of homogeneous catalysis.

Both iron(II) ions and iron(III) ions catalyse this reaction as the reactions above can occur in either order as both $S_2O_8^{2-}$ and I^- ions are present in the solution.

Homogeneous autocatalysis

Autocatalysis occurs when the product in a reaction acts as a catalyst for the reaction. A common example of this is:

Reaction of manganate(VII) ions, MnO_4^-, with ethanedioate ions, $C_2O_4^{2-}$.

The equation for the reaction is:

$$5C_2O_4^{2-} + 2MnO_4^- + 16H^+ \rightarrow 10CO_2 + 2Mn^{2+} + 8H_2O$$

The reaction between MnO_4^- ions and $C_2O_4^{2-}$ ions is initially slow as the two negative ions repel each other.

However, the reaction is catalysed by the manganese(II) ions, Mn^{2+}, formed in the reaction. As these are formed the reaction rate increased dramatically.

The Mn^{2+} ions catalyse the reaction in two reactions:

$$4Mn^{2+} + MnO_4^- + 8H^+ \rightarrow 5Mn^{3+} + 4H_2O$$

$$2Mn^{3+} + C_2O_4^{2-} \rightarrow 2Mn^{2+} + 5CO_2$$

Because the Mn^{2+} ion is positively charged it is attracted to the MnO_4^- ion, therefore increasing the rate of the reaction.

The reaction can be monitored using a colorimeter as the MnO_4^- ion is purple in solution.

A graph of absorbance or concentration of the MnO_4^- ion against time looks like the one shown below. The graph has labels to explain the shape.

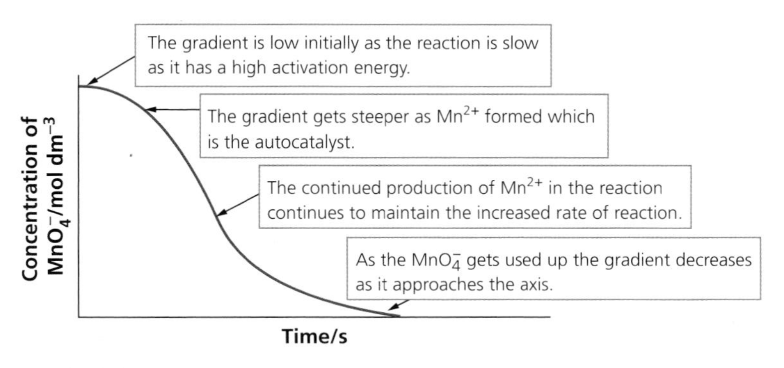

> **TIP**
> Make sure that you can sketch the shape of this curve or explain it. This is an example of homogeneous autocatalysis as the catalyst and reactant ions are all present in solution.

In both the examples above, the ions of manganese and iron are able to act as catalysts as they have variable oxidation states. The ability to change oxidation state is important to their role as a catalysts. Group 1 and 2 elements are not able to act as homogeneous catalysts as they do not have **variable oxidation states** in their compounds.

TEST YOURSELF 4

1 Write two equations to show the role of V_2O_5 in the contact process.

2 For the following reactions:

A: $N_2(g) + 3H_2(g) \rightleftharpoons 2NH_3(g)$

B: $S_2O_8^{2-} + 2I^- \rightarrow 2SO_4^{2-} + I_2$

C: $5C_2O_4^{2-} + 2MnO_4^- + 16H^+ \rightarrow 10CO_2 + 2Mn^{2+} + 8H_2O$

a) Name a catalyst for each reaction.

b) Which reaction(s) use a homogeneous catalyst?

c) Which reaction(s) use a heterogeneous catalyst?

d) Which reaction(s) show autocatalysis?

3 Explain why Group 1 metal compounds are not able to act as homogeneous catalysts but transition metal compounds can.

Practice questions

1 Which one of the following is the correct colour for a solution of the compound NH_4VO_3?

A yellow

B blue

C green

D violet *(1)*

2 Using the standard electrode potentials below, choose the oxidising agent capable of oxidising vanadium from the +2 to the +4 oxidation state but not to the +5 oxidation state.

	$E^\ominus$/V
$Cl_2(aq) + 2e^- \rightleftharpoons Cl^-(aq)$	+1.36
$I_2(aq) + 2e^- \rightleftharpoons 2I^-(aq)$	+0.54
$SO_4^{2-}(aq) + 2H^+(aq) + 2e^- \rightleftharpoons 2H_2O(l) + SO_2(g)$	+0.17
$VO^{2+}(aq) + 2H^+(aq) + e^- \rightleftharpoons V^{3+}(aq) + H_2O(l)$	+0.32
$VO_2^+(aq) + 2H^+(aq) + e^- \rightleftharpoons VO^{2+}(aq) + H_2O(l)$	+1.00
$V^{3+}(aq) + e^- \rightleftharpoons V^{2+}(aq)$	−0.26
$Cu^{2+}(aq) + 2e^- \rightleftharpoons Cu(s)$	+0.34

A chlorine

B Cu^{2+} ions

C iodine

D sulfate ions *(1)*

3 In which of the following lists are the transition metals all in the +6 oxidation state?

A	$KMnO_4$	K_2CrO_4	K_3CoO_4
B	$K_2Cr_2O_7$	KVO_3	K_2MnO_4
C	K_2MnO_4	$K_2Cr_2O_7$	K_2FeO_4
D	K_2CrO_4	K_3CoO_4	KVO_3

(1)

4 25.0 cm^3 of a solution of acidified hydrogen peroxide is titrated against 0.124 mol dm^{-3} potassium managate(VII) solution. 17.55 cm^3 of the potassium manganate(VII) solution are required. In this reaction the hydrogen peroxide is oxidised to oxygen gas.

a) Write an equation for the reaction between acidified hydrogen peroxide and potassium manganate(VII).

b) Calculate the concentration of the hydrogen peroxide solution in mol dm^{-3}. Give your answer to 3 significant figures. *(4)*

5 Manganate(VII) ions react with ethanedioate ions.

The half equations for the reactions are given below:

$$C_2O_4^{2-} \rightarrow 2CO_2 + 2e^-$$

$$MnO_4^- + 8H^+ + 5e^- \rightarrow Mn^{2+} + 4H_2O$$

a) Write an equation for the reaction between ethanedioate ions and manganate(VII) ions. *(1)*

b) Explain why this reaction may be followed by colorimetry. *(1)*

c) Sketch a graph of the concentration of manganate(VII) ions against time. Explain the shape of the graph. *(4)*

6 Vanadium(V) oxide, V_2O_5, is the catalyst used in the contact process. The equations showing its mode of action are given below:

$$SO_2 + V_2O_5 \rightarrow SO_3 + V_2O_4$$

$$2V_2O_4 + O_2 \rightarrow 2V_2O_5$$

a) Using these equations write an overall reaction for the reaction V_2O_5 catalyses in the contact process. *(1)*

b) Explain why V_2O_5 is a catalyst based on the equations. *(1)*

c) Explain why a compound of vanadium is able to act as a catalyst in this reaction *(1)*

7 A sample of iron(II) ethanedioate, FeC_2O_4, was dissolved in an excess of sulfuric acid. The volume was made up to 500 cm^3 and a 25.0 cm^3 sample of the solution titrated against 0.0148 mol dm^{-3} potassium manganate(VII) solution. 25.25 cm^3 of the potassium manganate(VII) solution were required.

The half equations involved are:

$MnO_4^- + 8H^+ + 5e^- \longrightarrow Mn_{2+} + 4H_2O$

$C_2O_4^{2-} \longrightarrow 2CO_2 + 2e^-$

$Fe^{2+} \longrightarrow Fe^{3+} + e^-$

a) Write an equation for the reaction between manganate(VII) ions and ethanedioate ions. *(1)*

b) Write an equation for the reaction between manganate(VII) ions and iron(II) ions. *(1)*

c) Using the equations you have written in (a) and (b) state the ratio of $MnO_4^- : FeC_2O_4$ *(1)*

d) Calculate the mass of iron(II) ethanedioate used. If you have been unable to calculate a ratio in (c), use a ratio of 2 $MnO_4^- : 7FeC_2O_4$. This is not the correct ratio. *(4)*

8 The graph below shows how the concentration of manganate(VII) ions changes against time during the reaction with ethanedioate ions, $C_2O_4^{2-}$.

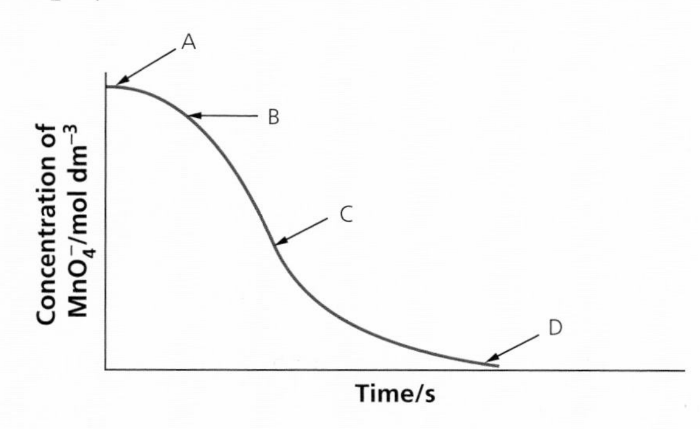

a) At which point on the graph (A, B, C or D) is the rate of the reaction fastest? *(1)*

b) Explain why the gradient of the graph is low at point A. *(2)*

c) Explain why the gradient of the line begin to increase between A and B. *(3)*

9 A sample of 4.483 g of hydrated iron(II) sulfate, $FeSO_4.7H_2O$, was dissolved in sulfuric acid and transferred to a volumetric flask. The volume was made up to 250 cm^3 and a 25.0 cm^3 sample was titrated against 0.0127 mol dm^{-3} potassium manganate(VII) solution.

a) Write an equation for the reaction between iron(II) ions and manganate(VII) ions. *(2)*

b) Calculate the volume of potassium manganate(VII) solution required for complete reaction. *(4)*

Stretch and challenge

10 Consider the half equations below and their associated redox potentials:

$VO_2^+ + e^- + 2H^+ \longrightarrow VO^{2+} + H_2O$	+1.00 V
$VO^{2+} + e^- + 2H^+ \longrightarrow V^{3+} + H_2O$	+0.34 V
$V^{3+} + e^- \longrightarrow V^{2+}$	−0.26 V
$Zn^{2+} + 2e^- \longrightarrow Zn$	−0.76 V
$SO_4^{2-}(aq) + 2H^+(aq) + 2e^- \rightleftharpoons 2H_2O(l) + SO_2(g)$	+0.17 V

a) Sulfur dioxide is mixed with a solution containing VO_2^+ ions.

i) Explain using the redox potentials which oxidation changes would occur. *(4)*

ii) Describe what would be observed. *(1)*

b) Zinc will reduce VO_2^+ ions to V^{2+} ions.

i) Explain using redox potentials why zinc will reduce vanadium from VO_2^+ to V^{2+}. *(3)*

ii) Write a half equation showing the overall reduction of vanadium from VO_2^+ to V^{2+}. *(2)*

iii) Write an ionic equation which shows the overall reaction between VO_2^+ and zinc where V^{2+} is the vanadium reduction product. *(2)*

10 Optical isomerism

COOH
C
H
OH
CH_3

PRIOR KNOWLEDGE

- Molecules which have the same molecular formula but a different structural formula are known as structural isomers.
- Stereoisomers are molecules which have the same structural formula but a different arrangement of atoms in space.
- E–Z isomers are stereoisomers, which exist due to restricted rotation about a C=C.

TEST YOURSELF ON PRIOR KNOWLEDGE 1

1 The compound C_4H_{10} exists as structural isomers.
  a) What is meant by the term structural isomers?
  b) How many structural isomers are there of C_4H_{10}? Name these isomers.
2 a) Draw and name a structural isomer of 1-chloropropane.
  b) Draw and name the stereoisomers of CHCl CHCl.
  c) Draw and name the stereoisomers of C_4H_8.
3 What is meant by the term stereoisomers?

There are two types of stereoisomers, E–Z which you met in Book 1 and **optical isomers**.

Carvone is an organic compound with the molecular formula $C_{10}H_{14}O$. It has two optical isomers, one that smells like caraway and the other that smells like spearmint. Both isomers have the same structural formula but are stereoisomers because they have a different arrangement of the atoms in space.

Optical isomerism

Optical isomerism occurs in molecules that have a carbon atom with four different atoms or groups attached to it tetrahedrally as shown in Figure 10.1.

Such compounds exhibit optical isomerism because the tetrahedral molecule is asymmetric, i.e. it has no centre, plane or axis of symmetry. As a result two tetrahedral arrangements occur in space; one is the mirror image of the other and they cannot be superimposed on each other. Figure 10.2 shows the optical isomers of 2-hyroxypropanoic acid. The mirror images are non-superimposable. All molecules have a mirror image, but usually it is the same molecule; non-superimposable mirror images and hence optical isomers only occur with chiral molecules. The two mirror images are optical isomers, and are often called **enantiomers**.

A molecule that is asymmetric and exhibits optical isomerism is said to be chiral. The carbon atom, with four different atoms or groups attached to it is called a **chiral** or **asymmetric centre**.

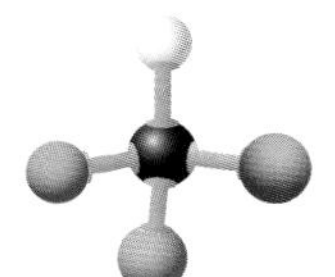

Figure 10.1 The carbon atom (black) has four different atoms attached to it tetrahedrally. The carbon atom is a chiral centre.

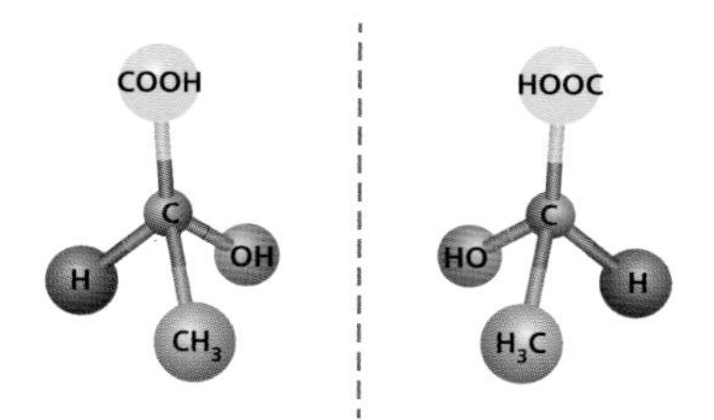

Figure 10.2

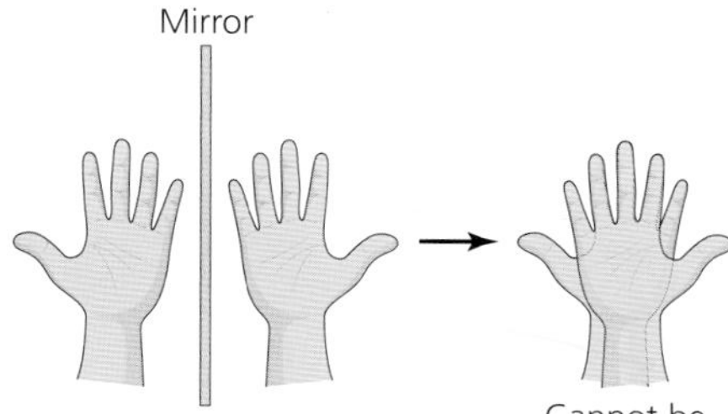

Figure 10.3 Your left and right hands are non-superimposable mirror images. A molecule is chiral if, like one of your hands, it cannot be superimposed on its mirror image. The word 'chiral' comes from the Greek word for hand.

When asked to draw optical isomers, a general method to follow is given below.

- Draw the displayed formula.
- Identify the chiral centre.
- Draw the 3d tetrahedral structure based on the chiral centre and insert the four different groups.
- Draw a dotted line to represent a mirror, and draw the second isomer as a mirror image.

EXAMPLE 1

Draw the optical isomers for 2-chlorobutane.

Answer

First draw the displayed formula.

```
     H    H    H    H
     |    |    |    |
H —  C —  C —  C —  C — H
     |    |    |    |
     H    Cl   H    H
```

Then identify the chiral centre. Looking closely at the molecule it is clear that the two end carbon atoms have 3 H atoms attached and the third carbon atom from the left has 2 H atoms so these cannot be chiral centres. The second carbon atom has the following groups attached to it:

1 H **2** Cl **3** CH_3 **4** C_2H_5

These are four different groups so it is a chiral centre and is marked with an asterisk as shown:

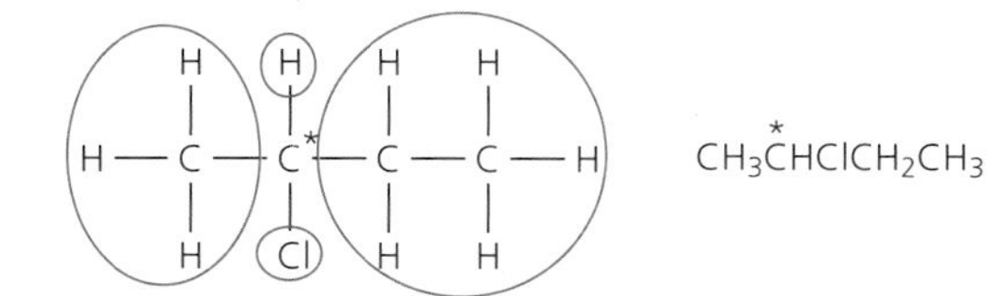

$CH_3\overset{*}{C}HClCH_2CH_3$

TIP

It is often useful to circle each of the four groups on the chiral carbon, this helps you remember what groups to place around the tetrahedron.

Finally to draw the optical isomers, simply draw a 3 dimensional tetrahedral arrangement and insert each of the four different groups at different points on the tetrahedron. Then place a dotted line to represent the mirror, and reflect the image as shown below.

These isomers cannot be superimposed on each other. They have the same molecular and structural formulae but differ only in the arrangement of groups around the chiral centre. You are expected to be able to draw optical isomers for molecules which have a single chiral centre.

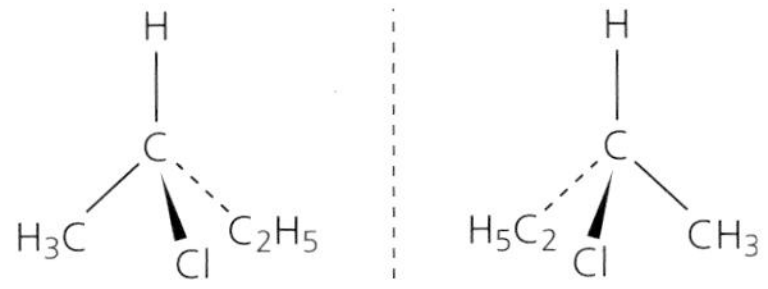

H, H_3C, Cl, C_2H_5 | H, H_5C_2, Cl, CH_3

TIP

Remember when drawing 3d molecules a normal line represents a bond in the plane of the paper, a dashed bond represents a bond extending backwards 'into' the paper and a wedged line means the bond protrudes forward, effectively 'out' of the paper.

EXAMPLE 2

Nineteen out of the 20 amino acids used in protein synthesis in the cells in your body are optically active – an example is the amino acid alanine. The one that is not optically active is glycine. State why glycine is not optically active and draw the optical isomers of alanine.

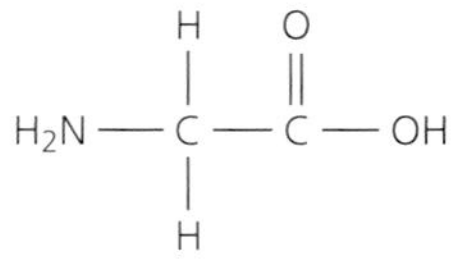

$H_2N-CH(H)-C(=O)-OH$ glycine

$H_2N-CH(CH_3)-C(=O)-OH$ alanine

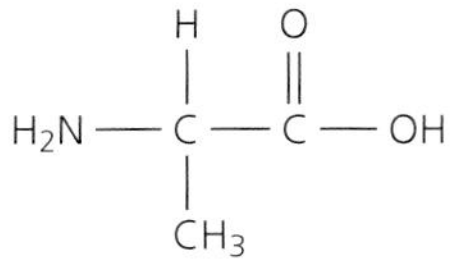

Answer

Glycine is not optically active as the central carbon atom does not have four different atoms or groups attached to it – the carbon has two H atoms attached.

Alanine has the following four groups attached to the central carbon atom.

1 H 2 CH_3 3 NH_2 4 COOH

The central carbon atom of alanine is a chiral centre.

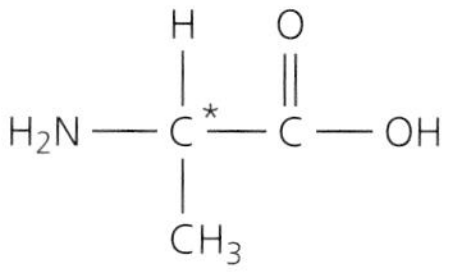

$H_2N-C^*H(CH_3)-C(=O)-OH$ $H_2N\overset{*}{C}H(CH_3)COOH$

The optical isomers can be drawn as:

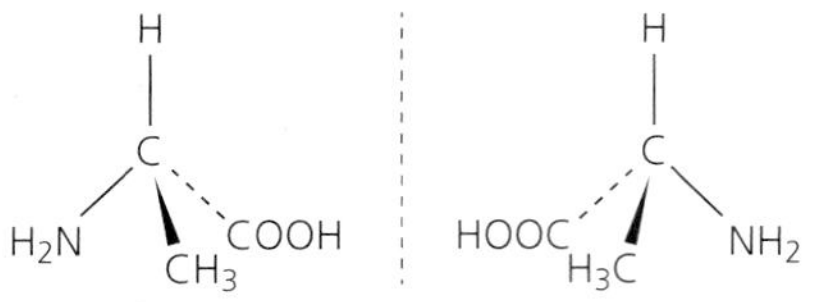

H, H_2N, CH_3, COOH | H, HOOC, H_3C, NH_2

TIP

Remember to correctly attach the atoms – the COOH must be drawn backwards in the mirror image.

TEST YOURSELF 1

1 Which of the following molecules have chiral centres?

a) $CH_3CH_2CH_2OH$
b) $CH_3CH_2CHCl\,CH_2CH_3$
c) $CH_3CH(CH_3)CO_2H$
d) $CH_3CH(OH)CO_2CH_3$
e) $CH_3CH_2CH_2C(OH)_2CH_2CH_2$
f) $CH_3CH_2CH_2CHClCH_3$

2 Lactic acid, $CH_3CHOHCOOH$, is a weak acid which builds up in muscles during exercise. The molecule has an asymmetric centre.

a) Explain what is meant by the term asymmetric centre.
b) Draw the structures of the two optical isomers of lactic acid.
c) What is the bond angle around the central carbon?
d) State the IUPAC name for lactic acid.

Optical activity

Unpolarised light Polariser Plane polarised light

A light beam consists of waves that vibrate in all planes at right angles to the direction in which the beam is travelling. A polaroid filter only allows light in one plane to pass through it, so when a light beam is passed through a polarising filter all the waves are absorbed apart from the ones vibrating in one particular plane. The light is said to be **plane polarised**.

An **optically active substance** is one that can rotate the plane of plane polarised light. When a beam of plane polarised light is passed through a solution of one optical isomer, the plane polarised light is rotated either to the left or to the right depending on which isomer is used.

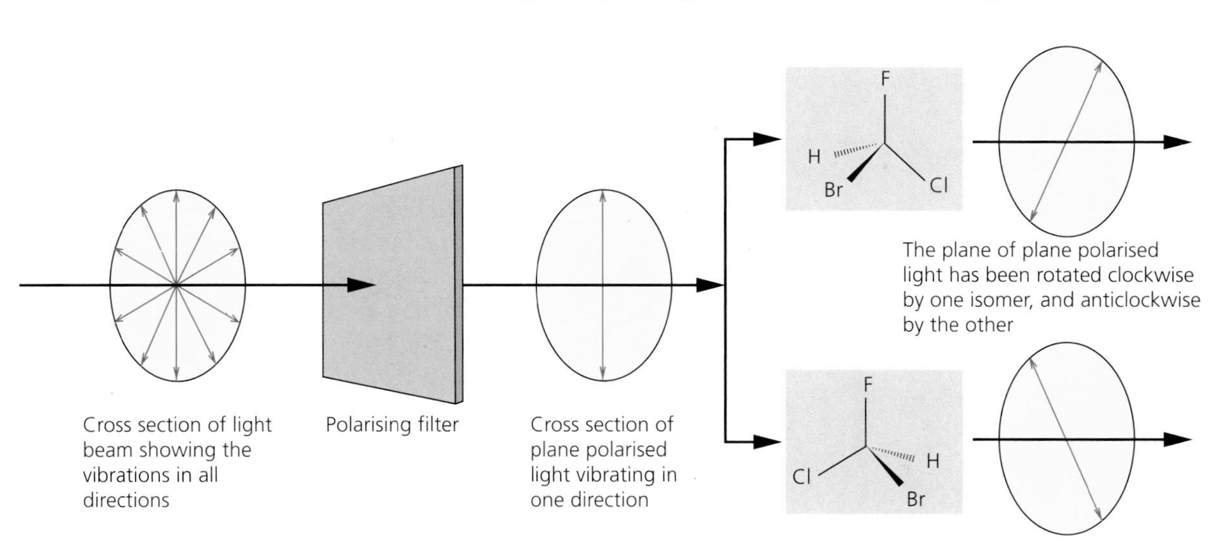

Figure 10.4 The diagram shows how light is polarised and the plane polarised light rotated in different directions by passing through the two optical isomers of bromochlorofluoromethane.

Optical isomers each *rotate the plane of plane polarised light in opposite directions* and hence they are optically active.

- If an optical isomer (enantiomer) rotates plane polarised light to the left, it is given the prefix – or L (which stands for laevorotatory)

- If an optical isomer (enantiomer) rotates plane polarised light to the right, it is given the prefix + or D (which stands for dextrorotatory). You will not be asked to assign a D or L prefix to an isomer simply from its structure.
- Mixing equal amounts of the same concentration of two enantiomers gives an **optically inactive mixture** which has no effect on plane polarised light because the two opposite effects cancel out. This mixture of equal amounts of each enantiomer is called a **racemic mixture** or **racemate**.

A **racemic mixture** (racemate) is an optically inactive mixture of equal amounts of enantiomers of the same concentration.

Plane polarised light is light in which all the waves vibrate in the same plane.

Optical isomers (enantiomers) are stereoisomers that occur as a result of chirality in molecules. They exist as non-superimposable mirror images and differ in their effect on plane polarised light.

An **asymmetric carbon** atom is chiral and has four different atoms or groups attached.

An **optically active substance** is one that can rotate the plane of plane polarised light.

Chiral: the structure and its image are non-superimposable.

EXAMPLE 3

Explain how you could distinguish between a racemate of lactic acid and one of the enantiomers of lactic acid?

Answer

Plane polarised light will be rotated by the single enantiomer but it will be unaffected by the racemate. The racemate is optically inactive as it contains equal amounts of each isomer, and one isomer rotates plane polarised light to the right, the other to the left and the two opposite effects cancel out.

When a chiral compound is synthesised in the laboratory, a mixture of optical isomers is often formed. Many drugs have optical isomers, and often complicated separation techniques are needed to separate the mixture. Separation is difficult because optical isomers have similar physical properties. Ibuprofen is a drug that targets muscle and bone pain, headaches and back pain. It is a chiral structure and has optical isomers. You can read more of the use of optical isomers in the pharmaceutical industry in Chapter 14.

TEST YOURSELF 2

1 How could you show that a solution of a drug was a racemate and not a single enantiomer?

2 2-bromobutane is optically active.

  a) Draw the displayed formula for 2-bromobutane.

  b) Mark the chiral carbon atom with an asterisk (*).

  c) Draw the two structures of 2-bromobutane that will form a racemic mixture.

  d) Explain why this racemic mixture will not rotate the plane of plane polarised light.

3 What is plane polarised light?

Practice questions

1 Which one of the following does not contain an asymmetric centre?

A $CH_3CH(CH_3)CH_2CH_3$

B $CH_3CH(OH)CH_2CH_3$

C $CH_3CHClCH_2CH_3$

D $CH_3CH(NH_2)CH_2CH_3$ *(1)*

2 Which one of the following displays optical isomerism?

A $CH_3CHOHCOOH$

B CH_2OHCH_2COOH

C $CH_2OHCOOCH_3$

D CH_2OHCH_2CHO *(1)*

3 3-hydroxybutanal, $CH_3CH(OH)CH_2CHO$ is optically active.

a) Explain the term **optically active**. *(1)*

b) Draw the three-dimensional structures for the two optical isomers of 3-hydroxybutanal. *(2)*

c) 3-hydroxybutanal may be dehydrated to form but-2-enal. Suggest a structure for but-2-enal. *(1)*

4 There are a number of structural isomers of C_4H_9OH. However, only one has an asymmetric centre and can exist as optical isomers.

a) What is meant by the term asymmetric centre? *(1)*

b) Explain in terms of structure the meaning of the term optical isomers. *(1)*

c) Draw the displayed formula of the structural isomer of C_4H_9OH which contains an asymmetric centre. *(1)*

d) Draw the two optical isomers of the molecule identified in (c). *(2)*

e) How can a solution of one optical isomer be distinguished from a solution of another? *(2)*

f) Only one isomer of C_4H_9OH can be oxidised to a ketone using acidified potassium dichromate(VI). Draw the alcohol and the corresponding ketone and state the colour change that occurs. *(4)*

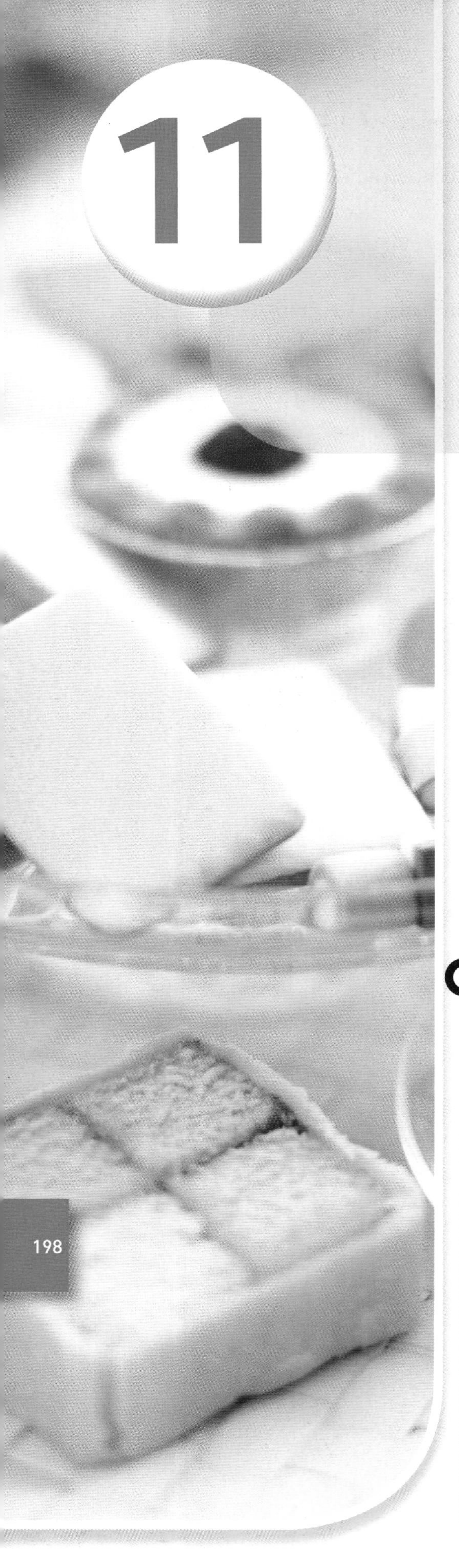

11 The carbonyl group

PRIOR KNOWLEDGE

- A functional group is a group of atoms that is responsible for the characteristic reactions of a molecule.
- A displayed formula shows how atoms are arranged and the bonds between them. Ionic parts of the molecule are shown using charges.
- A structural formula shows the arrangement of atoms in a molecule, carbon by carbon, with the attached hydrogens and functional groups.
- A skeletal formula shows the carbon skeleton with hydrogen atoms removed and functional groups present.
- Be able to apply the IUPAC rules for nomenclature to name organic compounds containing up to six carbon atoms.
- A homologous series is a series of compounds which have the same general formula. Successive members differ from each other by a $-CH_2$ group. They have similar chemical reactions and show a gradation in physical properties.
- Molecules which have the same molecular formula but a different structural formula are known as structural isomers
- Primary alcohols can be oxidised to aldehydes and then to carboxylic acids. Secondary alcohols can be oxidised to ketones, which are fairly resistant to oxidation. Tertiary alcohols are not easily oxidised.

The carbonyl group is found in many organic compounds including aldehydes, ketones, esters and acids. Benzaldehyde is an aldehyde that is used to make almond essence, an ingredient in Bakewell tarts. It is also added to give flavour to marzipan, which is commonly used on Christmas cakes and Battenberg cakes.

TEST YOURSELF ON PRIOR KNOWLEDGE 1

1 Name the following compounds.

a) $CH_3CH(OH)CH_2CH_3$

b) $CH_3CH_2CH{=}CHCH_2CH_2CH_3$

c)

d)

e)

```
     H   H   H
     |   |   |
 H — C — C — C — H
     |   |   |
     H   |   H
     H — C — H
         |
         H
```

f) $CHBrClCHClCH_3$

g)

```
     H   H   H   Br
     |   |   |   |
 H — C — C — C — C — Cl
     |   |   |   |
     H   H   H   H
```

2 Choose one compound from question 1 which is represented as

a) a skeletal formula.

b) a displayed formula.

3 Name the functional group present in 1(a) and in (b)

4 Draw the structure of an isomer of the compound in 1(e).

5 To use a disposable breathalyser, a motorist blows into a tube containing crystals of acidified potassium dichromate(VI). Alcohol on the motorist's breath is detected by a colour change in the crystals.

a) What colour change occurs when acidified potassium dichromate(VI) reacts with ethanol?

b) Write an equation for the oxidation of ethanol using [O] to represent the acidified potassium dichromate(VI).

c) Name the organic product formed.

d) What class of alcohol is ethanol?

Chanel No. 5 was the first perfume made with synthetically produced aldehydes and is one of the world's most iconic perfumes. Perfume is a mixture of fragrant compounds and solvents which together give a pleasant smell. The earliest perfumes used natural ingredients such as extracts from flowers, grasses, spices and fruit. Chanel No. 5, developed in 1912 by Earnest Beaux, was the first perfume to use a synthetically produced aldehyde, 2-methylundecanal, together with jasmine, rose and lily scents to create its strong, pleasant smell. Today, perfumes contain a mix of natural and synthetic chemicals, many of which are aldehydes, ketones and esters that contain the carbonyl group. The **carbonyl group** is the functional group in a range of organic molecules. In this chapter you will study the aldehydes, ketones, carboxylic acids, esters, acid anhydrides, acyl chlorides and amides. Some examples of these are shown in Figure 11.1.

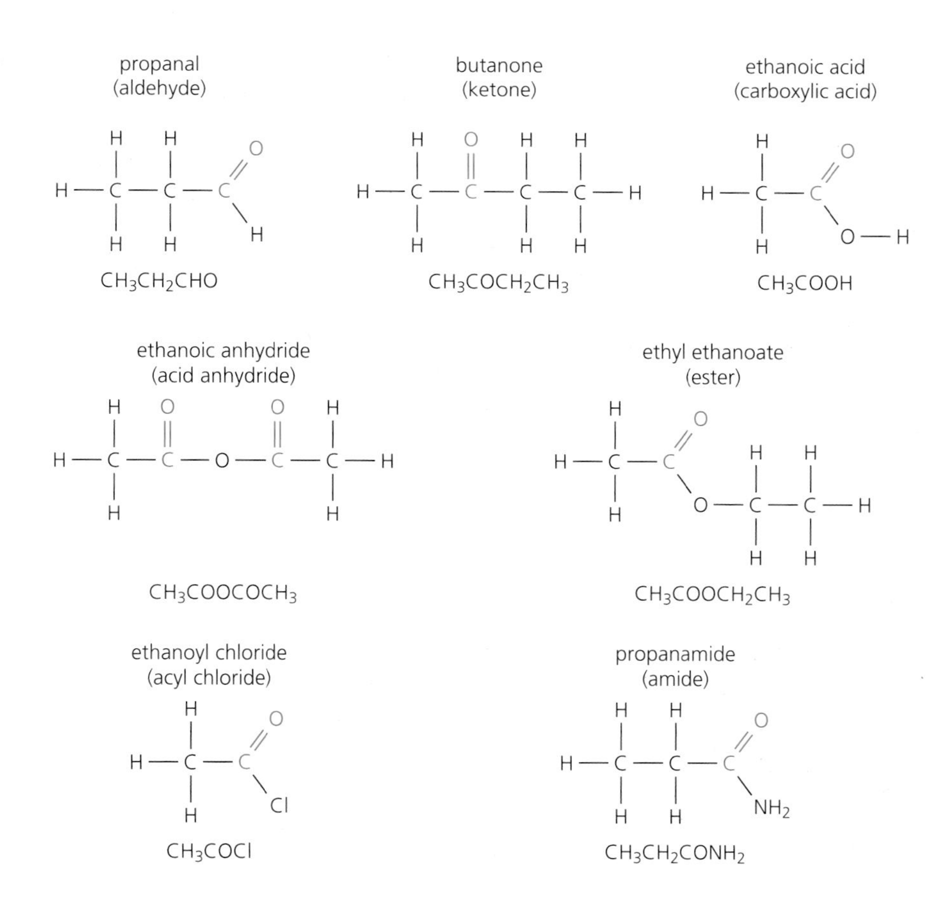

Figure 11.1 The carbonyl group, shown in red, is found in all of these different organic compounds.

Aldehydes and ketones

The carbonyl group

The carbonyl functional group is shown in Figure 11.2. It has a double bond and, as expected for a double bond, undergoes addition reactions. However the carbonyl double bond does not react in the same way as the carbon–carbon double bond in an alkene. This is because the C=O bond is polar. Oxygen is more electronegative than carbon and draws the electrons in the covalent bond towards itself resulting in a δ+ charge on the carbon and a δ– charge on the oxygen. The δ+ carbon is readily attacked by nucleophiles (a nucleophile is a lone pair donor) and hence the carbonyl group undergoes nucleophilic addition reactions.

Figure 11.2 The carbonyl functional group.

Naming aldehydes

Aldehydes have the general formula $C_nH_{2n}O$. The table below gives the names of the first four straight chain members of the aldehydes.

Name	Structural formula	Displayed formula
methanal	HCHO	
ethanal	CH_3CHO	
propanal	CH_3CH_2CHO	
butanal	$CH_3CH_2CH_2CHO$	

TIP

When writing structural formula always write the aldehyde as –CHO. Writing –COH is incorrect and easily confused with alcohols. Ethanal, for example, is written CH_3CHO not CH_3COH.

Aldehydes are named according to IUPAC rules. The names are based on the carbon skeleton with the ending changed from **–ane** to **–anal**.

Aldehydes do not need a positional number for the position of the carbonyl group as it is *always at the end of the chain* and this is carbon number one, if there are other substituents or groups on the chain.

Figure 11.3 Benzaldehyde is an aldehyde that is used to make almond essence, an ingredient in Bakewell tarts. It is also added to give flavour to marzipan, which is commonly used on Christmas cakes and Battenberg cakes.

EXAMPLE 1

The structure below shows an aldehyde which is present in coffee and burnt cocoa. Name this structure

TIP

When naming organic compounds, always put a dash between a number and a word, and a comma between two numbers, e.g. 2,2-dichloropentanal.

Answer

- Name the longest unbranched carbon chain in the molecule that includes the functional group – *there are four carbons so the stem is but.*
- The C=O is at the end of the chain, so the compound is an aldehyde – *butanal*
- Number the carbon atoms in the carbon chain, with the carbon of the carbonyl as carbon 1
- Name the side chains and/or substituents attached to the main carbon chain and give their position – *there is a methyl side chain on carbon 2 so it is 2-methyl*

The name of the compound is 2-methylbutanal.

EXAMPLE 2

Name the molecule shown below.

H–⁶C(H)(H)–⁵C(Cl)(H)–⁴C(CH₃)(H)–³C(H)(Cl)–²C(H)(H)–¹C(=O)H

Answer

- Name the longest unbranched carbon chain in the molecule that includes the functional group – *there are 6 carbons so the stem is hex.*
- The C=O is at the end of the chain, so the compound is an aldehyde – *hexanal*
- Number the carbon atoms in the carbon chain, with the carbon of the carbonyl as carbon 1
- Name the side chains and/or substituents attached to the main carbon chain and give their position – *there is a chloro side group on position 3 and on position 5 so this is 3,5-dichloro and a methyl side chain on carbon 4 so this is 4-methyl*

All substituent groups are written alphabetically

The name of the compound is 3,5-dichloro-4-methylhexanal

TIP

When naming, the substituents are written alphabetically. However, the alphabetical rule ignores 'di', 'tri' and 'tetra'. 1,2,3-trichloro-2 methylpentane is a correct name as the chloro is alphabetically before methyl and the 'tri' is ignored.

Aldehydes can have **structural isomers**, for example butanal is isomeric with 2-methylpropanal. Both have the same molecular formula (C_4H_8O), but each have a different structural formula.

H–C(H)(H)–C(H)(H)–C(H)(H)–C(=O)H butanal

H–C(H)(H)–C(CH₃)(H)–C(=O)H 2-methylpropanal

Naming ketones

Ketones also have the general formula $C_nH_{2n}O$. The table below gives the first two members of the ketones.

In ketones, the carbonyl group can be at any position on the chain, except for the end, so the first member of the ketones is propanone.

TIP

Remember in nomenclature there is a difference in numbering a molecule with a functional group rather than a substituted alkane. In substituted alkanes the chain is numbered from the end that will give the lowest locant numbers for the substituents BUT when there is a functional group, the position of the functional group dictates the numbering in the chain.

Name	Structural formula	Displayed formula
propanone	CH_3COCH_3	H H H—C—C—C—H H O H
butanone	$CH_3COCH_2CH_3$	H H H H—C—C—C—C—H H O H H

Ketones are named according to IUPAC rules. The names are based on the carbon skeleton with the ending changed from **–ane** to **–anone.**

EXAMPLE 3

This molecule shown below is important in the manufacture of various herbicides and fungicides. Name the molecule.

```
    H    H         H    H
    |    |         |    |
H—5C —4C —3C —2C —1C — H
    |    |    ||   |    |
    H    H    O   CH3   H
```

Answer

- Name the longest unbranched carbon chain in the molecule that includes the functional group – *there are 5 carbons so the stem is pent.*
- Number the carbon atoms in the longest carbon chain.
- The C=O is on carbon 3 and so the compound is a ketone, and the positon of the carbonyl group must be given – *pentan-3-one.*
- Name the side chains and/or substituents attached to the main carbon chain and give their position *– there is one methyl side chain on carbon 2 so this is 2-methyl, note that since the ketone group is on carbon 3, the numbering could be from left to right – but this gives 4-methyl – the lower number is chosen – 2-methyl.*

The name of the compound is 2-methylpentan-3-one.

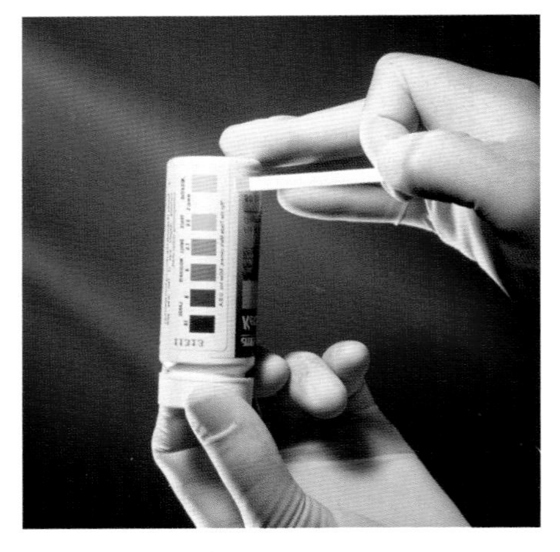

Figure 11.4 This is the Ketostix test for ketones, conducted on the urine of a diabetic patient. Patients with diabetes lack the hormone insulin, which breaks down glucose in the blood. The body uses fat as an alternative energy source, leading to a build up of ketones in the blood and urine that may eventually cause a diabetic coma.

Some ketones may have positional isomerism of the functional group; the molecular formula is the same but the structural formula is different due to the different positon of the carbonyl group. The pentanones are the first ketones to exhibit positional isomerism of the functional group as pentan-2-one and pentan-3-one exist.

```
   H         H    H    H
   |         |    |    |
H— C — C — C — C — C — H        pentan-2-one
   |   ||    |    |    |
   H   O     H    H    H

   H    H         H    H
   |    |         |    |
H— C — C — C — C — C — H        pentan-3-one
   |    |   ||    |    |
   H    H   O     H    H
```

The same is true for hexanone. Hexan-2-one and hexan-3-one have positional isomers.

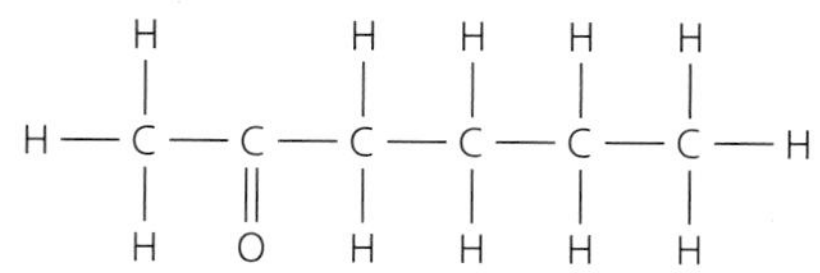

hexan-2-one

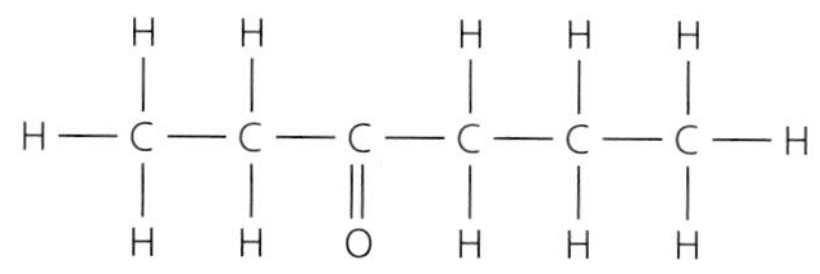

hexan-3-one

Aldehydes and ketones exhibit functional group isomerism. Propanone and propanal are functional group isomers; they have different structural formula but the same molecular formula.

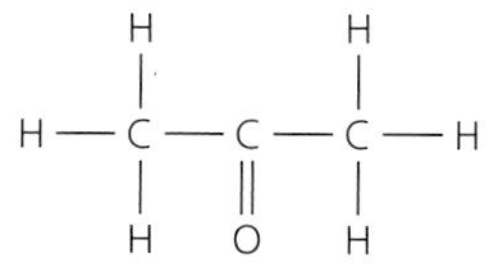

propanone
C_3H_6O

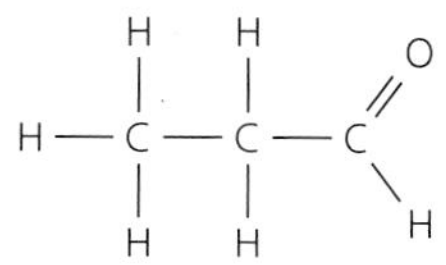

propanal
C_3H_6O

Physical properties of aldehydes and ketones

Boiling points

Van der Waals' forces exist between one aldehyde molecule and another, and in addition there are permanent dipole-permanent dipole attractions between polar carbonyl groups on different molecules. The same is true of ketones.

This means that the boiling points of aldehydes and ketones are higher than that of alkanes with a similar relative molecular mass (M_r).

The table below shows the boiling points of three organic molecules with similar M_r values.

Homologous series	Molecule	M_r	Boiling point/°C
Alkane	$CH_3CH_2CH_3$ (propane)	44	-42
Aldehyde	CH_3CHO (ethanal)	44	+21
Alcohol	CH_3CH_2OH (ethanol)	46	+79

- All three of these molecules have similar van der Waals' forces between their molecules, as there is a similar M_r and so a similar number of electrons.
- The boiling point of the aldehyde is much higher than the boiling point of the alkane due to permanent dipole attractions between the carbonyl groups on neighbouring aldehyde molecules.
- The boiling point of the alcohol is much higher than the boiling point of the aldehyde due to hydrogen bonds between the hydroxyl groups on neighbouring alcohol molecules.

The table below gives the boiling points of the first members of the aldehydes and ketones.

Aldehyde	Boiling point/°C	Ketone	Boiling point/°C
methanal	−21	propanone	56
ethanal	21	butanone	80
propanal	49	pentan-3-one	102

As the length of the carbon chain increases, the boiling point of the aldehydes increases. This is due to a greater number of electrons and hence stronger van der Waals' forces of attraction between the molecules. Methanal is a gas but the other aldehydes shown here are liquids.

The boiling point of the ketones increases similarly, again due to an increase in number of electrons.

Ketones tend to have a slightly higher boiling point than their isomeric aldehydes (compare propanone and propanal). This is due to the position of the carbonyl group. A carbonyl group at the end of the molecule (in aldehydes) gives a longer non-polar section meaning there are more effective van der Waals' forces and less effective permanent dipole attractions. The effect however is small and other factors play a part such as the symmetry of the ketone structure.

Branched chain aldehydes and ketones have lower boiling points than their straight chain isomers due to less effective permanent dipole attractions. The branching allows less interaction between the molecules. For example butanal ($CH_3CH_2CH_2CHO$) has a boiling point of 74 °C whereas 2-methylpropanal ($(CH_3)_2CHCHO$) has a boiling point of 63 °C.

Solubility in water

All short chain aldehydes and ketones mix well with water because the polar carbonyl groups are able to form **hydrogen bonds** with water molecules.

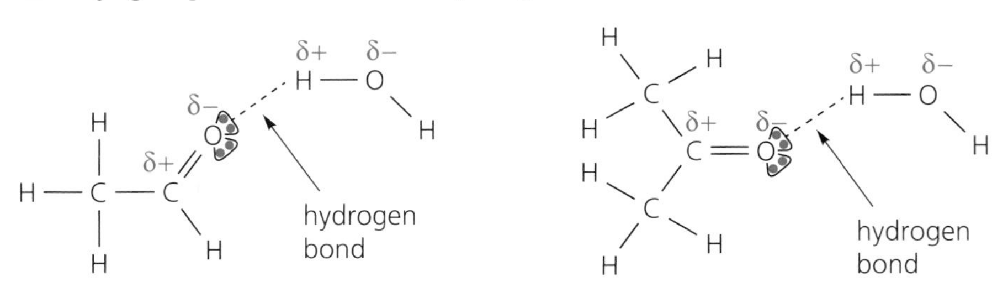

Figure 11.5 The hydrogen bond between ethanal and water and between propanone and water.

Oxidation

You will recall that aldehydes and ketones are the products of the oxidation of alcohols. The oxidising agent often used is a solution of potassium dichromate(VI) dissolved in dilute sulfuric acid – this is commonly referred to as acidified potassium dichromate(VI) and is represented in an equation by [O].

Aldehydes can be oxidised into carboxylic acids. Ketones can only be oxidised by very strong oxidising agents, and they are not oxidised by the reagents used in A2 chemistry.

primary alcohol $\xrightarrow{[O]}$ aldehyde $\xrightarrow{[O]}$ carboxylic acid

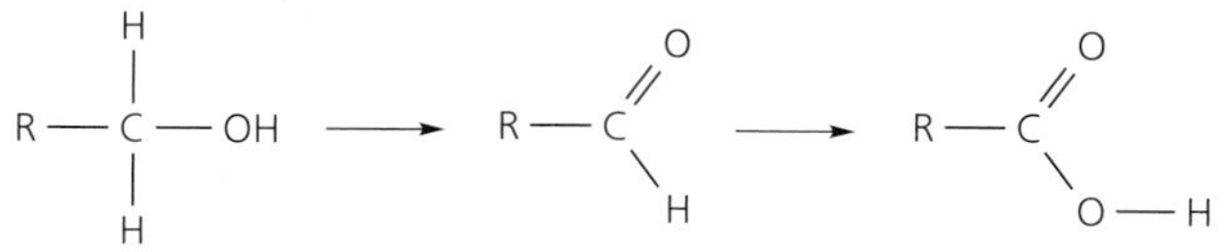

$$R{-}CH_2{-}OH \longrightarrow R{-}C(=O){-}H \longrightarrow R{-}C(=O){-}O{-}H$$

Equation:

ethanal $\xrightarrow{[O]}$ ethanoic acid

$CH_3CHO + [O] \rightarrow CH_3COOH$

Condition: warm with acidified potassium dichromate(VI)

Observation: orange solution changes to green.

The aldehyde is oxidised to the carboxylic acid and the **orange** dichromate ion is reduced to the **green** chromium(III) ion, Cr^{3+} according to the ionic equation:

$$\underset{\text{orange}}{Cr_2O_7^{2-}} + 14H^+ + 6e^- \rightarrow \underset{\text{green}}{2Cr^{3+}} + 7H_2O$$

If an aldehyde is to be prepared from a primary alcohol it is important to immediately distil it from the reaction mixture before it can be oxidised further to the acid. Prolonged heating of a primary alcohol under reflux with excess oxidising agent would cause complete oxidation to the carboxylic acid

secondary alcohol $\xrightarrow{[O]}$ ketone $\longrightarrow$ no further oxidation

Equation:

propan-2-ol $\xrightarrow{[O]}$ propanone

$CH_3C(OH)HCH_3 + [O] \rightarrow CH_3COCH_3 + H_2O$

Condition: warm with acidified potassium dichromate (VI)

Observation: orange solution changes to green.

Tertiary alcohols are not oxidised by acidified potassium dichromate(VI) solution. The oxidising agent removes hydrogen from the primary and secondary alcohols as shown below.

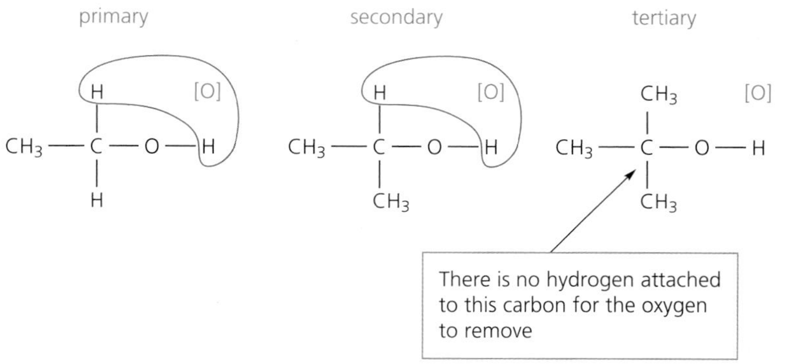

primary: $CH_3{-}CH_2{-}O{-}H$ [O]

secondary: $CH_3{-}CH(CH_3){-}O{-}H$ [O]

tertiary: $CH_3{-}C(CH_3)_2{-}O{-}H$ [O]

There is no hydrogen attached to this carbon for the oxygen to remove

This enables the C=O bond to form. Tertiary alcohols do not have a hydrogen atom attached to the carbon with the –OH and so cannot be easily oxidised.

Figure 11.6 Carrying out the Fehling's solution test using a water bath: The orange-red precipitate identifies an aldehyde.

TIP

If asked for a reagent to distinguish between aldehydes and ketones, give the correct name, e.g. Tollens' reagent. Writing $[Ag(NH_3)_2]^+$ is incorrect as it is not possible to have a reagent that only contains this ion; there must be other ions present.

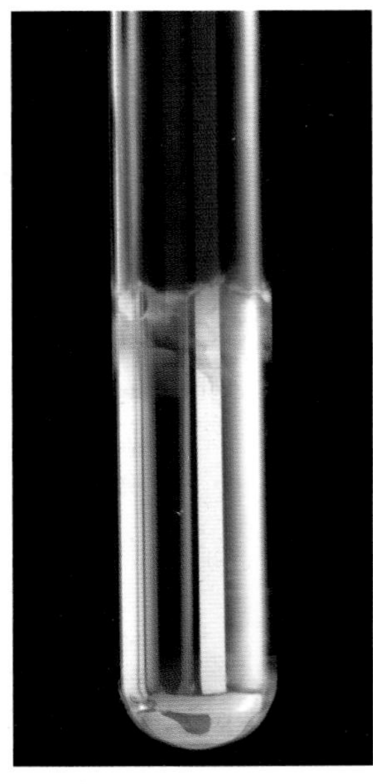

Figure 11.7 A silver mirror is deposited on the test tube when an aldehyde is warmed with Tollens' reagent.

Chemical test to distinguish between aldehydes and ketones

Fehling's solution and Tollens' reagent are mild oxidising agents that are used to distinguish between aldehydes and ketones. Aldehydes are oxidised and ketones are not.

Fehling's solution contains the copper(II) complex ion in alkaline solution. It is a blue solution.

Method

- Add a few drops of the unknown solution to 1 cm^3 of freshly prepared Fehling's solution in a test tube.
- **Warm** in a hot water bath.
- If the unknown is an aldehyde an **orange–red precipitate** will be produced, if it is a ketone, there is no reaction and the solution remains blue.

The Cu^{2+} ions in the oxidising agent (Fehling's solution) are reduced by the aldehyde to form an orange–red precipitate of copper(I)oxide (Cu_2O). The aldehyde is oxidised to the carboxylic acid.

$$2Cu^{2+} + H_2O + 2e^- \rightarrow Cu_2O + 2H^+$$

The equation for the reduction may be simply written as:

$$Cu^{2+} + e^- \rightarrow Cu^+$$

Tollens' reagent contains the complex ion, $[Ag(NH_3)_2]^+$. It is formed by adding sodium hydroxide solution to silver nitrate solution to form a brown precipitate of silver oxide and then adding ammonia solution drop wise until the precipitate dissolves to form a colourless solution. Tollens' reagent is often referred to as ammoniacal silver nitrate.

Method

- Add a few drops of the unknown solution to 2 cm^3 of freshly prepared Tollens' reagent in a test tube.
- **Warm** in a hot water bath
- If the unknown is an aldehyde, a **silver mirror** will be produced on the test tube, if it is a ketone, there is no reaction and the solution remains colourless.

The silver ions in the oxidising agent (Tollens' reagent) are reduced in this reaction to produce silver

$$Ag^+ + e^- \rightarrow Ag$$

The aldehyde is oxidised to the corresponding carboxylic acid.

Reduction

Aldehydes and ketones can be reduced to alcohols using the reducing agent sodium tetrahydridoborate(III) $NaBH_4$.The structure of this compound is shown below

Na^+ [H, B, H, H, H]$^-$

In equations a reducing agent is represented by [H].

aldehyde $\xrightarrow{[H]}$ primary alcohol

Equation:

$$CH_3CH_2CHO + 2[H] \rightarrow CH_3CH_2CH_2OH$$

propanal — propan-1-ol

$$H-CH_2-CH_2-CHO + 2[H] \rightarrow H-CH_2-CH_2-CH_2-OH$$

propanal aldehyde — propan-1-ol primary alcohol

Condition: heat under reflux with sodium tetrahydridoborate(III) in aqueous ethanol followed by acidification with dilute sulfuric acid.

ketone $\xrightarrow{[H]}$ secondary alcohol

Equation:

$$(CH_3)_2C{=}O + 2[H] \rightarrow CH_3-CH(OH)-CH_3$$

propanone ketone — propan-2-ol secondary alcohol

Condition: heat under reflux with sodium tetrahydridoborate(III) in **aqueous solution** followed by acidification with dilute sulfuric acid.

Sodium tetrahydridoborate(III) is a less reactive reducing agent than many others, and so it could be used to selectively reduce only the carbonyl group in a molecule which contained more than one functional group. It is widely used today in the production of certain antibiotics.

TEST YOURSELF 2

1 Draw the structure of the following aldehydes and ketones:
- **a)** butanal
- **b)** propanone
- **c)** ethanal
- **d)** methanal
- **e)** pentan-2-one
- **f)** 2-methylbutanal
- **g)** 4-chloropentanal

2 From the following list of aldehydes and ketones: ethanol, propanone, butanone, propanal
- **a)** Which can undergo reduction?
- **b)** Which can undergo oxidation?
- **c)** Which have the same molecular formula?
- **d)** Which would show a reaction with acidified potassium dichromate(VI) solution?
- **e)** Which would react with Tollens' reagent. State the observation which would occur.
- **f)** Which would be isomers of each other?

3 Write an equation for the reduction of methanal using [H] to represent the reducing agent.

4 Write an equation for the oxidation of propan-2-ol using [O] to represent the oxidising agent.

5 Write an equation for the oxidation of butanal using [O] to represent the oxidising agent.

6 Complete the table by placing a tick if a reaction occurs when the oxidising agent is warmed with each of the substances shown. Place a cross if the reaction does not occur.

Oxidising agent	Primary alcohol	Secondary alcohol	Tertiary alcohol	Aldehyde	Ketone
Acidified potassium dichromate					
Fehling's solution					
Tollens' reagent					

7 From the following list of organic chemicals, choose the correct answer for each question:

A ethanal	**B** propanone	**C** ethanoic acid
D propanal	**E** butanone	**F** butanoic acid
G propan-1-ol	**H** propan-2-ol	**I** methanol

a) Which of the substances can be oxidised to propanoic acid?
b) Which of the substances can be reduced to ethanol?
c) Which of the substances has the empirical formula C_2H_4O?
d) Which of the substances is a secondary alcohol?
e) State two substances which are isomeric.
f) Which of the substances produce a silver mirror with Tollens' reagent?

Reaction of aldehydes and ketones with hydrogen cyanide

Hydrogen cyanide reacts with aldehydes and ketones to produce compounds called hydroxynitriles.

$$CH_3CHO + HCN \longrightarrow CH_3CH(OH)CN$$

ethanal (aldehyde) + hydrogen cyanide → 2-hydroxypropanenitrile (hydroxynitrile)

TIP
When you name hydroxynitriles, remember that the carbon with the nitrogen attached is always counted as the first carbon in the chain. Also ensure that the side group is called *hydroxy* not hydroxyl, in the name.

The product of this reaction is a hydroxynitrile, with the functional group –CN. When naming this compound, the longest carbon–carbon chain must include the carbon in the –CN group, hence it is *propanenitrile*, and the OH is a side group on carbon 2, named 2-hydroxy. The name is 2-hydroxypropanenitrile.

The product of the reaction of a ketone with hydrogen cyanide is also a nitrile, with the functional group –CN. When naming this compound, the longest carbon–carbon chain must include the carbon in the –CN group, hence it is *propanenitrile*, and the OH is a side group on carbon 2, named 2-hydroxy. The name is 2-hydroxy-2-methylpropanenitrile.

$$(CH_3)_2C{=}O + HCN \longrightarrow (CH_3)_2C(OH)CN$$

propanone (ketone) + hydrogen cyanide → 2-hydroxy-2-methylpropanenitrile

The hazards of using KCN

The reactions of aldehydes and ketones with HCN are not usually carried out using hydrogen cyanide as it is an extremely toxic and flammable gas.

Instead it is usual to generate hydrogen cyanide in the reaction mixture by adding a dilute acid to an aqueous solution of **potassium cyanide**. Care must also be exercised when handling potassium cyanide as it is toxic when ingested and forms hydrogen cyanide when in contact with acid.

Nucleophilic addition reactions

The carbonyl group is unsaturated and can undergo addition reactions. An addition reaction is one in which the pi bond of the double bond is broken and species are added across the double bond.

The carbonyl group is also polar and the carbon δ+ is susceptible to attack by nucleophiles.

A **nucleophile** is a lone pair donor. It is an atom or group that is attracted to an electron deficient centre, where it donates the lone pair to form a new covalent bond.

Hence aldehydes and ketones take part in nucleophilic addition reactions. There are two different nucleophilic addition mechanisms which you need to be familiar with.

TIP
When you write mechanisms involving nucleophiles you must show the lone pair. In this example you must show both the lone pair and the negative sign on the hydride ion.

Nucleophilic addition mechanism for reduction reactions of aldehydes and ketones with $NaBH_4$

The BH_4^- ion in $NaBH_4$ is a source of hydride ions (H^-). A hydride ion is a negatively charged hydrogen ion with a lone pair of electrons, $H:^-$. The hydride ion acts as a nucleophile and attacks the carbon δ+. The lone pair of electrons on the hydride ion is donated and forms a bond with the carbon and at the same time the higher energy pi electrons in the carbonyl group move onto the oxygen, giving it a negative charge.

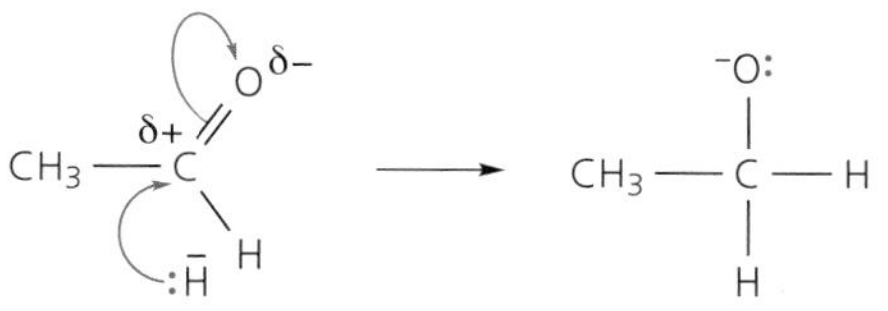

$CH_3-C^{\delta+}(H)=O^{\delta-}$ + $:\bar{H}$ ⟶ $CH_3-C(H)(H)-O:^-$

TIP
Be very careful when drawing curly arrows in a mechanism. Curly arrows represent the movement of electron pairs. The **formation** of a covalent bond is shown by a curly arrow that **starts from a lone electron pair or from another covalent bond.** Similarly the **breaking** of a covalent bond is shown by a curly arrow **starting** from the bond.

When acid is added to complete the reaction the lone pair on the negative ion forms a covalent bond with a hydrogen ion of the acid, producing an alcohol product.

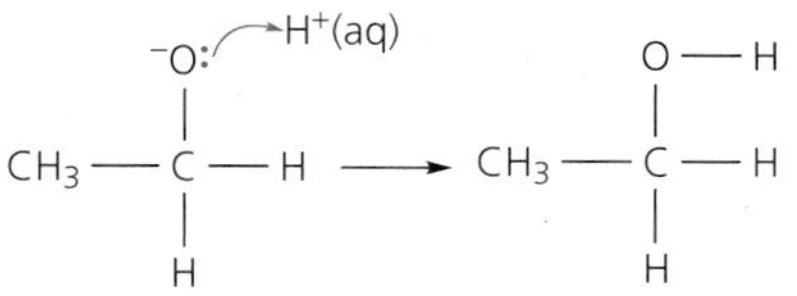

$CH_3-C(H)(H)-O:^-$ + $H^+(aq)$ ⟶ $CH_3-C(H)(H)-O-H$

This reaction is an addition reaction, because the C=O bond breaks and the nucleophile adds across the C=O bond.

EXAMPLE 4

Name and outline a mechanism for the reaction of propanone with $NaBH_4$.

Answer

When answering questions on mechanisms it is sufficient to simply draw the flow diagram, no words of explanation are necessary.

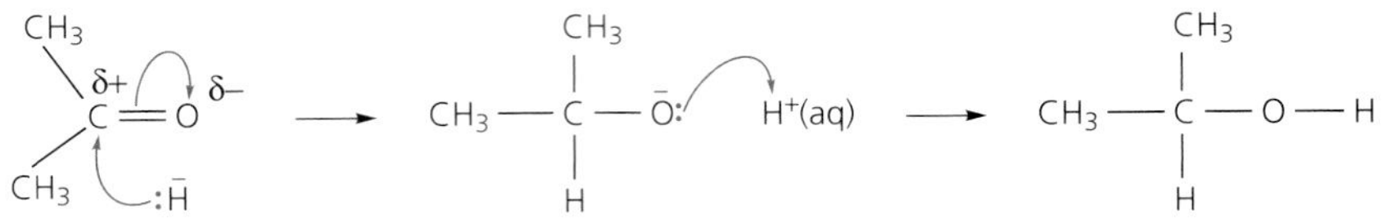

CH_3 CH_3 $C^{\delta+}=O^{\delta-}$ + :$\bar{H}$ ⟶ $CH_3—C(CH_3)(H)—\bar{O}$: + $H^+(aq)$ ⟶ $CH_3—C(CH_3)(H)—O—H$

Name: Nucleophilic addition

Nucleophilic addition mechanism for the reaction with KCN followed by dilute acid

The cyanide ion acts as a nucleophile and attacks the carbon δ+. The lone pair of electrons on the cyanide ion is donated and forms a bond with the carbon and at the same time the higher energy pi electrons in the carbonyl group move onto the oxygen, giving it a negative charge.

When acid is added to complete the reaction the lone pair on the negative ion forms a covalent bond with a hydrogen ion of the acid, producing a hydroxynitrile product.

The full mechanism for ethanal and potassium cyanide and dilute acid is shown below

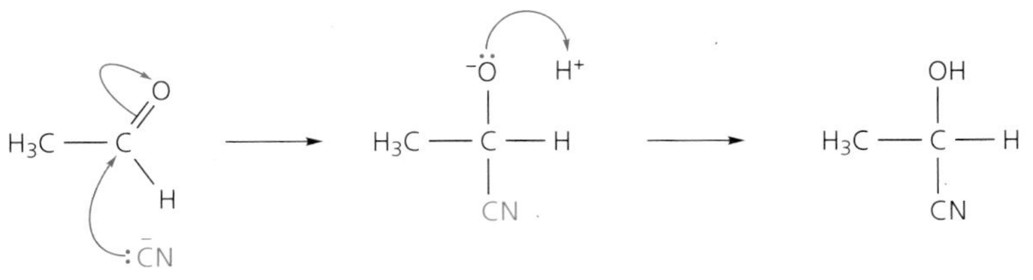

$H_3C—C(H)=O$ + :$\bar{C}N$ ⟶ $H_3C—C(\bar{O}$:$)(H)(CN)$ + H^+ ⟶ $H_3C—C(OH)(H)(CN)$

All aldehydes and ketones undergo similar mechanisms with $NaBH_4$ and KCN. You must be able to draw the mechanism for any carbonyl compound reacting with these reagents. Essentially all that changes is the structure of the carbonyl compound and the final product.

Nucleophilic addition reactions of KCN followed by dilute acid can produce a racemate

When potassium cyanide reacts with ethanal, 2-hydroxypropanenitrile is formed. This compound exhibits optical isomerism as it has an asymmetric centre with four different atoms or groups attached as shown below

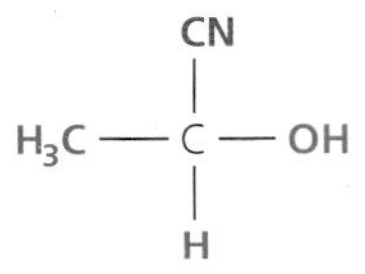

$H_3C—C(CN)(H)—OH$

In the reaction of ethanal with potassium cyanide, the product is optically inactive. This is because a racemic mixture – a 50/50 mixture of the two optical isomers – is formed. The carbonyl bond is planar and the cyanide ion could attack the carbon atom equally from either side. As a result, the formation of each enantiomer is equally likely and so a racemate is formed

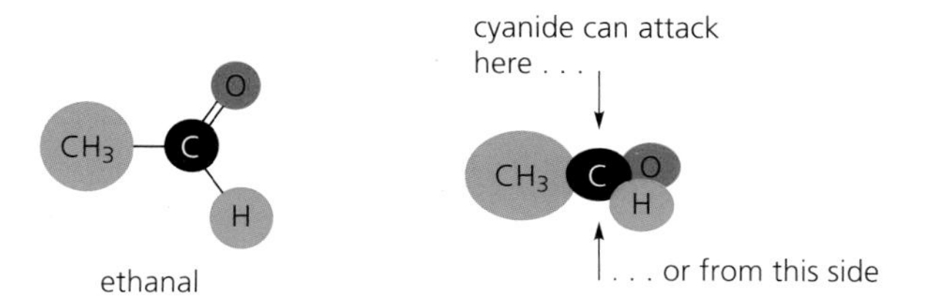

Figure 11.8 The nucleophile is equally likely to attack from above or below.

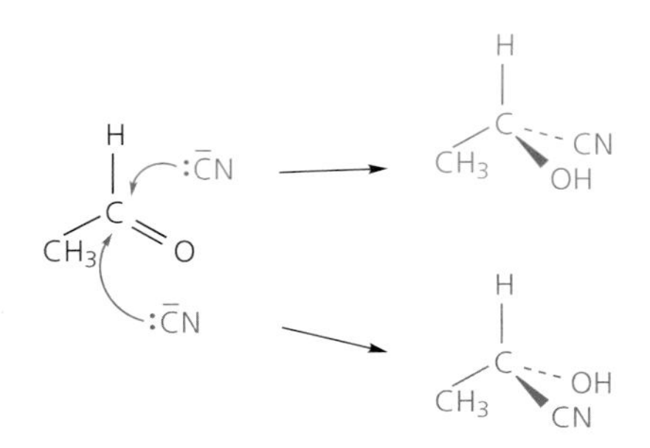

Figure 11.9 How a racemate is formed.

All aldeyhydes will produce a racemate.

Propanone, a symmetrical ketone, produces the product 2-hydroxy-2-methylpropanenitrile which does not have an asymmetric carbon atom and the product is optically inactive. Unsymmetrical ketones such as $CH_3COCH_2CH_3$ will produce a racemate, because the product has an asymmetric carbon, and the nucleophile is equally likely to attack it from either side.

TEST YOURSELF 3

1 a) Write the equation for the reaction of HCN with propanal.
b) Name the organic product.
c) Name the mechanism for this reaction.
d) Outline the mechanism for the reaction of KCN with propanal followed by dilute acid.
e) Suggest why the product of this reaction is optically inactive.

2 Propanal can be reduced.
a) Name a reducing agent for this reaction.
b) Outline a mechanism for the reduction of propanal and name the nucleophile.
c) Name the organic product.

3 Insects contain small organic molecules which they release to signal alarm. One alarm chemical is hex-2-enal $CH_3CH_2CH{=}CHCHO$.
a) Write an equation for the reaction of this compound with bromine water.
b) Write an equation for the reaction of this compound with HCN.

Carboxylic acids

Structure and functional group

Carboxylic acids have the general structure:

```
        O
       //
R — C
       \
        O — H
```

The functional group is called the **carboxyl group**. It is drawn as

```
      O
     //
— C
     \
      O — H
```

and is written as –COOH.

The carboxyl group is a combination of a carbonyl group (C=O) and a hydroxyl group (O–H).

Carboxylic acids are weak acids as they dissociate partially in solution; only a small fraction of the molecules are ionised in solution.

Many carboxylic acids are found in nature. For example ant stings contain methanoic acid, citrus fruits contain citric acid and the smell of goats and other farmyard animals is caused by hexanoic acid. The leaves of rhubarb contain ethanedioic (oxalic) acid. The level of acid in the leaves is so high that it is dangerous to eat them – the acid can lower the calcium ion concentration in the blood to dangerously low levels.

Naming carboxylic acids

Carboxylic acids are named according to IUPAC rules. The names are based on the carbon skeleton with the ending changed from **–ane** to **–anoic acid.**

The table gives the names, structural and displayed formulae of the first four carboxylic acids.

Name of acid	Structural formula	Displayed formula
methanoic acid	HCOOH	H—C(=O)—O—H
ethanoic acid	CH_3COOH	H—C(H)(H)—C(=O)—O—H
propanoic acid	CH_3CH_2COOH	H—C(H)(H)—C(H)(H)—C(=O)—O—H
butanoic acid	$CH_3CH_2CH_2COOH$	H—C(H)(H)—C(H)(H)—C(H)(H)—C(=O)—O—H

IUPAC nomenclature rules state that **the carboxyl carbon in the COOH functional group is always carbon number 1.** Any substituents are numbered based on this.

EXAMPLE 5

The acid shown below is found in very smelly cheeses. Name this structure.

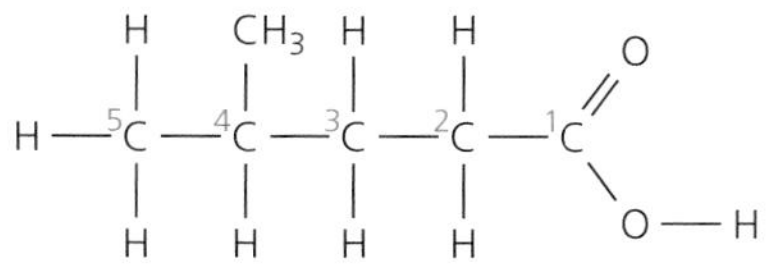

H—⁵C(H)(H)—⁴C(CH₃)(H)—³C(H)(H)—²C(H)(H)—¹C(=O)—O—H

Answer

- Name the longest unbranched carbon chain in the molecule that includes the carbon of the functional group – *there are 5 carbons so the stem is pent.*
- It has a COOH functional group so it is an acid – *pentanoic acid.*
- Number the carbon atoms with the carbon of the carboxyl group as carbon 1.
- Name the side chains and/or substituents attached to the main carbon chain and give their position – *there is a methyl side chain on carbon 4 – 4-methyl.*

The name of the structure is 4-methylpentanoic acid.

Physical properties

Solubility in water

Short chain carboxylic acids are **very soluble** in water. This is because the highly polar carbonyl and hydroxyl groups can form **hydrogen bonds** with water. As the number of carbon atoms in the chain increases, the solubility decreases due to the longer non-polar hydrocarbon chain.

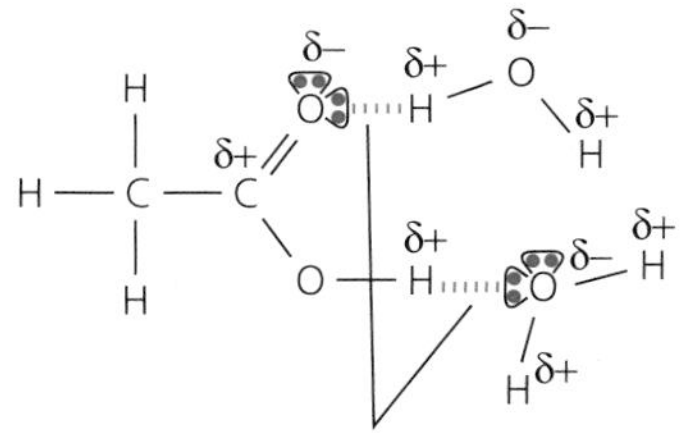

Boiling point

The boiling point of carboxylic acids is higher than corresponding alcohols.

Substance	Formula	M_r	Boiling point/°C
propan-1-ol	$CH_3CH_2CH_2OH$	60.0	97.2
ethanoic acid	CH_3COOH	60.0	118.0

As shown in the table, propan-1-ol and ethanoic acid have the same relative molecular mass and therefore the same number of electrons, resulting in similar van der Waals' forces between the molecules. Both have hydrogen bonds between their molecules. The higher boiling point of the carboxylic acid is because the hydrogen bonding occurs between two molecules of the acid to form a dimer. This doubles the size of the molecule and increases the van der Waals' forces between the dimers resulting in a higher boiling point.

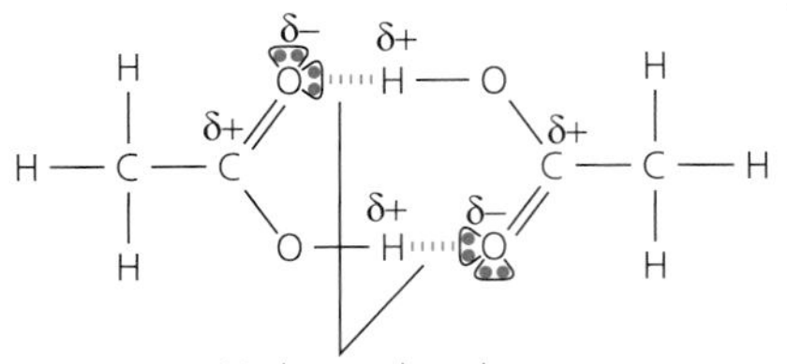

Hydrogen bonds

Salts of carboxylic acids

Like all acids, carboxylic acids can form salts in their reactions. The metal salt of a carboxylic acid is called a carboxylate. The sodium salt of ethanoic acid, sodium ethanoate, is formed by the reaction of ethanoic acid with sodium hydroxide, and has the structure shown below.

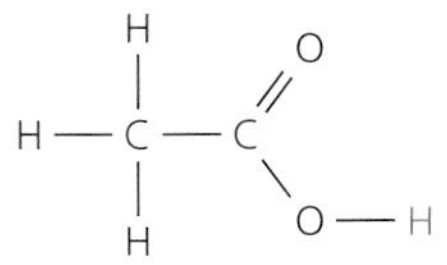

ethanoic acid
(acid)

sodium ethanoate
(salt)

TIP

When writing the formula of a salt you can put the charges in, stressing the ionic nature of the compound, e.g. $CH_3COO^-Na^+$ or you can leave them out, e.g. CH_3COONa. However, to put just one charge in and omit the other is incorrect.

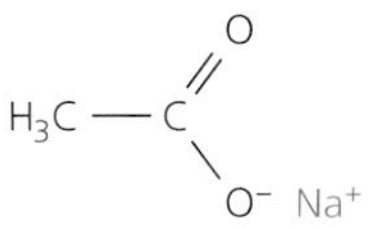

When drawing this structure, do not place a line between the oxygen and the sodium – it is not a covalent bond, it is ionic.

The names of the first four carboxylic acid salts are shown below

Carboxylic acid	Name of salt
methanoic	methanoate
ethanoic	ethanoate
propanoic	propanoate
butanoic	butanoate

Acid reactions of carboxylic acids

All carboxylic acids in aqueous solution act as acids dissociating to form $H^+(aq)$ (or $H_3O^+(aq)$) and the carboxylate ion. They are **weak acids** because they are partially dissociated in solution.

For example:

$$CH_3COOH(aq) \rightleftharpoons CH_3COO^-(aq) + H^+(aq)$$

ethanoic acid — ethanoate ion — hydrogen ion

or

$$CH_3COOH + H_2O \rightleftharpoons CH_3COO^- + H_3O^+$$

ethanoic acid — water — ethanoate ion — hydroxonium ion

Carboxylic acids take part in typical acid reactions – with carbonates, metals and bases to form salts.

With carbonates

A carboxylic acid reacts with carbonates according to the general equation

acid + carbonate $\longrightarrow$ salt + water + carbon dioxide

For example:

$$2CH_3COOH + Na_2CO_3 \rightarrow 2CH_3COONa + CO_2 + H_2O$$

ethanoic acid, sodium carbonate, sodium ethanoate, carbon dioxide, water

$$CH_3COOH + NaHCO_3 \rightarrow CH_3COONa + CO_2 + H_2O$$

ethanoic acid, sodium hydrogen carbonate, sodium ethanoate, carbon dioxide, water

Observations: There will be effervescence and the solid sodium carbonate will be used up, producing a colourless solution.

Figure 11.10 Carboxylic acids react with carbonates and produce effervescence. This is the test for a carboxylic acid.

Test for a carboxylic acid

Despite being weak acids, carboxylic acids are stronger than carbonic acid and release carbon dioxide when reacted with carbonates. This is the reaction used to test for carboxylic acids.

- Add a spatula of sodium carbonate or sodium hydrogen carbonate to the solution.
- If it is a carboxylic acid, effervescence is observed.
- The gas can be collected and bubbled into colourless limewater which should turn cloudy, proving that the gas produced is carbon dioxide.

With metals

A carboxylic acid reacts with reactive metals according to the general equation:

acid + metal → salt + hydrogen

For example:

$$2CH_3COOH + Mg \rightarrow (CH_3COO)_2Mg + H_2$$

ethanoic acid, magnesium, magnesium ethanoate, hydrogen

Observation: There will be effervescence and the solid magnesium will be used up producing a colourless solution.

With bases

A carboxylic acid reacts with bases according to the general equation

acid + base → salt + water

For example:

$$CH_3COOH + NaOH \rightarrow CH_3COONa + H_2O$$

ethanoic acid, sodium hydroxide, sodium ethanoate, water

Observations: There is release of heat and the colourless solution remains.

With ammonia

$$CH_3COOH + NH_3 \rightarrow CH_3COONH_4$$

ethanoic acid, ammonia, ammonium ethanoate

Observations: There is release of heat and the colourless solution remains.

TEST YOURSELF 4

1 a) Draw the structure of the following carboxylic acids:

i) 3-chloropentanoic acid

ii) 4-methylpent-2-enoic acid

iii) 2-hydroxypropanoic acid

iv) 3-hydroxypropanoic acid

b) Three of the acids in (a) can exist as stereoisomers. Which one cannot?

c) Name the acid in (a) which can exist as E-Z isomers.

2 a) Complete the equation for the dissociation of ethanoic acid in solution.

$CH_3COOH \rightleftharpoons$

b) Name the anion formed in this reaction.

c) Would you expect ethanoic acid to be a good electrolyte? Explain your answer.

3 Write equations for the formation of the following salts:

a) potassium ethanoate from potassium hydroxide

b) sodium ethanoate from sodium carbonate

c) magnesium propanoate from magnesium oxide

d) ammonium methanoate from ammonia

e) ammonium ethanoate may be dehydrated by heating. Write an equation for this dehydration and name the organic compound formed.

4 How would you test a substance and prove that it was a carboxylic acid?

Esters

Carboxylic acids react with alcohols in the presence of a strong acid catalyst, to produce esters. Esters have the general structure:

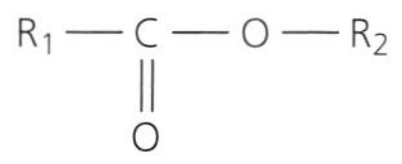

$$R_1-\underset{\underset{O}{\|}}{C}-O-R_2$$

R_1 is from the acid and R_2 is from the alcohol.

The formation of the ester above can be represented by the equation:

$$R_1-C(=O)-O-H \quad + \quad R_2OH \quad \rightleftharpoons \quad R_1-C(=O)-O-R_2 \quad + \quad H_2O$$

Water is eliminated from the reactants in this reaction, so the reaction can be described as a condensation reaction. It is a reversible reaction.

The functional group of an ester is the –COO– group which is called simply an ester group or an ester linkage.

Naming esters

The name of an ester is an **alkyl carboxylate**. When naming, the alcohol provides the **alkyl** part of the name and the carboxylic acid provides the **carboxylate** part of the name. For example the ester made from methanol and propanoic acid is methyl propanoate.

The table below gives the names of some common esters and their parent acids and alcohols.

Carboxylic acid	Alcohol	Ester name	Ester formula
ethanoic acid	ethanol	ethyl ethanoate	$CH_3COOCH_2CH_3$
ethanoic acid	methanol	methyl ethanoate	CH_3COOCH_3
propanoic acid	ethanol	ethyl propanoate	$CH_3CH_2COOCH_2CH_3$
butanoic acid	propan-1-ol	propyl butanoate	$CH_3CH_2CH_2COOCH_2CH_2CH_3$

TIP

It is best when drawing the structural formula of the ester to start with the acid end of the molecule.

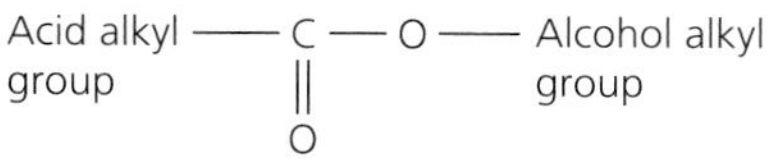

Acid alkyl group — C(=O) — O — Alcohol alkyl group

Esterification equations for formation of esters

Example 1:

$$CH_3COOH + CH_3CH_2OH \rightleftharpoons CH_3COOCH_2CH_3 + H_2O$$

ethanoic acid + ethanol ⇌ ethyl ethanoate + water

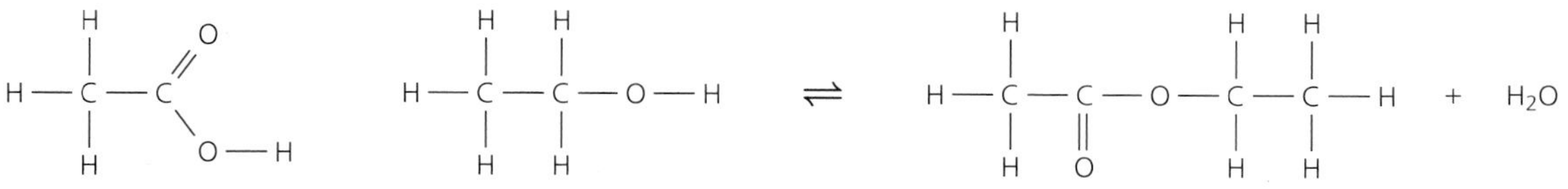

Condition: A catalyst of concentrated sulfuric acid is used and the mixture is heated.

Example 2:

$$HCOOH + CH_3CH_2CH_2OH \rightleftharpoons HCOOCH_2CH_2CH_3 + H_2O$$

methanoic acid + propan-1-ol ⇌ propyl methanoate + water

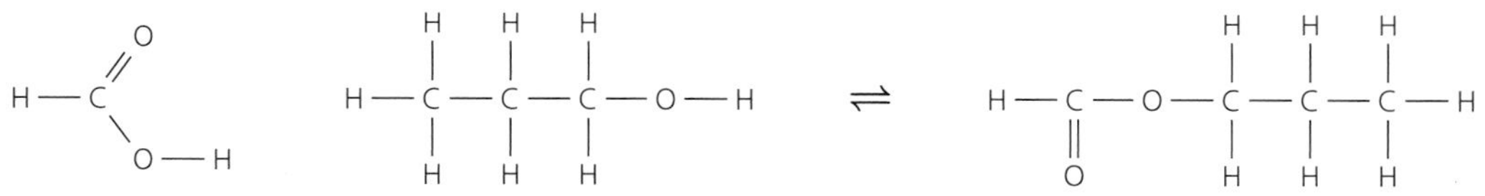

Questions can often be set which give the structure of the ester and expect you to determine the structure of the acid and alcohol from which they are formed

Uses of esters

Solvents

1 Esters are used as **plasticisers**, which are additives mixed into polymers to improve their flexibility. For example, PVC is a rigid, strong polymer used in drainpipes and window frames. If it is treated with up to 18% by mass of a plasticiser made from an ester it becomes cling film, a much more flexible polymer that is used to wrap non-fatty foods.

2 Esters are used as **solvents** for organic compounds. They are volatile and so are easily separated from the solute. Ethyl ethanoate is a common solvent due to its low cost and low toxicity. It is used in paints and nail varnish remover.

Ester	Aroma
benzyl ethanoate	jasmine
octyl ethanoate	orange
ethyl butanoate	pineapple
ethyl cinnamate	cinnamon

3 Esters have pleasant smells and are used in **perfumes**, together with other compounds. The table on the left shows the smell of some different esters.

4 Esters are used in **food flavourings.** Many artificial fruit flavourings contain esters, for example ethyl methanaote is used to give a raspberry flavour and pentyl pentanoate gives a pineapple flavour.

REQUIRED PRACTICAL 10

Preparation of an ester (an organic liquid)

The ester ethyl ethanoate was prepared from ethanol and ethanoic acid, using the method given below.

- Mix equimolar volumes of ethanoic acid and ethanol in a pear-shaped flask.
- Add 1 cm^3 of concentrated sulfuric acid *slowly with cooling and shaking.*
- Assemble the apparatus for reflux, add some anti-bumping granules and heat for 20 minutes.
- Distil off the ethyl ethanoate and collect the fraction between 74 and 79 °C.
- Place the crude ethyl ethanoate in a separating funnel and shake with sodium carbonate solution. Invert the funnel and open the tap occasionally.
- Allow the layers to separate and discard the lower aqueous layer.
- Add some calcium chloride solution to the ethyl ethanoate to remove any ethanol impurities, shake, and run off the lower aqueous layer.
- Place ester in a boiling tube.
- Add a spatula of anhydrous calcium chloride and stopper the boiling tube and shake. Repeat until the ester is clear.
- Filter to remove calcium chloride.
- Redistill to remove any remaining organic impurities, collecting the fraction at the boiling point.

1 Why must the concentrated sulfuric acid be added slowly and with cooling?
2 Why is concentrated sulfuric acid used in this reaction?
3 Write a balanced symbol equation for the reaction to prepare ethyl ethanoate.
4 What is the nature and purpose of anti-bumping granules?
5 What is the function of the sodium carbonate solution?
6 Why do you need to open the tap from time to time?
7 Why should the distillation flask be clean and dry?
8 Why is anhydrous calcium chloride added until the ester is clear?
9 Calculate the percentage yield if the actual yield is 50.0 g of ethyl ethanoate from 40.0 g of ethanol and 52.0 g of ethanoic acid.

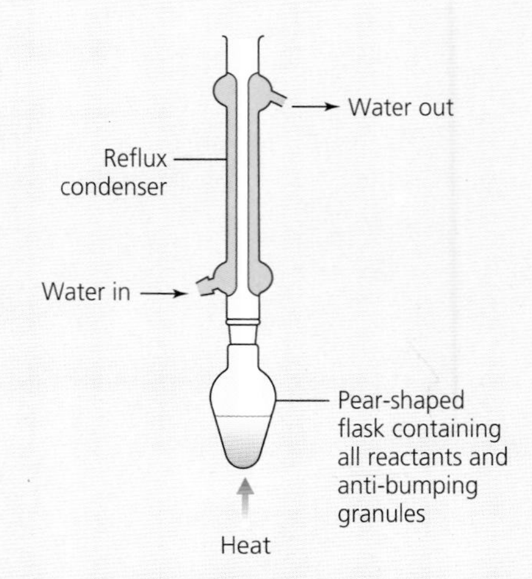

Figure 11.11 Apparatus for heating under reflux.

TIP

Reflux is the continuous boiling and condensing of a reaction mixture. Refer to Book 1 Chapter 13 on **Halogenoalkanes** to refresh your memory on reflux, distillation and the use of a separating funnel.

Hydrolysis of esters

Hydrolysis is the reverse of an esterification reaction. The ester is split (lysis) by the action of water (hydro) into the carboxylic acid and alcohol.

The hydrolysis of an ester needs heat and can be catalysed by either a dilute mineral acid such as hydrochloric acid or a solution of an alkali such as sodium hydroxide. There is a difference in the products of acid hydrolysis and alkaline hydrolysis.

TIP

It is important to remember that alkaline hydrolysis produces the sodium salt of the carboxylic acid and the alcohol, whereas acid hydrolysis produces the carboxylic acid and the alcohol.

Acid hydrolysis

In acidic conditions, esters are not completely hydrolysed – an equilibrium mixture is formed in which some ester is present.

$$\text{ester} + \text{water} \overset{H^+}{\rightleftharpoons} \text{carboxylic acid} + \text{alcohol}$$

For example:

$$\underset{\text{ethyl ethanoate}}{CH_3COOC_2H_5} + \underset{\text{water}}{H_2O} \overset{H^+}{\rightleftharpoons} \underset{\text{ethanoic acid}}{CH_3COOH} + \underset{\text{ethanol}}{C_2H_5OH}$$

Condition: heat under reflux with dilute sulfuric or hydrochloric acid.

Alkaline hydrolysis

In alkaline conditions, esters undergo complete hydrolysis forming the corresponding alcohol and the salt of the carboxylic acid. The reaction in alkaline solution is quicker.

$$\text{ester} + \text{water} \rightarrow \text{carboxylic acid salt} + \text{alcohol}$$

For example:

$$\underset{\text{ethyl ethanoate}}{CH_3COOC_2H_5} + \underset{\text{sodium hydroxide}}{NaOH} \rightarrow \underset{\text{sodium ethanoate}}{CH_3COONa} + \underset{\text{ethanol}}{C_2H_5OH}$$

Condition: heat under reflux with aqueous sodium hydroxide.

In order to liberate the free acid from its salt in alkaline hydrolysis, a dilute mineral acid such as dilute hydrochloric acid should be added.

$$CH_3COONa + HCl \rightarrow CH_3COOH + NaCl$$

This reaction is sometimes called saponification and is the basis of soap making.

Soap is manufactured today using a similar technique as in the 18th century. Molten tallow (beef fat) or other fat is heated with sodium hydroxide and saponified to glycerol and sodium salts of fatty acids. After saponification, the water layer containing glycerol is drawn off and the soap is boiled in water to remove excess sodium hydroxide and glycerol. Additives such as pumice, dyes and perfumes are added and the soap poured into a mould.

Vegetable oils and fats

Vegetable oils and animal fats are **esters** of propane-1,2,3-triol (a tri-alcohol) and a long chain carboxylic acid called a fatty acid.

Propane-1,2,3-triol (also known as **glycerol**) has the structure shown.

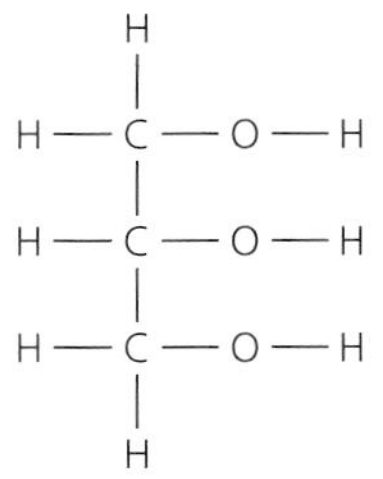

$$\begin{array}{c} H \\ | \\ H-C-O-H \\ | \\ H-C-O-H \\ | \\ H-C-O-H \\ | \\ H \end{array}$$

Fatty acids generally have the structure

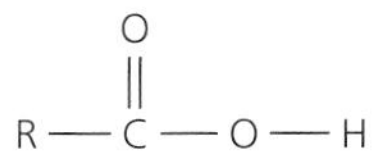

$$R-\overset{\overset{\large O}{||}}{C}-O-H$$

The R is hydrogen or an alkyl group.

Two common fatty acids are stearic acid, $CH_3(CH_2)_{16}COOH$, a saturated fatty acid containing 18 carbon atoms and oleic acid $CH_3(CH_2)_7CH{=}CH(CH_2)_7COOH$, an unsaturated fatty acid containing 18 carbon atoms.

The fats formed in the condensation reaction between fatty acids and glycerol are triesters and are often referred to as triglycerides. Propane-1,2,3-triol has three OH groups and so it reacts with three fatty acids to form a triglyceride.

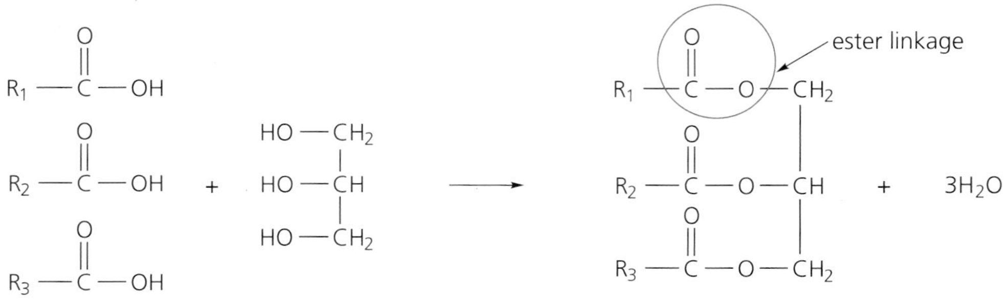

$$\begin{array}{l} R_1-\overset{\overset{O}{||}}{C}-OH \\ R_2-\overset{\overset{O}{||}}{C}-OH \\ R_3-\overset{\overset{O}{||}}{C}-OH \end{array} + \begin{array}{l} HO-CH_2 \\ HO-CH \\ HO-CH_2 \end{array} \longrightarrow \begin{array}{l} R_1-\overset{\overset{O}{||}}{C}-O-CH_2 \\ R_2-\overset{\overset{O}{||}}{C}-O-CH \\ R_3-\overset{\overset{O}{||}}{C}-O-CH_2 \end{array} + 3H_2O$$

3 fatty acids — glycerol — triglyceride ester (ester linkage circled)

A **triglyceride** is an ester of propane-1,2,3-triol (glycerol) and three fatty acid molecules. The fatty acids may or may not all be the same.

Fatty acids are naturally occurring long chain carboxylic acids.

Saturated fatty acids do not have a double bond in the hydrocarbon chain.

Unsaturated fatty acids have at least one C=C double bond in the hydrocarbon chain.

It is important to note the triester linkage in the triglyceride. The three fatty acids which form the lipid may be the same (e.g. three stearic acids) or they may be different. The triglyceride is a triester as it has three ester linkages.

Hydrolysis of vegetable oils and fats

Fats and oils are esters and can be hydrolysed using hot alkali. **Saponification** is the alkaline hydrolysis of fats into glycerol and the salts of the fatty acids present in the fat.

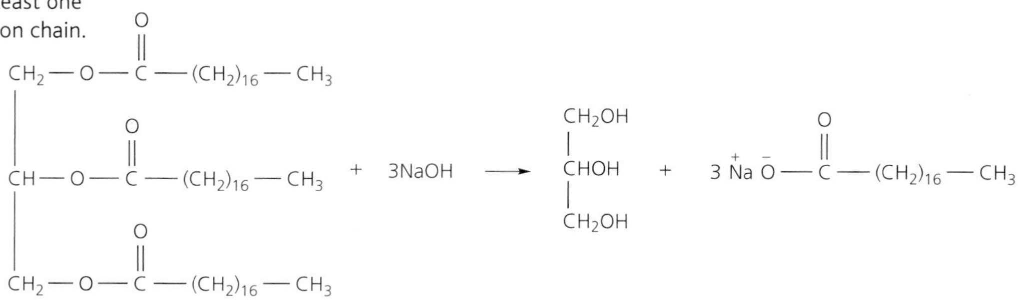

$$\begin{array}{l} CH_2-O-\overset{\overset{O}{||}}{C}-(CH_2)_{16}-CH_3 \\ CH-O-\overset{\overset{O}{||}}{C}-(CH_2)_{16}-CH_3 \\ CH_2-O-\overset{\overset{O}{||}}{C}-(CH_2)_{16}-CH_3 \end{array} + 3NaOH \longrightarrow \begin{array}{c} CH_2OH \\ | \\ CHOH \\ | \\ CH_2OH \end{array} + 3\,\overset{+}{Na}\,\overset{-}{O}-\overset{\overset{O}{||}}{C}-(CH_2)_{16}-CH_3$$

Condition: heat under reflux with aqueous sodium hydroxide

The sodium or potassium salts of the fatty acids are called **soaps**. Soaps generally contain C_{16} and C_{18} salts, but some also contain lower molecular mass carboxylates. Whenever soaps are manufactured, the glycerol produced is a useful by-product used in the pharmaceutical and cosmetic industry.

Biodiesel

Biodiesel is a renewable fuel made from vegetable oils such as rapeseed oil, palm oil and soybean oil. It can be used in normal diesel engines to power cars and buses.

Biodiesel consists of a mixture of methyl esters of long chain carboxylic acids (fatty acids).It is produced by heating vegetable oils (triglycerides) with methanol in the presence of an acid catalyst. The process can be called **trans-esterification** – reacting a ester with an alcohol to produce a different ester and a different alcohol.

$$\begin{array}{l} CH_2-O-C(=O)-R_1 \\ \quad | \\ CH-O-C(=O)-R_2 \\ \quad | \\ CH_2-O-C(=O)-R_3 \end{array} + 3\,CH_3OH \longrightarrow \begin{array}{l} CH_3-O-C(=O)-R_1 \\ CH_3-O-C(=O)-R_2 \\ CH_3-O-C(=O)-R_3 \end{array} + \begin{array}{l} CH_2-OH \\ \quad | \\ CH-OH \\ \quad | \\ CH_2-OH \end{array}$$

triglyceride — methanol — mixture of methyl esters of fatty acids (biodiesel) — glycerol

The alkyl groups R_1, R_2 and R_3 can be the same or different. The reaction is reversible so an excess of methanol is used to drive the equilibrium to the right, and under appropriate conditions this process can produce a 98% yield.

The glycerol produced is a by-product and can be used in pharmaceuticals and cosmetics. It has moisturising properties because of the three hydroxyl groups that can form hydrogen bonds with water and prevent its evaporation.

The use of biodiesel is increasing. However, there are concerns in some countries that using land for growing crops for biodiesel is in competition with land use for growing crops to produce food, and this could lead to food shortages.

TEST YOURSELF 5

1 Name the following esters:

a) $CH_3COOCH_2CH_3$

b) $HCOOCH_3$

c) $CH_3CH_2OOCCH_2CH_3$

d) $HCOOCH_2CH_3$

2 Write the equations for the formation of the following esters from alcohols and carboxylic acids.

a) methyl ethanoate

b) ethyl pentanoate

c) propyl butanoate

3 State the name of the catalyst used in the preparation of an ester from an alcohol and a carboxylic acid.

4 a) Write an equation for the hydrolysis of ethyl methanoate using sodium hydroxide.
   b) Write an equation for the hydrolysis of propyl methanoate using hydrochloric acid.

5 State the IUPAC name of glycerol.

6 What is saponification?

7 Trilinolein (rape seed oil) may be represented as:

$CH_2OOC(CH_2)_7CH{=}CHCH_2CH{=}CH(CH_2)_4CH_3$
|
$CHOOC(CH_2)_7CH{=}CHCH_2CH{=}CH(CH_2)_4CH_3$
|
$CH_2OOC(CH_2)_7CH{=}CHCH_2CH{=}CH(CH_2)_4CH_3$

   a) Write an equation for the formation of trilinolein from 3 molecules of the fatty acid linoleic acid and glycerol.
   b) Explain why trilinolein is described as an unsaturated fat.
   c) Rapeseed oil can be converted into biodiesel by reaction with methanol. Draw the structure of this biodiesel molecule.

Carboxylic acid derivatives

Acid derivatives are compounds that are related to carboxylic acids; the OH group has been replaced by something else.

Acyl chlorides (also known as acid chlorides) are derived from carboxylic acids by replacing the OH group by a chlorine atom.

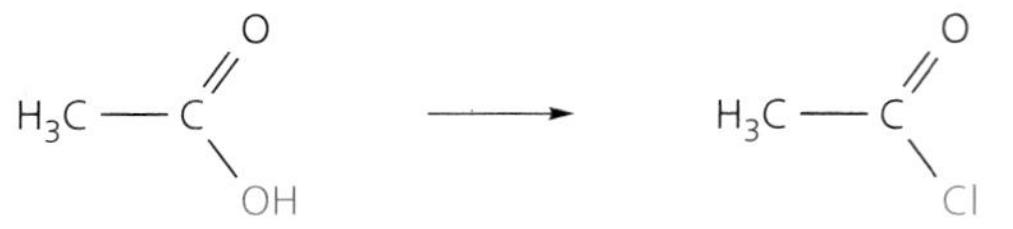

$H_3C{-}COOH \longrightarrow H_3C{-}COCl$

ethanoic acid (carboxylic acid) → ethanoyl chloride (acyl chloride)

Acid chlorides have the functional group:

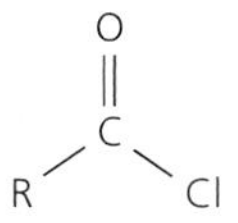

O
||
C
R Cl

and are named **–anoyl chloride.** For example, $CH_3CH_2CH_2COCl$ is butanoyl chloride.

Acid anhydrides are formed when two molecules of a carboxylic acid join together and water is eliminated.

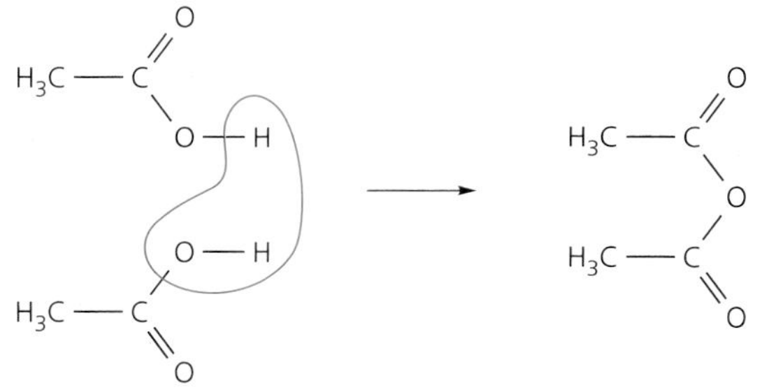

$2\ H_3C{-}COOH \longrightarrow (H_3C{-}CO)_2O$

2 molecules ethanoic acid (carboxylic acid) → ethanoic anhydride (acid anhydride)

The structure of ethanoic anhydride can be written as $(CH_3CO)_2O$. Anhydrides are derivatives of carboxylic acids as the –OH group is replaced by the carboxylate (RCOO–) group.

Acid anhydrides have the functional group

$$R-C(=O)-O-C(=O)-R$$

and are named **–anoic anhydride**. For example $(CH_3CH_2CO)_2O$ is propanoic anhydride.

Amides are derived from carboxylic acids by replacing the –OH with a $-NH_2$ group.

$$CH_3-C(=O)-OH \longrightarrow CH_3-C(=O)-NH_2$$

ethanoic acid (carboxylic acid) → ethanamide (amide)

Amides contain the amide functional group

$$R-C(=O)-NH_2$$

and are named using **–anamide**. For example $CH_3CH_2CH_2CONH_2$ is butanamide.

Acylation

An acyl functional group has the structure

$$R-C(=O)-$$

Acylation is the process of replacing a hydrogen atom in certain molecules by an acyl group (RCO–). Acylation can be carried out using acid derivatives such as acyl chlorides and acid anhydrides which act as acylating agents. In general an acylating agent can be represented by

$$R-C(=O)-X$$

where X = Cl in acid chlorides X = OCOR in anhydrides.

Acyl chlorides and acid anhydrides show similar reactions with water, alcohol, ammonia and primary amines. Acyl chlorides react vigorously with water, and so in these reactions, anhydrous conditions must be used to prevent hydrolysis of the acyl chloride. Acid anhydrides undergo slower reactions than acyl chlorides.

TIP

Notice that in these reactions the organic product is the same. The only difference is the other products formed, for example acyl chlorides form HCl and acid anhydrides form CH_3COOH.

1. *With alcohol*

Acyl chloride

CH_3COCl	+	CH_3CH_2OH	→	$CH_3COOCH_2CH_3$	+	HCl
ethanoyl chloride		ethanol		ethyl ethanoate		hydrogen chloride

Observation: A vigorous reaction which produces steamy fumes of HCl(g), heat

This is a suitable way of producing an **ester** from an alcohol because it occurs at room temperature, is irreversible and the **hydrogen chloride** is removed as a gas, forming pure ester. This method is not commonly used in the laboratory due to the volatile and poisonous nature of the acid chlorides. The normal laboratory preparation of an ester uses an alcohol and a carboxylic acid and needs heat, a catalyst and is reversible so it is difficult to get a high yield of ester.

Acid anhydride

$$(CH_3CO)_2O + CH_3CH_2OH \rightarrow CH_3COOCH_2CH_3 + CH_3COOH$$

ethanoic anhydride, ethanol, ethyl ethanoate, ethanoic acid

Observation: This is a slower and less vigorous reaction than that of an acid chloride and water.

The reaction needs warming. An **ester** is produced in addition to **ethanoic acid.**

Commercially, acid anhydrides are used preferentially to acyl chlorides in acylation reactions, as the reactions are easier to control

2. *With water*

Acyl chloride

$$R\text{–}COCl + H_2O \rightarrow R\text{–}COOH + HCl$$

$$CH_3CH_2COCl + H_2O \rightarrow CH_3CH_2COOH + HCl$$

propanoyl chloride, water, propanoic acid, hydrogen chloride

Observation: a vigorous reaction producing steamy fumes of hydrogen chloride.

Acid anhydride

$$R_1\text{-}COOCO\text{-}R_2 + H_2O \rightarrow R_1\text{-}COOH + R_2\text{-}COOH$$

In this case two molecules of carboxylic acid are produced.

$$(CH_3CH_2CO)_2O + H_2O \rightarrow 2CH_3CH_2COOH$$

propanoic anhydride, water, propanoic acid

Observation: A slower reaction occurs at room temperature. Two colourless solutions produce another colourless solution.

Figure 11.12 The pipetted addition of ethanoyl chloride to the water in the beaker has formed ethanoic acid and fumes (white) of hydrochloric acid, formed as hydrogen chloride (HCl) reacts with water in the air. An ammonia-soaked glass rod tests for HCl by forming white fumes of ammonium chloride.

3. *With ammonia*

Acyl chloride

The reaction happens in stages. First an **amide** and hydrogen chloride are formed.

$$CH_3COCl + NH_3 \rightarrow CH_3CONH_2 + HCl$$

However, because ammonia is basic it will react with the hydrogen chloride to form an ammonium salt

$$NH_3 + HCl \rightarrow NH_4Cl$$

As a result, two ammonia molecules react – this is usually shown in a combined equation as below:

$$CH_3COCl + 2NH_3 \rightarrow CH_3CONH_2 + NH_4Cl$$

CH_3COCl	$2NH_3$	CH_3CONH_2	NH_4Cl
ethanoyl chloride	ammonia	ethanamide	ammonium chloride

Observation: a violent reaction producing white smoke that is a mixture of solid ammonium chloride and ethanamide. Some of the mixture remains dissolved in water as a colourless solution.

Acid anhydride

This reaction is similar and also occurs in two stages. First an amide and hydrogen chloride are formed.

$$(CH_3CO)_2O + NH_3 \rightarrow CH_3CONH_2 + CH_3COOH$$

However, because ammonia is basic it will react with the ethanoic acid to form an ammonium salt

$$NH_3 + CH_3COOH \rightarrow CH_3COONH_4$$

As a result, two ammonia molecules react – this is usually shown in a combined equation as below:

$$(CH_3CO)_2O + 2NH_3 \rightarrow CH_3CONH_2 + CH_3COONH_4$$

$(CH_3CO)_2O$	$2NH_3$	CH_3CONH_2	CH_3COONH_4
ethanoic anhydride	ammonia	ethanamide	ammonium ethanoate

Observation: A slower reaction than that of ethanol and ethanoyl chloride. Heating may be needed.

4. *With primary amines*

Acyl chloride

Ethanoyl chloride reacts with amines eliminating hydrogen chloride and forming an **N-substituted amide** – this means that a methyl group has substituted one of the hydrogen atoms of the NH_2 group. The reaction occurs in two stages:

For example:

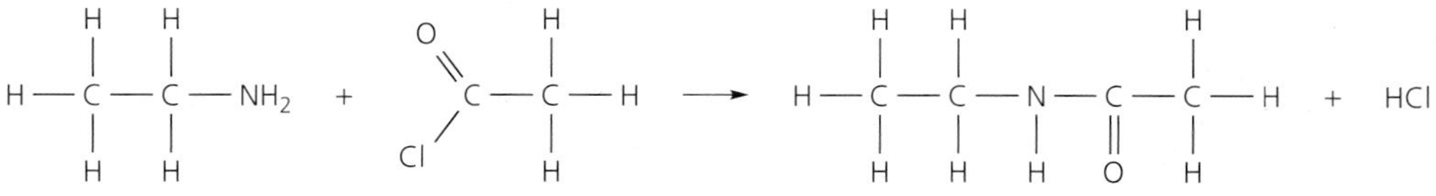

```
     H   H                O       H               H   H           H
     |   |                 \\     |               |   |           |
 H — C — C — NH2    +        C — C — H   ——>  H — C — C — N — C — C — H   +   HCl
     |   |                  /     |               |   |   |   ||  |
     H   H               Cl       H               H   H   H   O   H
```

ethylamine + ethanoyl chloride → N-ethylethanamide + HCl

The amine is basic and reacts with the hydrogen chloride to form an ammonium salt – methylammonium chloride

$$CH_3NH_2 + HCl \rightarrow CH_3NH_3Cl$$

As a result, two ammonia molecules react – this is usually shown in a combined equation as below:

$$CH_3CH_2COCl + 2CH_3CH_2NH_2 \rightarrow CH_3CH_2CONHCH_3 + CH_3CH_2NH_3Cl$$

propanoyl chloride   ethyl amine   N-ethylethanamide   ethylammonium chloride

Acid anhydride

$$(CH_3CH_2CO)_2O + CH_3NH_2 \rightarrow CH_3CH_2CONHCH_3 + CH_3COOH$$

propanoic anhydride   methyl amine   N-methylpropanamide   ethanoic acid

The amine is basic and will react with the ethanoic acid formed to produce a salt

$$CH_3COOH + CH_3NH_2 \rightarrow CH_3COONH_3CH_3$$

The salt is called methylammonium ethanoate. It is just like ammonium ethanoate, except that one of the hydrogens has been replaced by a methyl group.

Again a combined equation is often used

$$(CH_3CH_2CO)_2O + 2CH_3NH_2 \rightarrow CH_3CH_2CONHCH_3 + CH_3COONH_3CH_3H$$

	Nucleophile	Products
Acyl chloride RCOCl	Water HO-H	Carboxylic acid + hydrogen chloride RCOOH + HCl
Acid anhydride $(RCO)_2O$	Water HO-H	Carboxylic acid 2RCOOH
Acyl chloride RCOCl	Alcohol R_1OH	Ester + hydrogen chloride $RCOOR_1$ + HCl
Acid anhydride $(RCO)_2O$	Alcohol R_1OH	Ester + carboxylic acid $RCOOR_1$ + RCOOH
Acyl chloride RCOCl	Ammonia $H\text{-}NH_2$	Amide + hydrogen chloride $RCONH_2$ + HCl Then $HCl + NH_3 \rightarrow NH_4Cl$
Acid anhydride $(RCO)_2O$	Ammonia $H\text{-}NH_2$	Amide + carboxylic acid $RCONH_2$+ RCOOH Then $RCOOH + NH_3 \rightarrow RCOONH_4$
Acyl chloride RCOCl	Primary amine $R_1NH\text{-}H$	N-substituted amide + hydrogen chloride $RCONHR_1$ + HCl Then $HCl + RNH_2 \rightarrow RNH_3Cl$
Acid anhydride $(RCO)_2O$	Primary amine $R_1NH\text{-}H$	N-substituted amide + carboxylic acid $RCONHR_1$ + RCOOH Then $RCOOH + RNH_2 \rightarrow RCOONH_3R$

Nucleophilic addition–elimination reactions of acyl chlorides

Acyl chlorides are extremely reactive and react with nucleophiles such as water, alcohols, ammonia and amines. The mechanism is **nucleophilic addition–elimination** and occurs in two steps.

Addition: the δ+ carbon atom of the polar C=O bond attracts the lone pair of a nucleophile. The lone pair forms a new bond to the carbon, and a pair of electrons in the C=O double bond are transferred to the oxygen leaving it negatively charged. The nucleophile used below is H_2O.

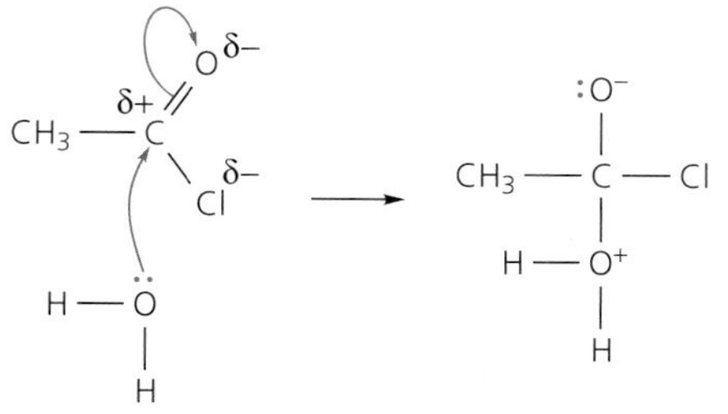

TIP
Remember, when asked to outline a mechanism you simply show a flow diagram with the correct curly arrows and lone pairs.

Elimination: the pair of electrons on the oxygen atom reform the double bond and the C–Cl bond of the acyl chloride breaks and releases a chloride ion, which reacts with a hydrogen atom on the protonated hydroxyl group to form HCl.

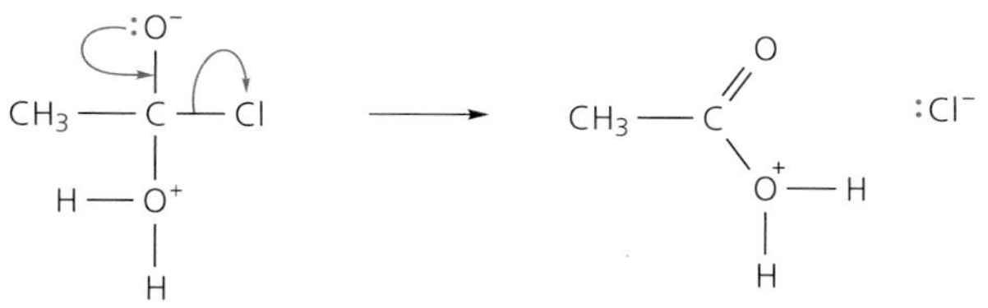

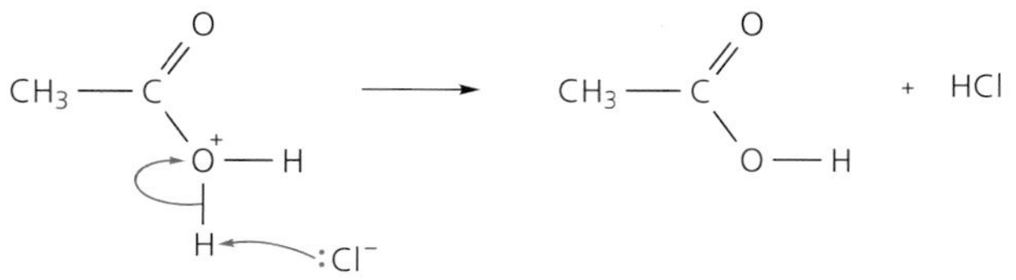

Often this elimination is shown in **one step** as in the examples shown below.

The following four mechanisms are all required.
Example 1: the reaction of ethanoyl chloride with water.

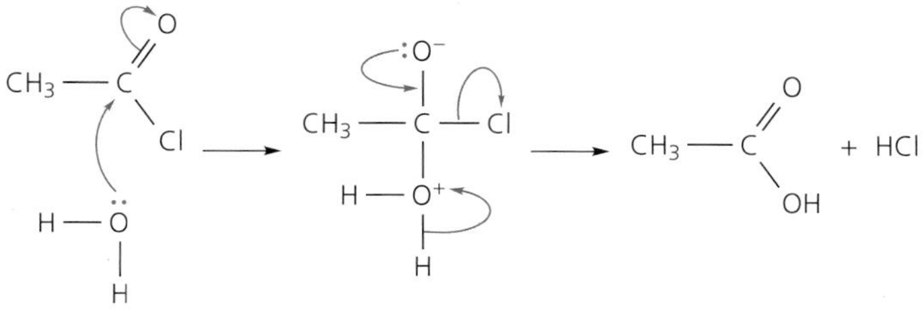

Example 2: the reaction of ethanoyl chloride with ethanol.

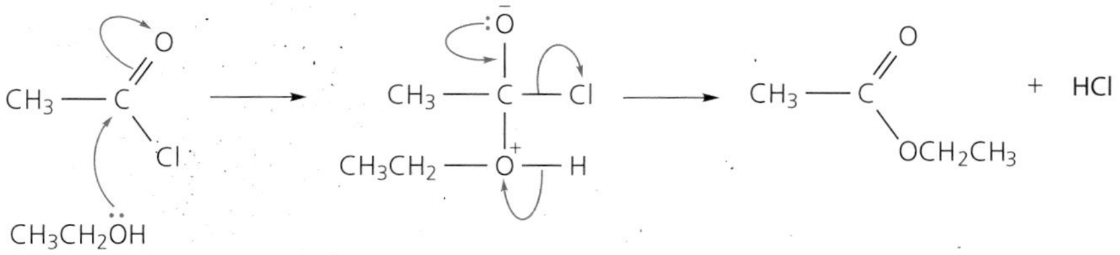

Example 3: the reaction of ethanoyl chloride with ammonia.

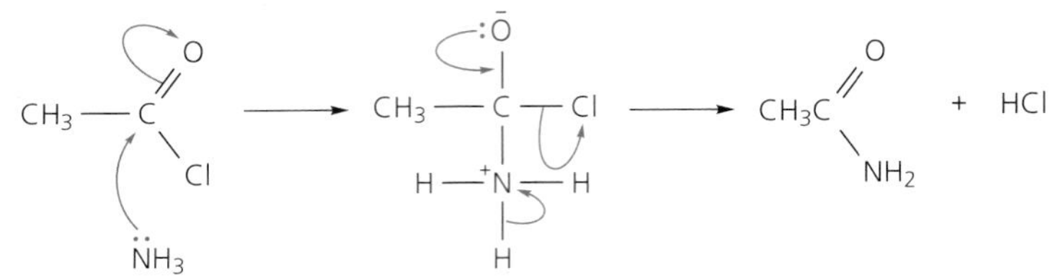

Remember that the HCl would react with another NH_3 to form an ammonium salt NH_4Cl.

Example 4: the reaction of ethanoyl chloride with primary amines, e.g. methyl amine.

TIP
The mechanism for nucleophilic addition-elimination needs to show 5 curly arrows as shown in examples 1-4.

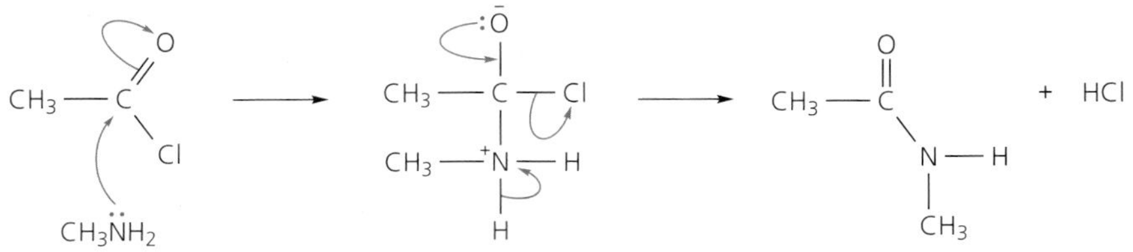

Remember that the HCl would react with another amine to form an ammonium salt.

Preparation of pure organic solids

Acylation reactions are used widely in the pharmaceutical industry. Two of the most commonly used drugs, paracetamol and aspirin are both manufactured using acylation reactions. Aspirin is manufactured by acylating 2-hydroxybenzenecarboxylic acid. The –OH group is esterified to form aspirin.

The **industrial advantages** of using ethanoic anhydride to acylate rather than ethanoyl chloride include

- it is less corrosive
- it is less vulnerable to hydrolysis
- it is less hazardous to use as it gives a less violent reaction
- it is cheaper than ethanoyl chloride
- it does not produce corrosive fumes of hydrogen chloride.

An equation for the preparation of aspirin from ethanoic anhydride and 2-hydroxybenzoic acid is shown below.

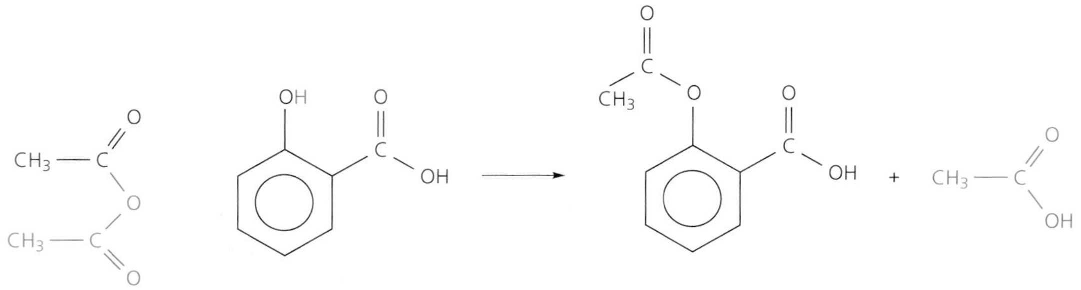

ethanoic anhydride — 2-hydrobenzoic acid — aspirin

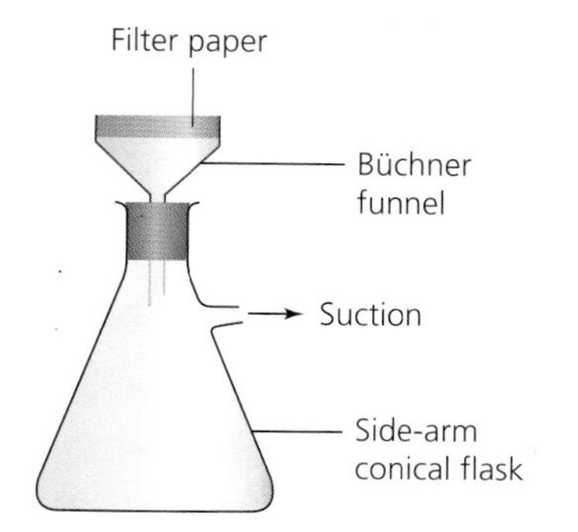

Figure 11.13 Suction filtration apparatus.

Organic solids such as aspirin must be produced in as pure a state as possible. Impurities can occur for many reasons; two of the most common are contamination with reactants due to an incomplete reaction and the presence of other compounds due to alternative competing reactions during the preparation.

Recrystallisation is a very important technique used to purify solids by removing unwanted by-products. The practical steps for the process of recrystallisation are:

- Dissolve the impure crystals in the **minimum volume of hot solvent**.
- Filter the hot solution by gravity filtration, using a hot funnel and fluted filter paper, to remove any insoluble impurities *(filtering through hot filter funnel and using fluted paper prevents precipitation of the solid)*.
- Allow the solution to cool and crystallise *(the impurities will remain in solution)*.
- Filter off the crystals using suction filtration *(suction filtration is faster than gravity filtration and gives a drier solid)*.
- Dry between two sheets of filter paper.

A minimum volume of hot solvent is used to dissolve the solid, making a saturated solution. The solution is cooled and the solubility of the compound drops causing it to recrystallise from solution. Impurities remain dissolved in the solution. A **minimum volume of hot solvent** is used to ensure that as much of the solute is obtained as possible.

TIP

The melting point tube can be attached to a thermometer with a rubber band and placed in an oil bath rather than in a melting point apparatus. Photos of both ways of finding a melting point are found in Practical skills, Chapter 17.

Checking the purity of an organic solid

A pure substance has a fixed melting point while an impure substance melts over a wide range of temperatures and at a lower temperature than the pure substance. The melting points of almost all substances are available in data tables. To check the purity of a solid, **a melting point can be determined** using the following method:

- Place some of the solid in a melting point tube.
- Place in melting point apparatus and heat slowly.
- Record the temperature at which the solid starts to melt and the temperature at which it finishes melting.
- Repeat and average the temperatures.
- Compare the melting point with known values in a data book.

The melting point of a substance is not the exact point at which it melts but rather the range of temperatures from when the sample starts to melt until it has completely melted. The greater the range the more impurities are present. A range of less than 2 °C indicates a fairly pure substance.

REQUIRED PRACTICAL 10

The preparation of aspirin (an organic solid) in the laboratory

Aspirin (2-ethanoylhydroxybenzenecarboxylic acid) is an antipyretic drug (reduces fever by lowering body temperature) and an analgesic (relieves pain).

Today, over one hundred companies hold patents for processes that generate aspirin. Each is slightly different but all end with the same molecule. The following method can be used to prepare aspirin in the laboratory.

- Place 20.0 g of 2-hydroxybenzoic acid in a pear-shaped flask and add 40 cm^3 of ethanoic anhydride ($(CH_3CO)_2O$).
- Safely add 5 cm^3 of conc. phosphoric(v) acid and heat under reflux for 30 minutes.
- Add water to hydrolyse any unreacted ethanoic anhydride to ethanoic acid, and pour the mixture into 400 g of crushed ice in a beaker.
- The product is removed by suction filtration, recrystallised from water and dried in a desiccator.
- The melting point is then determined.

The reaction can be represented as follows.

$HOOCC_6H_4OH + (CH_3CO)_2O \rightarrow HOOCC_6H_4OCOCH_3 + CH_3COOH$

1 Suggest the role of concentrated phosphoric(v) acid in this preparation.
2 Explain how the concentrated phosphoric(v) acid is added safely.
3 Draw a labelled diagram of the apparatus set up for reflux.
4 Assuming a 70% yield, calculate the mass of 2-hydroxybenzoic acid required to form 5.0 g of pure aspirin.
5 Write an equation for the hydrolysis of ethanoic anhydride.
6 Why is the mixture poured onto crushed ice?
7 Why is suction filtration used rather than gravity filtration?
8 Describe how the impure product is recrystallised.
9 Calculate the atom economy for the preparation of aspirin by this method.
10 An alternative preparation, using ethanoyl chloride rather than ethanoic anhydride, has a higher atom economy. Why is this reaction not used in industry?
11 The melting point of aspirin was determined. From the result, state how you would determine if the aspirin was pure.

TEST YOURSELF 6

1 Name the following:
   a) CH_3CH_2COCl
   b) $CH_3CH_2CH_2CONH_2$
   c) $(CH_3CH_2CO)_2O$
   d) CH_3COCl
2 Write equations for the reaction of:
   a) ethanoyl chloride and propanol
   b) butaonyl chloride and ammonia
   c) propanoyl chloride and methylamine
   d) ethanoic anhydride and ethanol
3 Name the mechanism for the reaction between ethanoyl chloride and water and state the name of the organic product.
4 Name the method used to purify solids.
5 A pure sample of solid ethanamide is prepared in the laboratory. How would you test its purity?

Practice questions

1 Which one of the following statements about the formation of an ester from ethanoyl chloride and propan-1-ol is correct?

A Concentrated sulfuric acid is required.
B Heat is required.
C The ester produced is called ethyl propanoate.
D The reaction goes to completion. *(1)*

2 Which one of the following compounds is optically active and incapable of reducing Fehling's solution?

A $CH_3CH(CH_3)CH_2CHO$
B $CH_3CH(C_2H_5)COCH_3$
C $CH_3CHClCH_2CHO$
D $CH_3CH(CH_3)COCH_3$ *(1)*

3 Butan-1-ol was reacted with an excess of propanoic acid in the presence of a small amount of concentrated sulfuric acid. 6.0 g of the alcohol produced 7.4 g of the ester. Which one of the following is the percentage yield of the ester?

A 57% **B** 70%
C 75% **D** 81% *(1)*

4 The scheme below summarises the formation and reactions of ethanoic acid.

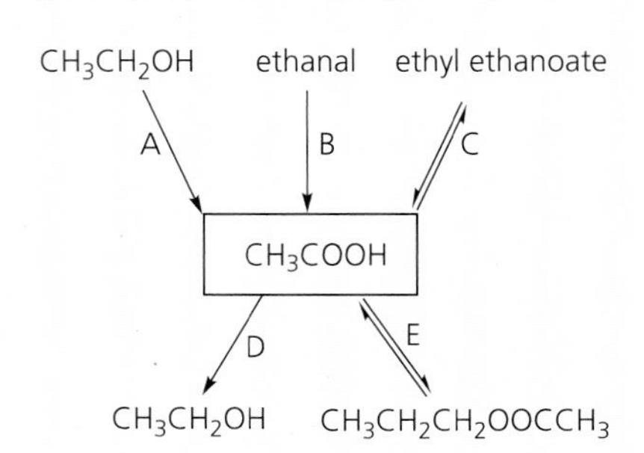

a) Give suitable reagents and conditions for each of the reactions A–E. *(10)*

b) Give the **type of reaction** occurring in the reactions A–E. *(5)*

5 Nucleophiles react with carbonyl compounds to form addition products.

a) Define the term nucleophile. *(1)*

b) Explain why the carbonyl group is susceptible to attack by a nucleophile. *(2)*

c) Write an equation for the reaction of propanal with HCN and name the product. *(2)*

d) Outline a mechanism for the reaction of propanal with HCN. *(3)*

e) Explain why the product is optically inactive. *(2)*

f) Aldehydes can be distinguished from ketones using the following reagents.

i) Complete the following table. *(4)*

Reagent	Formula of metal/ion before test	Formula of metal/ion after test
Tollens' reagent		
Fehling's solution		

ii) Which reagent(s) will give a positive test for ethanal? *(1)*

6 Carboxylic acids, esters and fats are all found in nature and contain the carbonyl group. They have different functions in nature, for example carboxylic acids are found in ant stings, esters flavour fruits, and fats are used for energy storage.

a) i) Ethanoic acid has a boiling point of 119 °C and ethyl methanoate, an ester has a boiling point of 56 °C. Explain the difference in boiling points. *(3)*

ii) Ethanoic acid reacts with solid potassium hydrogen carbonate. Write an equation for this reaction and state one observation. *(2)*

b) i) Ethanol reacts with methanoic acid to form ethyl methanoate. Write an equation for this equilibrium reaction and explain one way in which the equilibrium yield of the ester could be increased. *(2)*

ii) Draw and name two structural isomers of ethyl methanoate. *(4)*

iii) Write an equation for the preparation of ethyl methanoate from an acid anhydride and an alcohol. *(1)*

c) Fats may be produced by the equilibrium reaction between a fatty acid and glycerol.

$$\begin{array}{l}CH_2OH\\ |\\ CHOH + 3RCOOH\\ |\\ CH_2OH\end{array} \rightleftharpoons \begin{array}{l}CH_2OOCR\\ |\\ CHOOCR \quad + 3H_2O\\ |\\ CH_2OOCR\end{array}$$

i) State the IUPAC name for glycerol. *(1)*

ii) The fat may alternatively be produced by the reaction of acyl chlorides (RCOOCl) with glycerol. Write the equation for the reaction. *(4)*

iii) The vegetable oil shown below reacts with methanol in the presence of potassium hydroxide to form biodiesel. Write the equation for this reaction. *(1)*

$$\begin{array}{l}CH_2OOCC_{17}H_{31}\\ CHOOCC_{17}H_{33}\\ CH_2OOCC_{17}H_{29}\end{array}$$

7 Complete the diagram below by inserting the organic product formed in each case. *(3)*

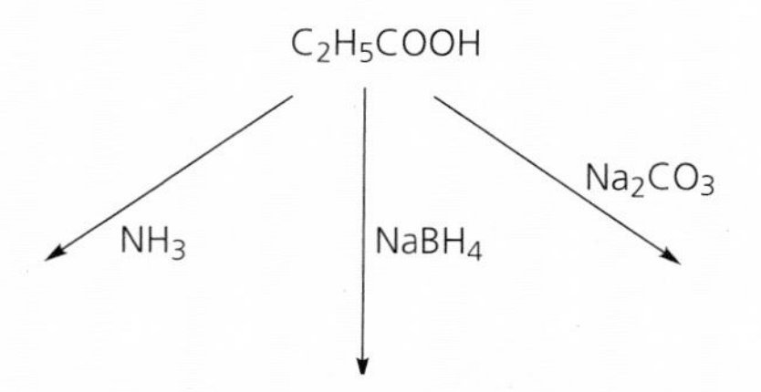

8 Which one of the following would not be affected by boiling with aqueous sodium hydroxide?

A ethyl ethanoate

B glycerol

C olive oil

D ethanoic acid *(1)*

9 a) Name and outline a mechanism for the reaction of CH_3CH_2COCl with CH_3NH_2. (3)

b) Write an equation for the reaction of CH_3CH_2COCl with ammonia. (1)

10 The flow scheme below shows a series of organic reactions.

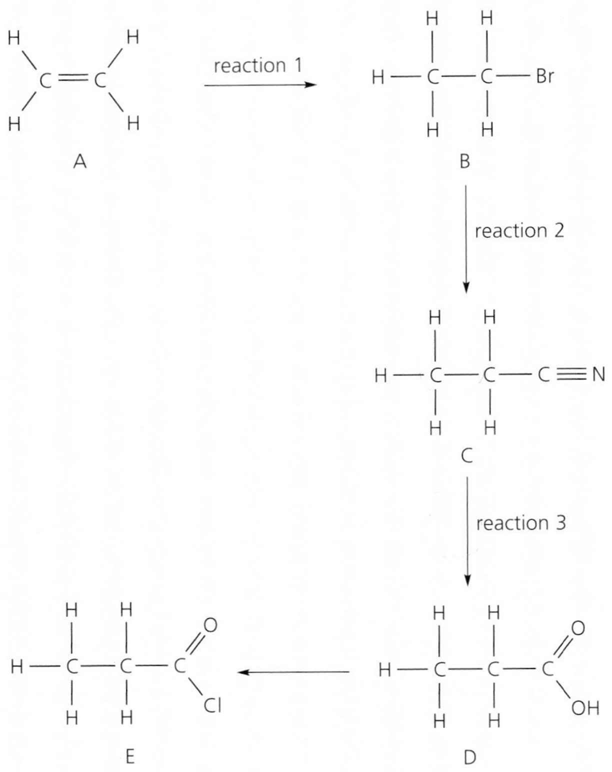

A → (reaction 1) → B → (reaction 2) → C → (reaction 3) → D → E

a) Name compounds A to E. (5)

b) State the type of reaction which is taking place in reactions 1 and 2. (2)

c) Name the mechanism by which reaction 1 occurs. (2)

d) Name the mechanism by which reaction 2 occurs. (2)

e) Name the reagents required for each of reactions 1 and 2. (2)

11 The following compounds can be distinguished by observing what happens in test tube reactions. For each pair, suggest a suitable reagent or reagents that could be added separately to each compound in order to distinguish between them. Describe what you would observe with each compound.

a) methyl ethanoate and propanoic acid (3)

b) butanone and propanoyl chloride (3)

c) butan-2-ol and 2-methylpropan-2-ol (3)

12 Dodecanoic acid, $C_{11}H_{23}COOH$, is an acid found in coconut oil and human breast milk. It is a white solid at room temperature with a melting point of 45 °C, and is insoluble in water.

a) Explain why ethanoic acid is soluble in water whereas dodecanoic acid is insoluble. (2)

b) Describe a chemical test to prove that dodecanoic acid is an acid. (2)

c) Describe how you would experimentally determine that a sample of solid dodecanoic acid is pure. (3)

d) Dodecanoic acid can be reduced to the corresponding alcohol. Write an equation for the reduction using [H] to represent the reducing agent and name a suitable reducing agent for the reaction. (2)

e) The main constituent of coconut oil is the triester formed from dodecanoic acid and glycerol. Write an equation for the formation of this triester. (1)

Stretch and challenge

13 Propanone can be converted into 2-bromopropane by a three-step synthesis as shown below. It is first reduced into compound X. Deduce the structure and names of compounds X and Y and for each of the three steps suggest a reagent that could be used and name the mechanism. (6)

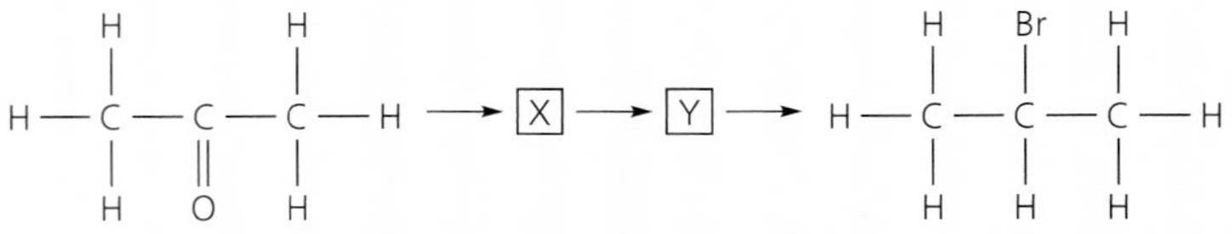

CH_3COCH_3 → X → Y → $CH_3CHBrCH_3$

12 Aromatic chemistry

PRIOR KNOWLEDGE

- Alkenes undergo addition reactions. The mechanism of these reactions is electrophilic addition.
- Alkenes decolourise orange bromine water.
- Understand the terms structural formula, displayed formula, molecular formula, empirical formula and skeletal formula

TEST YOURSELF ON PRIOR KNOWLEDGE 1

1 a) Write an equation for the reaction of bromine with ethene.
  b) State the colour change in this reaction.
  c) Name the product in this reaction.

2 Rubber is a polymer of isoprene. The structure of isoprene is shown below.

$$H_2C{=}C(CH_3){-}CH{=}CH_2$$

  a) What is the molecular formula for isoprene?
  b) What is the empirical formula for isoprene?
  c) What is the skeletal formula for isoprene?
  d) What is the IUPAC name for isoprene?
  e) Isoprene reacts with excess bromine. Write an equation for this reaction and name the product.
  f) What is observed in the reaction of isoprene with bromine?

The word aromatic comes from the Latin word 'aroma' meaning fragrance. Many of the first organic compounds produced from natural substances such as cinnamon bark and vanilla beans, were found to give off pleasant smells. When these compounds were analysed they were found to contain benzene rings. Today's definition of an **aromatic compound** is one which contains a benzene ring. The smell generated by a compound has nothing to do with the presence of a benzene ring and aromatic compounds do not all have a pleasant smell!

Aromatic chemicals are very important in the production of synthetic compounds such as drugs, dyes, explosives and plastics.

The history of the structure of benzene

In 1825 Michael Faraday discovered benzene, a colourless, sweet smelling liquid. It was found to have the empirical formula CH and relative molecular mass 78.0 which gave a molecular formula of C_6H_6. He first extracted benzene from whale oil, which was commonly used for lighting homes.

Scientists speculated over the structure of benzene for many years. Initially it was suggested that it had a structure with several double bonds, like that shown in Figure 12.1.

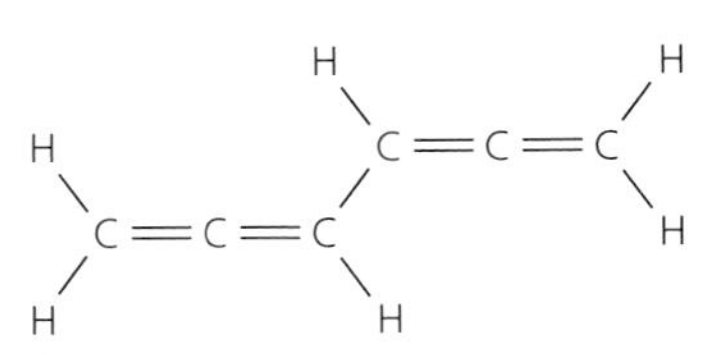

Figure 12.1 A linear structure was initially suggested for benzene.

However, experimental evidence showed that benzene was rather unreactive for such an unsaturated molecule and this structure was rejected.

In 1865 Friedrich August Kekulé proposed a hexagonal structure of six carbon atoms joined by alternate single and double bonds as shown in Figure 12.2 This was the first time that it was considered that carbon atoms could join together in rings rather than chains.

displayed formula or skeletal formula

Figure 12.2 The Kekulé structure of benzene.

Alkenes are reactive molecules due to their double bonds, and since Kekulé's structure contained double bonds, benzene was expected to react in a similar way to alkenes, and to decolourise bromine water. Benzene, however, did not decolourise bromine water and did not readily undergo addition reactions. To account for this lack of reactivity, Kekulé proposed that his structure had two forms that existed in a rapid equilibrium. An approaching bromine molecule could not be attracted to a double bond before the structure changed, and so the bromine could not react with the double bond. In this model, the structure of benzene is called a resonance hybrid and it is often represented as shown below.

The carbon–carbon bond lengths

The structure of benzene suggested by Kekulé had alternating single and double bonds. In 1922 Kathleen Lonsdale used X-ray diffraction to measure the bond lengths in benzene. She found all the carbon–carbon bonds were the *same length*, at an intermediate value between that of a double and single carbon–carbon bond. This suggested that the Kekulé structure of benzene was incorrect.

Bond	Bond length
carbon–carbon bond in benzene	0.140 nm
C–C	0.153 nm
C=C	0.134 nm

Thermochemical data: enthalpy of hydrogenation

When cyclohexene, with one carbon–carbon double bond is hydrogenated, the enthalpy change of hydrogenation is $-120\,kJ\,mol^{-1}$.

cyclohexene + H_2 ⟶ cyclohexane

Enthalpy of hydrogenation is defined as the change in enthalpy, which occurs when 1 mol of an unsaturated compound reacts with an excess of hydrogen to become fully saturated.

A **sigma bond** (δ) is formed by direct overlap of two orbitals on adjacent carbons, each providing one electron, so that the electron density is concentrated between the two nuclei.

A **pi bond** (π) is formed by the sideways overlap of two p orbitals on adjacent carbon atoms, each providing one electron.

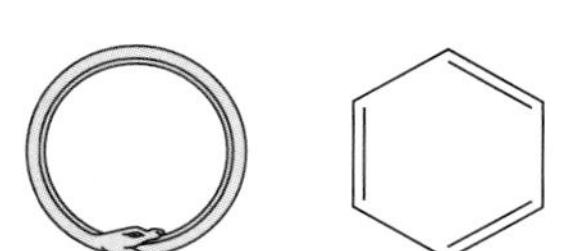

Figure 12.3 Kekulé spent years studying the nature of carbon–carbon bonds. He said that he discovered the ring shape of the benzene molecule after having a daydream of a snake seizing its own tail. This is an ancient symbol known as the ouroboras.

Kekulé's benzene structure has three double bonds, so it would be expected to have a enthalpy change of $-360\,kJ\,mol^{-1}$, three times that of cyclohexane because three double bonds are hydrogenated. A theoretical cyclic compound with three double bonds is cyclohexa-1,3,5-triene.

cyclohexa-1,3,5-triene + $3H_2$ ⟶ cyclohexane

However, when benzene is hydrogenated the enthalpy change is only $-208\,kJ\,mol^{-1}$, this is *152 kJ mol⁻¹ less than expected*. This means that the actual structure of benzene is more stable than the Kekulé structure (cyclohexa-1,3,5-triene)

The delocalised model of benzene

Today's accepted structure for benzene is of a delocalised model which has the following features:

- it is a *planar hexagonal* molecule of six carbon atoms.
- all *carbon–carbon bond lengths are intermediate in length* between that of a single C–C and a double C=C.
- each carbon uses three of its outer electrons to form three sigma bonds to two other carbon atoms, and one hydrogen atom. This leaves each carbon atom with one electron in a p orbital.

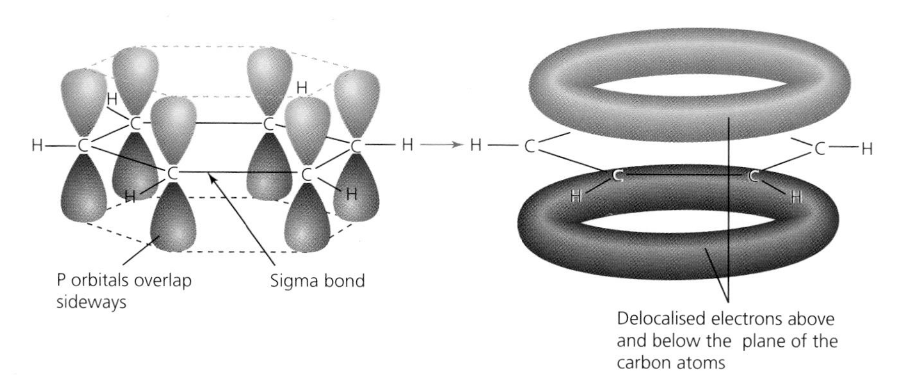

Figure 12.5 The formation of the delocalised electron structure of benzene.

- The lobes of the p orbitals do not pair up to form three carbon–carbon double bonds as in the Kekulé structure, but overlap sideways to form a ring of charge above and below the plane of the molecule. This overlap produces a system of pi bonds. The six p electrons spread over the whole ring – they *delocalise* and give regions of electron density above and below the ring as shown in Figure 12.5.
- Delocalisation of p electrons makes benzene *more stable* than the theoretical molecule cyclohexa-1,3,5-triene. The enthalpy of hydrogenation is $152\,kJ\,mol^{-1}$ less than expected due to the extra stability of the delocalised electrons that are more spread out and so have fewer electron–electron repulsions.
- The delocalisation of electrons also accounts for the intermediate carbon–carbon bond length.

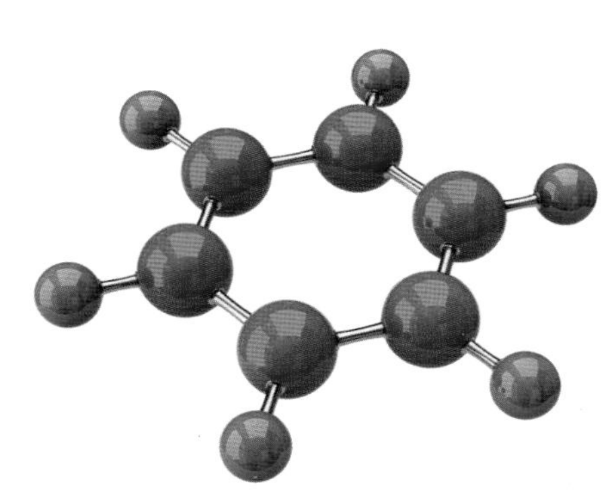

Figure 12.4 Computer-generated model of a molecule of benzene. It is a planar molecule.

- The bond angles in the ring are 120°, meaning there is no strain in this arrangement, another factor contributing to the stability of the molecule.
- A circle is used to represent the ring of delocalised electrons, as shown below.

Delocalised electrons are bonding electrons that are not fixed between two atoms, but shared by three or more atoms.

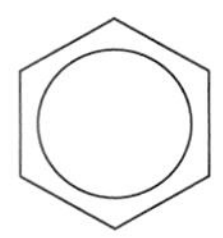

TEST YOURSELF 2

1. Explain why the bond lengths of the carbon–carbon bonds in benzene prove that Kekulé's structure is not correct.
2. State the shape of benzene and give the bond angle around a carbon atom.
3. How does a delocalised ring of electrons form in benzene?

Benzene and addition and substitution reactions

A **substitution** reaction is one in which an atom or group of atoms is replaced with a different atom or group of atoms.

An **addition** reaction is one in which a reactant is joined to an unsaturated molecule to make a saturated molecule.

Benzene and alkenes are similar in that they have regions of high electron density – the delocalised ring of pi electrons in benzene and the double bond in alkenes. As a result of this they both react with electrophiles.

Alkenes undergo addition reactions but, as a result of its stability benzene is fairly *resistant to addition* reactions and instead *substitution reactions* tend to occur. To undergo addition reactions, the electrons from the delocalised system in benzene would need to bond with the atom or group being added. The delocalised electron system would be broken, resulting in a loss of stability. Instead benzene undergoes substitution reactions, where one or more of the hydrogen atoms are replaced by another atom or group. The organic product formed retains the delocalised ring of electrons and hence the stability of the benzene ring.

Bromine water is not decolourised by benzene, as it does not undergo an addition reaction with bromine because the delocalised ring of electrons would be broken.

+ Br — Br

Br
Br

Figure 12.6 Two test tubes containing an organic layer floating on top of an aqueous layer. The aqueous layer contains bromine water (orange). The organic layer on the right (cyclohexene) decolourised the water because cyclohexene contains a reactive double bond which undergoes addition reactions. The organic layer on the left (benzene) did not decolourise the bromine water because it does not undergo addition reactions due to the stable delocalised electron ring.

Naming compounds based on benzene

If one of the hydrogen atoms on a benzene ring is replaced by an atom or group then a benzene derivative is formed. Some common benzene derivatives are shown below.

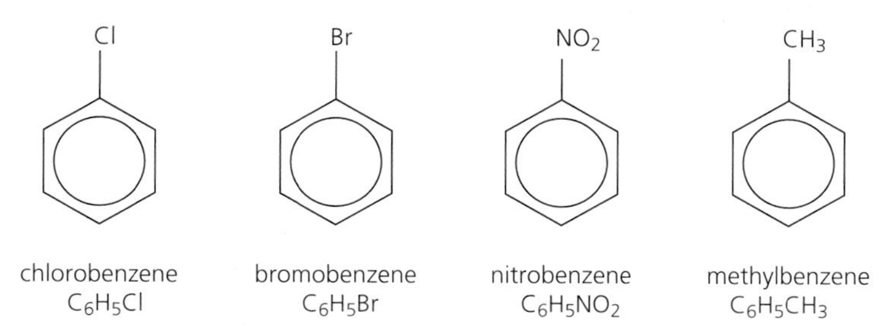

If there is more than one atom or group on the benzene ring, the position of the group must be given. Carbon atoms in the benzene are numbered 1–6 to give the smallest possible position of each group. For example,

CH_3

Cl

could be 1-chloro-3-methylbenzene or 1-chloro-5-methylbenzene. The correct name is 1-chloro-3-methylbenzene, using the *smallest* number for the position of the methyl group.

Figure 12.7 shows three structural isomers of C_7H_7Br.

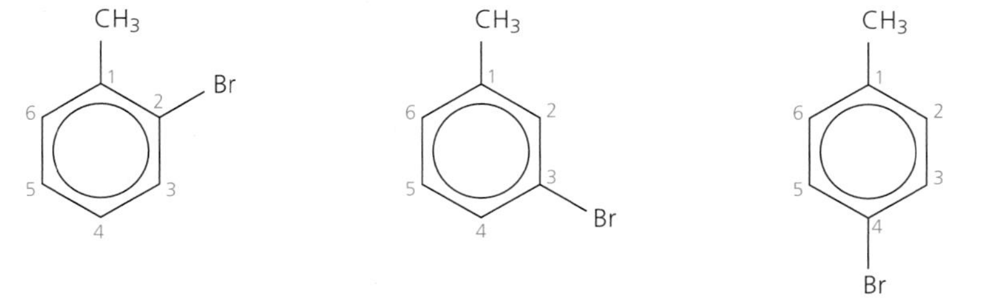

Figure 12.7 The structural isomers of C_7H_7Br.

Alkyl groups are formed by removing a hydrogen atom from an alkane, for example a methyl group CH_3 is formed from methane CH_4 by removing a hydrogen. Removing a hydrogen group from a benzene ring forms a *phenyl* group C_6H_5. Some compounds containing a benzene ring have their name based on phenyl. Some examples are shown below.

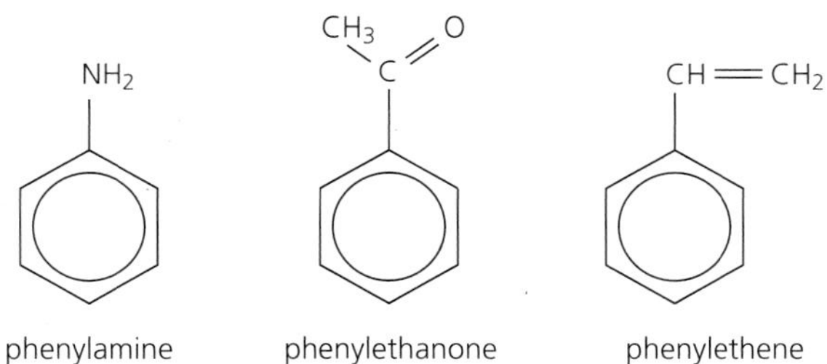

TEST YOURSELF 3

1 Name the following structures.

a)

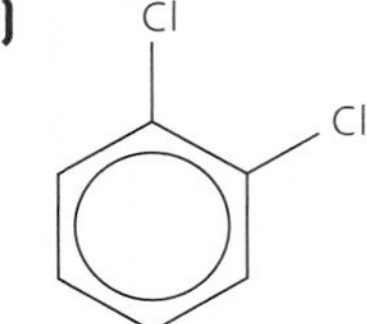

b) CH_3, Cl

c)

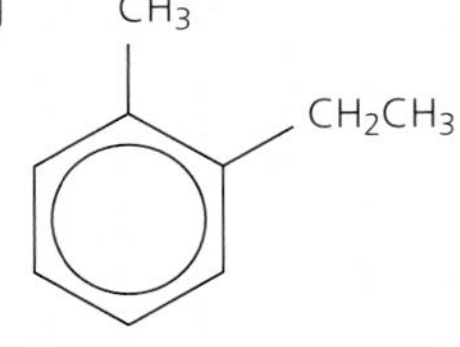

d)

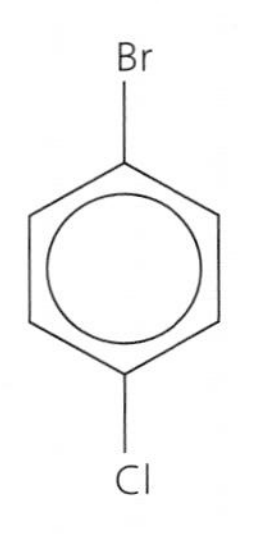

2 a) Name a reagent that could be used to distinguish between benzene and cyclohexene.

b) State what is observed when this reagent is shaken with benzene and then with cyclohexene in separate test tubes.

3 Why does benzene undergo substitution reactions rather than addition reactions?

4 Draw a possible structure for an aromatic compound which has the empirical formula C_7H_8.

5 Is this structure named 1-bromo-2-chlorobenzene or is it 1-chloro-2-bromobenzene? Explain your answer.

Br, Cl

Electrophilic substitution

The region of high electron density above and below the plane of the molecule results in benzene being attacked by electrophiles. Substitution reactions occur rather than addition reactions as this preserves the stability of the benzene ring. The mechanism is described as electrophilic substitution.

Mechanism of electrophilic substitution

A general mechanism for *electrophilic substitution* is shown in Figure 12.8.

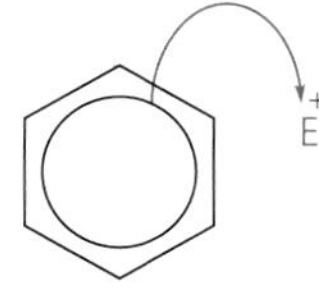

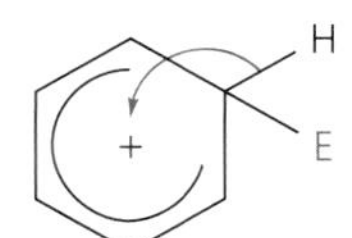

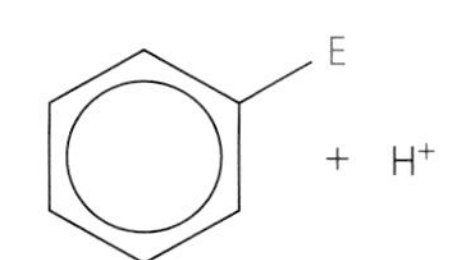

Figure 12.8 The mechanism of electrophilic substitution.

- In the first step the high electron density of the delocalised ring of electrons attracts the electrophile (E^+) and a pair of electrons from the ring of delocalised pi electrons form a bond with the electrophile (E^+), breaking the electron ring.
- This produces a highly unstable intermediate which has only a partially delocalised electron system containing four delocalised electrons.
- A C–H bond breaks in the unstable intermediate and the two electrons in the bond move back into the pi electron system, reforming the stable delocalised electron ring. The hydrogen is lost as H^+.

TIP

If you need to outline the mechanism, you simply need to draw a flow scheme with curly arrows as shown in Figure 12.8 – words of explanation are not necessary.

An electrophile is an **electron pair acceptor**. It is usually is an atom or group of atoms that is attracted to an electron rich centre where it accepts a pair of electrons to form a new covalent bond.

Benzene reacts with different types of electrophiles. Generally these electrophilic substitution reactions employ heat, concentrated reagents and catalysts. This is because of the high stability and low reactivity of benzene. Two electrophilic substitution reactions which you must study in detail are nitration and acylation. Only monosubstitution, where one atom or group is replaced is required.

Nitration

In nitration of benzene, a nitro group (NO_2) replaces one of the hydrogen atoms. To nitrate benzene a mixture of concentrated nitric and concentrated sulfuric acid is used at 50 °C. The mixture of concentrated nitric and concentrated sulfuric acid is often referred to as the nitrating mixture.

$$\text{benzene } C_6H_6 + HNO_3 \xrightarrow[50°C]{\text{conc } H_2SO_4 \text{ catalyst}} \text{nitrobenzene } C_6H_5NO_2 + H_2O$$

Condition: concentrated sulfuric acid and concentrated nitric acid at 50 °C.

The concentrated sulfuric acid is a stronger acid than nitric acid and donates a proton to it, forming an intermediate $H_2NO_3^+$ that decomposes to produce NO_2^+, the *nitronium* ion that is the electrophile.

$$H_2SO_4 + HNO_3 \rightleftharpoons HSO_4^- + H_2NO_3^+$$

$$H_2NO_3^+ \rightarrow NO_2^+ + H_2O$$

$$H_2SO_4 + H_2O \rightarrow HSO_4^- + H_3O^+$$

The overall equation for the generation of the electrophile is

$$HNO_3 + 2H_2SO_4 \rightarrow NO_2^+ + 2HSO_4^- + H_3O^+$$

The mechanism for mononitration is shown in Figure 12.9.

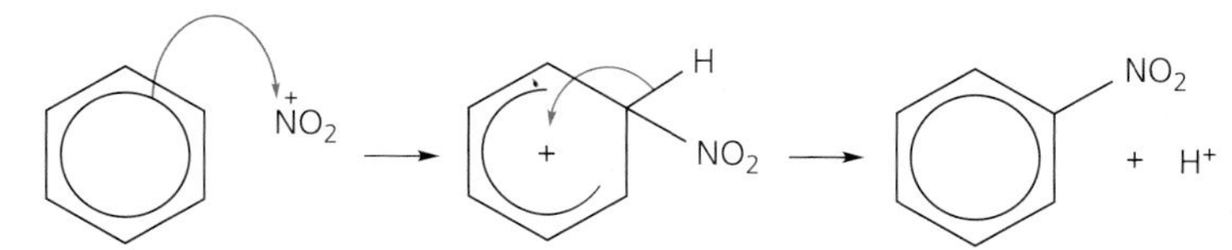

Figure 12.9 The electrophilic substitution mechanism for the nitration of benzene.

TIP

The first curly arrow must come from the delocalised ring to the N or the + on the N of the electrophile. In the intermediate the broken electron ring must not extend beyond carbon 2 or carbon 6. It must be approximately centred on carbon 1.

The second curly arrow must come from the middle of the bond into the hexagon.

The concentrated sulfuric acid acts as a catalyst in the reaction as it is regenerated in the last step when H^+ ion is released in the mechanism and combines with HSO_4^- to reform sulfuric acid.

Uses of nitration

The nitration of benzene and other aromatic compounds is useful in industry

1 **In the manufacture of explosives**. Many nitro compounds formed by nitration are used in explosives – for example TNT 2-methyl-1,3,5-trinitrobenzene otherwise known as 2,4,6-trinitrotoluene is a well-known

explosive used to demolish buildings and to blast mines and quarries. It is made by nitrating methyl benzene.

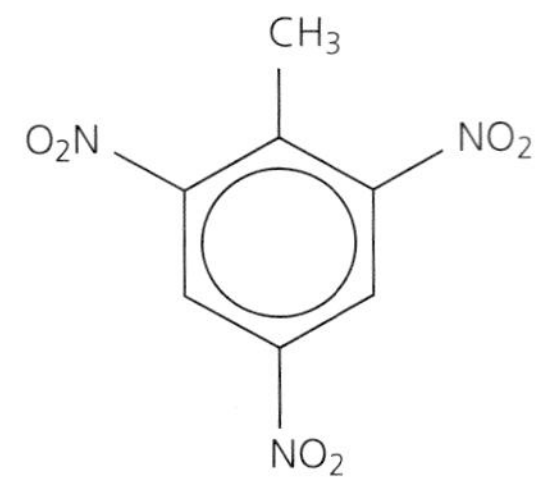

TNT

The presence of the methyl group which is electron releasing makes the benzene more susceptible to attack by electrophiles, so three nitro groups substitute rather than just one.

2 **In the formation of amines**. Nitrobenzene can be reduced to phenylamine (see Chapter 13) which is an important chemical in the production of dyes.

3 **In organic synthesis**. In organic chemistry the synthesis of a particular compound is often a multistage process. Often nitration is one of the steps in synthesis. Organic synthesis will be looked at in detail in Chapter 15.

TEST YOURSELF 4

1 The first step in the nitration mechanism of benzene is shown below. Both benzene and the intermediate structure formed have pi bonds.

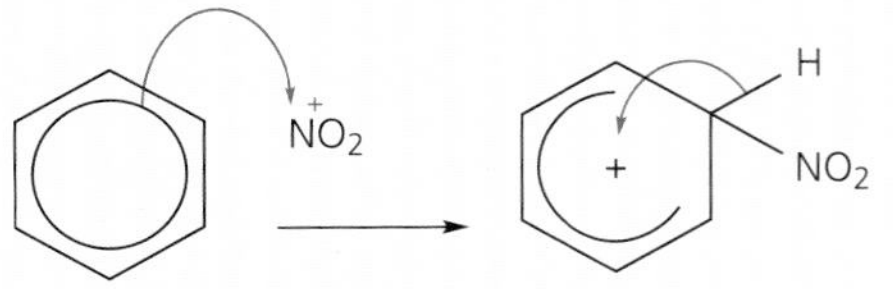

a) Name this mechanism.
b) What does the curly arrow represent?
c) State how many electrons are involved in the pi bonding in each structure.
d) Describe in words what happens in the next step of this mechanism.
e) Name two types of commercially important materials whose manufacture involves the nitration of benzene.
f) State the conditions required for the nitration of benzene.

Acylation

Friedel-Crafts reactions were developed in 1877 by the French chemist Charles Friedel and the American chemist James Mason Craft. They are electrophilic substitution reactions in which an acyl group is attached to the ring, replacing a hydrogen atom. They are useful as the benzene forms a bond with a carbon, producing a side chain.

Benzene can be acylated using an acyl chloride, in the presence of a catalyst, to form an aromatic ketone. The catalyst used is aluminium chloride, and the conditions must be anhydrous to prevent its hydrolysis.

Acylation is the process of replacing a hydrogen atom in certain molecules by an acyl group (RCO–).

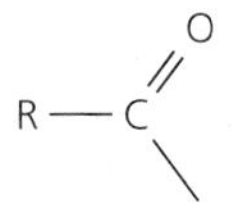

an acyl group

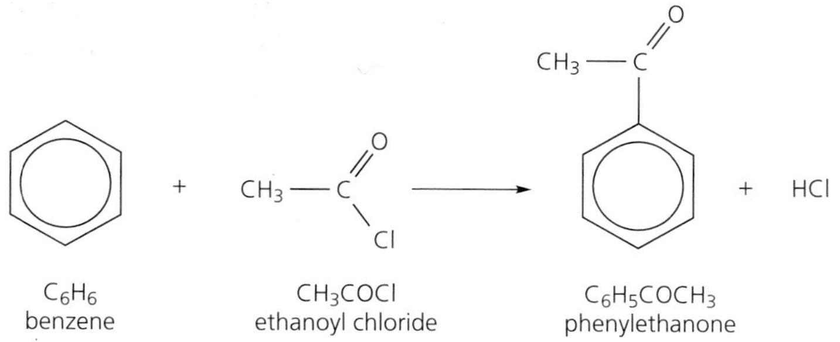

TIP

Remember that phenylethanone is written $C_6H_5COCH_3$. Writing the formula as $C_6H_5CH_3CO$ is incorrect. The -one in the name phenylethanone shows that it is a ketone and has a C=O in the chain, the phenyl indicates that a C_6H_5 is attached.

The required electrophile is the **acylium** ion $CH_3—\overset{+}{C}=O$ It is formed by reaction between the ethanoyl chloride and the aluminium chloride catalyst. The equation for the formation of the electrophile is

$$CH_3COCl + AlCl_3 \rightarrow CH_3\overset{+}{C}O + AlCl_4^-$$

The mechanism for acylation is shown in Figure 12.10.

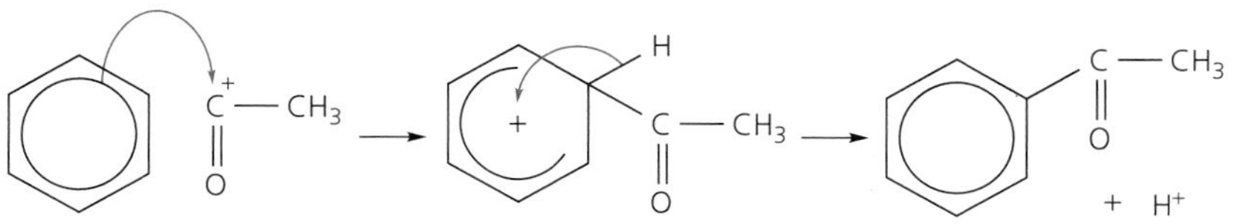

Figure 12.10 The electrophilic substitution mechanism for the acylation of benzene.

The catalyst is regenerated:

$$H^+ + AlCl_4^- \rightarrow AlCl_3 + HCl$$

When methyl benzene reacts with ethanoyl chloride in the presence of aluminium chloride a similar electrophilic substitution reaction occurs but the reaction is faster than the reaction of benzene. This is because the methyl group increases the electron density on the benzene ring which means the electrophile is attracted more, and so the reaction is faster.

An alternative acylation agent is ethanoic anhydride. Instead of HCl, ethanoic acid is produced.

$$C_6H_6 + (CH_3CO)_2O \rightarrow C_6H_5COOCH_3 + CH_3COOH$$

Importance of Friedel-Crafts acylation

Friedel-Crafts are important in synthesis of other organic compounds as they allow carbon–carbon bonds to form. Acylation introduces a reactive carbonyl functional group to the ring. This can undergo the reactions of a carbonyl group and so act as an intermediate in the synthesis of other compounds.

TEST YOURSELF 5

1 a) Write an equation for the reaction of butanoyl chloride with benzene.
   b) Explain why this reaction is an acylation.
   c) Name the catalyst for this reaction and write an equation to show how it reacts with butanoyl chloride molecules to produce an electrophile.
   d) At the end of this reaction the catalyst is regenerated. Write an equation to show the regeneration of the catalyst.
   e) Why must the reaction mixture be kept completely dry in this reaction?

2 Phenylpropanone is used in the manufacture of the drug amphetamine.
   a) Draw the structural formula of phenylpropanone.
   b) Write an equation to show how phenylpropanone is produced from benzene and state the conditions and reagent used.
   c) Write the empirical formula of phenylpropanone.
   d) Name and draw the skeletal formula of a phenylaldehyde which is an isomer of phenylpropanone.

REQUIRED PRACTICAL 10

Preparation of methyl 3-nitrobenzoate (an organic solid)

Dogs are extremely sensitive to smell and can be trained to detect the smell of different types of chemicals. The drug cocaine hydrochloride hydrolyses in moist air to give methyl benzoate. Drug-sniffing dogs are trained to detect the smell of methyl benzoate.

Nitrobenzene is not prepared in the laboratory due to the high toxicity of benzene. To illustrate nitration in the laboratory methyl benzoate, a pleasant-smelling organic compound of low toxicity is used. The procedure followed is detailed below.

- Weigh out 2.7 g of methyl benzoate in a small conical flask and then dissolve in 5 cm^3 of concentrated sulfuric acid. When the solid has dissolved cool the mixture in ice.
- Prepare the nitrating mixture by carefully adding 2 cm^3 of concentrated sulfuric acid to 2 cm^3 of concentrated nitric acid and then cool this mixture in ice.
- Add the nitrating mixture drop by drop to the solution of the methylbenzoate, stirring with a thermometer and keeping the temperature below 10 °C. When the addition is complete allow the mixture to stand at room temperature for 15 minutes.
- Pour the reaction mixture onto 25 g of crushed ice and stir until all the ice has melted and crystalline methyl 3-nitrobenzoate is formed.
- Filter the crystals using Buchner filtration, wash with cold water, recrystallise and obtain the melting point.

1 Write an equation for the nitration of methyl benzoate to produce methyl-3-nitrobenzoate.
2 What homologous series does methyl benzoate belong to?
3 What is the molecular and the empirical formula of methyl benzoate?
4 What conditions were used during this preparation to prevent further nitration of the product to the dinitro derivative?
5 How do the conditions for the nitration of methyl benzoate differ from those for the nitration of benzene?
7 Why were the crystals washed with cold water?
6 Explain why recrystallisation was carried out.
8 Calculate the theoretical yield of methyl 3-nitrobenzoate based on the mass of methyl benzoate used in this preparation.
9 If 2.1 g of methyl 3-nitrobenzoate are obtained, calculate the percentage yield.

Practice questions

1 How many electrons are there in the delocalised pi electron system in benzene?

A 3

B 6

C 9

D 12 *(1)*

2 What is the total number of isomers of dichlorobenzene, $C_6H_4Cl_2$?

A 2

B 3

C 4

D 5 *(1)*

3 Which one of the following statements about benzene is **incorrect**?

A A total of six electrons per molecule are delocalised.

B All of the carbon-carbon bonds are the same length.

C The bond angles are all 120°.

D The empirical formula is C_6H_6. *(1)*

4 Cyclohexa-1,3,5-triene is a hypothetical molecule. It can be hydrogenated to form cyclohexane

a) Write an equation for the hydrogenation of cyclohexa-1,3,5-triene. *(1)*

b) Use the data below to state and explain the stability of benzene compared with the hypothetical cyclohexa-1,3,5-triene molecule. *(2)*

	Enthalpy of hydrogenation/$kJ\,mol^{-1}$
cyclohexene	−120
benzene	−208

c) How would you experimentally distinguish between a sample of benzene and a sample of cyclohexene? *(2)*

5 Methylbenzene $C_6H_5CH_3$ is used to make artificial sweeteners, pharmaceuticals and explosives. Monobromination occurs when one bromine atom substitutes a hydrogen in the aromatic ring in methylbenzene, resulting in three possible structural isomers.

a) Draw and name two of the structural isomers. *(4)*

b) Explain why methylbenzene undergoes substitution reactions rather than addition reactions with bromine. *(2)*

6 **a)** Benzene may be converted to methylbenzene (toluene) by reaction with chloromethane in the presence of aluminium chloride. The mechanism is similar to that for the nitration of benzene.

$$C_6H_6 + CH_3Cl \rightarrow C_6H_5CH_3 + HCl$$

i) What is the function of the aluminium chloride? *(1)*

ii) Suggest the formula of the electrophile. *(1)*

b) The explosive trinitrotoluene (TNT) is prepared by the nitration of methylbenzene (toluene).

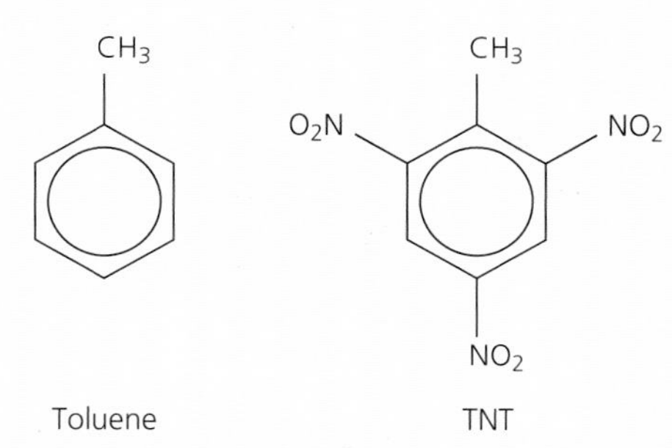

Toluene TNT

i) What is the IUPAC name for TNT? *(1)*

ii) TNT burns to form a mixture of carbon dioxide, nitrogen and water. Write an equation for the complete combustion of TNT. *(2)*

c) TNT is prepared from toluene by using the same nitrating mixture as is used to nitrate benzene.

i) Name the reagents present in the nitrating mixture. *(2)*

ii) Write an equation for the formation of the nitrating species. *(1)*

iii) What name is given to the nitrating species? *(1)*

d) The mechanism for nitrating toluene is similar to that for nitrating benzene.

i) What name is given to this mechanism? *(1)*

ii) Outline a mechanism for the mononitration of toluene. *(3)*

iii) Name the organic product in the mononitration of toluene. *(3)*

7 Ethanoyl chloride reacts with benzene.

a) Write an equation for this reaction and name the organic product. *(2)*

b) Name the catalyst for this reaction and write an equation to show how the catalyst reacts with ethanoyl chloride to produce a reactive intermediate. *(1)*

c) Outline the mechanism for the reaction of benzene with the reactive intermediate. *(1)*

d) Name the mechanism. *(1)*

8 a) Draw a dot and cross diagram to show the bonding in the nitronium ion NO_2^+. *(1)*

b) Explain why there is a positive charge on the nitronium ion. *(1)*

c) Explain if the nitronium ion is an electrophile or nucleophile. *(1)*

d) Write an equation for the generation of the nitronium ion. *(1)*

e) The nitronium ion reacts with benzene. Explain if this reaction is addition or substitution. *(2)*

f) How does the mechanism for this reaction show that the sulfuric acid is acting as a catalyst. *(1)*

Stretch and challenge

9 Benzene reacts with electrophiles. In a reaction with an electrophile the compound Y, shown below was formed.

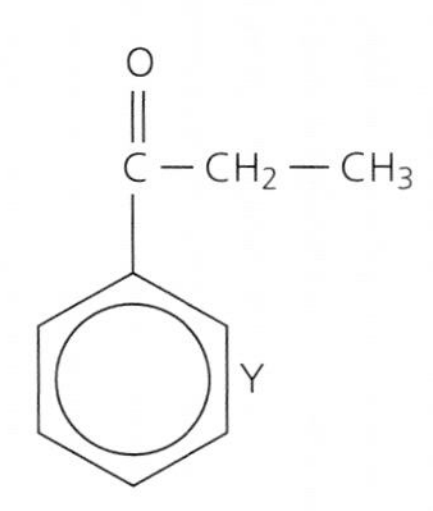

a) Name the two substances that react to generate the electrophile, and write an equation to show the formation of this electrophile. *(3)*

b) Outline a mechanism for the reaction of this electrophile with benzene to form compound Y. *(3)*

c) What is observed when compound Y is warmed with Fehling's solution? *(1)*

d) Another compound X is an isomer of Y. X produces a silver mirror when warmed with the reagent Z. Compound X is optically active. Identify reagent Z and suggest a structure for X. *(3)*

13 Amines

PRIOR KNOWLEDGE

- Ammonia has the formula NH_3 and consists of a nitrogen atom covalently bonded to three hydrogen atoms. The nitrogen atom has a lone pair of electrons.
- Due to its lone pair ammonia can act as a nucleophile, an electron pair donor.
- A Brønsted-Lowry base is a proton acceptor, ammonia is a weak base.
- Halogenoalkanes (e.g. chloromethane CH_3Cl) have a polar carbon–halogen bond due to the halogen atoms having a greater electronegativity than carbon.
- Halogenoalkanes undergo substitution reactions with the nucleophile NH_3 to form amines.

TEST YOURSELF ON PRIOR KNOWLEDGE 1

1 Draw a dot and cross diagram to show the bonding in NH_3 and label the lone pair.

2 Why is ammonia a Brønsted-Lowry base?

3 Name the following structures:

a)
```
    H   H   H   Br
    |   |   |   |
H — C — C — C — C — H
    |   |   |   |
    H   Br  Cl  H
```

b)
```
    H   H   H
    |   |   |
F — C — C — C — F
    |   |   |
    H   F   H
```

c)
```
    H   Cl  H
    |   |   |
H — C — C — C — H
    |   |   |
    H   H   H
```

d)
```
    H   CH3
    |   |
H — C — C — Br
    |   |
    H   H
```

e)
```
  Br
  |
 / \_/
```

4 Explain why ammonia is a nucleophile.

Amines are a group of nitrogen containing organic compounds that are derived from ammonia. As with ammonia, they are characterised by a distinctive unpleasant smell similar to rotting fish.

Amines play an important role in the survival of life – they are involved in the creation of amino acids, the building blocks of proteins and they are found in many chemicals in the body. Adrenaline, a hormone that helps the body to deal with sudden stress, is an amine. Serotonin, a neurotransmitter for the brain that controls the feeling of hunger and helps regulate sleeping patterns is also an amine. Amines are classified as primary, secondary or tertiary depending on the number of hydrogen atoms substituted.

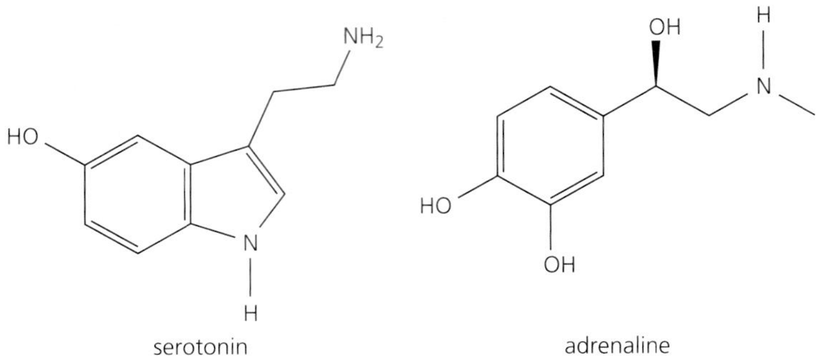

serotonin adrenaline

Figure 13.1 Serotonin and adrenaline are amines. Note the NH_2 and NH groups.

The global demand for amines is increasing yearly and is expected to rise to an average of 4501 kilo tonnes by 2020. This increase in demand for amines reflects their wide ranging applications.

- Many analgesics (medicines that relieve pain) are amines. Morphine is an example.
- Salbutamol, the active ingredient in asthma inhalers, is an amine.
- Many synthetic dyes are made from amines. Methyl orange is made from an amine.
- Amines are used in anaesthetics such as Novocaine, which is a local anaesthetic used in dentistry. It is an aromatic amine.

The structure of amines

> **TIP**
> You came across the terms primary, secondary and tertiary in Book 1 when you studied alcohols. Note that the terms do not have the same meaning with amines.

Amines are compounds based on ammonia, where hydrogen atoms have been replaced by alkyl or aryl (C_6H_5–) groups. Primary amines contain the $-NH_2$ functional group, which is called the **amino group**.

- A **primary amine** contains **one** alkyl or aryl group attached to the nitrogen atom as only one hydrogen atom in ammonia has been replaced. This means there is only one carbon chain attached to the nitrogen.
- A **secondary amine** contains **two** alkyl or aryl groups attached to the nitrogen atom as two hydrogen atoms in ammonia have been replaced.
- A **tertiary amine** contains **three** alkyl or aryl groups attached to the nitrogen atom as three hydrogen atoms in ammonia have been replaced.

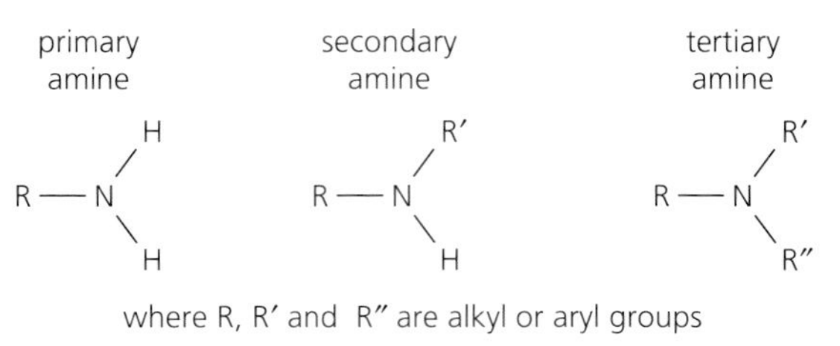

$CH_3CH_2NH_2$ — ethylamine

CH_3—N(CH_2CH_3)(H) — ethylmethylamine

CH_3—N(CH_3)(CH_3) — trimethylamine

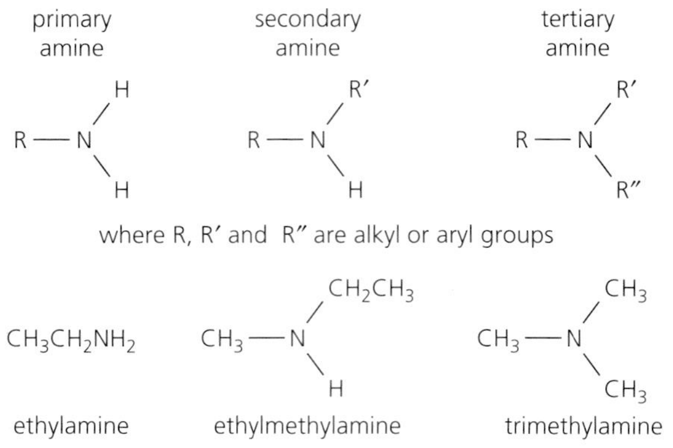

Figure 13.2 The structure of a primary, secondary and tertiary amine.

Quaternary ammonium compounds

These are produced from tertiary amines when the nitrogen's lone pair of electrons forms a dative covalent bond to a fourth alkyl group. Hence a quaternary ammonium salt has four alkyl groups attached to the nitrogen atom.

Like ammonium salts they are crystalline ionic solids.

For example,

$(CH_3)_4N^+I^-$ $CH_3-N^+(CH_3)_2-CH_3\ \ I^-$

is tetramethylammonium iodide, a quaternary salt.

TIP
Remember that the nitrogen in ammonia can form a dative covalent bond with another hydrogen producing a quaternary ammonium salt.

Naming amines

Amines are organic derivatives of ammonia. They are usually named according to the alkyl or aryl chain followed by **–amine.** Naming amines can be confusing as there are many variations on the names. The following examples will illustrate how to name different amines.

TIP
The easiest way to recognise a primary, secondary or tertiary amine is to count the number of H atoms on the N atom.

- **2** H atoms on N atom – primary amine, e.g. $C_2H_5N\mathbf{H_2}$
- **1** H atom on N atom – secondary amine, e.g. $(C_2H_5)_2N\mathbf{H}$
- **0** H atoms on N atom – tertiary amine, e.g. $(C_2H_5)_3N$

EXAMPLE 1

Name the compound:

$H-CH_2-CH_2-CH_2-NH_2$ (displayed: H–C(H)(H)–C(H)(H)–C(H)(H)–N(H)(H))

Answer

- Identify if it is a primary, secondary or tertiary amine – it is a primary amine as it has only one alkyl group attached to the N.
- Name the alkyl chain – there are 3 carbons so the stem is *propyl.*

The compound is *propylamine* (you may also see this named as 1-aminopropane).

EXAMPLE 2

Name the compound: $CH_3-CH(NH_2)-CH_3$

Answer

- Identify if it is a primary, secondary or tertiary amine – it is a primary amine as it has only one alkyl group attached to the nitrogen.
- Name the alkyl chain – there are 3 carbons so the stem is *prop.*
- In this case the amino group is attached to the second carbon so the number must be included in the name.

The compound is *2-aminopropane.*

TIP
When naming amines, if you need to give the position of the carbon to which the NH_2 group is attached, use the 'amino' form of naming.

EXAMPLE 3

Name the compound:

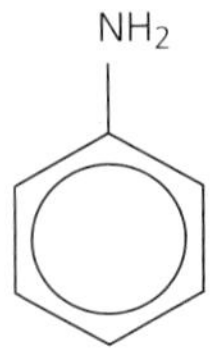

NH_2 attached to a benzene ring

Answer

- Identify if it is a primary, secondary or tertiary amine – it is a primary amine as it has only one aryl group attached to the nitrogen.
- Name the aryl group – *phenyl*.

The compound is *phenylamine*.

EXAMPLE 4

Name the compound $(CH_3)_2NH$:

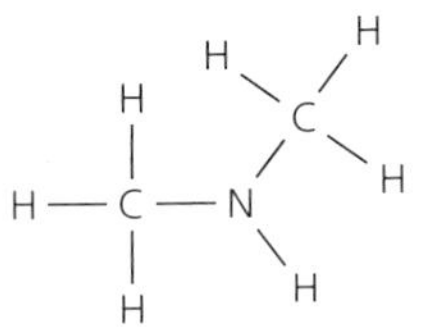

H–C(H)(H)–N(H)–C(H)(H)(H)

Answer

- Identify if it is a primary, secondary or tertiary amine – it is a secondary amine as it has two alkyl groups attached to the nitrogen.
- Name the longest alkyl chain – it contains one carbon – *methyl*.
- Name the other alkyl chain – it contains one carbon – *methyl*.
- The prefix di and tri are used if there are two or three of the same alkyl group – there are two methyls so it is named dimethyl.

The compound is *dimethylamine*.

TIP

The prefix N- is used to show that the alkyl groups are attached to the main chain via the nitrogen atom.

EXAMPLE 5

Name the compound:

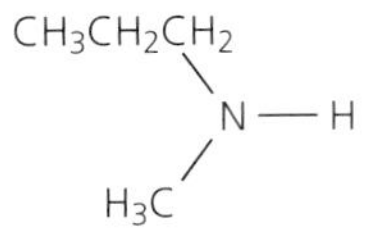

$CH_3CH_2CH_2$–N(–H)–H_3C

Answer

- Identify if it is a primary, secondary or tertiary amine – it is a secondary amine as it has two alkyl groups attached to the nitrogen.
- Name the longest alkyl chain – it contains three carbons – *propyl*.
- Name the other alkyl chain – it contains one carbon – *methyl*. This second alkyl chain is attached to the nitrogen, so a prefix N- is added to show this.

The compound is *N-methylpropylamine*.

TIP
The prefix diamino is used if a compound contains two amino groups, for example 1,2-diaminoethane is $H_2NCH_2CH_2NH_2$.

EXAMPLE 6

Name the compound:

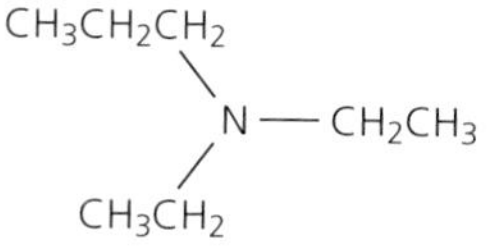

Answer

- Identify if it is a primary, secondary or tertiary amine – it is a tertiary amine as it has three alkyl groups attached to the nitrogen.
- Name the longest alkyl chain – it contains three carbons – *propyl*.
- Name the other alkyl chain – it contains two carbons – *ethyl*. The second alkyl chain is attached to the nitrogen, so a prefix N- is added to show this.
- Name the third alkyl chain – it contains two carbons – *ethyl* and it also is attached to the nitrogen – so the prefix N is added.
- The prefix di and tri are used if there are two or three of the same alkyl group. There are two ethyl groups so it is N,N-diethyl.

The compound is *N,N-diethylpropylamine*.

Compounds that contain an amino group and other functional groups are named based on the hierarchy in organic chemistry which states that the order of priority from highest to lowest is as follows:

1 carboxyl (–COOH)

2 aldehyde (–CHO)

3 ketone (–CO–)

4 hydroxyl (–OH)

5 amines ($-NH_2$, –RNH, $-R_2N$)

6 alkene (–C=C–)

The following groups are always named as prefixes in alphabetical order: alkyl (R–), phenyl (C_6H_5–), chloro (Cl–), bromo (Br–), iodo (I–), nitro ($-NO_2$).

For example:

H_2NCH_2COOH	aminoethanoic acid (also known as glycine)
$H_2NC_6H_4OH$	4-aminophenol
$H_2NCH_3CH_2CHO$	3-aminopropanal

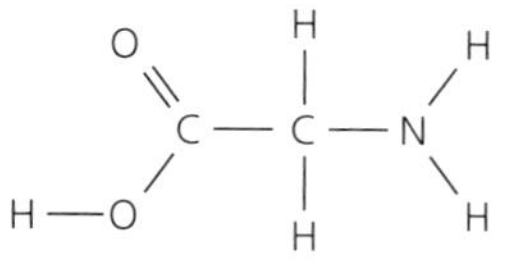

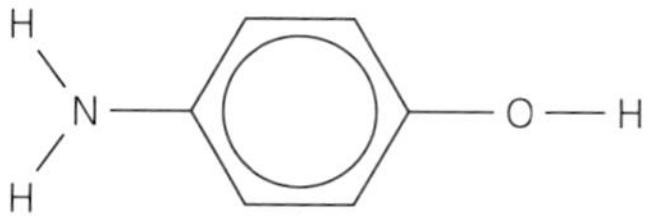

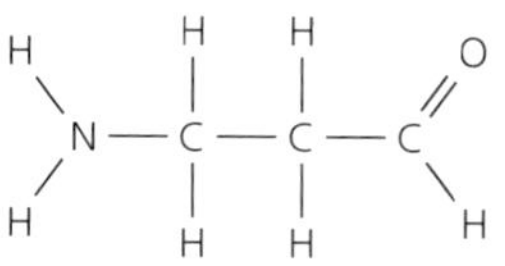

TIP
2-aminoethanoic acid may be called aminoethanoic acid; the 2 is not essential as the amino group cannot be attached to any other carbon.

TEST YOURSELF 2

1 Name the following compounds

a) $CH_3CH_2NH_2$

b) $C_6H_5NH_2$

c) $(CH_3)_2NH$

d)

H — C(H)(H) — N(CH₃)(CH₃) structure: a central N bonded to three C atoms, each C carrying three H atoms

e) $CH_3CH(NH_2)\ CH_2CH_3$

f) $CH_3—CH(NH_2)—CH_2—CH(NH_2)—CH_3$

g) $CH_3CH_3CH_2CH_2NH_2$

h) skeletal structure with H_2N, OH and O labels

2 Classify the amines in the table below as primary, secondary or tertiary amines by placing a tick in the table below.

Amine	Primary	Secondary	Tertiary
methylamine			
ethylamine			
dimethylamine			
phenylamine			
triethylamine			

3 Isomers of $CH_3CH_2CH_2NH_2$ include another primary amine, a secondary amine and a tertiary amine. Draw the structures of these three isomers. Label each structure as primary, secondary or tertiary.

Preparation of primary amines

From halogenoalkanes

Halogenoalkanes react with ammonia to produce an amine.

R–X	+	$2NH_3$	→	$R–NH_2$ + NH_4X
halogenoalkane		ammonia		amine ammonium halide

For example:

CH_3CH_2Cl	+	$2NH_3$	→	$CH_3CH_2NH_2 + NH_4Cl$
chloroethane		ammonia		ethylamine ammonium chloride

Condition: heat in a sealed flask with **excess** ammonia in ethanol. A sealed glass tube is used because the ammonia would escape as a gas if reflux was implemented.

The ammonia has a lone pair of electrons and acts as a nucleophile attacking the $C^{\delta+}$ of the polar C–X bond. The mechanism is **nucleophilic substitution**

The amine produced also has a lone pair of electrons on the nitrogen atom so it can act as a nucleophile. This can attack another molecule of chloroethane, causing further substitution and so continuing the reaction.

$$CH_3CH_2Cl + CH_3CH_2NH_2 \rightarrow (CH_3CH_2)_2NH + HCl$$

TIP

You have come across this reaction and the mechanism before in Book 1, and you will revise the mechanism later on in this chapter.

If **excess ammonia** is used then a primary amine is the major product. If **excess** of the **halogenoalkane** is used then successive substitution is more likely to occur and the reaction produces ethylamine, diethylamine and triethylamine. We will look at this in more detail on page 255. Due to a mixture of different types of amine being produced, this method is not generally used for preparing primary amines. A better method is by reduction of a nitrile.

From nitriles by reduction

Amines can be formed by the reduction of nitriles using hydrogen in the presence of a nickel catalyst or by using a reducing agent such as lithium tetrahydridoaluminate(III) ($LiAlH_4$) in ether.

$$R-C\equiv N + 2H_2 \rightarrow R\text{–}CH_2NH_2$$

$$\underset{\text{ethanenitrile}}{CH_3CN} + 2H_2 \rightarrow \underset{\text{ethylamine}}{CH_3CH_2NH_2}$$

Condition: hydrogen in the presence of a nickel catalyst

$$R-C\equiv N + 4[H] \rightarrow R-CH_2NH_2$$

or using a reducing agent of $LiAIH_4$

$$\underset{\text{propanenitrile}}{CH_3CH_2CN} + 4[H] \rightarrow \underset{\text{propylamine}}{CH_3CH_2CH_2NH_2}$$

Condition: [H] is lithium tetrahydridoaluminate(III) in dry ether

The reduction reaction gives a better yield than the preparation of amines from halogenoalkanes, and there are no other products.

Preparation of aromatic amines

By reduction of nitro compounds

Aromatic amines such as phenylamine are prepared by reduction of nitrocompounds. Phenylamine is prepared by reduction of nitrobenzene using tin and concentrated hydrochloric acid as reducing agent. Overall this reduction can be written as:

$$RNO_2 + 6[H] \rightarrow RNH_2 + 2H_2O$$

$$\underset{\text{nitrobenzene}}{C_6H_5-NO_2} + 6[H] \longrightarrow \underset{\text{phenylamine}}{C_6H_5-NH_2} + 2H_2O$$

Condition: heat under reflux with tin and excess concentrated hydrochloric acid, followed by adding concentrated sodium hydroxide.

Since excess acid is used (rather than getting phenylamine directly), the protonated form of phenylamine, i.e. the **phenylammonium ion,** is formed.

$$C_6H_5-NH_3^+$$

phenylammonium ion

Concentrated sodium hydroxide is added to remove the hydrogen ion from the NH_3^+ group. The addition of concentrated alkali liberates the free amine.

$$C_6H_5NH_3^+ + OH^- \longrightarrow C_6H_5NH_2 + H_2O$$

phenylammonium salt — phenylamine

When writing an equation for this reaction, it is usual to write the equation in one step:

$$C_6H_5NO_2 + 6\,[H] \longrightarrow C_6H_5NH_2 + 2H_2O$$

nitrobenzene — phenylamine

TIP

It is important to start trying to link your organic chemistry. A useful scheme to remember is:

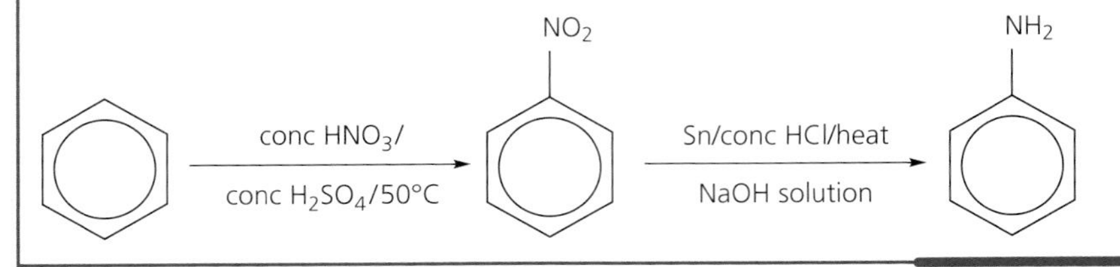

The nitrobenzene used to produce phenylamine is first produced by nitration of benzene (see page 239).

Questions may often involve **two nitro groups** being reduced.

TIP

The command word '**Identify**' allows the you to choose to use either the name or the formula of a reagent in their answer.

EXAMPLE 7

Write a balanced equation for this reaction using molecular formula and using [H] for the reducing agent. Identify a reducing agent for this conversion.

(1,4-dinitrobenzene, NO_2 / NO_2 → 1,4-diaminobenzene, NH_2 / NH_2)

Answer

The general equation for the reduction of a nitro compound is:

$$RNO_2 + 6[H] \rightarrow RNH_2 + 2H_2O$$

Note that in this case there are **two** nitro groups hence the equation is:

$$C_6H_4N_2O_4 + 12[H] \rightarrow C_6H_8N_2 + 4H_2O$$

The reducing agent is tin and concentrated hydrochloric acid

Use of aromatic amines

Aromatic amines prepared by the reduction or nitro compounds are used **in the manufacture of dyes.** These dyes, known as azo dyes were the first synthetic dyes to be produced towards the end of the 19th century. Prior to this, vegetable dyes, which easily faded were used. Azo dyes fade much more slowly and are not removed by water or detergent. These dyes account for about 60% of all the dyes used in food and textiles.

Basic properties of amines

Amines, like ammonia, are weak bases. A base is a **proton** (hydrogen ion) **acceptor.** The lone pair of electrons on the nitrogen atom of ammonia and amines can accept a proton as shown below. A dative covalent bond forms between the lone pair of the nitrogen and the proton.

$$H_3C{-}\ddot{N}H_2 + H^+(aq) \longrightarrow [H_3C{-}NH_2{\rightarrow}H]^+$$

methylamine, proton, methylammonium ion (Dative covalent bond)

Basic reactions of amines

With dilute mineral acids

In general, bases react with acids. Ammonia and amines are similar in their reaction with acids:

ammonia + acid → ammonium salt

amine + acid → alkyl ammonium salt

With dilute hydrochloric acid:

$$CH_3NH_2 + HCl \rightleftharpoons CH_3NH_3^+Cl^-$$

methylamine — methylammonium chloride

$$C_6H_5NH_2 + HCl \rightleftharpoons C_6H_5NH_3Cl$$

phenylamine — phenylammonium chloride

TIP

Remember when writing the formula of alkyl ammonium salts you can put BOTH charges in (as in the methylammonium chloride example) to show the ionic character, or you can leave them both out (as in the phenylammonium chloride example). Never put just one charge in – it is incorrect.

With dilute sulfuric acid:

$$2CH_3CH_2NH_2 + H_2SO_4 \rightleftharpoons (CH_3CH_2NH_3)_2SO_4$$

ethylamine — ethylammonium sulfate

These reactions with acids can be reversed using alkali. The amine can be reproduced again from its salt using a dilute alkali such as sodium hydroxide solution.

With water

When amines react with water, they accept a hydrogen ion from water to produce an alkylammonium ion and hydroxide ions.

$$CH_3NH_2 + H_2O \rightleftharpoons CH_3NH_3^+ + OH^-$$

methylamine methylammonium ion

The solution formed is weakly basic because the equilibrium lies to the left as methylamine is only partly ionised, and as a result little of it has reacted with the water resulting in a solution with low $[OH^-]$.

Comparison of base strength

The basicity of amines depends upon the availability of the lone pair on the N atom, which is used to bond with the proton. Different types of amines have different basic strengths as shown in the figure below:

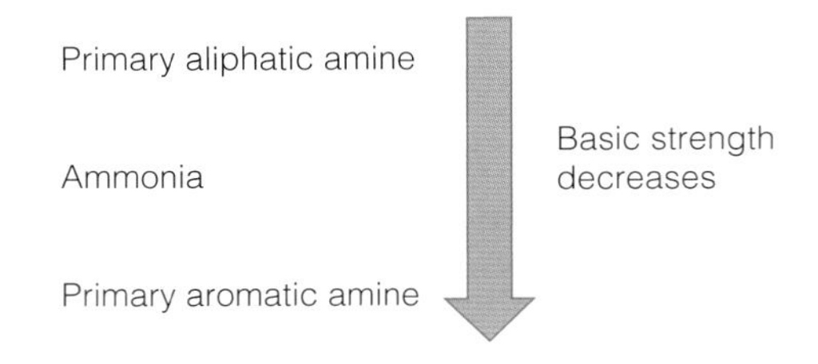

Primary aliphatic amines are stronger bases than ammonia because of the alkyl group attached to the nitrogen. The alkyl group is said to be **electron donating** – it releases electrons meaning there is slightly more electron density on the nitrogen atom. As a result, the **lone pair is more available** and so has an increased ability to accept a proton. Aliphatic amines generally increase in base strength as the number of alkyl groups attached to the nitrogen atom increases.

The trend in basicity for the first three primary amines is shown in Figure 13.3.

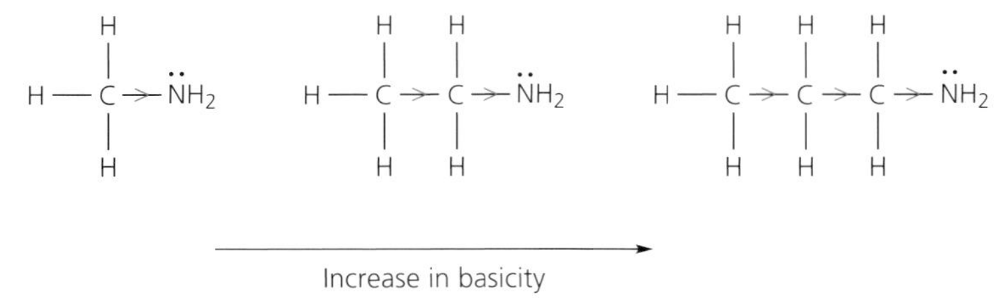

Figure 13.3 The diagram shows that as the size of the alkyl group increases the electron donating ability increases, and so the lone pair is more available resulting in an increase in basicity.

Primary aromatic amines such as phenylamine, are weaker bases than ammonia because nitrogen's lone pair of electrons can overlap with the delocalised pi electrons in the benzene ring. The lone pair is delocalised into the pi system, the electron density on the nitrogen is decreased and the **lone pair is less available** for accepting a proton.

The relative strength of the primary amines can be compared by looking at the pK_a values in the table below. The pK_a values given are for the conjugate RNH_3^+. The stronger the base, the weaker the conjugate acid. As a result, the stronger base will have a higher pK_a value for its conjugate base – in this table propylamine is the strongest base.

Name	Formula	pKa	
propylamine (primary aliphatic)	$CH_3CH_2CH_2NH_2$	10.84	Decreasing basicity ↓
ethylamine (primary aliphatic)	$CH_3CH_2NH_2$	10.73	
methylamine (primary aliphatic)	CH_3NH_2	10.64	
ammonia	NH_3	9.25	
phenylamine (primary aromatic)	$C_6H_5NH_2$	4.62	

TEST YOURSELF 3

1 Ammonia and ethylamine are examples of weak Brønsted–Lowry bases.

a) State the meaning of the term Brønsted–Lowry base.

b) i) Write an equation for the reaction of ethylamine ($CH_3CH_2NH_2$) with water to form a weakly alkaline solution.

ii) In terms of this reaction, state why the solution formed is weakly alkaline.

c) State which is the stronger base, ammonia or ethylamine. Explain your answer.

2 a) Methylamine reacts with hydrochloric acid as shown below.

$CH_3NH_2 + HCl \rightarrow CH_3NH_3Cl$

i) Name the product of this reaction.

ii) Explain how the free amine may be liberated.

b) Place the following basic compounds in order of basicity from 1 (most basic) to 4 (least basic) by placing the number in the table below.

Compound	Basicity
ammonia	
dimethylamine	
methylamine	
phenylamine	

c) Explain how ethylamine reacts with water to form an alkaline solution using an equation.

d) Write an equation for the reaction of methylamine with sulfuric acid.

3 For the flow scheme

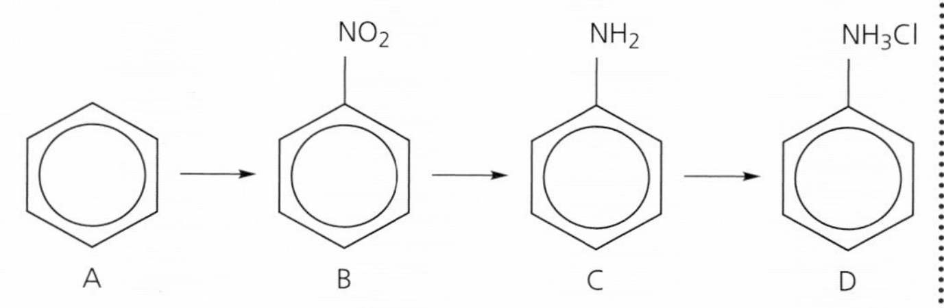

a) Name structures A–D

b) Name the reagents for each step in the conversion

c) Name the mechanism for the conversion of A to B

4 Explain why X is a weaker base than Y.

NH2 NH2

X Y

Nucleophilic properties of amines

Nucleophiles are lone pair donors. All amines contain a lone pair of electrons on the nitrogen atom, so they act as nucleophiles.

Nucleophilic substitution reaction with halogenoalkanes

The reaction of a halogenoalkane with ammonia and amines forms primary, secondary, tertiary amines and quaternary ammonium salts.

Making a primary amine

This reaction happens in two stages. Initially a salt is formed:

$$CH_3CH_2Br + NH_3 \rightarrow CH_3CH_2NH_3Br$$

ethylammonium bromide

Then the salt reacts with the excess ammonia in the mixture and removes a hydrogen ion forming a primary amine.

$$CH_3CH_2NH_3Br + NH_3 \rightleftharpoons CH_3CH_2NH_2 + NH_4Br$$

ethylamine (primary amine)

Making a secondary amine

Further substitution reactions may occur. This is because the primary amine produced has a lone pair so it can act as a nucleophile and continue to react with any unused halogenoalkane, in the same stages as before.

$$CH_3CH_2Br + CH_3CH_2NH_2 \rightarrow (CH_3CH_2)_2NH_2Br$$

$$(CH_3CH_2)_2NH_2Br + NH_3 \rightleftharpoons (CH_3CH_2)_2NH + NH_4Br$$

diethylammonium bromide; diethylamine (secondary amine)

This successive substitution results in the formation of a secondary amine.

Making a tertiary amine

The secondary amine produced also has a lone pair so it too can act as a nucleophile and continue to react with any unused halogenoalkane, in the same stages as before. This further substitution results in a tertiary amine.

$$CH_3CH_2Br + (CH_3CH_2)_2NH \rightarrow (CH_3CH_2)_3NHBr$$

$$(CH_3CH_2)_3NHBr + NH_3 \rightleftharpoons (CH_3CH_2)_3N + NH_4Br$$

triethylamine (tertiary amine)

Making a quaternary ammonium salt

Finally, the tertiary amine reacts with the halogenoalkane to form a quaternary ammonium salt. There is no longer a lone pair on the nitrogen so the quaternary ammonium salt cannot act as a nucleophile and the reaction stops.

$$CH_3CH_2Br + (CH_3CH_2)_3N \rightarrow (CH_3CH_2)_4NBr$$

quaternary ammonium salt

Choosing the conditions

In this reaction, due to the repeated nucleophilic substitution secondary amines, tertiary amines and quaternary ammonium salts may all be present depending on the conditions. The product of each reaction is a better nucleophile than the starting material as it contains more electron donating alkyl groups.

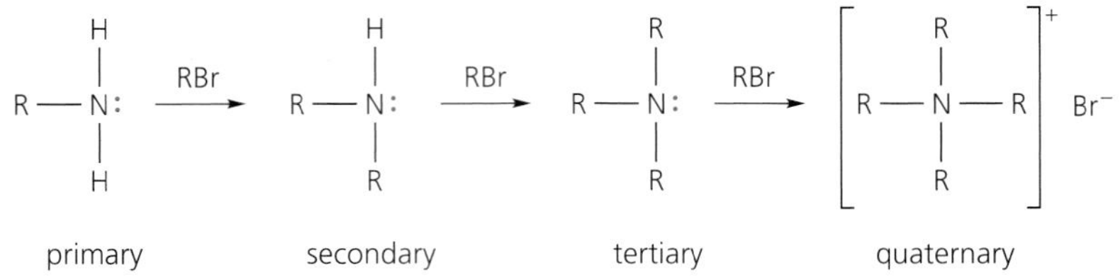

$$RNH_2 \xrightarrow{RBr} R_2NH \xrightarrow{RBr} R_3N \xrightarrow{RBr} [R_4N]^+Br^-$$

primary — secondary — tertiary — quaternary

Figure 13.4 The successive substitutions of an amine.

The initial conditions can be adjusted to favour production of a particular type of amine.

- **Excess ammonia** favours the production of **primary amines** as it is less likely that another halogenoalkane molecule will react with an amine, when there is a large number of unreacted ammonia molecules available.

- **Excess halogenoalkane** favours the production of the **quaternary ammonium salts** as it ensures that each ammonia reacts with four halogenoalkanes molecules.

The use of quaternary ammonium salts

Quaternary ammonium salts are used in the production of **cationic surfactants** that are found in **detergents, fabric softeners** and **hair conditioners.** They coat the surface of the cloth or hair with positive charges and reduce the static due to negatively charged electrons. A quaternary ammonium salt found in detergent is shown in Figure 13.5.

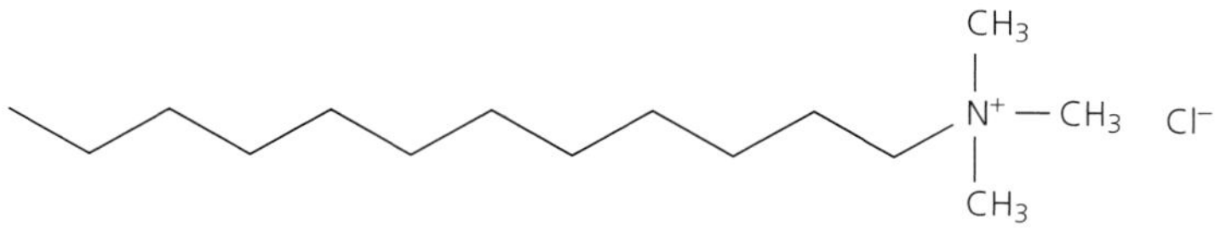

Figure 13.5 Quaternary ammonium salts are found in fabric softener/water repellents and hair conditioners.

Mechanism of the nucleophilic substitution

You have studied the mechanism of nucleophilic substitution in Book 1, for the reaction of halogenoalkanes with ammonia.

$$CH_3CH_2Br + 2NH_3 \rightarrow CH_3CH_2NH_2 + NH_4Br$$

bromoethane + ammonia → ethylamine + ammonium bromide

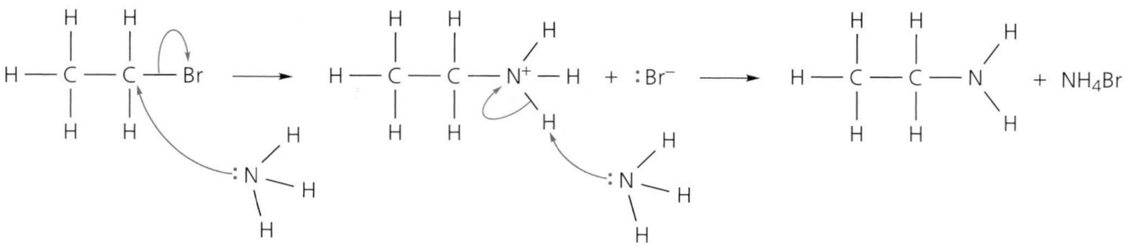

TIP

The balanced symbol equation for the overall reaction between ammonia and a halogenoalkane contains two ammonia molecules. The first acts as a nucleophile and substitutes the halogen. The second acts as a base and accepts a proton, a hydrogen ion, from the reactive intermediate.

Further substitution is possible as the product, the primary amine, is also a nucleophile. The mechanism below illustrates how further substituted amines are formed.

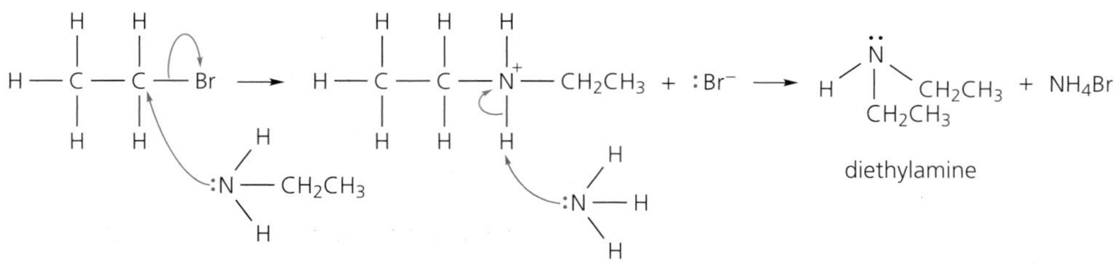

The mechanism for the formation of a quaternary ammonium ion is shown below.

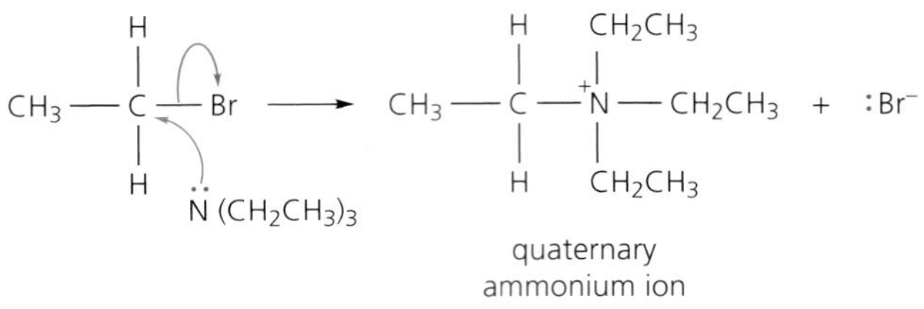

$$CH_3CH_2Br + :N(CH_2CH_3)_3 \longrightarrow CH_3CH_2\overset{+}{N}(CH_2CH_3)_3 + :Br^-$$

quaternary ammonium ion

Nucleophilic addition–elimination reactions of ammonia and primary amines with acyl chlorides and acid anhydrides

Primary amines and ammonia both react with acyl chlorides and acid anhydrides to form substituted amides. You have already studied these reactions on page 225 of this book. You must also be familiar with the nucleophilic addition–elimination mechanism outlined for these reactions on page 227.

TEST YOURSELF 4

1 Draw the structural formula of the two compounds formed in the reaction of CH_3NH_2 with ethanoic anhydride.

2 **a)** Write a equation for the reaction of $CH_3CH_2NH_2$ with propanoyl chloride.
  b) Name the mechanism for this reaction.
  c) Outline the mechanism for this reaction.
  d) Name the product.

3 **a)** Name the compound $(CH_3)_2NH$.
  b) $(CH_3)_2NH$ can be formed by the reaction of an excess of CH_3NH_2 with CH_3Br. Name and outline a mechanism for this reaction.
  c) Name the type of compound produced when a large excess of CH_3Br reacts with CH_3NH_2.
  d) Give a use for this type of compound.

ACTIVITY

Paracetamol

Paracetamol and aspirin are the largest selling pain relief tablets available without a prescription. Paracetamol can be made by reacting 4-aminophenol with ethanoyl chloride.

1 Suggest a structure for 4-aminophenol (4-aminohydroxybenzene).

2 Paracetamol has the IUPAC name 4-hydroxy-(N-ethanoylaminobenzene).
  a) Suggest a structure for paracetamol.
  b) Write an equation for the reaction of 4-aminophenol with ethanoyl chloride to produce paracetamol.
  c) Name the mechanism for this reaction.

3 State two reasons why ethanoic anhydride is used in industry instead of ethanoyl chloride.

Practice questions

1 Which one of the following gives the order of increasing basic strength for ammonia, ethylamine and phenylamine?

A ammonia, phenylamine, ethylamine

B ethylamine, ammonia, phenylamine

C phenylamine, ammonia, ethylamine

D phenylamine, ethylamine, ammonia *(1)*

2 The mechanism for the reaction of methyl amine with chloroethane is described as

A electrophilic substitution

B nucleophilic addition-elimination

C nucleophilic addition

D nucleophilic substitution *(1)*

3 Iodoethane may be converted to propylamine by reaction with

A ammonia

B ammonia followed by $LiAlH_4$

C potassium cyanide followed by ammonia

D potassium cyanide followed by $LiAlH_4$ *(1)*

4 Amines can be prepared from halogenoalkanes or from nitriles.

a) 1-Bromobutane is heated with excess ammonia in a sealed tube.

i) Draw the structure and name the amine formed in this reaction. *(2)*

ii) Ammonia is a nucleophile and the mechanism is described as nucleophilic substitution. What is a nucleophile? *(1)*

iii) Explain what is meant by a substitution reaction. *(1)*

b) Propanenitrile is formed from 1-bromoethane by reaction of potassium cyanide.

i) Write an equation for the reaction of 1-bromoethane with potassium cyanide. *(1)*

ii) What reagent is used to convert propanenitrile to an amine? *(1)*

iii) Name the amine formed when propanenitrile is converted to an amine. *(1)*

iv) What type of reaction occurs when a nitrile is converted to an amine? *(1)*

v) Draw a structural equation showing all bonds for the reaction when propanenitrile is converted to an amine. *(1)*

5 **a) i)** Give the meaning of the term Brønsted–Lowry base. *(1)*

ii) State and explain which of ammonia or butylamine is the stronger base. *(2)*

b) Draw the structure of the tertiary amine which is an isomer of butylamine. *(1)*

c) i) Draw the structure of the species formed when the amine $CH_3(CH_2)_{17}NH_2$ reacts with an excess of CH_3Br. *(1)*

ii) Name the type of compound formed in part (c) (i) and give a use for such compounds. *(2)*

d) Name and outline a mechanism for the reaction in which N-methylethanamide is formed from CH_3NH_2 and CH_3COCl. *(3)*

6 Amines have characteristic properties and can form hydrogen bonds due to the polarity of the N–H bond.

a) Show the polarity of the N–H bond. *(1)*

b) The table below gives the melting and boiling points of some amines.

Amine	Melting point/°C	Boiling point/°C	State at 25 °C
methylamine	−93	−61.8	
ethylamine	−81	17	
propylamine	−83	49	
dimethylamine	−92	7	
trimethylamine	−117	3	

i) Complete the table giving the physical state of each amine at room temperature (25 °C). *(5)*

ii) Explain why propylamine is miscible with water. *(1)*

c) i) Name the type of mechanism for the reaction between phenylamine and bromomethane. *(1)*

ii) Name the type of substance formed when phenylamine reacts with a large excess of bromomethane. *(1)*

Stretch and challenge

7 Propylamine can be prepared in two different ways.

One method involves starting with CH_3CH_3Br and forms an intermediate before forming propylamine.

The second method involves starting from $CH_3CH_2CH_2Br$ and involves just one step.

a) Name the intermediate compound in method one and give the reagents and conditions for all steps in the process. In addition give the reagent and condition for the second method. *(6)*

b) State a disadvantage of each method of preparation of propylamine. *(2)*

14 Polymers, amino acids and DNA

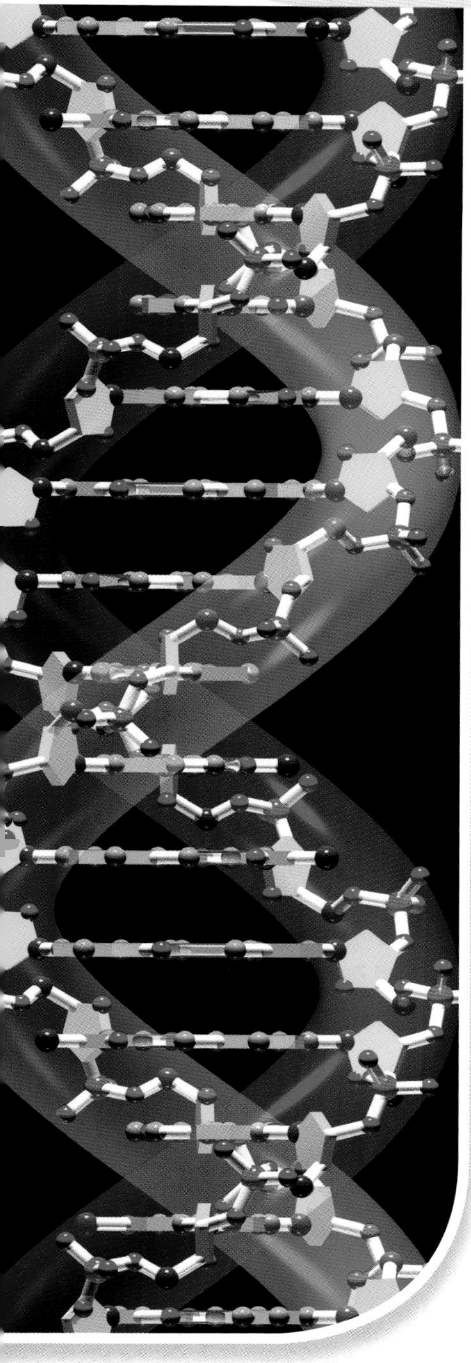

PRIOR KNOWLEDGE

- Addition polymers are formed from alkenes in an addition reaction.

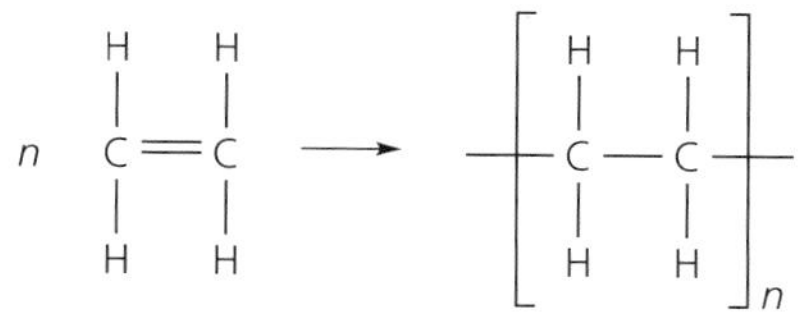

$$n\ \mathrm{CH_2{=}CH_2} \longrightarrow \left[\!-\mathrm{CH_2}-\mathrm{CH_2}-\!\right]_n$$

Figure 14.1 The equation for the formation of polythene from ethene. Ethene is the monomer and poly(ethene) is the polymer, often called polythene.

- Polythene chains only have van der Waals' forces between chains so are not good for forming fibres or weaving.
- The groups on the alkene may differ creating different polymers.
- If an H is replaced with a Cl, the monomer is chloroethene commonly called vinyl chloride.
- The polymer formed is called poly(chloroethene) or polyvinylchloride (PVC). 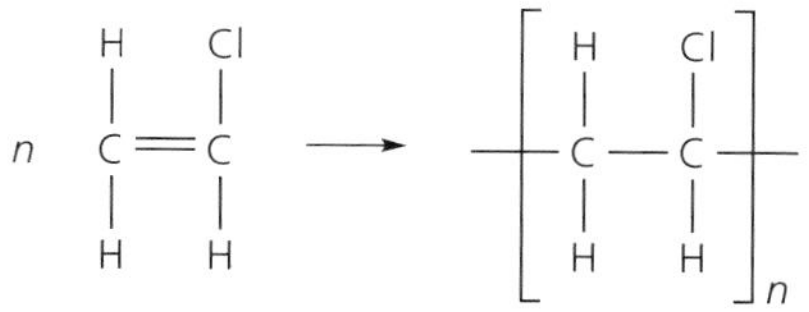
- Two repeating units of the polymer may be shown as:

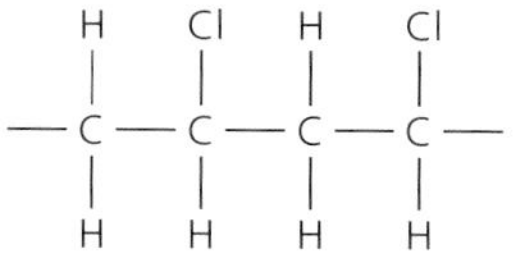

$$n\ \mathrm{CHCl{=}CH_2} \longrightarrow \left[\!-\mathrm{CH_2}-\mathrm{CHCl}-\!\right]_n$$

$$-\mathrm{CH_2}-\mathrm{CHCl}-\mathrm{CH_2}-\mathrm{CHCl}-$$

- If the Cl is replaced with a benzene ring the monomer is called phenylethene (commonly called styrene) and the polymer is poly(phenylethene) or polystyrene.

TEST YOURSELF ON PRIOR KNOWLEDGE 1

1 Name the addition polymer formed from the following monomers
  - **a)** ethene
  - **b)** chloroethene
  - **c)** styrene

2 Using the following monomer:

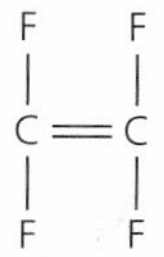

  - **a)** Name the monomer.
  - **b)** Draw the structure of two repeating units of the polymer formed from this monomer.
  - **c)** Name the type of polymerisation.
  - **d)** The polymer is commonly called PTFE. Explain the name.

Condensation polymerisation

Molecules which possess COOH groups and OH group or COOH groups and NH_2 groups can form condensation polymers. During the formation of a condensation polymer, water or HCl is eliminated.

A condensation polymer is a polymer formed when monomers join together and eliminate a small molecule such as water (or hydrogen chloride).

There are two types of condensation polymers considered here:

- polyesters
- polyamides.

Polyesters

A polyester is a condensation polymer formed between a diol and a dicarboxylic acid (or between molecules of a single compound that contain both a COOH group and an OH group). The reaction is described as condensation polymerisation as it eliminates water.

This is in contrast to the addition polymerisation of ethene to form polythene.

One polyester is formed from the reaction between ethane-1,2-diol and benzene-1,4-dicarboxylic acid (terephthalic acid). The polyester is called poly(ethylene terephthalate) or PET (sometimes PETE).

PET is a polyester as it contains many ester (COO) groups.

$$n\,HO-CH_2-CH_2-OH + n\,HOOC-C_6H_4-COOH \longrightarrow \left[-O-CO-C_6H_4-CO-O-CH_2-CH_2-\right]_n + 2n\,H_2O$$

ethane-1,2-diol; benzene-1,4-dicarboxylic acid (terephthalic acid); poly(ethylene terephthalate)

Two molecules of water are eliminated for the formation of one repeating unit, as one repeating unit has two ester groups (COO) and the formation of each one eliminates a molecule of water.

A polyester could be formed using an diacyl chloride (a molecule containing two acid chloride groups) and an diol. In that case 2HCl would be eliminated.

Most plastic bottles used for soft drinks are now made of PET (polyethylene terephthalate), a plastic that is recyclable. At the start of the recycling process, the bottles are compacted into bales, and then shredded. Discarded plastic drinks bottles are compacted at a recycling facility. The plastic chips formed are cleaned, and then sent to be used in new bottles.

Polyamides

Nylon is a condensation polymer formed from the reaction between 1,6-diaminohexane and adipic acid (hexane-1,6-dioic acid).

Nylon is a polyamide as it contains many amide groups (CONH).

$$n\,H_2N-(CH_2)_6-NH_2 + n\,HOOC-(CH_2)_4-COOH \longrightarrow \left[-NH-CO-(CH_2)_4-CO-NH-(CH_2)_6-\right]_n + 2n\,H_2O$$

hexane-1,6,diamine; hexane-1-6-dioic acid; nylon

This form of nylon is often called nylon-6,6 as it is made from a diamine containing 6 carbon atoms and a dioic acid containing 6 carbon atoms. Nylon-6,10 also exists where the dioic acid contains 10 carbons.

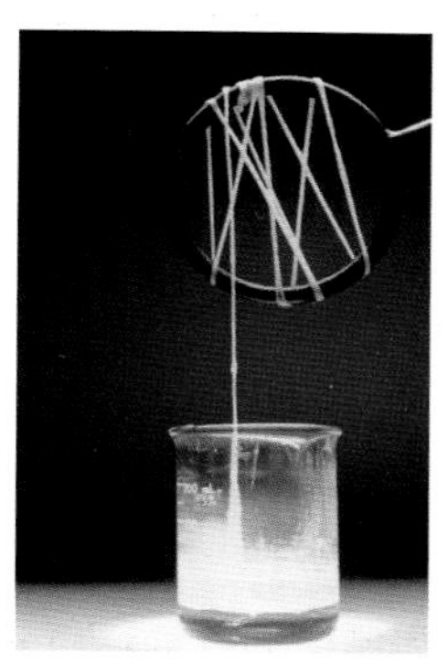

Figure 14.2 The synthetic fibre, nylon is made by the condensation polymerisation of amine-containing molecules and carboxylic acid-containing molecules to form long chains (polymers). Nylon is used to make fabrics, and also ropes and some medical prosthetics.

A diacyl chloride (molecule containing two acid chloride (COCl) groups may be used in place of the dioic acid and the elimination product is HCl rather than water.

$$n\ H_2N{-}(CH_2)_6{-}NH_2 + n\ ClOC{-}(CH_2)_4{-}COCl \longrightarrow \left[{-}NH{-}CO{-}(CH_2)_4{-}CO{-}NH{-}(CH_2)_6{-}\right]_n + 2n\ HCl$$

hexane-1,6,diamine; hexanedioyl dichloride; nylon

Hexanedioyl dichloride is often called adipoyl chloride from the common name of the dioic acid. Sebacic acid is the common name for decane-1,10-dioic acid and the acid chloride is called sebacoyl chloride. Sebacoyl chloride is often used in the nylon rope trick forming nylon from the boundary of two layers of sebacoyl chloride and 1.6-diaminohexane.

Kevlar is a polyamide polymer formed from benzene-1,4-dicarboxylic acid and benzene-1,6-diamine.

$$n\ HOOC{-}C_6H_4{-}COOH + n\ H_2N{-}C_6H_4{-}NH_2 \longrightarrow \left[{-}CO{-}C_6H_4{-}CO{-}NH{-}C_6H_4{-}NH{-}\right]_n + 2H_2O$$

There are hydrogen bonds between the chains of polymers in polyamides due to the presence of C=O bonds and N–H bonds. Polyesters have permanent dipole–dipole attractions between the polymer chains due to the presence of the polar C=O bonds. There are greater forces of attraction between the polymer chains in polyamides and polyesters compared to addition polymers, so polyamides like nylon and Kevlar and polyesters like Terylene can be used to create clothing. The stronger forces of attraction between the polymer chains mean that the fibres can be woven together or knit together. Formula 1 racing drivers use helmets made of carbon fibre covered with a layer of Kevlar which gives protection, but is still very light. Kevlar can also be spun into sheets which are five times stronger than steel, and are used in bullet proof vests.

Most addition polymers cannot be used to create clothing as there are only weak van der Waals' forces of attraction between the polymer chains.

TIP

Make sure that you can draw the repeating unit in a polymer based on the monomers or the structures of the monomers from a section of the polymer.

EXAMPLE 1

Lactic acid, $CH_3CH(OH)COOH$, can form a condensation polymer called PLA, poly(lactic acid). Draw a section of PLA showing two repeating units.

Answer

The COOH and OH groups of the single monomer allow it to form a condensation polymer. It is best to draw the monomer with the COOH and OH groups facing each other to be able to draw a section of the polymer.

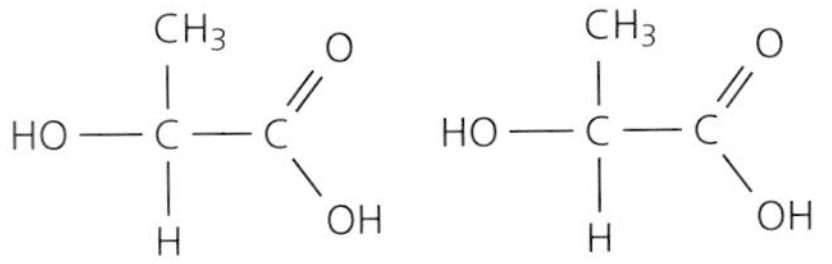

$HO—C(CH_3)(H)—C(=O)OH \quad HO—C(CH_3)(H)—C(=O)OH$

The polymer is formed from the reaction between the OH group and the COOH group. The structure of the polymer showing two repeating units is:

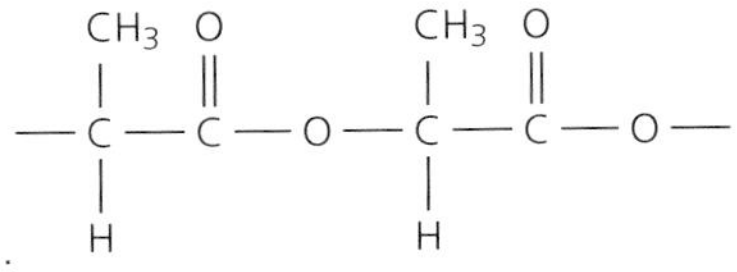

$—C(CH_3)(H)—C(=O)—O—C(CH_3)(H)—C(=O)—O—$

This polyester condensation polymer shows two repeating units, but you can start from anywhere in the polymer as long as two complete repeating units are shown. The polymer could be drawn as shown below:

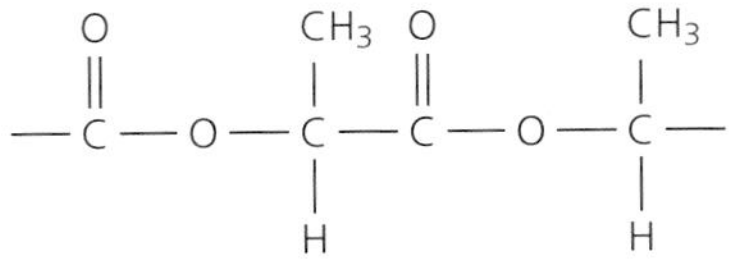

$—C(=O)—O—C(CH_3)(H)—C(=O)—O—C(CH_3)(H)—$

TIP

Examples will include being asked to draw a certain number of repeating units of a polymer.

EXAMPLE 2

The structure below shows the repeating unit of a polymer.

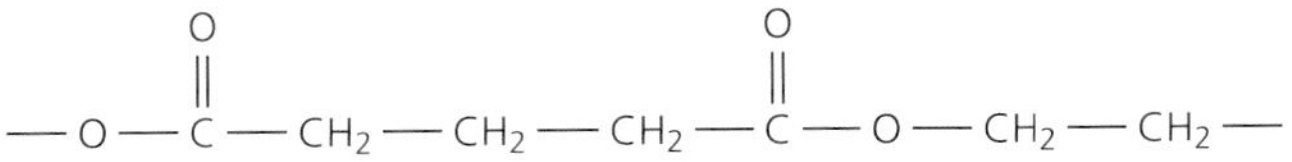

$—O—C(=O)—CH_2—CH_2—CH_2—C(=O)—O—CH_2—CH_2—$

1 Give the IUPAC name for the carboxylic acid used to form this polymer.
2 Give the IUPAC name of the alcohol used to form this polymer.
3 Name the type of polymer formed.

Answers

1 Take the polymer and use the C=O groups to identify the acid from which the polymer was formed.

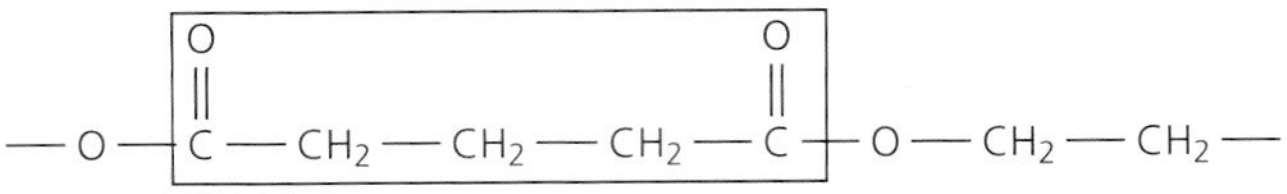

$—O—[C(=O)—CH_2—CH_2—CH_2—C(=O)]—O—CH_2—CH_2—$

The acid is: $HOOCCH_2CH_2CH_2COOH$. This acid is pentanedioic acid.

2 The alcohol structure and name can be determined by looking at the O atoms. They might be on different end of the polymer; one O atom in this polymer is on the left but the structure of the alcohol is $HOCH_2CH_2OH$. Its name is ethane-1,2-diol.

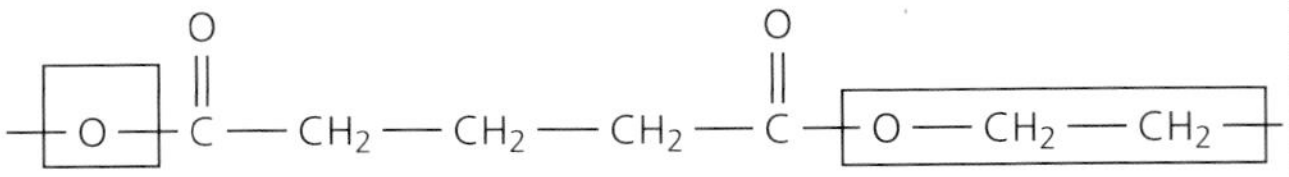

$[O]—C(=O)—CH_2—CH_2—CH_2—C(=O)—[O—CH_2—CH_2]—$

3 Condensation polymer or polyester.

TIP

In dioic acids the carbon atoms that are part of the carbon chain are counted as part of the chain, HOOC–COOH is ethanedioic acid. $HOOC–CH_2–COOH$ is propanedioic acid.

TIP

Diols and triols require a number for the position of the OH groups in the molecules. They also require an extra 'e' in the name.

TIP

If this example had stated that the polymer is a condensation polymer, you would be expected to give the type of polymer as a polyester.

Biodegradability of polymers

A biodegradable polymer is one that can be broken down in the environment by the action of micro-organisms. Polyalkenes are chemically inert and non-biodegradable because they are saturated and have no bond polarity.

Condensation polymers can be hydrolysed and so are now being used more than addition polymers. This means that they are biodegradable in the environment. Many drink bottles (especially water bottles) are now made from PET.

Addition polymers, such as polythene, will remain for thousands of years in the environment, but PET can be hydrolysed by micro-organisms in the environment so it is biodegradable.

However, condensation polymers break down more quickly on a compost heap than in landfill. There is less water available in landfill to hydrolyse the polymer.

There is also less oxygen available in landfill and it is colder. There is less UV light to assist breakdown of the polymers. A landfill site has fewer bacteria than would be present in a compost heap.

Disposal of waste

A large proportion of household and industrial waste in the UK is plastic. Plastic waste or polymer waste is particularly worrying as it does not biodegrade and stays in the ecosystem for hundreds of years. Some 7% of household waste is plastic as shown in the chart below.

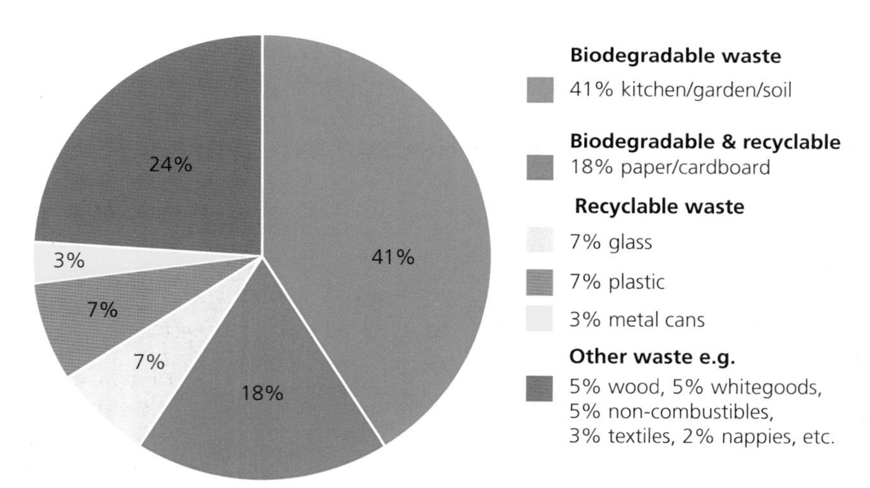

Figure 14.3 Composition of UK household waste.

The methods of disposing of polymers are landfill and incineration. Both have their advantages and disadvantages.

Landfill

Disadvantages

- Landfill wastes land and often pollutes the land with polymers which will take hundreds of years to decompose.
- Some polymers can also leach compounds into the soil.
- Landfill sites are often an eyesore.
- Landfill releases methane, which is a very effective greenhouse gas.

Advantages
- Landfill is the most cost-effective method of waste disposal.

Incineration

Disadvantages
- Incineration releases greenhouse gases into the air and also some toxic gases depending on the polymer being incinerated.
- Incineration can also produce gases that cause acid rain.
- Incineration still produces waste, which has to be sent to landfill, but this is 90% less than waste sent directly to landfill.
- More expensive than landfill (but can be linked with energy recovery where the heat of combustion is used to power an electricity generator).

Advantages
- Incineration saves money with regard to transport as waste can be incinerated locally.
- Incineration prevents unsightly landfill sites.

The EU has guidelines about how to handle waste materials, including polymers. EU policy ranks waste management strategies in the following order:

1 Prevention of waste

2 Recycling and re-use of material

3 Safe disposal of waste that is not recyclable or re-useable – in this ranking order:

a) Incineration with Energy Recovery
b) Incineration
c) Landfill

Reduce, re-use, recycle

Reduce, re-use and recycle are referred to as the 3Rs. Waste management has become more and more important and we have been encouraged to reduce the amount of waste we produce. Manufacturers must limit the amount of packaging and as much packaging as possible must be able to be recycled. Reducing the amount of waste saves the Earth's resources. We are encouraged to re-use material such as carrier bags over and over again. At the end of the lifetime of an item, then it should be recycled. All this saves resources.

Strategies to reduce, control and manage polymer waste

Many strategies are already in place to reduce, control and manage polymer waste.

- Reduction in the use of polymers in packaging.
- Reduction in use of carrier bags.
- Dedicated polymer recycling (blue bins).
- Use of biodegradable polymers.
- Businesses must recycle polymers.

TEST YOURSELF 2

1 Name the monomers used to make the following condensation polymers:
  a) PET **b)** Nylon-6,6 **c)** Kevlar

2 Draw the structure of the repeating unit formed from condensation polymerisation between the following two monomers.
  $H_2NCH_2CH_2NH_2$ $HOOCCH_2CH_2CH_2COOH$

3 The following polymer is a condensation polymer formed from a diacyl chloride and a diol.

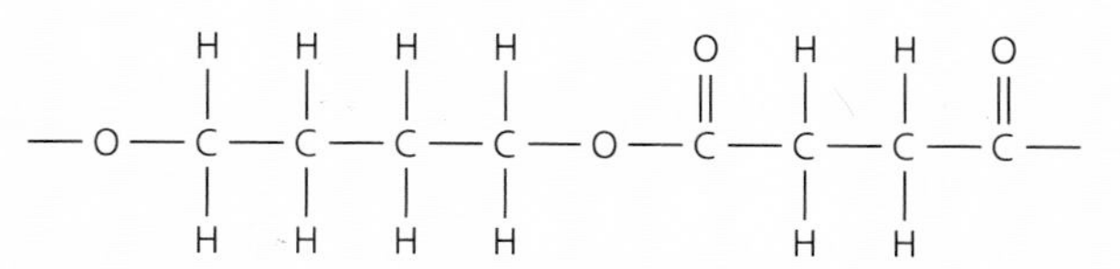

  a) Name the diacyl chloride from which it is formed.
  b) Name the diol from which it is formed.
  c) Name the type of polymer.

Amino acids, proteins and DNA

Proteins are compounds of the elements carbon, hydrogen, oxygen and nitrogen and usually contain some sulfur as well. Proteins perform many different roles within the body such as:

1 Enzymes are proteins.

2 Muscle fibres are made of several proteins.

3 Antibodies in the immune system are proteins.

4 Some hormones are proteins.

5 Bone is a matrix of protein and calcium phosphate.

6 Haemoglobin and chlorophyll are mainly protein.

There are many other roles of proteins within the body. In fact, the word protein comes from the Greek word *proteios* meaning first. They are often described as the fundamental material in living organisms.

Structure

The structure of proteins is the most complex all the nutrients and indeed among the most complex of any molecule within any organism. Their structure is divided into three levels:

1 **Primary structure** of a protein is concerned with the order in which the amino acids are joined together in the molecule and corresponds to the normal structural formulae that have been considered for simpler molecules. → E.g - polypeptide

2 **Secondary structure** of a protein is concerned with the folding, coiling and puckering of the chain of atoms of the protein molecule. → E.g. - α Helix

3 **Tertiary structure** of a protein is concerned with the final folding of the protein molecule to form a globular protein as dictated by the interaction between groups attached to the main body of the chain.

Primary structure

Proteins are often called natural polymers, but unlike synthetic ones, they are made up from a selection of 20 different subunits called amino acids. The sequence of these amino acids in the protein chain joined together by peptide bonds makes up the primary structure of the protein molecule.

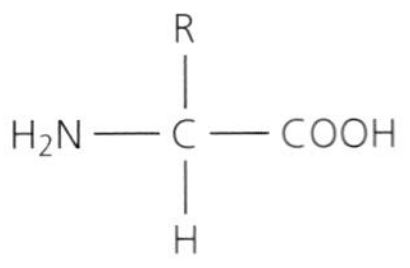

$H_2N-C(R)(H)-COOH$

Many amino acids are **α-aminocarboxylic acids**. Variation in the structure of these monomers occurs in the side chain, denoted by R in the structure on the left.

Amino acids do not always behave like organic compounds. They have melting points well in excess of 200 °C, whereas most organic compounds of similar M_r are liquids at room temperature. Amino acids are soluble in water and other polar solvents, but are insoluble in non-polar solvents such as ether or benzene. They are less acidic than most carboxylic acids and less basic than most amines.

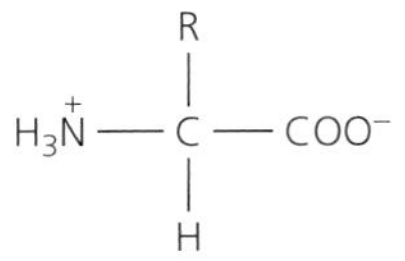

$H_3\overset{+}{N}-C(R)(H)-COO^-$

The reason for the odd properties of amino acids is that they contain an acidic carboxyl group and a basic amino group on the same molecule. Amino acids undergo an internal acid–base reaction to yield a dipolar ion (called a zwitterion). In the solid state, an amino acid has the structure shown on the left.

Solid amino acids contain ionic bonds, which leads to the higher than expected melting point.

TIP
Read over the section on pH to make sure you understand this.

This zwitterion structure also explains the weakly acidic and basic properties of an amino acid as it has the weakly basic COO^- group and the weakly acidic NH_3^+ group.

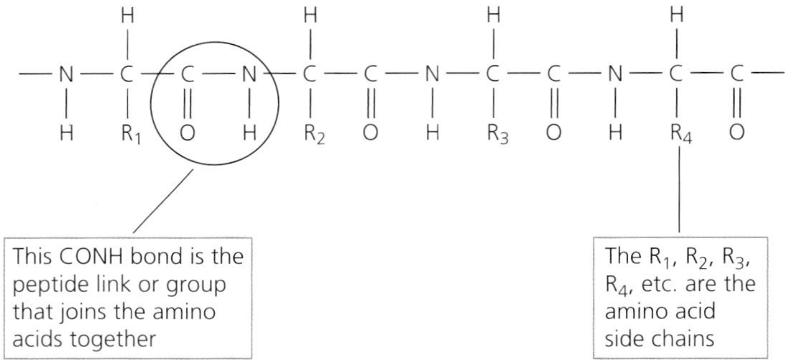

$-N(H)-C(H)(R_1)-C(=O)-N(H)-C(H)(R_2)-C(=O)-N(H)-C(H)(R_3)-C(=O)-N(H)-C(H)(R_4)-C(=O)-$

This CONH bond is the peptide link or group that joins the amino acids together

The R_1, R_2, R_3, R_4, etc. are the amino acid side chains

Proteins are polyamides. This is due to the condensation reaction which can occur between the carboxyl group on one amino acid and the amino group on the next. The bond linking these two amino acids is called a peptide group or peptide link.

Protein molecules would have the form shown on the left.

Peptide groups are formed in a condensation reaction (a reaction in which a small molecule is eliminated, in this case water) between the carboxyl end of one amino acid and the amino end of another.

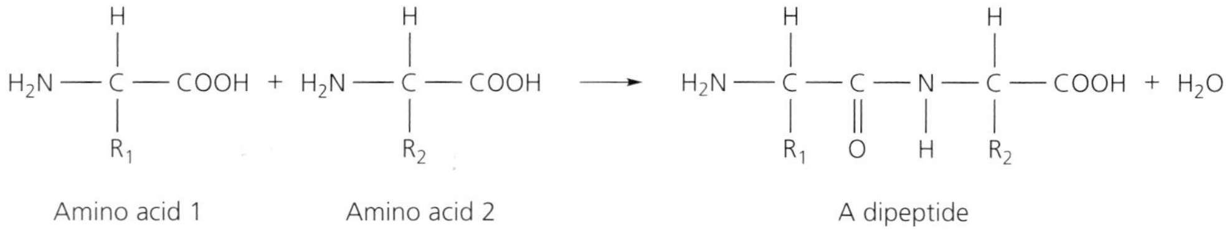

$$H_2N-C(H)(R_1)-COOH + H_2N-C(H)(R_2)-COOH \longrightarrow H_2N-C(H)(R_1)-C(=O)-N(H)-C(H)(R_2)-COOH + H_2O$$

Amino acid 1 | Amino acid 2 | A dipeptide

Many amino acids joined in a chain creates a polypeptide and longer polypeptides are called proteins.

TIP
The structures of six amino acids are given in your Data Booklet.

Amino acids
There are 20 naturally occurring amino acids: These vary in the R side chain. The properties of the R side chain also vary and this allows a convenient way of classifying the 20 amino acids in groups.

These groups are:

1 Non-polar aliphatic R group amino acids

2 Non-polar aromatic R groups amino acids

3 Polar uncharged R groups amino acids

4 Acidic R group amino acids

5 Basic R group amino acids.

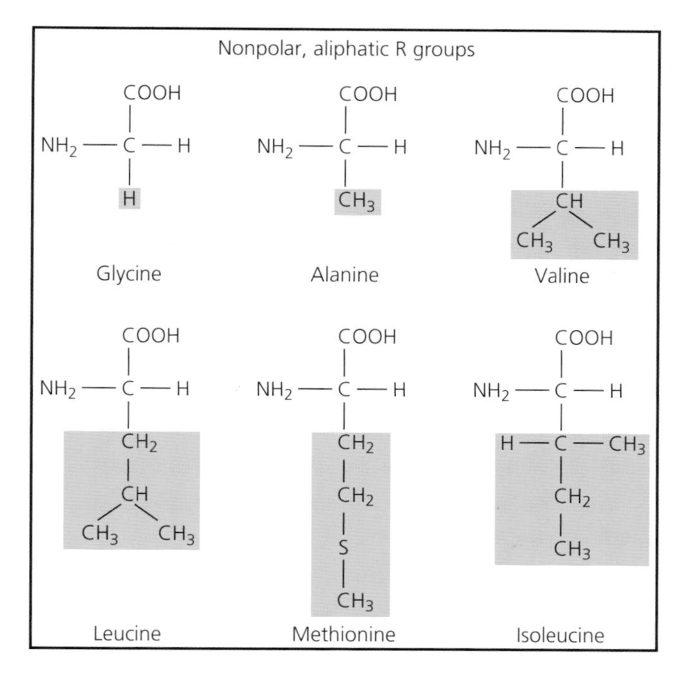

Nonpolar aromatic R groups

Phenylalanine

Tyrosine

Tryptophan

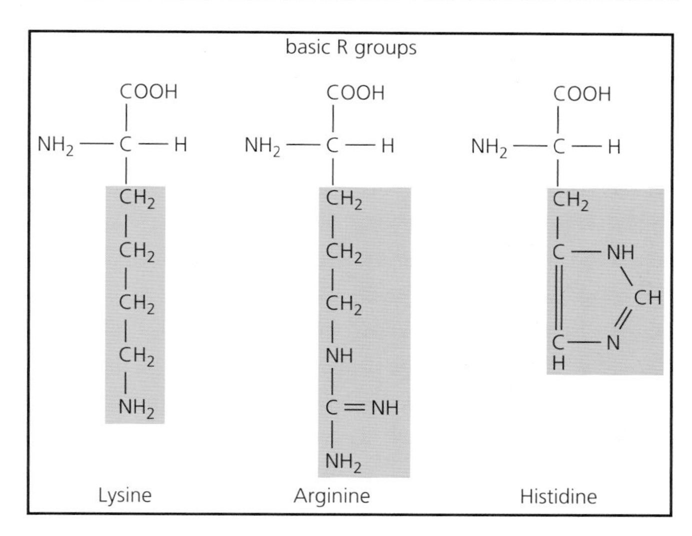

Polar, uncharged R groups

Serine

Threonine

Cysteine

Proline

Asparagine

Glutamine

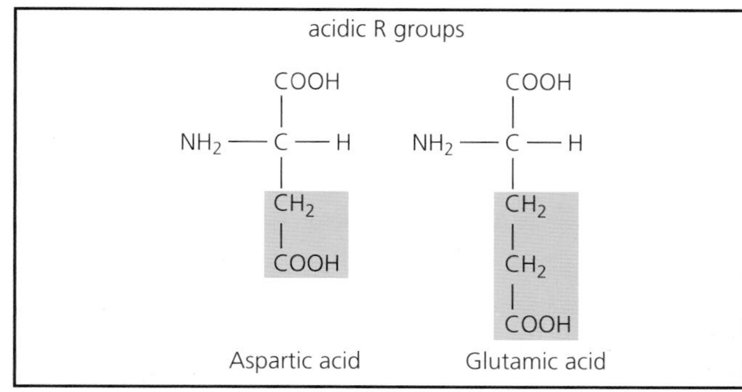

Figure 14.4 The 20 amino acids present in human proteins.

TIP

You do not need to remember the structure of each amino acid but make sure you can write the correct IUPAC names for some common amino acids.

EXAMPLE 3

Name the following amino acids:

1 H_2NCH_2COOH

2 $NH_2-C(H)(COOH)-C(H)(CH_3)-CH_2-CH_3$

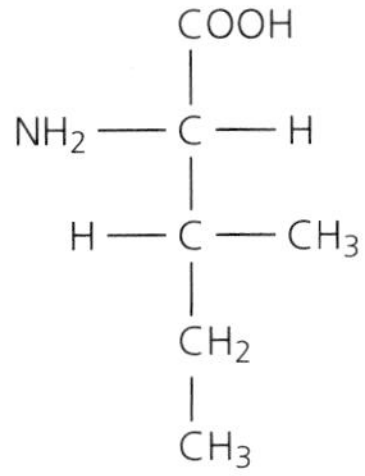

Answers

1 aminoethanoic acid

2 2-amino-3-methylpentanoic acid

Amino acids in acidic and alkaline solution

Amino acids contain the COOH group and the NH_2 group. Both of these groups are affected by changes in pH.

The COOH group at high pH deprotonates to form the COO^- group. At low pH in an excess of H^+ ions, the COOH group is protonated as COOH.

The NH_2 group can be protonated at low pH to form NH_3^+. At high pH the NH_2 group is deprotonated so NH_2 is formed.

For amino acids that contain other COOH groups or NH_2 group in the R group, these groups will also be protonated or deprotonated depending on the pH.

TIP

In a structure the + should be shown on the N atom. The – charge on COO^- should be shown on the oxygen atom at the end of the single bond.

EXAMPLE 4

Alanine, leucine and lysine have the following structures:

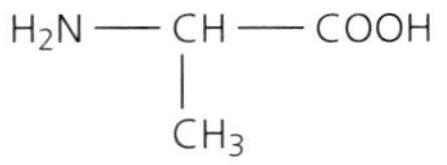

alanine: $H_2N-CH(CH_3)-COOH$

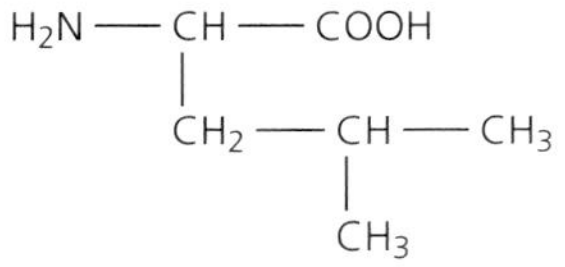

leucine: $H_2N-CH(CH_2-CH(CH_3)-CH_3)-COOH$

lysine: $H_2N-CH(CH_2-CH_2-CH_2-CH_2-NH_2)-COOH$

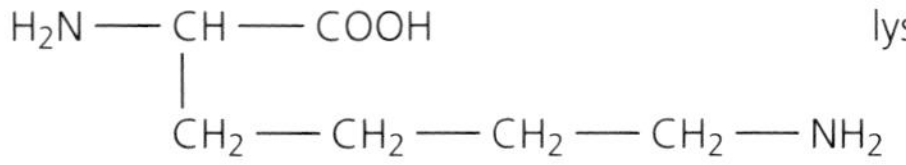

1 Draw the structure of alanine at high pH.

2 Draw the structure of lysine at low pH.

3 Give the IUPAC name for leucine.

Answer

1 $H_2N-C(CH_3)(H)-COO^-$

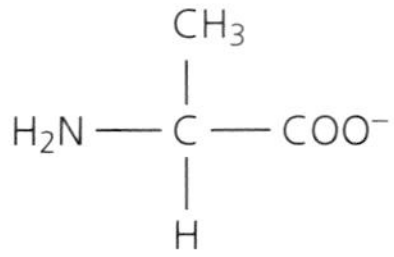

At high pH the COOH group is deprotonated to form COO^-. The amino acid is negatively charged at high pH.

2 $H_3\overset{+}{N}-C(H)(CH_2-CH_2-CH_2-CH_2-\overset{+}{N}H_3)-COOH$

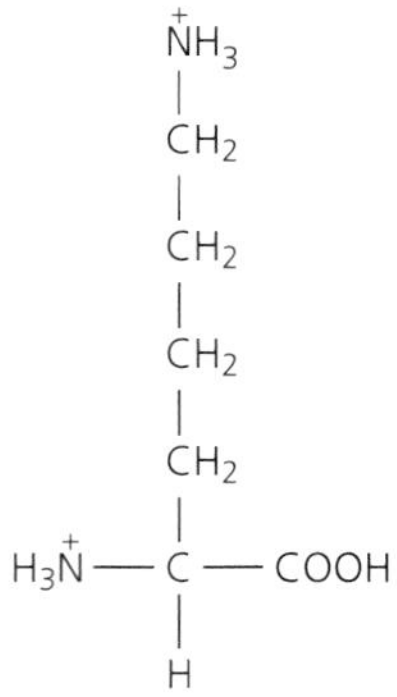

At low pH the two NH_2 groups in lysine are both protonated. The amino acid is positively charged at low pH.

3 2-amino-4-methylpentanoic acid

Amino acids from peptides

You may also be expected to extract the structure of an amino acid from a section of a polypeptide or protein.

EXAMPLE 5

The tripeptide below shows three different amino acids.

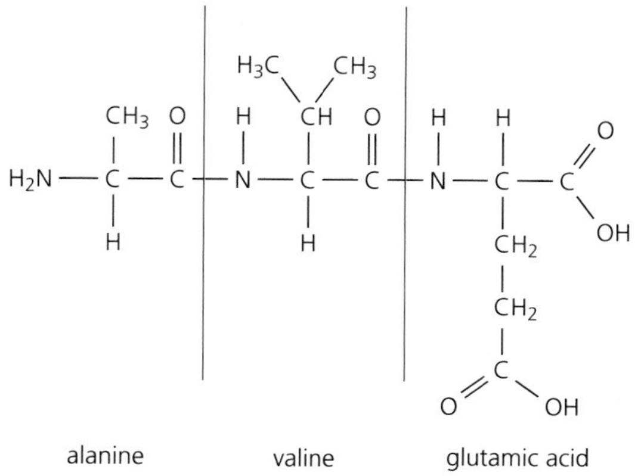

alanine | valine | glutamic acid

1 Draw the structure of the zwitterion formed by alanine.
2 Give the IUPAC name for valine.
3 Draw the structure of species formed by glutamic acid at high pH.
4 Draw the structure of the dipeptide formed from two glutamic acid molecules.
5 Valine reacts with ethanol in the presence of concentrated sulfuric acid. Draw the structure of the compound formed.

Answers

1

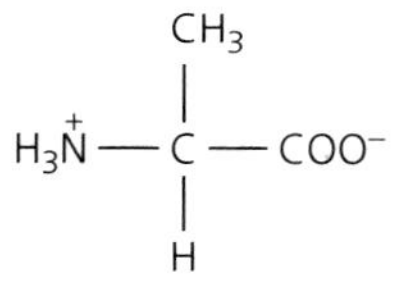

2 2-amino-3-methylbutanoic acid

3

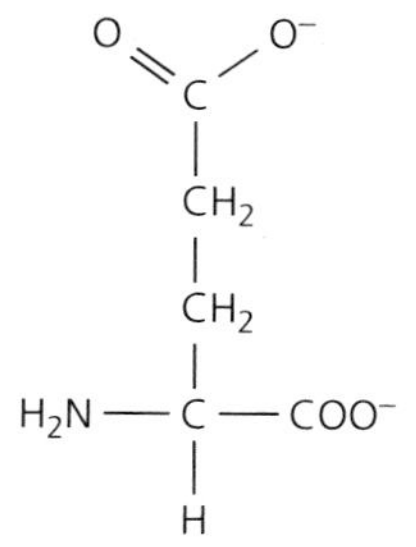

4

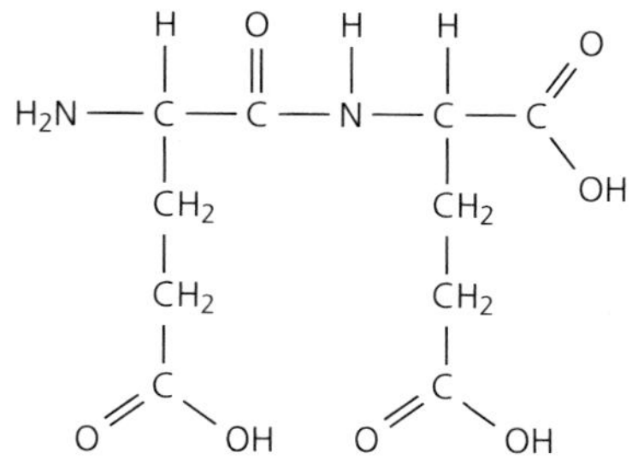

5 This is simply the ester formed from the COOH group in valine with the OH group of ethanol.

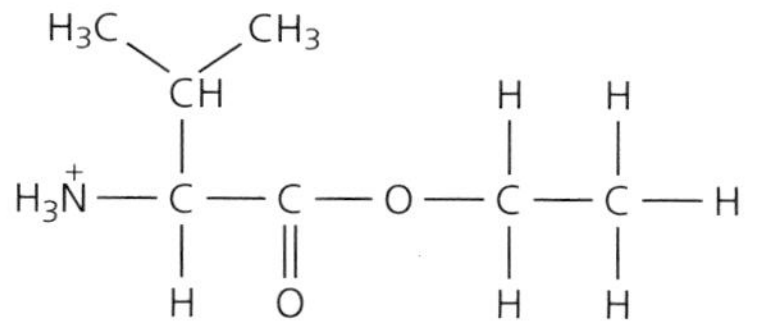

TIP
As concentrated sulfuric acid is present it would be expected that the NH_2 group is protonated as shown.

ACTIVITY

Lysine

Today about 500 different amino acids are known, of which nine are called 'essential amino acids' because they cannot be made by the human body, and so must be taken in as food. Lysine is an essential amino acid. Good dietary sources of lysine include high protein foods such as eggs, meat, soy beans, Parmesan cheese and salmon.

Lysine has the formula $H_2N(CH_2)_4CH(NH_2)COOH$. It is an optically active molecule with a melting point of 196 °C.

1 Give the IUPAC name for lysine.
2 What is meant by the term optically active?
3 Draw the two optical isomers of lysine, labelling the asymmetric carbon with an asterisk (*).
4 Explain why lysine has a relatively high melting point.
5 What is meant by the term zwitterion?
6 a) Draw the structure of the species formed by lysine at low pH.
  b) Write the formula of the organic ion present when lysine is dissolved in an alkaline solution.

7 Draw the structure of a dimer formed when two molecules of lysine react.

8 An amino acid was found to have the following composition by mass:

Element	% composition
N	10.5
C	36.1
H	5.3
O	48.1

Deduce the empirical formula for this amino acid.

Secondary structure

The secondary structure of a protein is the arrangement of the peptide-linked backbone of the amino acid chain. It usually involved coiling of the amino acid chain into one of two structures. This interaction between the polar C=O groups and the N–H groups hold the amino acid backbone in certain orientations. The two main secondary structure organisations of a protein are α-helices and β-pleated sheets. These involve hydrogen bonding between C=O and N–H groups.

α-Helix

The α-helix is a right-handed helix. It is the most common type of secondary structure and forms when the amino acid R groups are small. Keratin, a protein found in fur and feathers contains a high percentage of α-helices. The structure of the helix is maintained by the hydrogen bonding between the polar C=O and N–H groups of different peptide bonds in the amino acid chain. Each turn of the helix contains 3.6 amino acid residues. The hydrogen bond occurs for example from the C=O group in the peptide bond after amino acid 1 in the chain to the N–H group in the peptide bond after amino acid 4 in the chain. In fact due to the order of C=O and N–H in the peptide chain, it is actually 3.6 amino acids away.

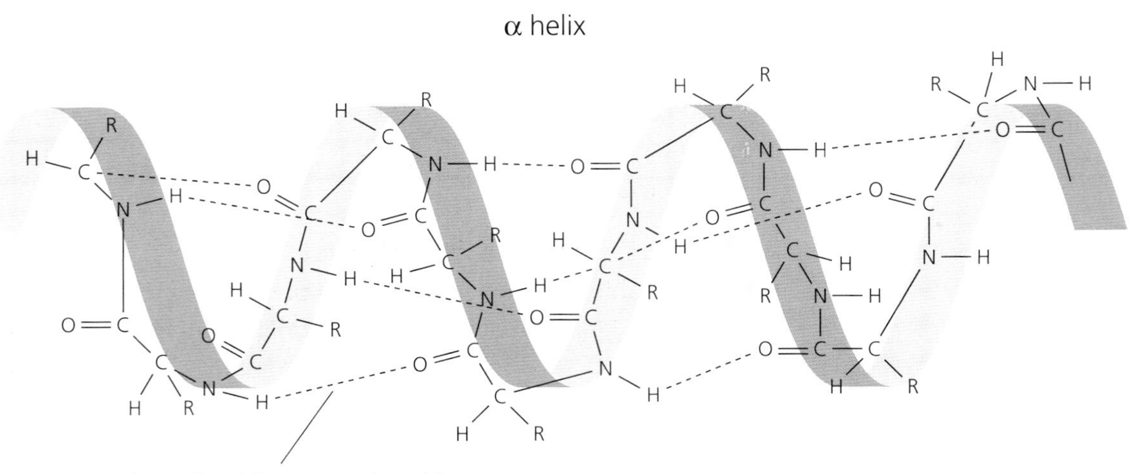

β-Pleated sheet

In a β-pleated sheet the amino acid chain is lined up side by side with another section of the same chain and the hydrogen bonds occur again between the C=O and N–H groups on peptide bonds which are facing each other. The protein found in silk (fibroin) contains a substantial proportion of β-pleated sheet structure.

β-pleated sheet

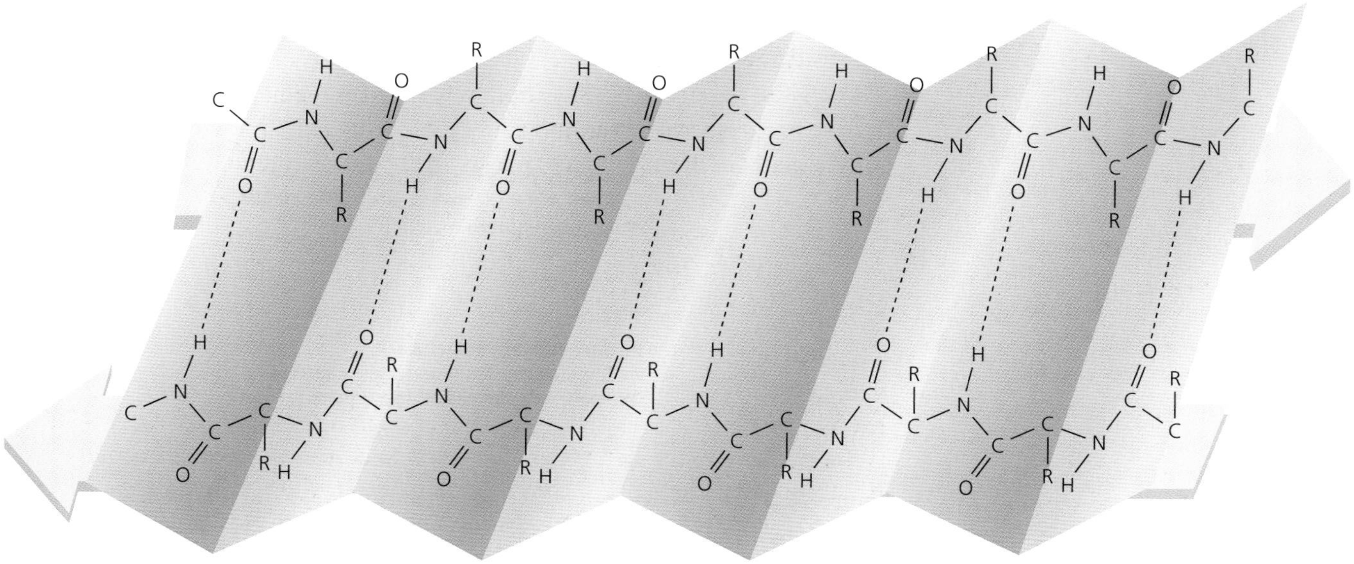

Tertiary structure

The tertiary structure of a protein is the way in which the coiled chain of amino acids is folded. Many different interactions contribute to the tertiary structure of a protein. These interactions usually involve amino acid side chain group which are well separated in the amino acid sequence. Interactions include:

1 Hydrogen bonding between polar side chain groups.

2 Disulfide bridges (–S–S–) between cysteine amino acid residues in the protein chain.

3 Salt bridges (ionic interaction between $RCOO^-$ and RNH_3^+ side chain groups).

4 Hydrophobic and hydrophilic interactions. (Polar and non-polar side chains localised in certain sections of the protein molecule depending on its function.)

5 Side chains may be too large to be packed closely together. The folding of the protein chain minimises steric hindrance.

The most stable tertiary structure is the one with the greatest number of stabilising interactions. Given a particular primary structure, a protein naturally assumes its most stable secondary and tertiary structure. For example, every insulin protein synthesised in the body has the same structure as it has the same amino acid sequence (primary structure).

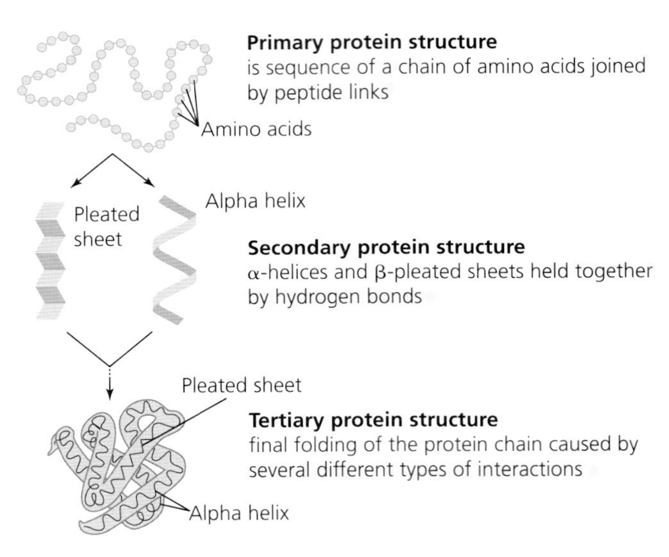

Figure 14.5 The primary, secondary and tertiary structure of a protein.

Fibrous and globular proteins

Proteins that have mainly secondary structure are fibrous in nature. Proteins with a secondary and tertiary structure are globular.

Fibrous proteins (also called structural proteins) form skin, muscle, the walls of the arteries and hair are composed of long thread-like molecules that are tough and insoluble in water. They are also resistant to the action of acid and alkali.

Globular proteins are small proteins, somewhat spherical in shape because of the folding of the protein chains upon themselves. Globular proteins are water soluble (due to the presence of hydrophilic side chains on the surface) and perform various functions within the organism. For example, haemoglobin transports oxygen to the cells; insulin aids in carbohydrate metabolism; antibodies render foreign proteins inactive; fibrinogen (soluble) can form insoluble fibres that results in blood clotting; and hormones carry messages throughout the body.

Enzymes in our diet

Enzymes are proteins which act as biological catalysts. They are named according to the reaction that they catalyse. In nutrition, enzymes are vitally important as the three main food groups, fats, proteins and carbohydrates are generally large insoluble molecules which need to be broken down to give smaller soluble molecules which can then be absorbed into the blood in the small intestine.

Mode of action of enzymes

The substrate is the reactant in the reaction that is catalysed by the enzyme.

The substrate fits into the active site in the enzyme and forms an enzyme–substrate complex. The bonds in the substrate are strained by interaction with the enzyme, and the enzyme–substrate complex becomes an enzyme–product complex. The product is not the same shape as the substrate and so the fit between the enzyme and the product is no longer perfect and this causes the product to dissociate from the enzyme.

Most enzymes have an M_r of between 12,000 and 120,000 and sometimes even higher. Most substrates, for example, an amino acid, are much smaller molecules. The specific location on the enzyme where the substrate binds is called the **active site**. The active site of the enzyme contains amino acid residues which can interact with the substrate and hold it. There are also residues within the active site which catalyse the reaction which the substrate undergoes.

The shape of the active site matches exactly the shape of the substrate and the mechanism for this is often described as a lock and key mechanism. The key being the substrate, and the lock being the enzyme.

Figure 14.6a shows the substrate binding to the active site of the enzyme, forming the enzyme–substrate complex, and then the products being released from the enzyme.

The active site of an enzyme is stereospecific. If two enantiomers of a molecule exist, the enzyme can only catalyse the reaction for one of the enantiomers.

Molecules can be designed that can fit into the active site and block it. These molecules are called enzyme inhibitors and are useful as drugs in metabolic and other diseases where it is important to block enzyme activity. Computer modelling of the enzyme and its active site can be used to help design enzyme inhibitors.

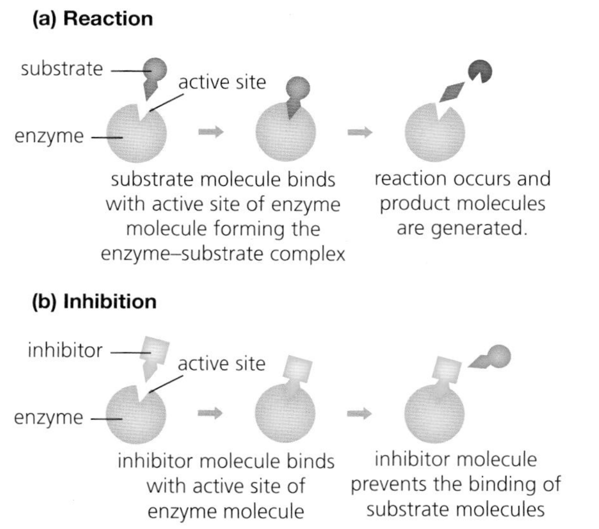

Figure 14.6 The reaction and inhibition processes in enzymes.

Figure 14.7 Thalidomide is a sedative drug, which was prescribed as the racemate to treat nausea in pregnant women in the early 1960s. It was later found to cause foetal abnormalities involving limb malformation, as shown by the deformed hand and forearm of the baby. Thalidomide is optically active: the (+) isomer was an effective drug but the (−) isomer caused deformation. The drug was banned in 1962. Today, the optical isomers of chiral drugs are separated before testing.

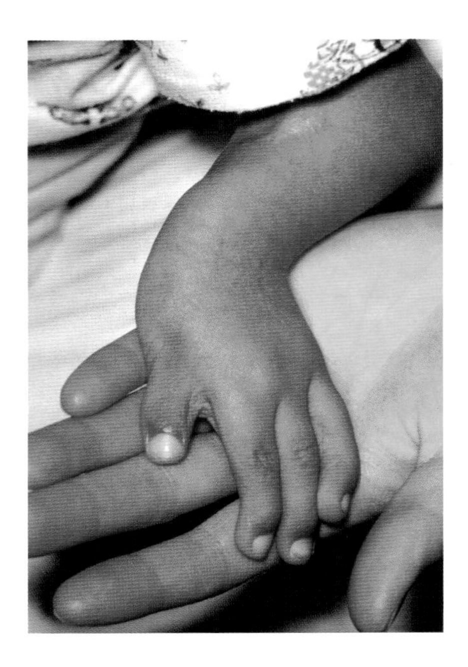

Effect of temperature and pH on enzyme activity

Fibrous proteins are insoluble in water, resistant to dilute acid and dilute alkali and are unaffected by moderate changes in temperature. Globular proteins, on the other hand, are relatively soluble and are affected by extremes of pH and temperature that disrupts the hydrogen bonding which maintains their tertiary structure.

The specific action of enzymes is due to their tertiary structure, which is maintained by hydrogen bonds. Consequently, enzymes are very sensitive to changes in temperature and pH. All enzyme action is destroyed on boiling.

The optimum temperature range for the activity of an enzyme is between 35 °C and 40 °C.

In general, the optimum temperature for the activity of plant enzymes is 5 °C. Enzymes in warm-blooded animals have an optimum temperature of 37 °C. An increase in temperature usually leads to an increase in the rate of a chemical reaction, but in the case of enzyme catalysed reactions, it will eventually lead to inactivation of the enzyme.

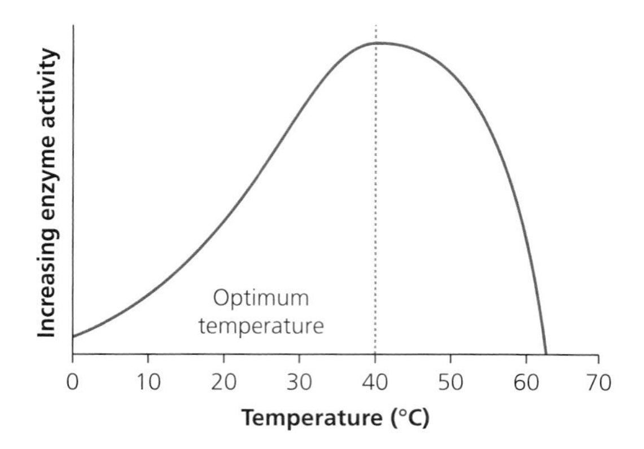

Enzyme activity is also dependent on the pH of the medium in which it acts. Most enzymes operate in environments of pH 7. Some like pepsin (a stomach protease enzyme) can only operate in acidic conditions.

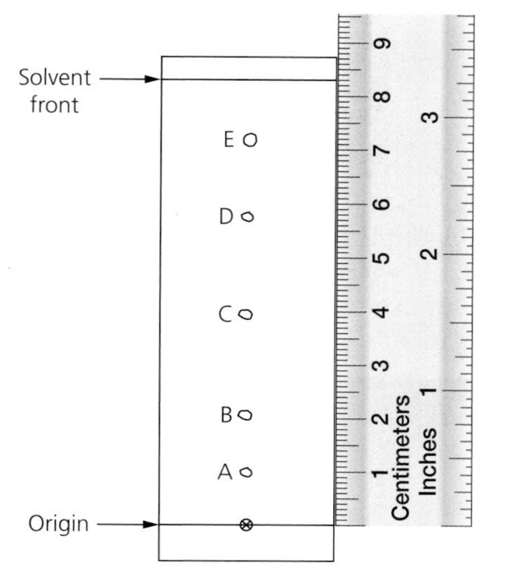

Hydrolysis of proteins

Proteins can be hydrolysed to the constituent amino acids by boiling with a dilute mineral acid such as hydrochloric acid. The amino acids can be analysed by thin-layer chromatography (TLC). The details of thin-layer chromatography will be considered in Chapter 15.

Amino acids are colourless and can be located on the chromatogram by staining with ninhydrin or using ultraviolet light. The retardation factor (R_f value) is determined by dividing the distance moved by a particular amino acid by the distance moved by the solvent. R_f values for amino acids in TLC with a particular solvent may be used to identify the amino acids present in the protein.

The diagram shows a representation of a thin-layer chromatogram of five amino acids that was run in a solvent of propanone and water mixed in a ratio of 1 : 3 and developed using ninhydrin.

The ruler allows the distance moved by the amino acids and the solvent to be measured, and is used to calculate the R_f values:

- The solvent has moved 8.3 cm or 83 mm.
- The centre of spot A has moved 1.0 cm or 10 mm. The R_f value for the amino acid at spot A is $\frac{10}{83} = 0.120$
- The centre of spot B has moved 2.1 cm or 21 mm. The R_f value for the amino acid at spot B is $\frac{21}{83} = 0.253$ and so on.

Amino acids with low R_f values have been less soluble in the solvent and have a greater affinity for the support on the TLC plate. This will be examined in greater detail later.

TEST YOURSELF 3

1 The structure below is of isoleucine, an amino acid.

```
           CH3
            |
           CH2
            |
  H3C —— CH
            |
  H2N —— C —— COOH
            |
            H
```

a) Identify any chiral centres in isoleucine and label each with an asterisk (*).

b) Draw the structure of the species obtained from isoleucine at low pH.

2 Name two feature of the secondary structure of a protein.

3 Explain how the R_f value is determined from a chromatogram.

4 Draw the structure of the zwitterion formed from alanine, $H_2NCH(CH_3)COOH$.

ACTIVITY

Urea

Urea is a waste product of many living organisms, and is the major organic component of human urine. This is because it is produced at the end of a chain of reactions which break down the amino acids that make up proteins. The average person excretes about 30 grams of urea a day, mostly through urine, but a small amount is also secreted in perspiration

Urea is an organic compound with the following structural formula:

$$H_2N-\overset{\overset{\displaystyle O}{\|}}{C}-NH_2$$

Urea was first synthesised in 1828 by Friedrich Wohler and this made urea the first *organic* compound to be synthesised from wholly *inorganic* starting materials.

Urea can be prepared in the laboratory by reacting lead(II) cyanate, $Pb(CNO)_2$, with ammonia and water to produce lead(II) hydroxide and ammonium cyanate, NH_4CNO. The ammonium cyanate then rearranges on heating to form urea, which has a melting point of 133 °C.

1 Write an equation for the reaction of lead(II) cyanate with ammonia and water.

2 Assuming an 80% yield, calculate the mass of lead(II) cyanate required to produce 420 g of ammonium cyanate.

3 a) The crude product in the preparation is purified by dissolving in the minimum volume of hot ethanol, filtering to remove insoluble impurities, and cooling. What name is given to this purification process?

b) What practical considerations determine the choice of solvent used?

c) Why is the minimum amount of hot solvent used?

d) How is the pure dry product obtained from the filtrate?

e) Giving practical details, describe how you would determine whether or not the crystals of urea produced are pure.

Urea is produced in industry from ammonia and carbon dioxide. In this process ammonium carbamate $H_2N-COONH_4$ forms and it then decomposes into urea and water. More than 90% of the world's industrial production of urea – approximately 184 million tonnes is used in nitrogen-release fertiliser.

Figure 14.8 Urea has the highest nitrogen content of all solid nitrogenous fertilisers in common use.

4 a) Write an equation for the reaction of ammonia and carbon dioxide to form ammonium carbamate.

b) Write an equation for the decomposition of ammonium carbamate into urea and water.

c) Calculate the percentage by mass of nitrogen in urea.

Urea is also used in the production of the polymer urea-methanal which is a thermosetting plastic resin used as adhesive to glue wood and MDF. It is produced in the condensation reaction between urea and methanal in the presence of heat and a base. Water is eliminated as the hydrogen atoms from one amino group on each of two urea molecules combine with the oxygen atom from a methanal molecule. The remaining $-CH_2-$ group from the methanal molecule then forms a bridge between two neighbouring urea molecules. This process, repeated many thousands of times, forms a long chain of urea-methanal.

5 Write an equation to show the condensation reaction between methanal and urea, to produce one unit of the polymer urea-methanal.

DNA

DNA is deoxyribonucleic acid. It is a double-stranded polymer that is found in the nucleus of cells and carries the genetic code in chemical form.

DNA is made up of a chain of 2-deoxyribose units connected by a phosphate ion. Bases are bonded to the 2-deoxyribose and hydrogen bonds between the bases hold the two strands together, forming a double-helix structure.

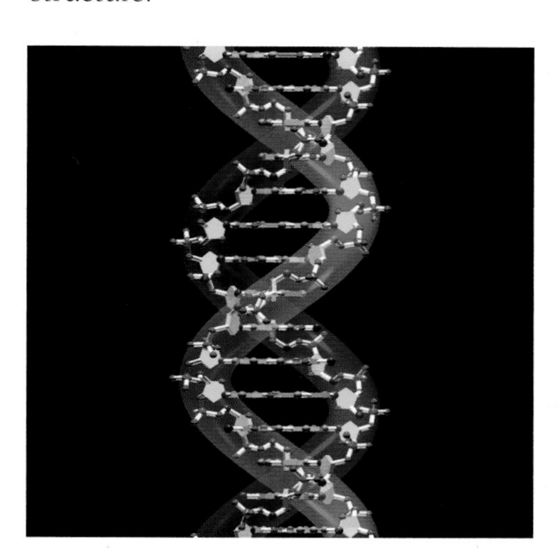

Figure 14.9 The yellow pentagons represent the 2-deoxyribose units which are connected by small phosphate ions. The two strands (highlighted in red and green) are often referred to as a sugar-phosphate backbone in DNA. The strands are connected by flat planar molecules which are the bases. The entire structure is a double helix arranged around the same axis.

Components of DNA

The components which make up the backbone sugar–phosphate strand in DNA are given below:

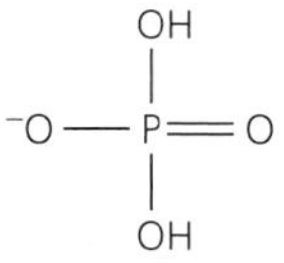

phosphate

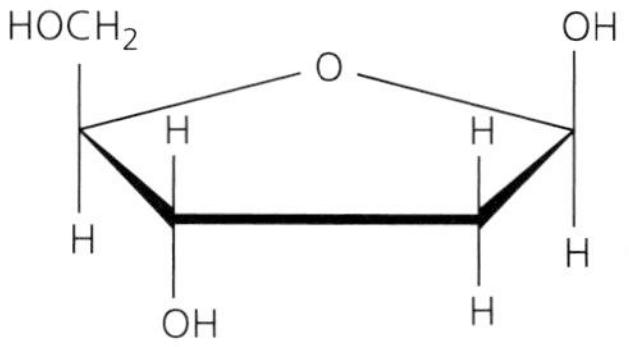

2-deoxyribose

The carbon atoms in 2-deoxyribose are numbered as shown in the diagram below. This numbering is very important as the other components are bonded at specific carbon atoms. Ribose is a pentose sugar with the formula $C_5H_{10}O_5$ but 2-deoxyribose is $C_5H_{10}O_4$. There is an oxygen atom removed from carbon number 2 in deoxyribose compared to ribose. The structure of ribose is also shown below.

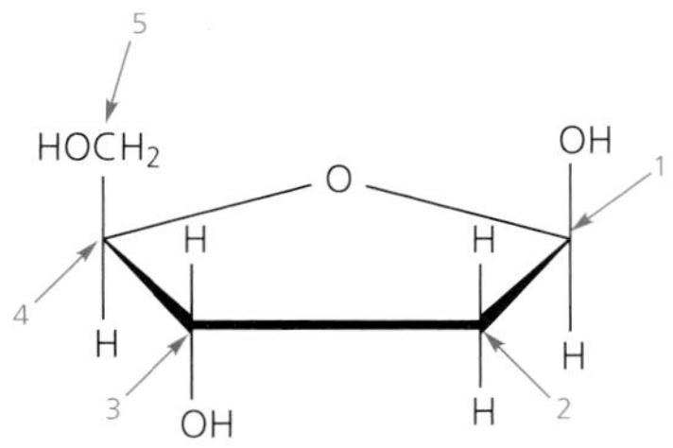

numbering of carbon atoms in 2-deoxyribose

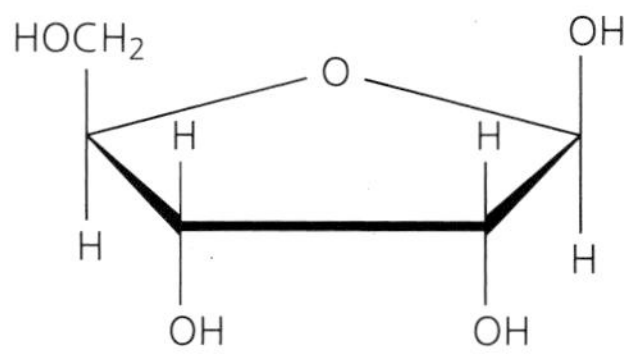

ribose

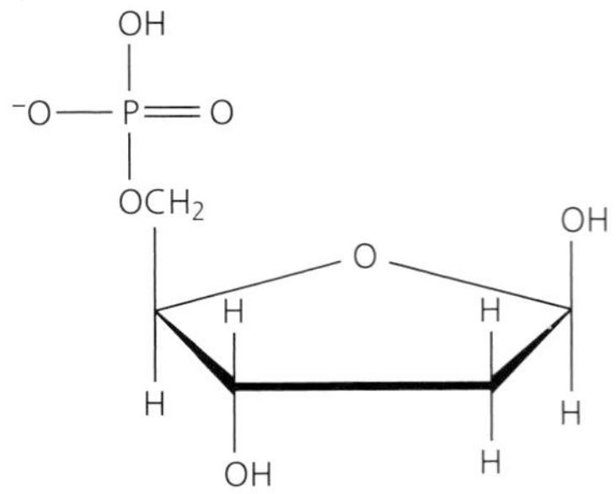

A phosphate ion is bonded to 2-dexoyribose at carbon number 5 as shown on the left:

The sugar-phosphate backbone is built from these units. The phosphate ion on carbon 5 bonds with carbon 3 on the next 2-deoxyribose unit.

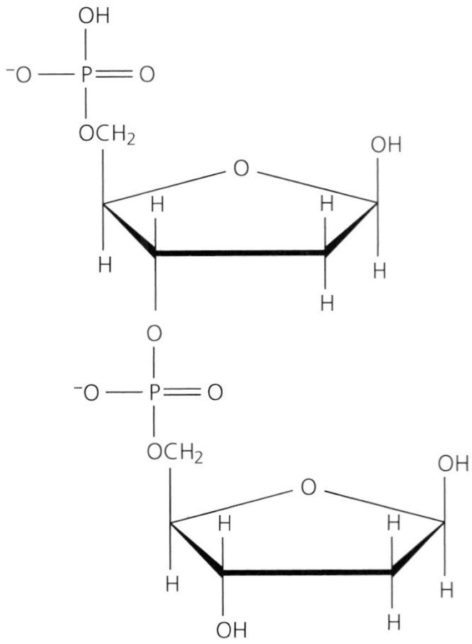

This continues creating the sugar-phosphate backbone of the DNA polymer.

A base is bonded at carbon 1 on each 2-deoxyribose molecule of the sugar–phosphate backbone. There are four bases in DNA. The structure of the bases in DNA are shown below:

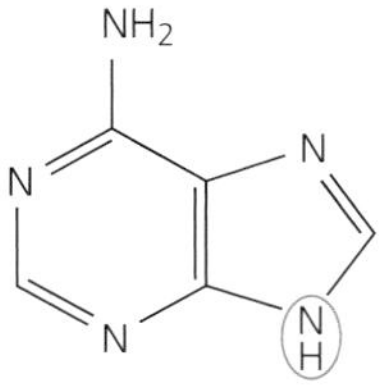

adenine

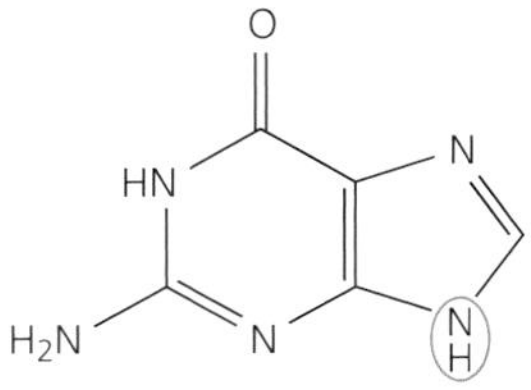

guanine

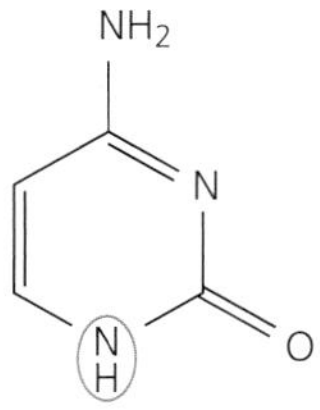

cytosine

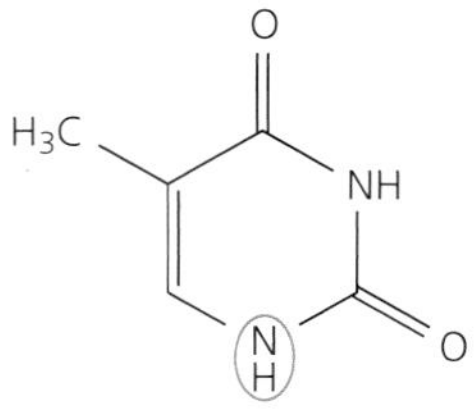

thymine

Adenine and guanine are purine bases and cytosine and thymine are pyrimidine bases.

Adenine, guanine, cytosine and thymine are referred to as the bases as they all contain amino groups.

The nitrogen atom, which bonds to carbon 1 of the 2-deoxyribose unit, is circled in green on the molecules above. Water is eliminated in this reaction.

A single unit of a base bonded to a 2-deoxyribose at carbon 1 with a phosphate ion bonded at carbon 5 is called a **nucleotide**.

There are four nucleotides, one for each of the bases. These nucleotides are shown in Figure 14.10. The nucleotides are called adenosine monophosphate, guanosine monophosphate, cytidine monophosphate and thymidine monophosphate.

adenosine monophosphate

guanosine monophosphate

cytidine monophosphate

thymidine monophosphate

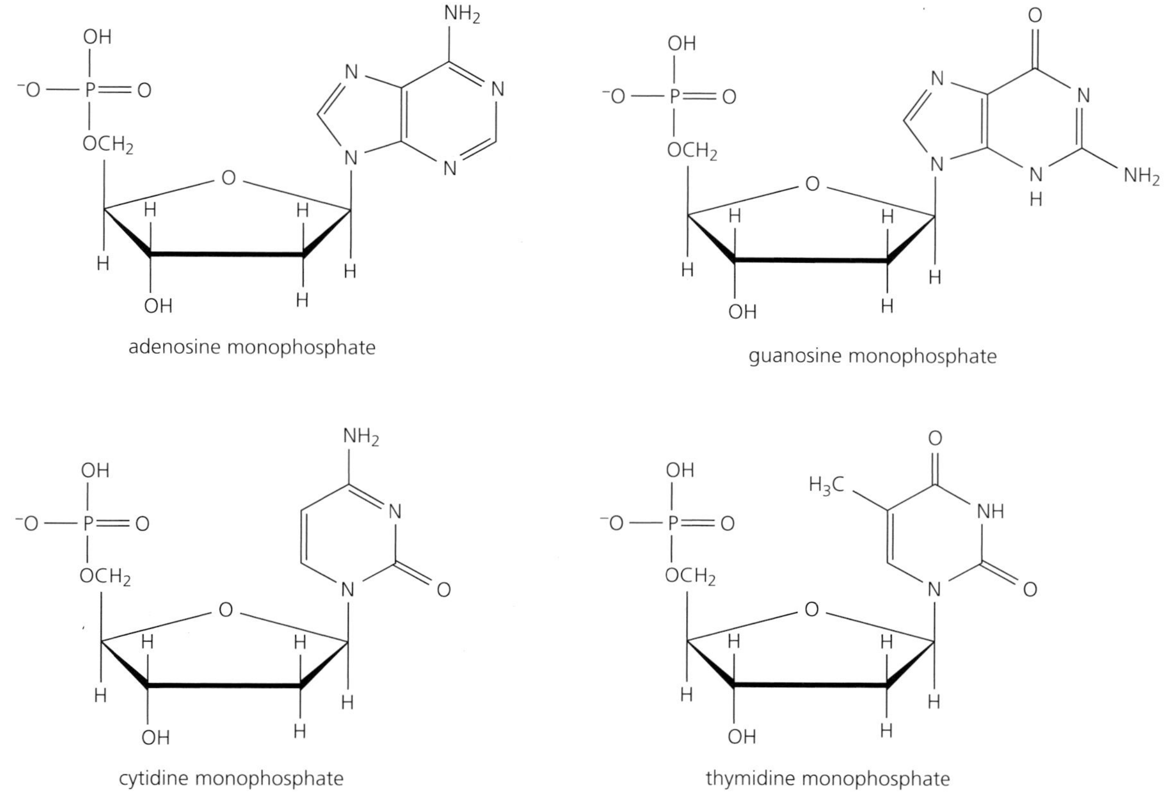

Figure 14.10 Adenosine monophosphate, guanosine monophosphate, cytidine monophosphate and thymidine monophosphate

TIP

Make sure that you are able to put together the components of DNA to form the nucleotides or recognise the bases and other components of DNA from a given structure.

DNA structure

The full structure of DNA is based on the connection of nucleotides to form a sugar-phosphate-sugar-phosphate polymer chain. The nucleotides are joined in a condensation reaction between the OH of the carbon 3 and the OH of the phosphate ion group. Water is eliminated. Another chain of nucleotides runs parallel and the bases are joined together by hydrogen bonds. One chain runs from its carbon 5 end (often called 5′) to carbon 3 (often called 3′) and the other runs from 3′ to 5′.

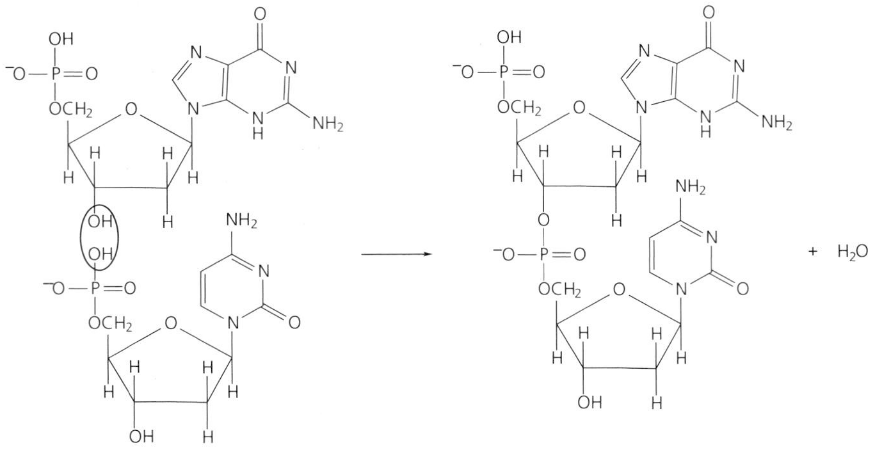
H2O
two mononucleotides
Dinucleotide

The condensation reaction continues forming a chain of nucleotides.

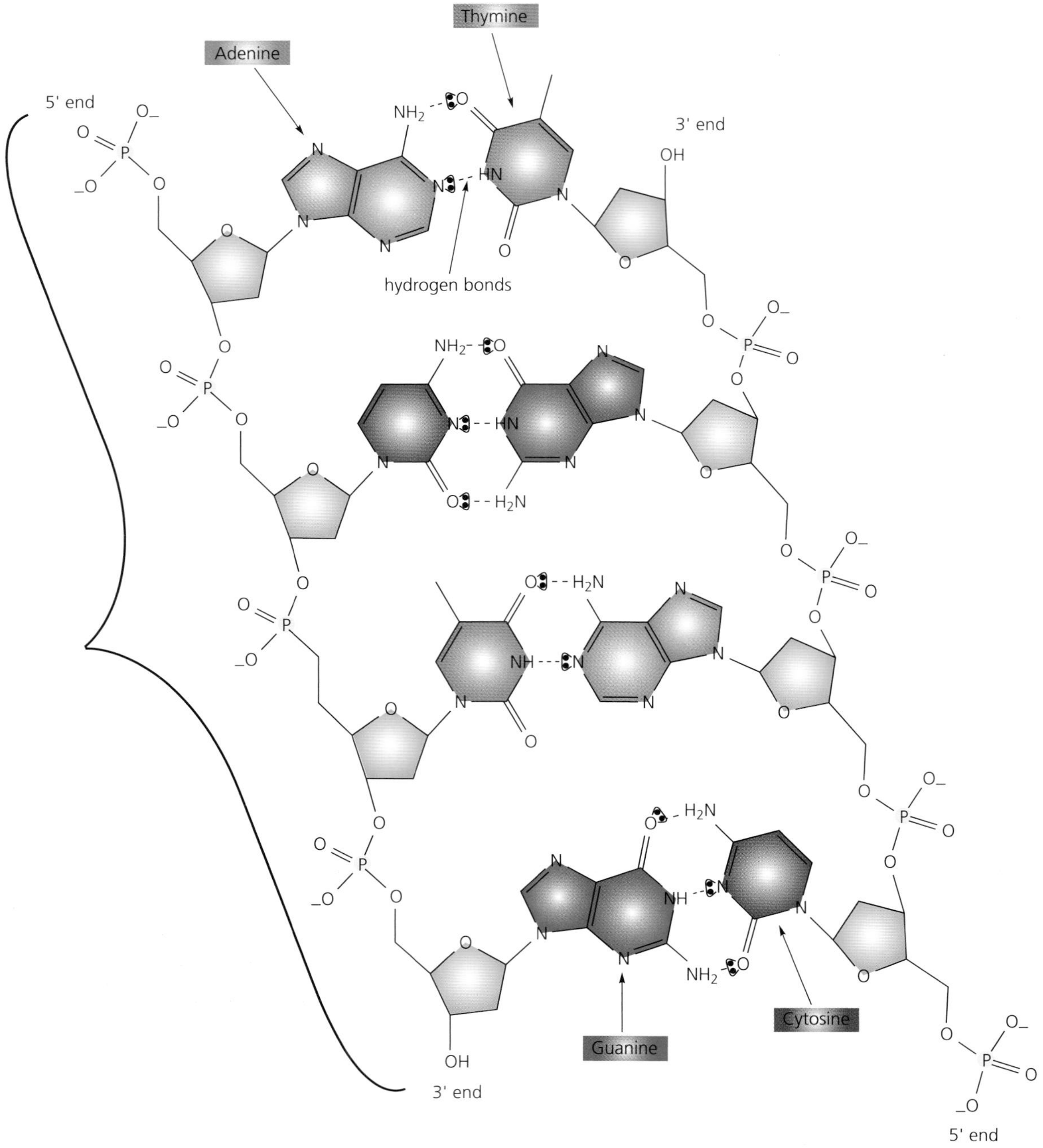
Thymine
Adenine
5' end
3' end
hydrogen bonds
Guanine
Cytosine
3' end
5' end

The bases are always in the same pairs. Adenine pairs up with thymine and guanine pairs up with cytosine. So an adenosine monophosphate nucleotide in one strand will be hydrogen bonded to a thymine monophosphate nucleotide in the other strand. Similarly a guanosine monophosphate nucleotide in one strand will be hydrogen bonded to a cytidine monophosphate nucleotide in the other strand.

The pairing of the bases guanine and cytosine is often referred to as a GC base pair and the pairing of the bases adenine and thymine is often referred to as an AT base pair.

There are three hydrogen bonds between the bases guanine and cytosine but only two hydrogen bonds between the bases adenine and thymine. This can alter the properties of DNA. A section of DNA which is rich in GC base pairs will require a higher temperature to separate the strands.

The diagram below shows the hydrogen bonds between thymine and adenine and also between cytosine and guanine. The hydrogen bonds form between a hydrogen atom bonded to a nitrogen atom in the base and a lone pair of electrons on either a nitrogen atom or an oxygen atom.

thymine H_3C O NH NH O — NH_2 N N N NH adenine

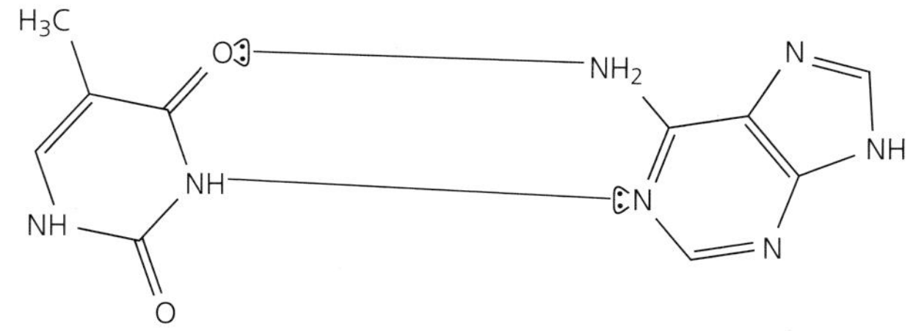

cytosine NH_2 N NH O — O HN N H_2N N NH guanine

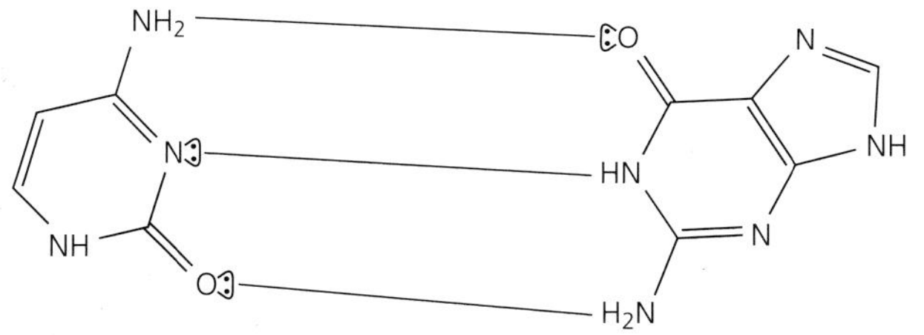

Short form of DNA

Often DNA may be represented as a double chain with G, C, A and T used to represent the bases as shown below.

G	T	G	A	C	A	T	A
C	A	C	T	G	T	A	T

You may be given part of a section like this and be asked to complete the bases.

EXAMPLE 6

Complete the sequence of bases in this representation of a DNA molecule.

A	G	A	C	C	G	T	C
T	C						

Answer

This question is testing your knowledge of the base pairs in DNA. It is important to remember that the base pairs are AT and GC.

A	G	A	C	C	G	T	C
T	C	T	G	G	C	A	G

Action of anti-cancer drugs

A cancerous cell is a cell in which cell division is not controlled and it continues to divide to form a tumour. For the cell to divide, its DNA must be replicated. This involves separating the DNA strands in the double helix, and enzymes moving along the strand copying the DNA. The enzymes catalyse the connection of the nucleotides forming two new double helices, one formed from each of the original strands.

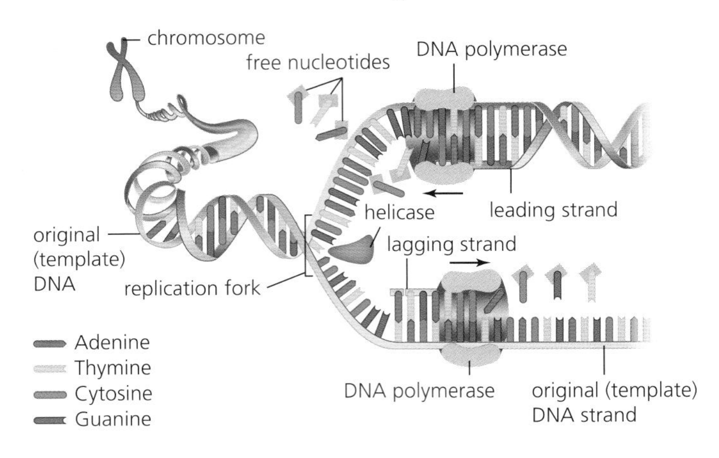

Figure 14.10 The helicase enzyme unzips the DNA double helix and DNA polymerase enzymes copy each strand.

Some anti-cancer drugs inhibit the enzymes that are involved in the synthesis of the nucleotides. Others inhibit enzymes which catalyse the DNA replication process. Cisplatin is a platinum(II) complex with 2 chloride ions and 2 ammonia molecules. It was introduced in the chapter on transition metals as a square planar complex which can cross-link DNA so preventing replication.

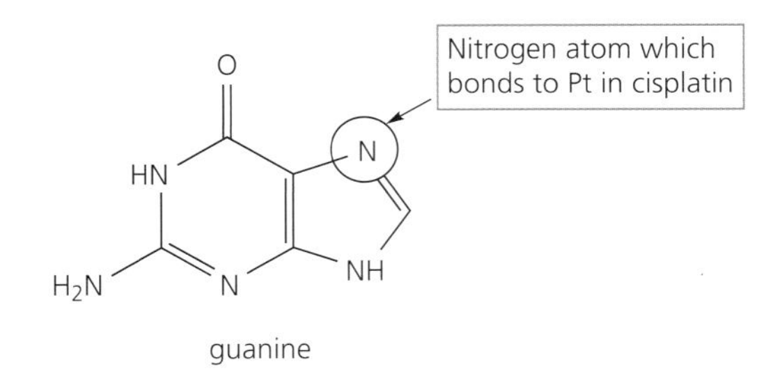

Cisplatin acts by allowing a coordinate bond to form between a nitrogen atom in guanine and the platinum in cisplatin. Cisplatin can also bond to other guanine residues in the same strand or between the strands. DNA cannot be replicated.

In adults, most cells are no longer dividing. However, some cells like sex cells, intestinal lining cells, hair and skin cells are still undergoing cell division. The use of anti-cancer drugs destroys these dividing cells as well. Patients taking these drugs lose their hair, have skin and digestive problems and are often left unable to have children.

Doctors will advise patients of the potential adverse risks and benefits of the use of anti-cancer drugs. Patients make the decision whether to take the treatment.

TEST YOURSELF 4

1. Name four bases in DNA.
2. Which two bases in DNA form three hydrogen bonds between them?
3. Describe the structure of a nucleotide.
4. Complete the sequence of bases in this section of DNA.

T C G A T A C G

A G

Practice questions

1 Which one of the following is the correct formula of hexanedioic acid used in the production of nylon-6,6?

A $CH_3(CH_2)_4COOH$

B $CH_3(CH_2)_2CH(COOH)_2$

C $HOOC(CH_2)_4COOH$

D $HOOC(CH_2)_6COOH$ *(1)*

2 Which one of the following is not a condensation polymer?

A Polyethylene terephthalate

B Nylon-6,6

C Polythene

D Kevlar *(1)*

3 Three amino acids are given below:

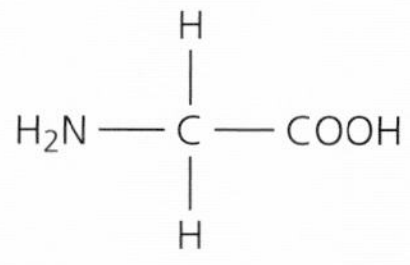

glycine

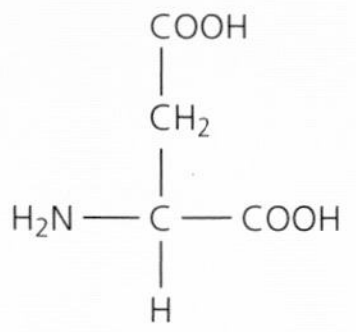

aspartic acid

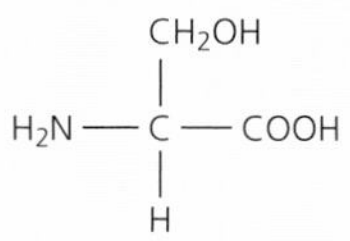

serine

a) Which one of the amino acids is not optically active? Explain your answer. *(2)*

b) Draw the structure of a dipeptide formed from one aspartic acid molecule and one serine molecule. *(2)*

c) Give the IUPAC name for aspartic acid. *(1)*

d) Give the IUPAC name for serine. *(1)*

e) Draw the species formed from aspartic acid at high pH. *(1)*

f) Draw the species form from serine at low pH. *(1)*

g) Serine reacts with methanol in the presence of concentrated sulfuric acid. Draw the structure of the compound formed. *(2)*

4 The IUPAC name for the amino acid, threonine is 2-amino-3-hydroxybutanoic acid.

a) Draw the structure of threonine.

b) Draw the structure of the species formed from threonine at low pH.

c) Threonine reacts with ethanoic acid in the presence of concentrated sulfuric acid. Draw the structure of the compound formed.

5 The diagram below shows a representation of a nucleotide.

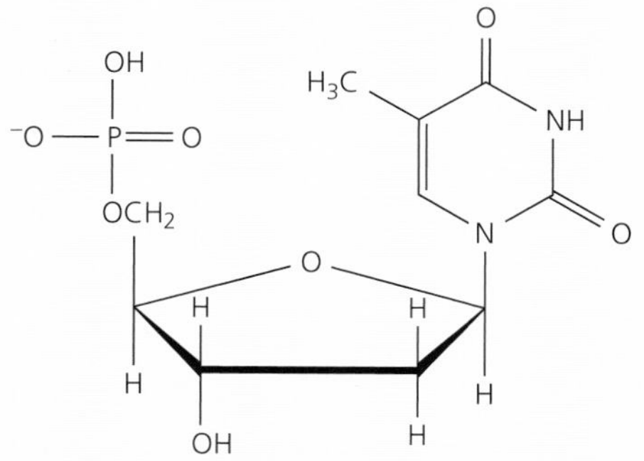

a) i) Name the base in this nucleotide. *(1)*

ii) Write the molecular formula for the base. *(1)*

b) Name the sugar component of the nucleotide. *(1)*

c) i) With which base does this base hydrogen bond? *(1)*

ii) On the base show the point at which hydrogen bonds form with the other base. *(1)*

d) On the sugar show the point of attachment of the next nucleotide. *(1)*

6 Cisplatin has the formula $[Pt(NH_3)_2Cl_2]$.

a) What is the oxidation state of platinum in cisplatin? *(1)*

b) State the shape of the complex. *(1)*

c) Cisplatin is used as an anti-cancer drug.

i) Explain how cisplatin acts as an anti-cancer drug. (3)

ii) State two adverse effects of the use of cisplatin. (2)

7 Polyesters are formed from the reaction between polyhydric alcohols such as ethane-1,2-diol and dicarboxylic acids such as terephthalic acid (benzene-1,4-dicarboxylic acid).

a) Draw the structural formula of the following compounds:

i) ethane-1,2-diol (1)

ii) terephthalic acid (1)

b) Give the polymer structure of the polyester formed in this reaction showing two repeating units and circle an ester group. (3)

c) Name the type of polymer formed in this polymerisation. (1)

d) Polyesters may also be formed using diacyl chlorides.

i) Draw the structure of the diacyl chloride formed from terephthalic acid. (1)

ii) Write an equation for the reaction of one molecule of the diacyl chloride with two molecules of ethanol. (2)

e) Terephthalic acid is formed from the oxidation of 1,4-dimethylbenzene. Write and equation for this oxidation using [O] to represent the oxidising agent. (2)

8 The following polymer is a condensation polymer formed from a diacyl chloride and a diol.

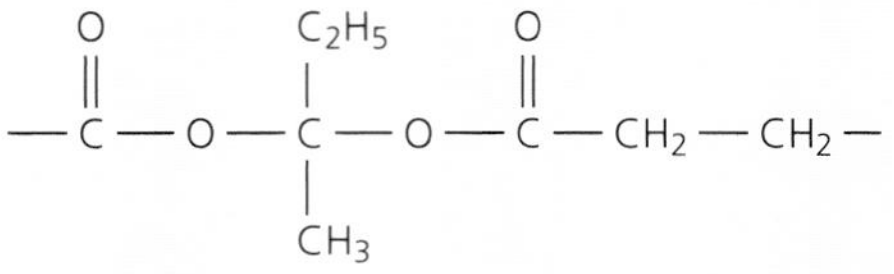

```
  O         C2H5        O
  ||         |          ||
—C —O—C—O—C—CH2—CH2—
             |
            CH3
```

a) Draw the diacyl chloride from which it is formed. (1)

b) Name the diol from which it is formed. (1)

c) Name the type of polymer. (1)

Stretch and challenge

9 Aspartame is an artificial sweetener used in many products. Its structure is shown below. It is the methyl ester of a dipeptide.

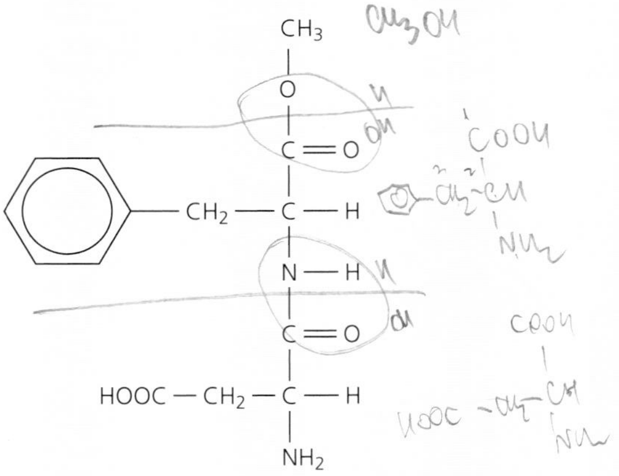

```
                   CH3
                   |
                   O
                   |
                   C=O
                   |
  C6H5 — CH2 — C — H
                   |
                   N — H
                   |
                   C=O
                   |
   HOOC — CH2 — C — H
                   |
                   NH2
```

a) Name the two amino acids present in aspartame. (2)

b) Draw the structure of the dipeptide at high pH. (1)

c) Draw the structure of the dipeptide at low pH. (1)

d) Place an asterisk (*) at any chiral centres on the structure of aspartame. (1)

e) A sample of aspartame was hydrolysed using dilute hydrochloric acid and thin-layer chromatography carried out on the products of hydrolysis. Ninhydrin was used to develop the chromatogram. Two organic products were detected using ninhydrin.

i) Draw the structures of all the organic products which would be formed on hydrolysis of aspartame using dilute hydrochloric acid. (2)

ii) Which of these products are detected using ninhydrin? (1)

15 Organic synthesis, NMR spectroscopy and chromatography

PRIOR KNOWLEDGE

- A homologous series is a family of organic compounds containing the same functional group.
- A functional group in an organic compound may be changed into another functional group.
- Organic chemistry requires knowledge of the reagents used and conditions required for a particular type of reaction.
- Infrared spectroscopy and mass spectrometry may be used to analyse and identify an organic compound.
- Infrared spectroscopy may be used to determine the bonds present in an organic compound and the fingerprint region used to identify the compound.
- Mass spectrometry may be used to determine the molecular formula of a compound.

TEST YOURSELF ON PRIOR KNOWLEDGE 1

1 Name the following aliphatic organic compounds and name the homologous series to which they belong.
   a) CH_3COOH
   b) CH_3CH_2CN
   c) $CH_3CH_2CH_2COCH_3$
   d) CH_3COOCH_3
   e) CH_3CH_2CHO
   f) $CH_2{=}CHCH_2CH_3$
2 Which groups in an organic compound would absorb at the following wavenumbers in an infrared spectrum?
   a) $1680–1750\,cm^{-1}$
   b) $3230–3550\,cm^{-1}$
   c) $2220–2260\,cm^{-1}$
3 Identify the alkene which has a peak, due to its molecular ion, at m/z =42 in its mass spectrum.

Organic synthesis

Organic synthesis is a branch of chemistry concerned with producing organic compounds by chemical reactions. The American chemist Robert Burns Woodward is regarded as the father of modern organic synthesis and was awarded the Nobel Prize in 1965 for synthesis of many natural products such as cholesterol, chlorophyll, quinine and cortisone.

The synthesis of organic compounds requires functional groups to be changed into other functional groups. The diagram below shows the connections between different homologous series and the table gives the details of reagents, conditions, mechanism and type of reaction for each conversion.

Questions on synthesis focus on conversion between different organic compounds, often as a one-step or a two-step process. However, synthetic routes with up to four steps are possible.

A chemist will always try to limit the number of steps in a conversion as there is loss at each step as the percentage yield is not 100% for each reaction. There may also be waste in terms of atom economy.

Chemists try to avoid the use of auxiliary chemicals such as solvents if they are not needed. Gas phase reactions are preferred as they do not require a solvent. Water may be used where a solvent is required. Organic solvents such as chlorinated compounds and ether are volatile and it is hard to control their escape into the atmosphere.

Chemists also try to minimise the use of hazardous starting materials. These may be toxic, corrosive or flammable. Often the use of potassium cyanide is avoided as it is toxic. Other synthetic routes will be used.

Summary of organic reactions

(The numbers on the arrows are references to the reaction details in the table on pages 289–300.)

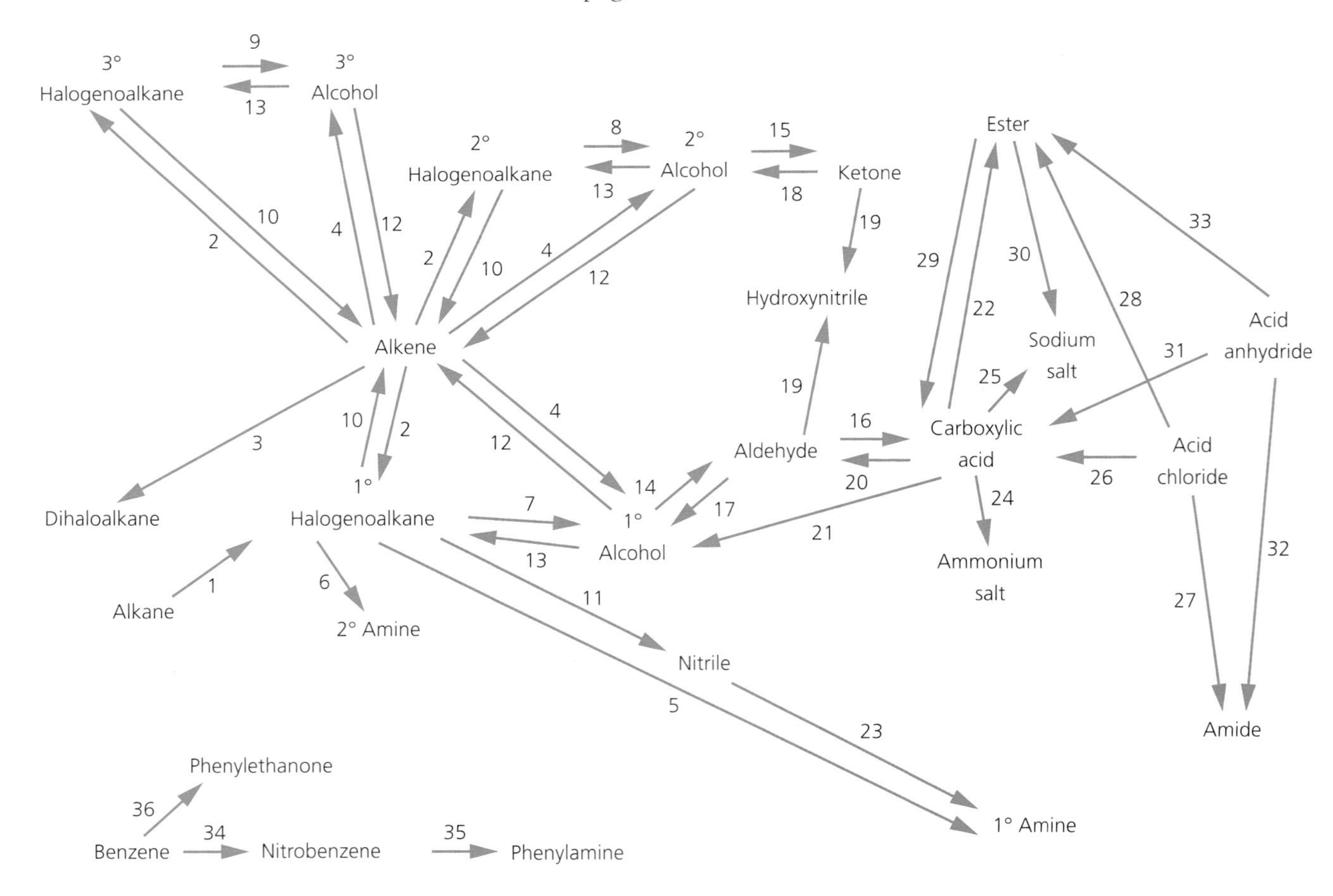

No	Example	Reagents	Conditions	Mechanism	Type of reaction
1	alkane → halogenoalkane $CH_4 + Cl_2 \rightarrow CH_3Cl + HCl$	Cl_2	UV light	Free-radical substitution	Substitution
2	alkene → halogenoalkane $C_2H_4 + HBr \rightarrow C_2H_5Br$	HX	N/A	Electrophilic addition	Addition
3	alkene → dihaloalkane $C_2H_4 + Br_2 \rightarrow CH_2BrCH_2Br$	Halogen, e.g. Br_2	N/A	Electrophilic addition	Addition
4	alkene → alcohol $C_2H_4 + H_2O \rightarrow C_2H_5OH$ 1°, 2° and 3° alcohols can be produced depending on the position of the C=C and the actual alkene used	H_2O	Concentrated H_2SO_4 or concentrated H_3PO_4 300 °C 7 MPa	Electrophilic addition	Addition followed by hydrolysis
5	1° halogenoalkane → 1° amine $C_2H_5Br + NH_3 \rightarrow C_2H_5NH_2 + HBr$	Concentrated NH_3	Excess NH_3 dissolved in ethanol	Nucleophilic substitution	Substitution
6	1° halogenoalkane → 2° amine $C_2H_5Br + C_2H_5NH_2 \rightarrow (C_2H_5)_2NH + HBr$	1° amine	Amine dissolved in ethanol	Nucleophilic substitution	Substitution
7	1° halogenoalkane → 1° alcohol $C_2H_5Br + NaOH \rightarrow C_2H_5OH + NaBr$	NaOH(aq)	Heat under reflux	Nucleophilic substitution	Substitution
8	2° halogenoalkane → 2° alcohol $CH_3CHBrCH_3 + NaOH \rightarrow CH_3CH(OH)CH_3 + NaBr$	NaOH(aq)	Heat under reflux	Nucleophilic substitution	Substitution
9	3° halogenoalkane → 3° alcohol $(CH_3)_3CBr + NaOH \rightarrow (CH_3)_3C(OH) + NaBr$	NaOH(aq)	Heat under reflux	Nucleophilic substitution	Substitution
10	1° or 2° or 3° halogenoalkane → alkene $C_2H_5Br + KOH \rightarrow C_2H_4 + KBr + H_2O$ The position of the halogen atom and the actual halogenoalkane will give different alkenes or even a mixture	KOH (dissolved in ethanol)	Heat under reflux	Elimination	Elimination
11	1° halogenoalkane → nitrile $C_2H_5Br + KCN \rightarrow C_2H_5CN + KBr$	Potassium cyanide (dissolved in ethanol)		Nucleophilic substitution	Substitution
12	alcohol → alkene $C_2H_5OH \rightarrow C_2H_4 + H_2O$ 1° or 2° or 3° alcohols can undergo this reaction to form differing alkenes or a mixture of alkenes	Concentrated H_2SO_4 or concentrated H_3PO_4 (or Al_2O_3 catalyst)	170 °C for acid dehydration	Elimination	Elimination Dehydration
13	alcohol → halogenoalkane $C_2H_5OH + HX \rightarrow C_2H_5X + H_2O$	HX (prepared in situ from NaX and conc H_2SO_4)	Heat under reflux	N/A	Substitution
14	1° alcohol → aldehyde $C_2H_5OH + [O] \rightarrow CH_3CHO + H_2O$	Acidified potassium dichromate(VI) solution	Heat and distil	N/A	Oxidation
15	2° alcohol → ketone $CH_3CH(OH)CH_3 + [O] \rightarrow CH_3COCH_3 + H_2O$	Acidified potassium dichromate(VI) solution	Heat under reflux	N/A	Oxidation
16	aldehyde → carboxylic acid $CH_3CHO + [O] \rightarrow CH_3COOH$	Acidified potassium dichromate(VI) solution	Heat under reflux	N/A	Oxidation
17	aldehyde → 1° alcohol $CH_3CH_2CHO + 2[H] \rightarrow CH_3CH_2CH_2OH$	$NaBH_4$	Aqueous solution	Nucleophilic addition	Reduction

No	Example	Reagents	Conditions	Mechanism	Type of reaction
18	ketone → 2° alcohol $CH_3COCH_3 + 2[H] \rightarrow CH_3CH(OH)CH_3$	$NaBH_4$	Aqueous solution	Nucleophilic addition	Reduction
19	aldehyde/ketone → hydroxynitrile $CH_3CHO + HCN \rightarrow CH_3CH(OH)CN$	KCN followed by dilute acid	N/A	Nucleophilic addition	Addition
20	carboxylic acid → aldehyde $CH_3COOH + 2[H] \rightarrow CH_3CHO + H_2O$	$LiAlH_4$	In dry ether	Nucleophilic addition	Reduction
21	carboxylic acid → 1° alcohol $CH_3COOH + 4[H] \rightarrow CH_3CH_2OH + H_2O$	$LiAlH_4$	In dry ether	Nucleophilic addition	Reduction
22	carboxylic acid → ester $CH_3COOH + C_2H_5OH \rightarrow CH_3COOC_2H_5 + H_2O$	Alcohol	Concentrated sulfuric acid	Nucleophilic addition–elimination	Elimination or condensation
23	nitrile → 1° amine $C_2H_5CN + 4[H] \rightarrow C_2H_5CH_2NH_2$	Lithal/$LiAlH_4$	In dry ether	Nucleophilic addition	Reduction
24	carboxylic acid → ammonium salt $CH_3COOH + NH_3 \rightarrow CH_3COONH_4$	Ammonia solution	Room temperature	N/A	Neutralisation
25	carboxylic acid → sodium salt $CH_3COOH + NaOH \rightarrow CH_3COONa + H_2O$	NaOH(aq) or Na_2CO_3	Room temperature	N/A	Neutralisation
26	acid chloride → carboxylic acid $CH_3COCl + H_2O \rightarrow CH_3COOH + HCl$	H_2O	Room temperature	Nucleophilic addition-elimination	Hydrolysis
27	acid chloride → amide $CH_3COCl + NH_3 \rightarrow CH_3CONH_2 + HCl$	Ammonia	Acid chloride added to concentrated ammonia	Nucleophilic addition-elimination	Substitution
28	acid chloride → ester $CH_3COCl + C_2H_5OH \rightarrow CH_3COOC_2H_5 + HCl$	Alcohol added to acid chloride	Room temperature	Nucleophilic addition-elimination	Esterification or elimination
29	ester → carboxylic acid $CH_3COOC_2H_5 + H_2O \rightarrow CH_3COOH + C_2H_5OH$	Dilute hydrochloric acid	Heat under reflux	N/A	Acid hydrolysis
30	ester → salt of carboxylic acid $CH_3COOC_2H_5 + NaOH \rightarrow CH_3COONa + C_2H_5OH$	Sodium hydroxide solution (or any alkali)	Heat under reflux	N/A	Base hydrolysis
31	acid anhydride → carboxylic acid $(CH_3CO)_2O + H_2O \rightarrow 2CH_3COOH$	H_2O	Room temperature	N/A	Hydrolysis
32	acid anhydride → amide $(CH_3CO)_2O + NH_3 \rightarrow CH_3CONH_2 + 3CH_3COOH$	Concentrated NH_3	Room temperature	N/A	N/A
33	acid anhydride → ester $(CH_3CO)_2O + CH_3CH_2OH \rightarrow CH_3COOCH_2CH_2 + CH_3COOH$	Alcohol	Room temperature	N/A	Elimination or condensation
34	benzene → nitrobenzene $C_6H_6 + HNO_3 \rightarrow C_6H_5NO_2 + H_2O$	Concentrated HNO_3 Concentrated H_2SO_4	Low temperature to prevent further nitration	Electrophilic substitution	Substitution
35	nitrobenzene → phenylamine $C_6H_5NO_2 + 6[H] \rightarrow C_6H_5NH_2 + 2H_2O$	Sn HCl	Heat under reflux and add NaOH(aq) to liberate the free amine	N/A	Reduction
36	benzene → phenylethanone $C_6H_6 + CH_3COCl \rightarrow C_6H_5COCH_3 + HCl$ $C_6H_6 + (CH_3CO)_2O \rightarrow C_6H_5COCH_3 + CH_3COOH$	CH_3COCl or $(CH_3CO)_2O$	$AlCl_3$ catalyst with CH_3COCl	Electrophilic substitution	Substitution

EXAMPLE 1

Ethanal may be converted into bromoethane using a three-step synthesis.

Step 1: Ethanal is reduced to compound **A**.
Step 2: Compound **A** is converted into compound **B**.
Step 3: Compound **B** is converted into bromoethane.
Give the names of compound **A** and **B**.

For each step, suggest a reagent which could be used and name the mechanism.

Answer

Firstly identify the homologous series to which each named substance belongs:

Ethanal is an aldehyde

Bromoethane is a 1° halogenoalkane

Another clue to the route is the word 'reduced' in step 1. Aldehydes are reduced to 1° alcohols (see reaction 17).

Extracting a three-step route from the diagram that starts with an aldehyde being reduced to a 1° alcohol could give the following:

Aldehyde →(17) 1° Alcohol →(12) Alkene →(2) 1° Halogenoalkane

> **TIP**
> Knowing the type of reaction is important. It can give a clue to the synthetic route.

Step 1: Ethanal is reduced to ethanol
Compound **A** is ethanol
Using reaction 17 in the table, the reagent required is $NaBH_4$.
The mechanism for reduction is nucleophilic addition.

Step 2: Ethanol is converted into ethene (an alkene)
Compound **B** is ethene
Using reaction 12 in the table, the reagent used in concentrated H_2SO_4.
The mechanism for the reaction is elimination.

Step 3: Ethene is converted into bromoethane
Using reaction 2 in the table, the reagent is HBr.
The mechanism for the reaction is electrophilic addition.

> **TIP**
> This type of question is synoptic as it uses all of your knowledge of organic chemistry and the reactions you have met. It is a good way to revise organic chemistry.

EXAMPLE 2

Compound **C** is synthesised in a three-step process from propene as shown below.

propene →(Step 1) compound A →(Step 2) compound B →(Step 3) $CH_3CH_2C(OH)(H)—C{\equiv}N$

compound C

Name compounds A, B and C.

For each step, suggest a reagent which could be used and any conditions required.

Name the mechanism for step 3.

Answer

Compound **C** is a hydroxynitrile. Hydroxynitriles are formed from aldehydes and ketones so compound **B** must be the corresponding aldehyde.

Propene is an alkene.

Aldehydes are formed on oxidation of a 1° alcohol.

Alkenes can be converted to 1° alcohols.

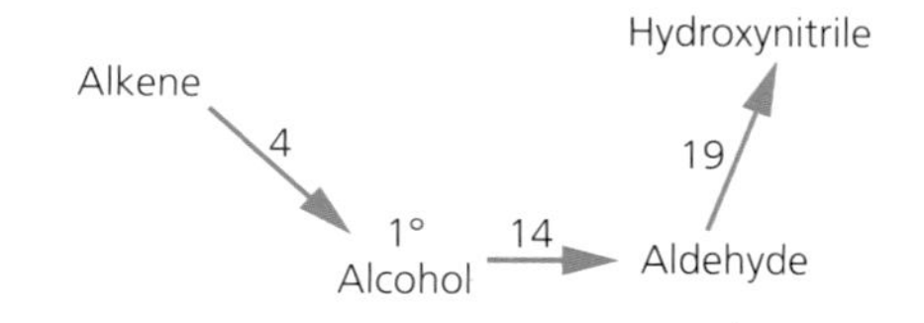

Step 1: Compound **A** is propan-1-ol.
Using reaction 4 in the table, possible reagents are concentrated H_2SO_4 or concentrated H_3PO_4 at a temperature of 300 °C and 7 MPa pressure.

Step 2: Compound **B** is propanal.
Using reaction 14, the reagent is acidified potassium dichromate(VI) solution.
The reaction should be heated and the product distilled off.

Step 3: Compound **C** is 2-hydroxybutanenitrile.
Using reaction 19, the reagents are KCN followed by dilute acid.
The mechanism is electrophilic addition.

TEST YOURSELF 2

1 Name the reagent used to carry out the following reactions:
  a) 1-bromobutane to butylamine
  b) 1-bromobutane to pentanenitrile
  c) nitrobenzene to phenylamine
  d) ethanenitrile to ethylamine

2 a) What is the intermediate in the two-step conversion of ethanoyl chloride to ethanal?
  b) Name the reagents for each step.

3 1-Chloropropane can be converted into butylamine in a two-step process.
  a) Name the intermediate.
  b) What are the reagents for each step?

Nuclear magnetic resonance spectroscopy

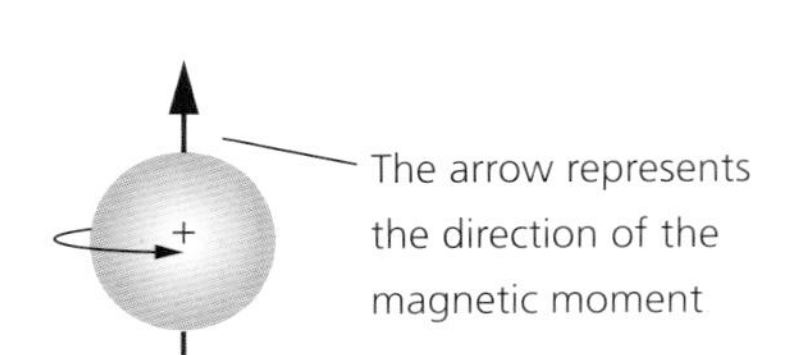

Nuclear magnetic resonance (NMR) spectroscopy is a method of examining the ^{1}H and ^{13}C nuclei within an organic molecule. ^{1}H and ^{13}C nuclei are referred to as spin-half nuclei and so they can act as a mini magnet.

If these nuclei are placed in a strong external magnetic field, they will align themselves with the magnetic field as shown in the diagram.

If a strong magnetic field is applied to this hydrogen nucleus, there are two possibilities. The magnetic moment can align with the external magnetic field or against it.

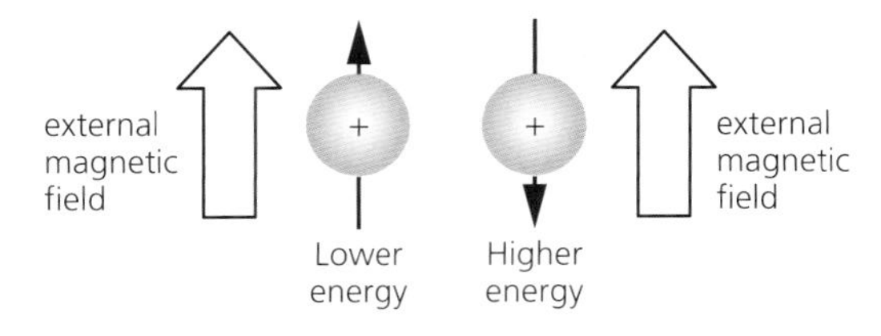

When the magnetic moments of the nuclei are aligned against the magnetic field, this is the higher energy state of the nuclei. The amount of energy required to flip the nuclei between the lower and the higher energy state is in the radio wave region of the electromagnetic spectrum and the energy absorbed can tell us a great deal about the environment in which the nuclei in the molecule are found.

Chemical shift

The chemical shift is a comparative measure of the energy required to flip the spin of the nuclei and is often represented by the Greek letter δ and is measured in parts per million (ppm). This forms the horizontal axis of the NMR spectrum and unusually the scale increase from right to left; the 0 value being on the right-hand side. The chemical shift values are given in the Data Booklet provided with your exam papers. These are given below:

Table B

^1H NMR chemical shift data

Type of proton	δ/ppm
RO**H**	0.5 – 5.0
RC**H**$_3$	0.7 – 1.2
RN**H**$_2$	1.0 – 4.5
R_2C**H**$_2$	1.2 – 1.4
R_3C**H**	1.4 – 1.6
R—C(=O)—C(—**H**)—	2.1 – 2.6
R—O—C(—**H**)—	3.1 – 3.9
RC**H**$_2$Cl or Br	3.1 – 4.2
R—C(=O)—O—C(—**H**)—	3.7 – 4.1
R(—)C=C(—)**H**	4.5 – 6.0
R—C(=O)—**H**	9.0 – 10.0
R—C(=O)—O—**H**	10.0 – 12.0

Table C

^{13}C NMR chemical shift data

Type of carbon	δ/ppm
—**C**—**C**—	5 – 40
R—**C**—Cl or Br	10 – 70
R—C(=O)—**C**—	20 – 50
R—**C**—N<	25 – 60
—**C**—O— alcohols, ethers or esters	50 – 90
>**C**=**C**<	90 – 150
R—**C**≡N	110 – 125
benzene ring	110 – 160
R—**C**(=O)— esters or acids	160 – 185
R—**C**(=O)— aldehydes or ketones	190 – 220

These chemical shift give information about the environment in which the ^1H and ^{13}C nuclei are found in the molecule. Those nuclei in the same environment appear at the same chemical shift on an NMR spectrum. A higher chemical shift usually indicates that the ^1H or ^{13}C nuclei are closer to an electronegative group or atom or an electron withdrawing group such as a benzene ring.

All chemical shifts are measured relative to a standard. The standard used in NMR spectroscopy is tetramethylsilane (TMS), $(CH_3)_4Si$, the structure of which is shown on the left.

$Si(CH_3)_4$: H_3C—Si—CH_3 with CH_3 above and below

tetramethylsilane

Figure 15.1 A nuclear magnetic resonance (NMR) spectrometer

Due to the higher electronegativity of carbon compared to silicon, the electrons in the four Si–C bonds are closer to the carbon atoms. This causes a knock-on effect for the electrons in the C–H bonds and hence the electrons are closer to any ^{13}C and ^1H nuclei than they would be in any other organic molecule. The proximity of these electrons to the nuclei shields the spinning particles in the nuclei.

Other ^1H and ^{13}C nuclei in organic molecules are said to be deshielded compared to those in TMS, and as the chemical shift is higher they appear to the left of the TMS peak in the NMR spectrum. The more electronegative the groups bonded to the nuclei (or even those attached to a carbon with hydrogen(s) attached) are, the more deshielded a hydrogen nucleus becomes and the larger the chemical shift.

The solvent used for ^1H NMR spectroscopy is tetrachloromethane or any deuterated solvent such as $CDCl_3$, CD_2Cl_2 or C_2D_6. These solvents do not contain ^1H nuclei which would give peaks on the spectrum.

^1H NMR spectroscopy

The spectrum below shows the low resolution ^1H NMR of ethanal, CH_3CHO. The three ^1H nuclei in the CH_3 group are all in the same molecular environment in a molecule compared to a hydrogen nucleus attached to an oxygen atom in an O–H bond. The spectrum shows these peaks. The structure of ethanal is shown on the left:

ethanal

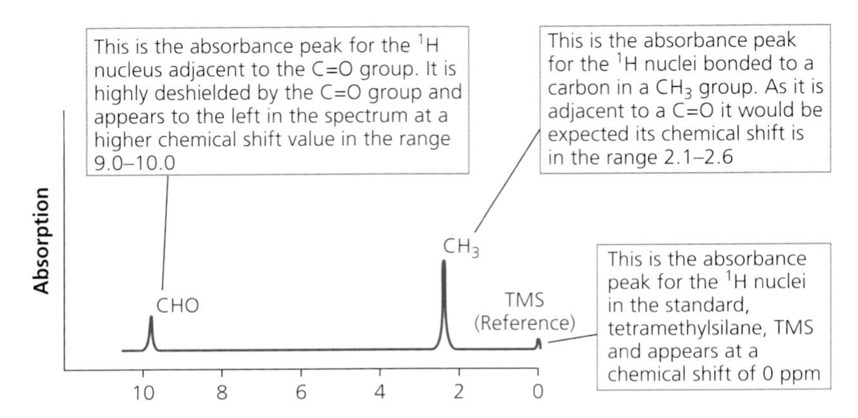

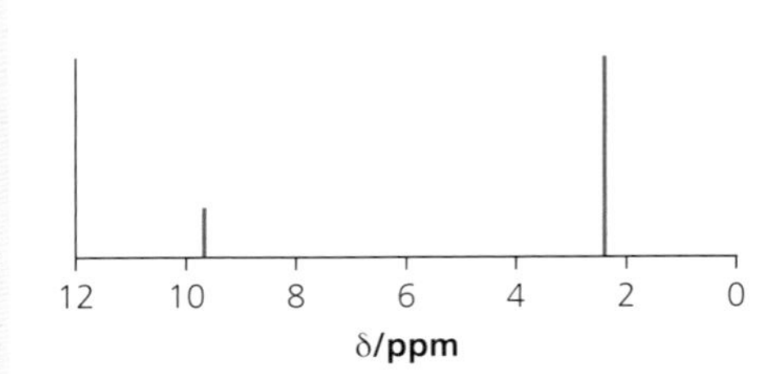

The low resolution ^1H NMR spectrum is often drawn more simply as a series of spikes.

Sometimes the peak for TMS is omitted as shown to make it clearer how many environments of ^1H nuclei are present in the molecule.

The chemical shift for ^1H nuclei in a CH_3 group is not always the same, as the other groups attached to the carbon may affect its degree of deshielding. The above spectrum for ethanal shows the CH_3 peak at 2.3 ppm as would be expected for ^1H adjacent to a C=O in the range 2.1–2.6 from the table on page 293. The chemical shift of ^1H nuclei vary depending on the adjacent groups. This is due to the oxygen atom attached to the carbon pulling the electrons in the C–H bonds away from the ^1H nuclei and so deshielding them to a greater extent than would normally be expected. However, it is true to say that the peaks at the lowest chemical shift in an aliphatic organic molecule would normally be those of ^1H nuclei in a CH_3 group, if they are present.

Hydrogen atoms that appear at the same chemical shift on the spectra and are therefore in the same environment within the molecule are said to be chemically equivalent hydrogen atoms.

Chemically equivalent hydrogen atoms are not distinguishable on an NMR spectrum. In the ethanal spectrum, there are two distinct types of ^1H nuclei. The three CH_3 ^1H nuclei are chemically equivalent and the single CHO ^1H nucleus is another environment. Two CH_3 groups may have all their ^1H nuclei in the same environment. for example in propanone, CH_3COCH_3, all 6 ^1H nuclei are in the same environment due to the symmetry of the molecule.

propanone

The ^1H NMR spectrum of propanone is shown below:

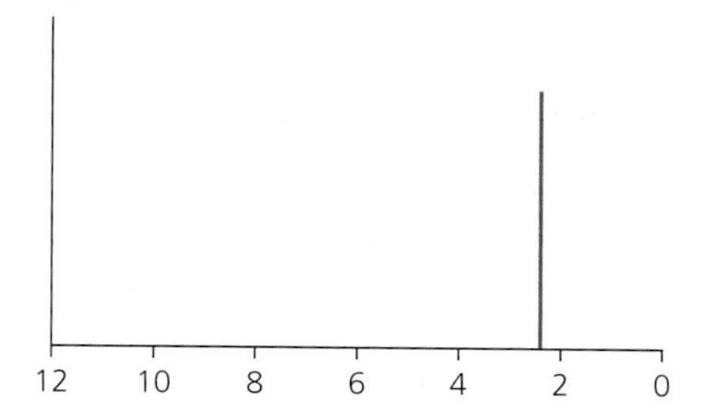

Peak integration

The ratio of the areas under the peaks in a ^1H NMR spectrum indicates the ratio of the numbers of chemically equivalent ^1H nuclei or equivalent protons. This is often given in a table with the chemical shift value. For example for the spectrum of ethanal, the table would look like the one below.

δ/ppm	2.3	9.8
Integration ratio	3	1

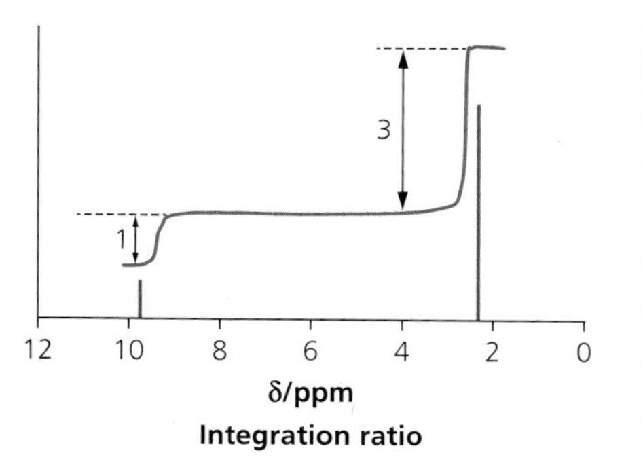

δ/ppm

Integration ratio

The ratio is calculated by integration of the area under the peaks as the spectrum is produced. It can also be given as a trace over the spectrum where the difference in the horizontal levels of the trace give the integration ratio.

The ratio of the difference in heights (measured using a ruler) can give the integration ratio.

Spin-spin splitting patterns

Neighbouring ^1H nuclei have an effect on each other, which can be seen in a high resolution ^1H NMR spectrum. Equivalent ^1H nuclei will appear at the same chemical shift value on the spectrum, but the peak for these ^1H nuclei will be split into a number of peaks depending on the number of non-equivalent ^1H nuclei bonded to the adjacent carbon atoms.

This splitting of the peaks into a set of peaks is called the spin-spin splitting pattern. The splitting pattern follows the **n+1 rule**. If there are 5 peaks (=n+1) on the spectrum, this is caused by 4 (=n) hydrogen nuclei on adjacent carbon atoms. The table shows the names and the ratios of the heights of the peaks in the various splits.

Number of hydrogen nuclei on adjacent carbon atoms [*n*]	Number of peaks [*n+1*] (Ratio of peak heights)	Name of pattern
0	1	Singlet
1	2 (1:1)	Doublet
2	3 (1:2:1)	Triplet
3	4 (1:3:3:1)	Quartet
4	5 (1:4:6:4:1)	Quintet
5	6 (1:5:10:10:5:1)	Sextet

If there is one non-equivalent proton bonded adjacent carbon atoms, the peak will be split into two peak of equal height. These peaks will be smaller, as the number of peaks in the environment is unchanged so the integration must be the same. These two peaks together are called a **doublet**.

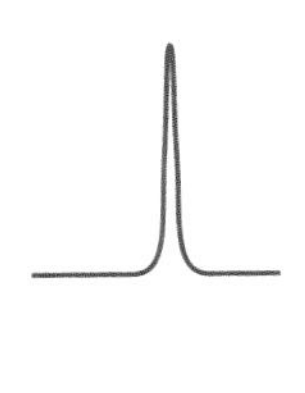

The ^{1}H nuclei bonded to the carbon atoms in the peak on the left have no hydrogen nuclei (protons) bonded to adjacent carbon atom(s) to split them. However, on the right the spectrum shows two peaks of approximately equal height (a doublet), which indicates one proton bonded to an adjacent carbon atom

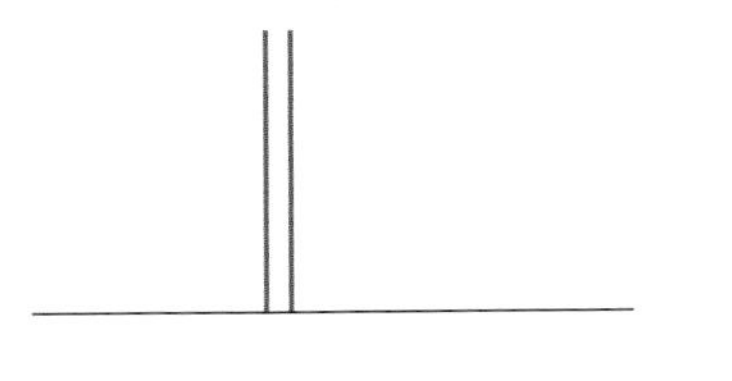

A doublet is often shown more simply as two spikes of approximately equal height.

If there are two equivalent ^{1}H nuclei or protons bonded to the carbon atom(s) adjacent to the carbon atom in question, the single peak we would have expected is split into three peaks, the heights of which are in the ratio 1 : 2 : 1. Again the total area under the three peaks is the same as the total area under the single unsplit peak so the peaks are smaller. The three peaks together are called a **triplet**.

Triplet caused by 2 non-equivalent protons bonded to adjacent carbon atoms

The triplet is often shown like this as a series of spikes in the spectrum

This splitting pattern is the result of the interactions of the spins of the ^{1}H nuclei hence the name spin-spin coupling is used. It is the spin-spin coupling which gives rise to the spin-spin splitting pattern observed in ^{1}H NMR spectra.

^{1}H NMR spectroscopy is a powerful tool to help determine structure and is often used with other analytical techniques such as infrared spectroscopy and mass spectrometry.

EXAMPLE 3

The ^{1}H NMR spectrum for chlorinated alkane is shown below:

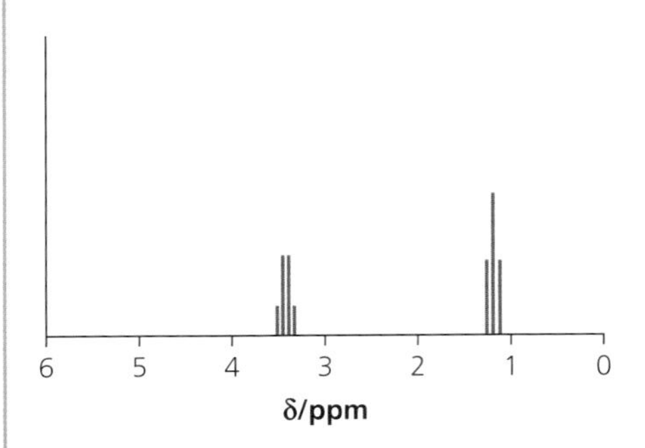

The integration table is shown below:

δ/ppm	1.2	3.4
Integration ratio	3	2

The peak at a chemical shift value of 1.2 in the spectrum would suggest a CH_3 group as it is within the range 0.7–1.2.

The integration ratio for this peak is 3, which would again suggest a CH_3 group.

The peak is split into a **triplet**, which indicates that it is split by 2 hydrogen nuclei (i.e. a neighbouring CH_2).

The peak at a chemical shift value of 3.4 would be a $RC\mathbf{H}_2Cl$ or Br as it is in the range 3.1–4.2. This peak is deshielded by the electronegative chlorine atom attached to the carbon.

The integration ratio for this peak is 2 which could again suggest a CH_2 group.

The peak is split into a **quartet** which indicates that it is split by 3 hydrogen nuclei (i.e. the $C\mathbf{H}_3$).

The evidence would suggest that the molecule is chloroethane.

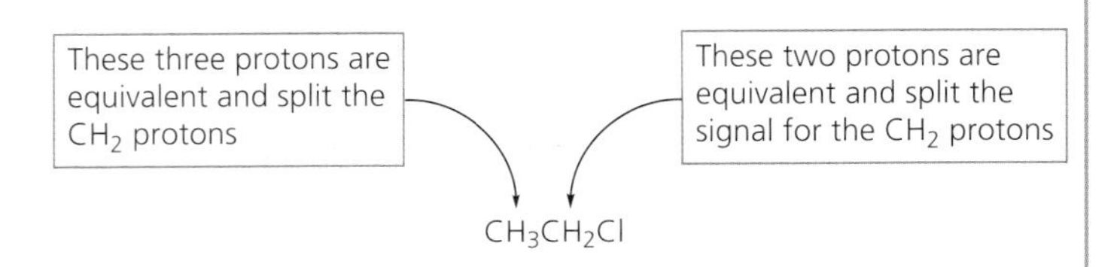

EXAMPLE 4

The ^1H NMR spectrum of a compound with molecular formula $C_6H_{12}O_2$ is given below.

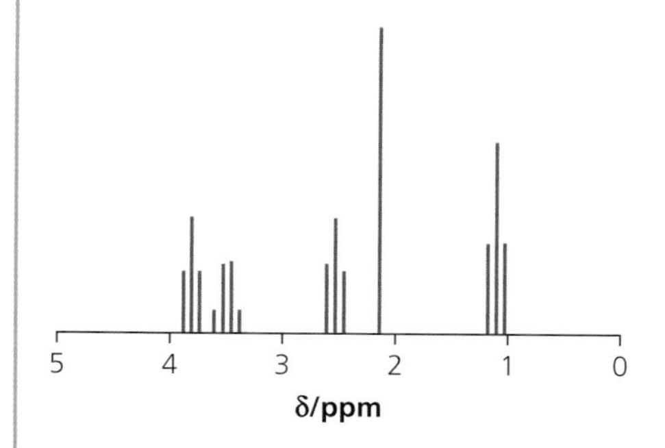

The integration gave the following data:

δ/ppm	1.2	2.2	2.6	3.5	3.8
Integration ratio	3	3	2	2	2
Spin-spin splitting	triplet				

Use Table B shown on page 293 and the information above to answer the following questions:

a) Explain why this is not the spectrum of a carboxylic acid.
b) Complete the table giving the spin-spin splitting patter observed in the spectrum.
c) Suggest which structural part of the molecule gives rise to the peaks at δ = 1.2 and 3.5.
d) Suggest which structural part of the molecule gives rise to the peak at δ = 2.2.
e) Suggest which structural part of the molecule gives rise to the peak at δ = 2.6 and 3.8.
f) Deduce the structure of this compound.

Answers

a) No peak at δ = 10.0–12.0 for a COOH.
b)

δ/ppm	1.2	2.2	2.6	3.5	3.8
Integration ratio	3	3	2	2	2
Spin-spin splitting	triplet	singlet	triplet	quartet	triplet

c) δ = 1.2 and an integration of 3 would suggest a CH_3 group.
The triplet would suggest that there is a CH_2 group adjacent.
As the CH_3 would cause the signal for the CH_2 group to be a quartet, it has to be the CH_2 at δ = 3.5 which is adjacent to the CH_3.
The chemical shift of the CH_2 would suggest it is adjacent to an oxygen atom (in the range 3.1–3.9).
The structural part here is:

```
    H   H
    |   |
H — C — C — O —
    |   |
    H   H
```

d) δ = 2.2 and an integration of 3 would again suggest a CH_3 group.
The singlet would suggest that there are no protons bonded to adjacent carbon atoms.
The chemical shift would suggest a C=O group adjacent (2.1–2.6) which would explain the lack of splitting.
The structural part here is:

```
    H   O
    |   ||
H — C — C —
    |
    H
```

e) δ = 2.6 and an integration of 2 would suggest a CH_2 group.
The triplet would suggest a CH_2 group is adjacent.
The chemical shift of 2.6 would suggest the CH_2 group is adjacent to a C=O group.
δ = 3.8 and an integration of 2 would suggest another CH_2 group.
The triplet would again suggest a CH_2 group is adjacent.
The chemical shift would suggest an oxygen atom (3.1–3.9) or an ester group (3.7–4.1) adjacent.
However the rest of the structure would suggest it is not an ester so the CH_2CH_2 must come between the other parts so this structural part is:

```
  O   H   H
  ||  |   |
— C — C — C — O —
      |   |
      H   H
```

f) The entire compound has the structure:

```
    H   O   H   H       H   H
    |   ||  |   |       |   |
H — C — C — C — C — O — C — C — H
    |       |   |       |   |
    H       H   H       H   H
```

TIP

The absence of a peak in (a) can be as much of a clue as its presence. This compound must be an ester or a molecule containing a C=O and a C-O-C. It could also contain a C=C and two OH groups.

Analysis of a given molecule

EXAMPLE 5

In another type of question, you could be given the molecule with the environments labelled and you have to match them to the peaks in the spectrum, state the number of peak in the ^1H NMR spectrum and/or explain the chemical shift, peak integration or spin-spin splitting.

$$\underset{a}{CH_3}—\overset{\overset{O}{\|}}{C}—O—\underset{b}{CH_2}—\underset{c}{CH_2}—\underset{d}{CH_3}$$

1 Identify the protons which will have the highest chemical shift in the ^1H NMR spectrum.
2 Give the splitting pattern of the protons at **d**.
3 Which protons would appear as a singlet on the spectrum?

Answers

1 Protons **a** and **b** are closest to the electronegative oxygen atoms.
 a should give a chemical shift in the range 2.1–2.6.
 b should give a chemical shift in the range 3.1–3.9
 The answer is **b**.
2 The protons in **d** have 2 protons bonded to adjacent carbons (a CH_2 on one side). This means that the splitting for d is a triplet.
3 A singlet appears when protons have no protons bonded to adjacent carbon atoms. The only protons for which this is the case is **a**, the CH_3 with a C=O adjacent.

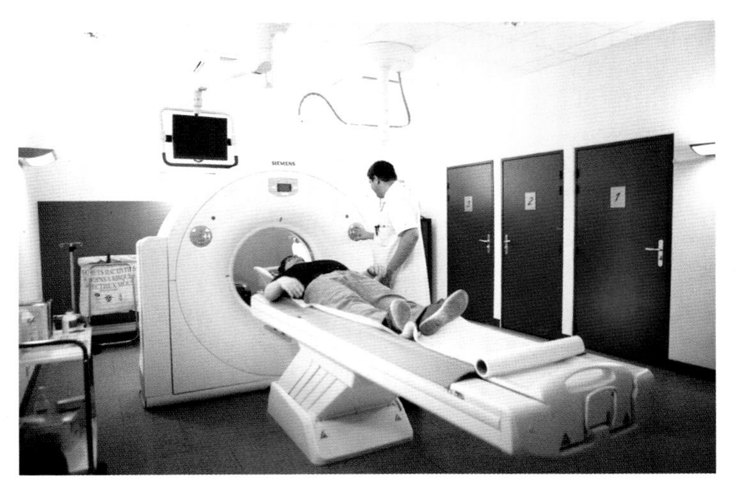

Figure 15.2 Nuclear magnetic resonance has been developed into the medical scanning technique known as magnetic resonance image MRI. The photograph shows a patient moving into a full body MRI scanning machine

^{13}C NMR spectroscopy

^{13}C NMR spectroscopy is much simpler than ^1H NMR spectroscopy.

^{13}C atoms occur as about 1% of all carbon atoms. ^{13}C NMR spectroscopy relies on the magnetic properties of the nuclei of these atoms.

TMS is again used as a standard for ^{13}C NMR spectroscopy and the solvent used is often $CDCl_3$. $CDCl_3$ will give a peak on a ^{13}C NMR spectrum, but this is easily identified and removed from the final spectrum. The peak at δ = 0 ppm for TMS is often removed as well.

There is no spin-spin splitting to worry about in ^{13}C NMR spectroscopy and you are simply looking for chemical shift and the number of types of equivalent ^{13}C nuclei in the molecule.

A spectrum is simply a number of peaks. Each peak corresponds to an environment of equivalent ^{13}C nuclei.

A ^{13}C NMR spectrum for 1-chloropropan-2-ol is shown below.

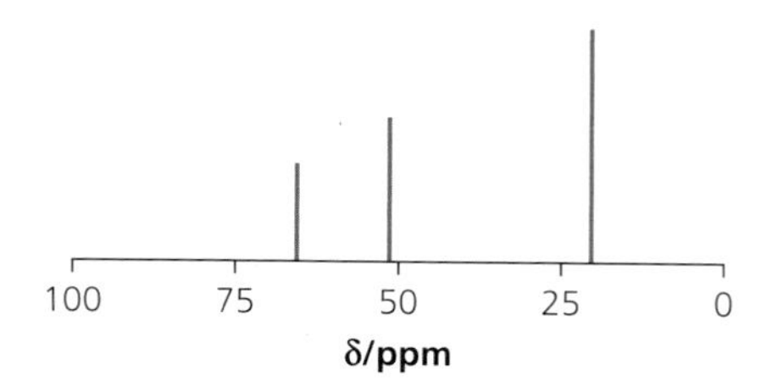

There are three peaks in the spectrum which correspond to the three ^{13}C nuclei.

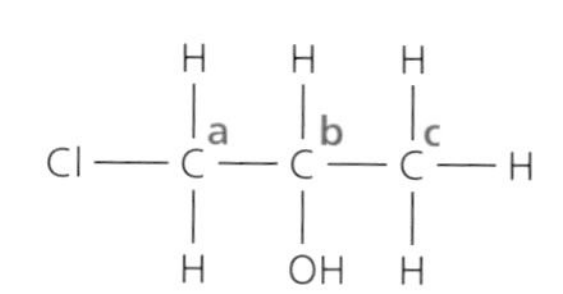

The structure of 1-chloropropan-2-ol is shown below with the carbon atoms labelled a, b and c.

Using Table C from the Data Booklet (see page 293 of this chapter):

- The peak at δ = 20 is caused by c (range 5–40)
- The peak at δ = 51 is caused by a (range 10–70)
- The peak at δ = 67 is caused by b (range 50–90)

There may be some confusion with a and b as they are both in the range, but the oxygen atom is more electronegative than the chlorine atom and so it more likely that the CH(OH) ^{13}C is the peak at 67.

Determining the number of peaks

The number of peaks expected in a ^{13}C NMR spectrum is a common question. This can be applied to aliphatic as well as aromatic molecules. Again the environments are determined based on the structural symmetry of the molecule.

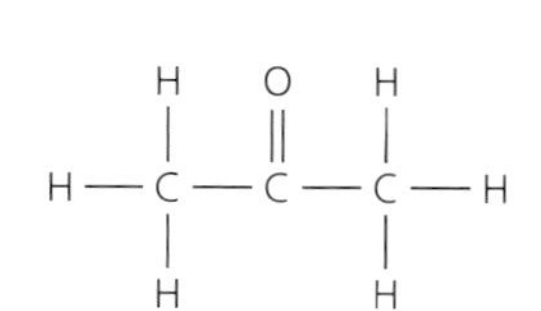

Two CH_3 groups bonded to the same carbon in a symmetrical molecule such as propanone.

The C in the two CH_3 groups are equivalent and so propanone will show two peaks on its ^{13}C NMR spectrum.

Propanal is shown below:

There are three peaks in the ^{13}C NMR spectrum of propanal. CH_3, CH_2 and the CHO will all give individual peaks.

Benzene gives a single peak as all the ^{13}C are in the same environment. The same is true of cyclohexane.

Substituted benzene compounds are treated differently.

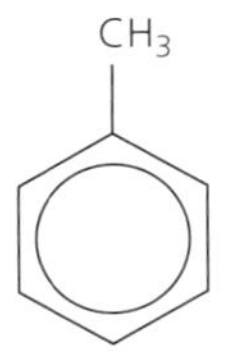

Methyl benzene (or toluene) shows 5 peaks in its ^{13}C NMR spectrum. This can be explained by examining the molecule.

The numbers represent the carbon atoms that are equivalent, i.e. both labelled 3 are equivalent and both labelled 4 are equivalent. The CH_3 group breaks the symmetry of the benzene ring so 2 and 5 are no longer equivalent.

However, 1,4-dimethylbenzene (shown below) has 3 peaks on its ^{13}C NMR spectrum. The carbon atoms are labelled as before. The presence of 2 CH_3 groups gives the molecule more symmetry so the top and bottom carbon atom in benzene are now equivalent and the other carbon atoms in the benzene ring are all equivalent (being equally removed from the CH_3 groups.

1,2-Dimethylbenzene (shown below) has 4 peaks in its ^{13}C NMR spectrum.

methylbenzene (toluene)

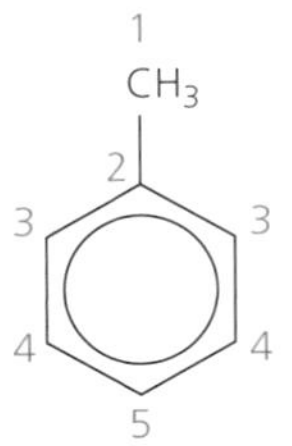

methylbenzene

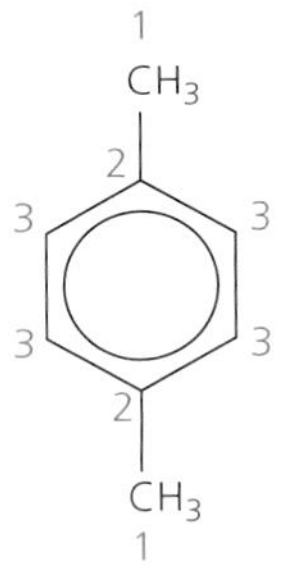

1,4-dimethylbenzene

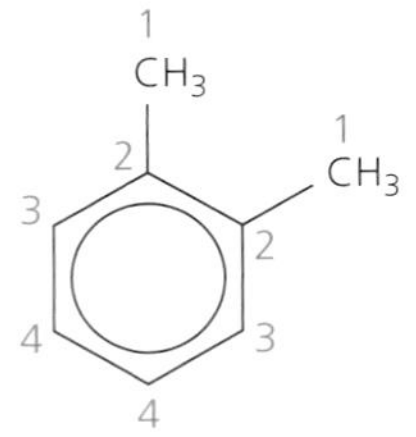

1,2-dimethylbenzene

EXAMPLE 6

Deduce the number of peaks in the ^{13}C NMR spectrum of the following molecules.

1

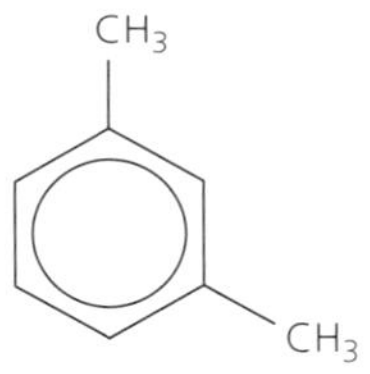

2

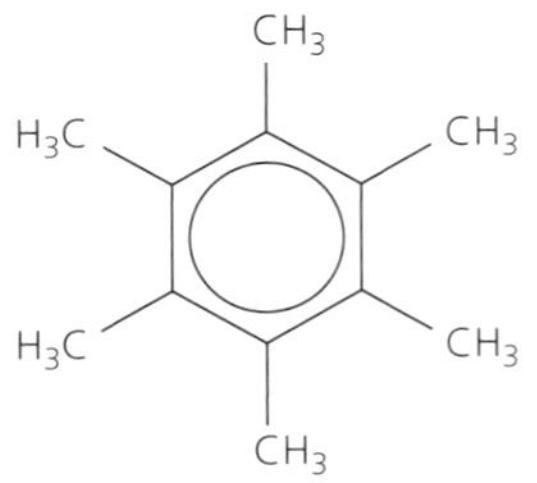

3

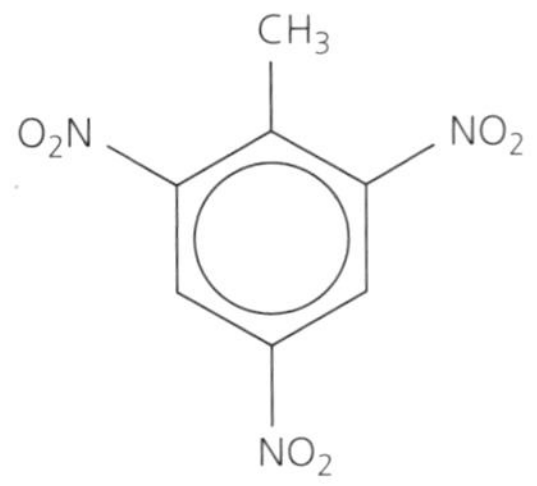

4 $CH_3COOCH(CH_3)_2$

Answers

1 5

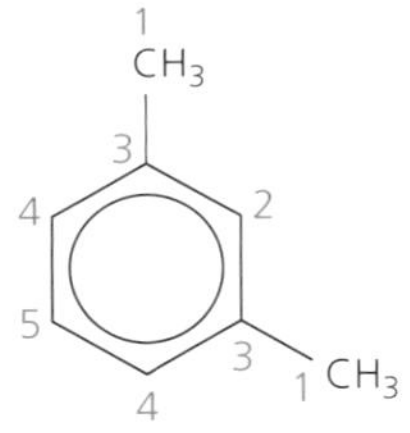

2 Answer is 2 as every CH_3 carbon atom is equivalent and all the carbon atoms in benzene are equivalent.

3 5

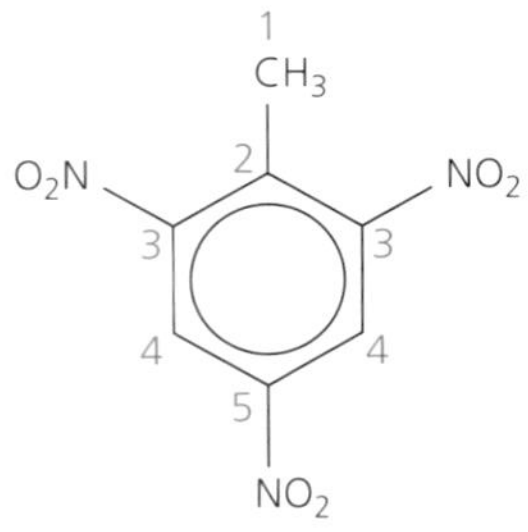

4 $\overset{1}{C}H_3\overset{2}{C}OO\overset{3}{C}H(\overset{4}{C}H_3)_2$ – the two carbon atoms in CH_3 groups are equivalent. The answer is 4.

TEST YOURSELF 3

1 Name the standard reference used for NMR spectroscopy.
2 Describe the spin-spin splitting pattern for ethanol, CH_3CH_2OH in its high resolution ^1H NMR spectrum.
3 How many peaks would be present in the ^{13}C NMR spectrum of propan-2-ol $CH_3CH(OH)CH_3$?
4 For the following compounds:
$CH_3CH_2COCH_3$ $CH_3CH_2COCH_2CH_3$ $CH_3CH_2CH_2CHO$
a) Name all the compounds.
b) Which one would give the least number of peaks in a ^{13}C NMR spectrum?

Chromatography

Chromatography is a method of separating soluble substances by their partition between two different phases. This was also studied on page 277.

All types of chromatography have a mobile phase and a stationary phase. The substances to be separated are initially in the mobile phase, but as they move through the chromatogram they may partition themselves into a liquid stationary phase or adsorb onto a solid stationary phase.

We will examine three different types of chromatography: thin-layer chromatography (TLC), column chromatography (CC) and gas chromatography (GC).

The three methods of chromatography we are looking at are described as having a liquid mobile phase (TLC or CC) or a gas mobile phase (GC). The stationary phase for paper and TLC and CC can be considered to be solid. The stationary phase for GC is a liquid.

The time the substances stay in the stationary phase depends on:

1 Size/mass of the solute molecules/ions.

2 Solubility in the mobile phase vs solubility in liquid stationary phase (partition)

3 Binding to the solid stationary phase (caused by adsorption).

- Particles that have a greater attraction to a *solid stationary phase* in TLC and CC will adsorb more onto this phase and will move more slowly through the chromatogram.
- Particles that are more soluble in the *liquid stationary phase in GC* will partition themselves into this phase and move along the chromatogram more slowly.
- Larger particles pass along a chromatogram more slowly too.
- Particles that are very soluble in the mobile phase compared to the liquid stationary phase will move quickly through the chromatogram.

The mobile and stationary phases of the three types of chromatography are shown in the table below.

Type of chromatography	Mobile phase	Stationary phase
Thin-layer chromatography	Liquid solvent e.g. water or organic solvents	Solid silica gel paste on a microscope slide or plastic plate or solvent/ water in the gel
Column chromatography	Liquid solvent e.g. water or organic solvents	Solid silica gel
Gas chromatography	Inert carrier gas, e.g. N_2, Ne	Microscopic film of liquid on a solid support

Thin-layer chromatography (TLC)

Thin-layer chromatography is carried out in almost an identical fashion to paper chromatography. The thin-layer plates are prepared from silica gel. Silica gel is a polymer form of silicic acid that has many –OH groups, making the surface of the gel/paste very polar. This allows polar substances to be held by the gel making their movement much slower than for non-polar substances.

1 Prepare a TLC plate using silica gel paste or use a commercial one.

2 Mark a fine pencil line about 0.5 cm from one end. Put a cross in the centre of the line. This line is called the **origin**.

3 Using a capillary tube, lift some of the sample dissolved in the appropriate solvent and spot onto the cross.

4 Place the thin-layer chromatogram in the same solvent (the end with the spot just dipping in).

5 Allow to run for about 20–30 minutes as solvent moves up.

6 Check every few minutes to ensure that the solvent has not run off the top end of the chromatogram.

7 As the solvent moves, the substances in the spot also move up and separate.

8 Once finished, draw a pencil line to show where the solvent has reached on the chromatogram. This is the solvent front.

9 The distance the solvent has moved in centimetres is called the R_s value.

10 $R_f = \frac{\text{distance moved by spot (to centre of spot)}}{\text{distance moved by solvent } (R_s)}$

11 Within a solvent system, a substance should always have the same R_f value, hence allowing comparison.

Figure 15.3 Thin-layer chromatography plate of black ink.

Developing chromatograms

Most chromatograms are not visible unless they are developed. The process of developing for amino acids with ninhydrin has been described previously. Some spots on chromatograms are only visible under UV light and can be marked with a pencil for viewing in normal light. Others require a solution of iodine to mark aromatic compounds.

Column chromatography (CC)

In this method, a solid such as silica gel is placed in a glass column with glass wool at the bottom to prevent the silica gel from blocking the column or coming out of the bottom. Glass wool is used as it is non-absorbent.

The silica gel is a polar solid stationary phase for the column chromatography. The solvent that will carry the substances to be separated is chosen to minimise time. Often small-scale tests are carried with TLC as it uses the same solid support to find the most appropriate solvent or mixture of solvents.

The solvent is run through the silica gel and from this point the column must not dry out.

The sample to be separated is dissolved in a solvent such as hexane. Hexane is a non-polar solvent and so polar substances are more likely to stay in the column longer as they will be more attracted to the polar silica gel. Non-polar substances will move quickly through the column as they will be more likely to remain in the non-polar solvent.

The eluent is the liquid that leaves the column at the bottom, and samples of the eluent may be taken for further analysis. The amount of time that a substance remains in the column is called the retention time.

Figure 15.4 Two bands of a red/brown colour can be seen in the column. These are different components of a mixture that is being separated – the different substances travel down the column of gel at different rates. The technique can be used to purify and isolate a required product.

The column is prepared and saturated with solvent. The mixture to be separated is dissolved in the solvent and placed on the top of the column. Fresh solvent is added to the column to stop it from drying out. The solvent moves through the column and the substances in the mixture separate based on their solubility in the solvent (mobile phase) and their retention by the silica gel (stationary phase). The liquid leaving the column is called the eluent. Substances are described as being eluted when they leave the column.

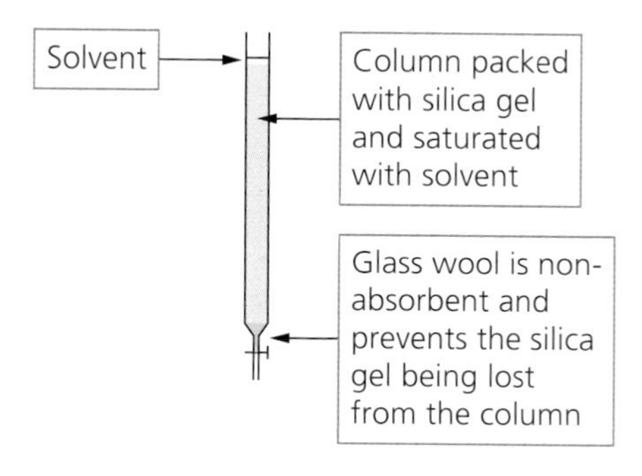

If the samples being eluted from the column are coloured, it is easy to see them being eluted. However many substances being used are not coloured and separate containers are used to collect 1 cm^3 or 5 cm^3 samples of eluent. The samples containing the separated substances may be analysed using TLC and developed using chemical developing agents or under UV light to determine which samples contained the individual substances.

EXAMPLE 7

A sample of ethanol was contaminated with ethanal. The sample was separated using column chromatography. The sample was dissolved in hexane and added to the column. Hexane continued to be added to the column.

Explain why ethanal was found to be present in samples of eluent collected first, whereas ethanol was collected later.

Answer

Ethanal is less polar than ethanol so it has a lower affinity for the polar stationary phase than ethanol. Ethanol being more polar remained bonded to the polar stationary phase longer and so stays longer in the column.

REQUIRED PRACTICAL 12

Thin-layer chromatography of amino acids in myoglobin

Myoglobin and haemoglobin are globular proteins that serve to bind and deliver oxygen in the body. Haemoglobin is the primary oxygen-carrying protein and is found in red blood cells. It consists of four connected polypeptide chains, each of which contains a non-protein haem group, the site at which oxygen is known to bind.

Myoglobin is the main oxygen-binding protein found in muscle cells. Myoglobin is only found in the bloodstream after muscle injury. Myoglobin also contains a haem group, but it consists of only one polypeptide chain containing 153 amino acid residues.

Myoglobin can be hydrolysed into its constituent amino acids, using acid hydrolysis. Thin-layer chromatography (TLC) can then be carried out on the resulting mixture and the R_f value calculated for each of the spots obtained on the chromatogram.

1 Why is it necessary to wear plastic gloves when holding a TLC plate?
2 Why is it necessary to draw a *pencil* base line 1.5 cm from the bottom of the plate.
3 How is a very tiny concentrated drop of amino acid solution added to the TLC plate?
4 Explain why the developing tank the solvent is at a depth of only 1 cm.
5 Explain why the developing tank is sealed with a lid when the TLC plate is placed in it.
6 Explain why the TLC plate is allowed to dry in a fume cupboard.
7 The substance used to develop the spots on the chromatogram is ninhydrin. State one safety precaution specific to the use of this substance, apart from the use of goggles.
8 The figure shows a developed chromatogram with three spots. The dotted line through each spot indicates its mid-point, which is used to measure the distance it has travelled. Using the figure, identify amino acid X on the chromatogram.

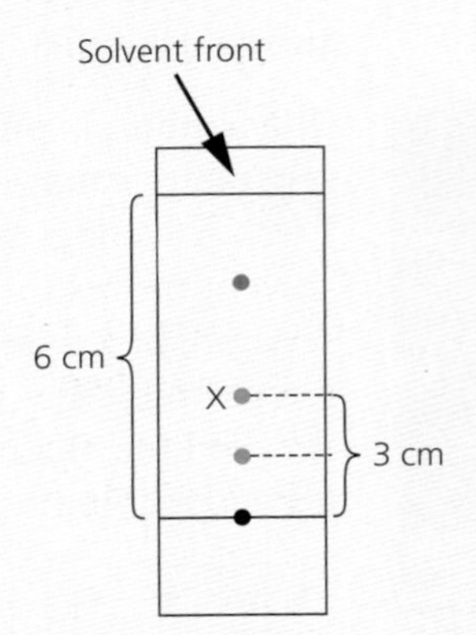

Table 15.1 The R_f value of some amino acids

Amino acid	R_f
Alanine	0.38
Arginine	0.20
Asparagine	0.50
Glutamic acid	0.30
Leucine	0.73
Lysine	0.14

9 Explain why different amino acids have different R_f values.

High concentrations of myoglobin in muscle cells allow organisms to hold their breath for a longer period of time. Diving mammals such as whales and seals have muscles with particularly high abundance of myoglobin. Muscle cells use myoglobin to accelerate oxygen diffusion and act as localised oxygen reserves for times of intense respiration.

Gas chromatography (GC)

Gas chromatography (or sometimes called gas-liquid chromatography – GLC) is carried out using a coiled tube containing a solid support that has a coating of liquid on its surface. The coiled tube is contained within an oven, which ensures that the sample when injected into the column will remain as a gas.

This type of chromatography works by partition. The substances in the gaseous mobile phase may dissolve in the stationary liquid phase and this retards their progress through the column. The more soluble the substances are in the liquid stationary phase, the more time they spend in the column.

The sample to be separated is heated to make it a gas and then mixed with an inert carrier gas such as nitrogen or helium. The sample then enters into the column and is allowed to pass through it.

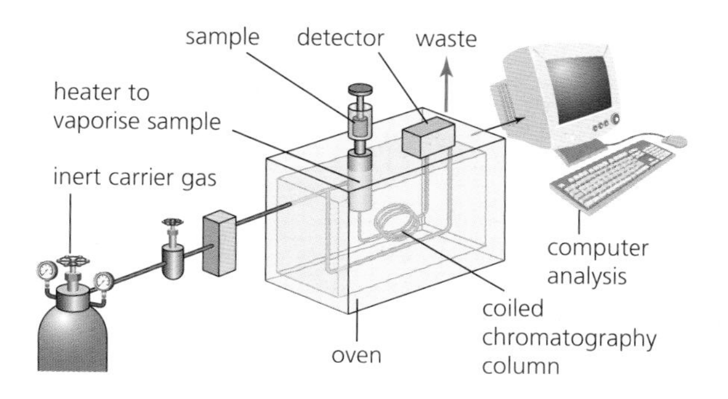

The computer detects substances being **eluted** from the column. A substance is eluted when it leaves the chromatography column. The time that each substance spends in the column is called its **retention time**. The computer analysis plots detector signal against retention time as shown below.

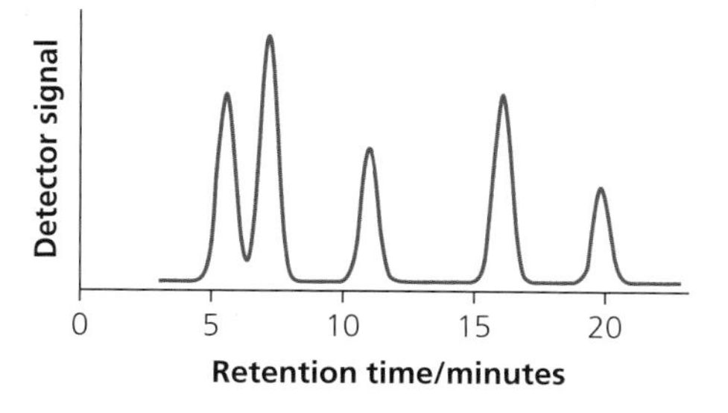

There are five components to the above sample as there are five separate peaks. The peaks at 5 and 7 minutes are close together so their properties must be similar. The peak at 20 minutes represents a substance that was very soluble in the liquid stationary phase as it was the last to elute from the column and therefore had the highest retention time.

Figure 15.5 A gas chromatography machine (left) connected to a mass spectrometer (right) in a forensic laboratory. This equipment is sensitive enough to detect minute quantities of illegal drugs in the hair of a suspect – weeks after any drugs were taken.

GC is also useful as the area under the peak gives a measure of the relative amounts of each substance present in the sample. This allows percentage composition of the mixture to be calculated. This can be useful in monitoring an equilibrium reaction. The reaction can be sampled and analysed using GC and the relative proportions of the reactants and products calculated from the area under the peaks.

GC is usually linked with mass spectrometry which will enable each peak to be identified. This combined technique is very powerful as it will identify each component of a mixture and calculate how much of each component is present.

If a non-destructive detector is used, GC can be used to separate the components of the mixture as they can be condensed.

EXAMPLE 8

The numbers above the peaks indicate the relative area under the peak which allows us to work out percentage composition.

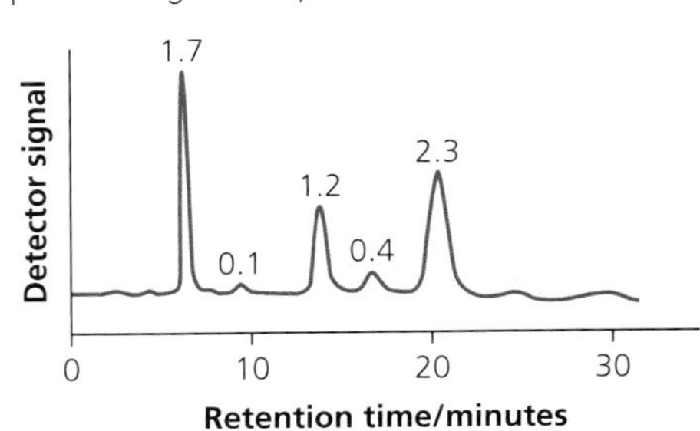

The total of the areas under the peaks should be calculated and then percentage composition can be worked out.

Answer

Total of all areas = 1.7 + 0.1 + 1.2 + 0.4 + 2.3 = 5.7

Peak at 7 minutes — percentage composition $= \frac{1.7}{5.7} \times 100 = 29.8\%$

Peak at 9 minutes — percentage composition $= \frac{0.1}{5.7} \times 100 = 1.8\%$

Peak at 14 minutes — percentage composition $= \frac{1.2}{5.7} \times 100 = 21.0\%$

Peak at 17 minutes — percentage composition $= \frac{0.4}{5.7} \times 100 = 7.0\%$

Peak at 20 minutes — percentage composition $= \frac{2.3}{5.7} \times 100 = 40.4\%$

TEST YOURSELF 4

1 Name a common stationary phase used in column chromatography.

2 Which one of ethanol, ethanal and ethanoic acid would you expect to be eluted first from silica gel column chromatography using a hexane solvent?

3 Name one way in which a TLC plate may be developed.

Practice questions

1 Which one of the following does not react with KCN?

A 1-bromopropane **B** propanal

C propanone **D** propanoic acid *(1)*

2 Which one of the following has 2 peaks on its ^{13}C NMR spectrum?

A 1,2-dibromobenzene

B 1,3-dibromobenzene

C 1,2,3-tribromobenzene

D 1,3,5-tribromobenzene *(1)*

3 Which one of the following is the correct splitting pattern in the ^{1}H NMR spectrum from lowest to highest chemical shift value for propanoic acid?

A quartet, triplet, singlet

B singlet, quartet, triplet

C triplet, quartet, singlet

D singlet, triplet, quartet *(1)*

4 Methoxymethane (CH_3OCH_3) and ethanol (CH_3CH_2OH) are isomers. The ^1H NMR spectrum are shown below.

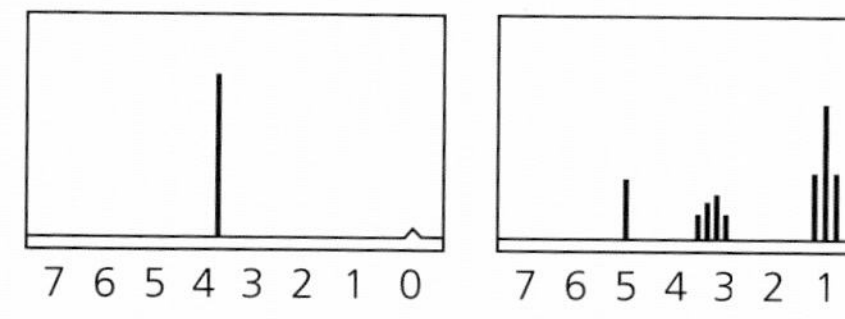

a) Explain why spectrum A is the spectrum of methoxymethane. *(2)*

b) Complete the table below giving the spin-spin splitting for spectrum B. *(2)*

δ/ppm	1.0	3.2	5.0
Integration ratio	3	3	1
Spin-spin splitting			

c) Ethanol contains an ethyl group, CH_3CH_2. Explain this from spectrum B. *(2)*

5 4-Methylpentan-2-one has the structure:

```
        CH3
        |         a              b
H3C — C — CH2 — C — CH3
        |              ||
        H              O
```

a) How many peaks are in the ^{13}C NMR spectrum of this compound? *(1)*

b) How many types of equivalent protons are in the ^1H NMR spectrum of this spectrum? *(1)*

c) Give the IUPAC name for this compound. *(1)*

d) What is the spin-spin splitting pattern for the protons labelled a? *(1)*

e) What is the spin-spin splitting pattern for the protons labelled b? *(1)*

f) Draw the structure of the compound formed on reaction of 4-methylpentan-2-one with $NaBH_4$. *(1)*

6 A ketone may be converted into a primary alcohol by a four-step process.

Step 1: C_4H_8O is reduced to compound A.

Step 2: Compound A is converted into compound B.

Step 3: Compound B is converted into compound C.

Step 4: Compound C is converted into a primary alcohol, D.

a) Give the structures and IUPAC name of A, B, C and D.

b) Name the reagents required for each step. *(12)*

7 The thin-layer chromatography plate shown below shows the separation of a mixture of organic compounds. Silica gel is the support and the solvent used was hexane. The mixture contained propan-2-ol, propanone and ethyl propanoate.

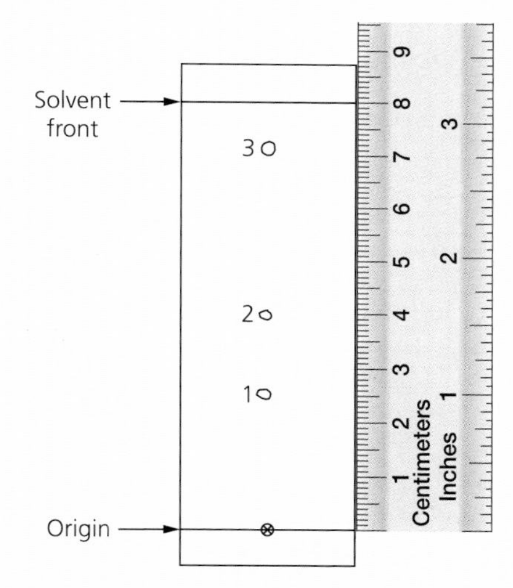

a) Calculate the R_f values for spots 1, 2 and 3. *(3)*

b) Which spot represents which substances in the mixture? *(2)*

c) Why is it important to wear gloves when preparing a TLC plate? *(1)*

8 The ^1H NMR spectrum below is for an organic compound which has the formula $C_4H_8O_2$ is shown below. The substance did not react with sodium carbonate. The integration ratios are given in the table below.

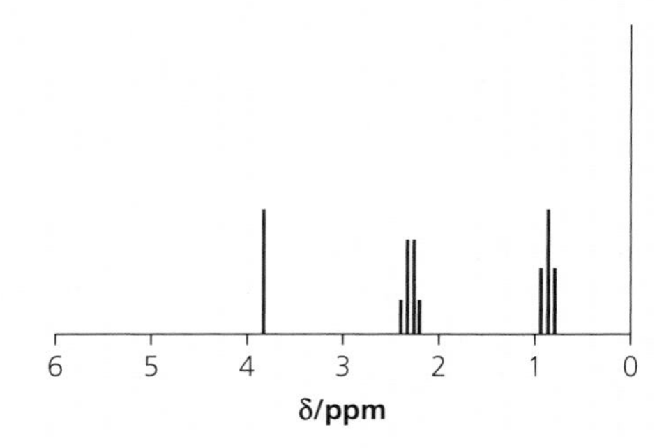

δ/ppm	0.9	2.4	3.8
Integration ratio	3	2	3

a) Suggest a fragment of the molecule which could be responsible for the peak at δ = 3.8. *(1)*

b) Suggest a fragment of the molecule which could be responsible for the peaks at δ = 0.9 and 2.4. *(1)*

c) Draw a possible structure of the molecule. *(1)*

d) Name a suitable solvent used in ^1H NMR spectroscopy. *(1)*

9 For the following organic compounds:

$CH_3CH_2CH_2OH$	$CH_3CH=CH_2$	CH_3CH_2COOH	$CH_3CH(OH)CH_3$
$CH_3CH_2CH_2Br$	$CH_3CH_2CH_3$	CH_3COCH_3	$CH_3CH_2CH_2NH_2$

a) Which one of the compounds would give 2 peaks on a ^{13}C NMR spectrum? *(1)*

b) Which of the compounds would react with acidified potassium dichromate giving an orange to green colour change? *(1)*

c) Which of the compounds would react with sodium hydroxide? *(1)*

d) Which one of the compounds has the empirical formula CH_3? *(1)*

e) Which of the substances can be reduced using $NaBH_4$ or $LiAlH_4$? *(1)*

f) $CH_3CH=CH_2$ can be converted to $CH_3CH_2CH_2NH_2$ in a two-step process.

i) Give the IUPAC name for $CH_3CH=CH_2$ and $CH_3CH_2CH_2NH_2$. *(2)*

ii) Give the reagent required to convert $CH_2CH=CH_2$ into $CH_3CH_2CH_2NH_2$ in a two-step process. State any conditions required and name the mechanism. *(7)*

Stretch and challenge

10 There are several isomers with the molecular formula $C_5H_{10}O$.

a) Draw the structure of one isomer which exhibits optical isomerism. Give the IUPAC name of the compound and label the chiral centre with an asterisk (*). *(3)*

b) Draw the structure of one isomer which exhibits E-Z isomerism. Draw the structure of the two stereoisomers and label them as E and Z. *(3)*

c) Draw the structure of one isomer which would react with water forming pentane-1,5-diol. *(1)*

d) Give the IUPAC name for the isomer which could be oxidised to 3-methylbutanoic acid using acidified potassium dichromate. *(1)*

16 Maths for chemistry

PRIOR KNOWLEDGE

- In calculation work you must be able to use an appropriate number of *significant figures*. If the raw data given is quoted to varying numbers of significant figures, the calculated result can only be reported to the limits of the least accurate measurement.
- *Standard form* expresses numbers in the form $A \times 10^n$, where A is a number between 1 and 10 and n is the number of places the decimal point moves. When converting between numbers in ordinary and standard form, the same number of significant figures must be kept.
- You must also be able to present your answers in calculations to a number of *decimal places* and to state the correct units. Conversion between units, e.g. cm^3 to dm^3 or kg to g, is also expected.
- The *arithmetic mean* is found by adding together all values and dividing by the total number of values. Outliers must be identified and ignored when calculating a mean.
- Per cent means 'out of 100' and is represented by the symbol %. Percentage error is found using the equation

  Percentage error = error × 100/quantity measured.

- To determine the percentage error when two burette readings are used to calculate a titre value, the error must be multiplied by two.
- You must be able to rearrange the subject of an equation, for example

  $$\text{moles} = \frac{\text{mass}}{M_r} \text{ and so } M_r = \frac{\text{mass}}{\text{moles}}.$$

- The independent variable is the variable for which values are changed by the investigator. The dependent variable is the variable of which the value is measured for each change in the independent. The controlled variable is one which must be kept constant, to prevent it affecting the outcome of the investigation. When plotting graphs the independent variable is placed on the x axis and the dependent variable is on the y axis. Appropriate scales must be devised for the axis, making the most effective use of the graph paper. Axes should be labelled with the name of the variable followed by a solidus (/) and the unit of measurement. A best-fit line or curve can be drawn by judging the approximate positon of the line so that there are approximately the same number of data points on each side of the line or curve.

Algebra

Changing the subject of an equation and substituting values

Book 1 outlined how to change the subject of a simple equation; however, you must now be able to change the subject of more complicated equations involving rates and equilibrium.

EXAMPLE 1

Rearrange the rate equation rate = $k[NO]^2[O_2]$ to make k the subject

Answer

Switch sides to get the new subject on the left:

$k[NO]^2[O_2] = \text{rate}$

You require k by itself on the left-hand side, hence $[NO]^2[O_2]$ must be moved to the right-hand side. Remember that k is multiplied by $[NO]^2[O_2]$ so when $[NO]^2[O_2]$ moves to the right it is divided.

$$k = \frac{\text{rate}}{[NO]^2[O_2]}$$

EXAMPLE 2

Rearrange the equation to make $[H^+]$ the subject.

$$K_a = \frac{[H^+]^2}{[\text{weak acid}]}$$

Answer

- Switch sides to get the new subject on the left.

$$\frac{[H^+]^2}{[\text{weak acid}]} = K_a$$

- On the left-hand side [weak acid] is divided so moving it to the right, it is multiplied, the inverse operation

$[H^+]^2 = K_a \times [\text{weak acid}]$

- To remove the power of two on the left-hand side, you must take the square root of the right-hand side.

$$[H^+] = \sqrt{K_a \times [\text{weak acid}]}$$

- To find $[H^+]$ simply substitute the values into this equation. If K_a is 1.74×10^{-5} mol dm^{-3} and it is a 0.20 mol dm^{-3} solution of ethanoic acid then:

$$[H^+] = \sqrt{K_a \times [\text{weak acid}]} = \sqrt{1.74 \times 10^{-5} \times 0.20}$$

$$= \sqrt{3.48 \times 10^{-6}}$$

$= 1.87 \times 10^{-3}$ mol dm^{-3}

TIP

To rearrange an equation with a squared power use the inverse operation, which is a square root. For example $3^2 = 9$, $\sqrt{9} = 3$. In general $a^2 = b$ so $a = \sqrt{b}$.

EXAMPLE 3

Rearrange the equilibrium expression to make [C] the subject.

$$K_c = \frac{[C]^2[D]}{[A][B]^2}$$

Answer

- Switch sides to get the new subject on the left:

$$\frac{[C]^2[D]}{[A][B]^2} = K_c$$

- If a quantity is divided on one side of the expression, it will be multiplied on the other side when moved:

$[C]^2[D] = K_c [A] [B]^2$

Both [A] and $[B]^2$ were divided, so when moved to the other side they are multiplied.

- If a quantity is multiplied on one side of the expression, it will be divided on the other side when moved:

$$[C]^2 = \frac{K_c[A][B]^2}{[D]}$$

- To remove the power of two on the left-hand side, you must take the square root of the right-hand side.

$$[C] = \sqrt{\frac{K_c[A][B]^2}{[D]}}$$

Units

For A2 you must be able to work out units for rate constants and equilibrium constants. To help with this it is useful to understand the laws of indices.

Multiplication

When multiplying numbers that have indices, add the powers:

$$x^2 \times x^3 = x^{(2+3)} = x^5$$

EXAMPLE 4

$mol\,dm^{-3} \times mol\,dm^{-3}$

Remember mol is really mol^1

$mol \times mol \times dm^{-3} \times dm^{-3} = mol^{(1+1)}\,dm^{(-3+-3)} = mol^2\,dm^{-6}$

The units $mol\,dm^{-3}$ are concentration units, so a simpler method is to replace the unit $mol\,dm^{-3}$ by the word concentration.

$mol\,dm^{-3} \times mol\,dm^{-3} = \text{concentration} \times \text{concentration}$

$= (\text{concentration})^{(1+1)} = (\text{concentration})^2$

This can then be converted back to units $(mol\,dm^{-3})^2 = mol^2\,dm^{-6}$

Division

When dividing numbers with indices, subtract the indices:

$$\frac{x^4}{x^3} = x^{(4-3)} = x^1 = x$$

Remember when dividing it is possible to cancel terms if the same term is on the top and bottom of the fraction.

$$\frac{\cancel{x^2} \times x^4}{\cancel{x^2}} = x^4$$

EXAMPLE 5

$$\frac{(mol\,dm^{-3})^2}{(mol\,dm^{-3})} = \frac{(\cancel{mol\,dm^{-3}})^{\cancel{2}\,1}}{(\cancel{mol\,dm^{-3}})} = mol\,dm^{-3}$$

$mol\,dm^{-3}$ is on the top and bottom of the fraction so it cancels

Or alternatively this can be written:

$$\frac{(\text{concentration})^2}{\text{concentration}} = \text{concentration}^{(2-1)} = \text{concentration} = mol\,dm^{-3}$$

EXAMPLE 6

$$\frac{(\text{mol dm}^{-3})^2}{(\text{mol dm}^{-3})^2} = \frac{(\cancel{\text{mol dm}^{-3}})^2}{(\cancel{\text{mol dm}^{-3}})^2} \text{ no units}$$

as the units top and bottom are the same, they cancel each other out.

$$\text{Units} = \frac{(\text{concentration})^2}{(\text{concentration})^2} = \text{concentration}^{(2-2)} = \text{concentration}^0 = \text{no units}$$

Remember $x^0 = 1$

Brackets

When taking the power of a number already raised to a power, multiply the powers.

$$(x^2)^3 = x^{(2\times3)} = x^6$$

EXAMPLE 7

$$(\text{mol dm}^{-3})^2 = \text{mol}^{(1 \times 2)} \times \text{dm}^{(-3 \times 2)} = \text{mol}^2\,\text{dm}^{-6}$$

Reciprocal

$$x^{-1} = \frac{1}{x}$$

EXAMPLE 8

$$\frac{(\text{mol dm}^{-3})^2}{(\text{mol dm}^{-3})^4} = \frac{(\cancel{\text{mol dm}^{-3}})^{\cancel{2}}}{(\cancel{\text{mol dm}^{-3}})^{\cancel{4}\,2}} = \frac{1}{(\text{mol dm}^{-3})^2} = (\text{mol dm}^{-3})^{-2} = \text{mol}^{-2}\,\text{dm}^6$$

EXAMPLE 9

$$\frac{\text{mol dm}^{-3}\text{s}^{-1}}{(\text{mol dm}^{-3})^3} = \frac{\cancel{\text{mol dm}^{-3}}\text{s}^{-1}}{(\cancel{\text{mol dm}^{-3}})^{\cancel{3}\,2}} = \frac{\text{s}^{-1}}{(\text{mol dm}^{-3})^2} = \frac{\text{s}^{-1}}{\text{mol}^{(2\times1)}\,\text{dm}^{(-3\times2)}} = \frac{\text{s}^{-1}}{\text{mol}^2\,\text{dm}^{-6}}$$

$$= \text{mol}^{-2}\,\text{dm}^6\,\text{s}^{-1}$$

Estimating results

Sometimes changing different experimental conditions has an effect on measureable values. Using mathematics, it is possible to estimate this effect.

For example, temperature is the only factor that affects the value of K_c for an equilibrium. This is because temperature affects the position of equilibrium and the concentrations of reactants and products will change.

An increase in the concentration of the products (and decrease in the concentration of the reactants) will increase the value of K_c. This is because the numerator of the K_c expression gets larger and the denominator gets smaller, giving a larger value of K_c.

For a reaction:

$$aA + bB \rightleftharpoons cC + dD$$

a decrease in the concentration of the products (and increase in the concentration of the reactants) will decrease the value of K_c when temperature is changed.

TIP

If the numerator (numbers on the top) increases and denominator (numbers on the bottom) decreases then K_c increases.

EXAMPLE 10

For the equilibrium:

$$N_2(g) + 3H_2(g) \rightleftharpoons 2NH_3(g) \qquad \Delta H = -92\,kJ\,mol^{-1}$$

From your knowledge of Le Chatelier's principle from Book 1, the forward reaction is exothermic (ΔH is negative) and so a decrease in temperature shifts the equilibrium position right in the forward exothermic direction which means more ammonia is produced.

In the equation for K_c

$$K_c = \frac{[NH_3]^2}{[N_2][H_2]^3}$$

If more ammonia is produced the numerator of the fraction gets bigger, and the denominator gets smaller, as there is a smaller concentration of nitrogen and hydrogen, and so the value of K_c will increase.

TIP

Remember that the equilibrium constant is affected in this way by changes in temperature only; it does not change when the pressure or concentration of the system changes – see page 75.

Graphs

The equation of a straight line graph

Every straight line can be represented by an equation: $y = mx + c$. The coordinates of every point on the line will solve the equation if you substitute them in the equation for x and y.

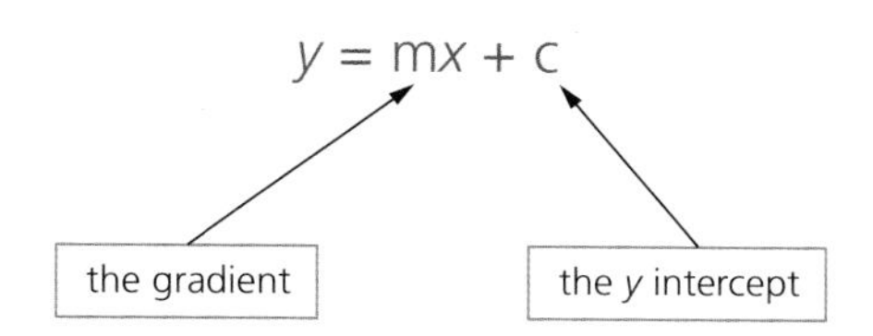

The y **intercept** is the point where the graph crosses the y-axis. It is the value for y when $x = 0$.

EXAMPLE 11

For the graph shown below, the equation of the line is:

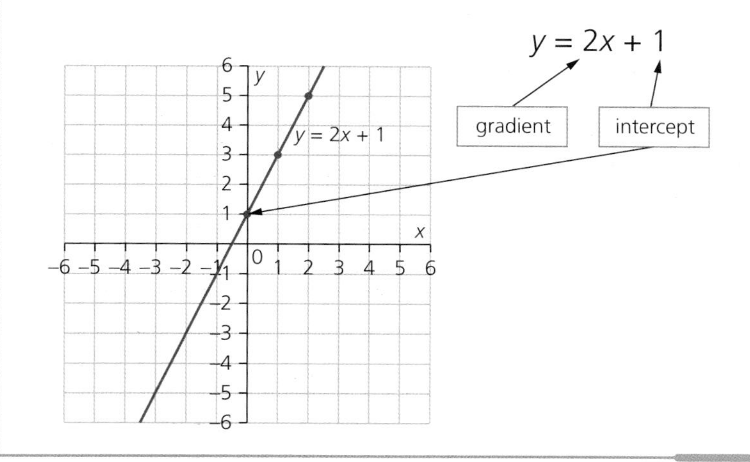

Gradient

Gradient is another word for 'slope'. The higher the gradient of a graph at a point, the steeper the line is at that point. A positive gradient means the line slopes up from left to right. A negative gradient means that the line slopes downwards from left to right. A zero gradient graph is a horizontal line. For a straight line graph the gradient is a constant value.

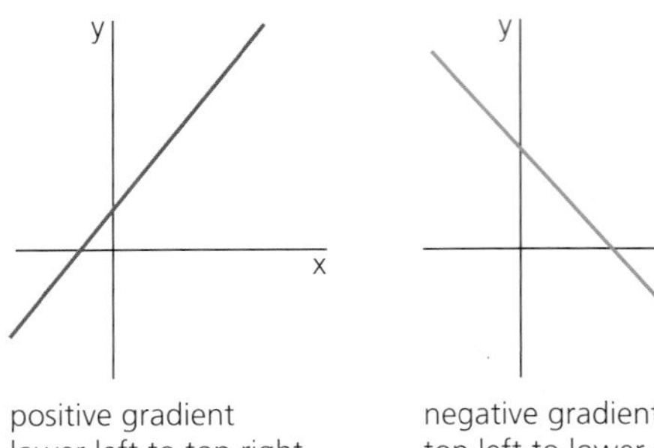

positive gradient lower left to top right.

negative gradient top left to lower right.

To find the gradient (m) of a straight line graph:

1 Choose any two points on the line.

2 At one end of the line chose a point and call it (x_1, y_1).

3 At the other end of the line chose another point and call it (x_2, y_2).

4 A more accurate answer is obtained when the points (x_1, y_1) and (x_2, y_2) are as far apart as possible.

5 Substitute your values into the equation:

$$m = \frac{y_2 - y_1}{x_2 - x_1}$$

where the numerator represents the vertical distance between the two points (the rise) and the denominator represents the horizontal distance between two points (the run).

A simpler equation to use may be gradient $(m) = \frac{\text{rise}}{\text{run}}$

or even gradient $(m) = \frac{\text{change in } y}{\text{change in } x}$

EXAMPLE 12

For this graph $y = mx + c$

The y intercept(c) is 1

The gradient (m) is calculated as:

$$m = \frac{y_2 - y_1}{x_2 - x_1} = \frac{3-1}{1-0} = \frac{2}{1} = 2$$

or

$$m = \frac{\text{rise}}{\text{run}} = \frac{2}{1} = 2$$

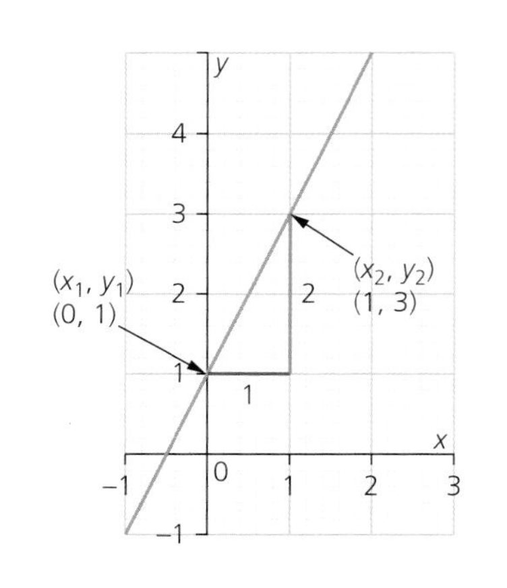

The equation for the line is:

$y = 2x + 1$

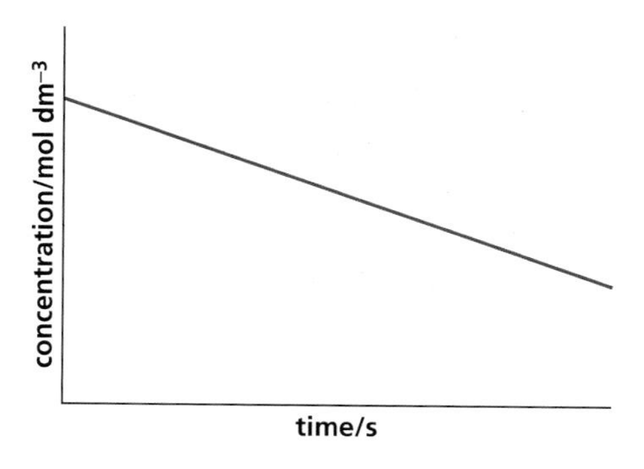

The rate equation for a zero order reaction is:

rate = $k[A]^0$.

The graph above is a concentration against time graph for a zero order reaction (see page 59) The gradient of this graph is equal to the rate constant. Gradients can have units. For the concentration against time graph the gradient does have units. This is because:

$$\text{Gradient (m)} = \frac{y_2 - y_1}{x_2 - x_1} = \frac{\text{concentration}}{\text{time}}$$

hence the units of the gradient of this graph are:

$$\frac{\text{mol dm}^{-3}}{\text{s}} = \text{mol dm}^{-3}\,\text{s}^{-1}$$

Tangent

The word tangent means 'touching' in Latin. The tangent is a straight line that just touches the curve at a given point and does not cross the curve.

To draw a tangent at a point (x,y):

1 Place your ruler through the point (x,y) on the curve.

2 Make sure your ruler does not touch the curve at any other point.

3 Draw a ruled pencil line passing through point (x,y).

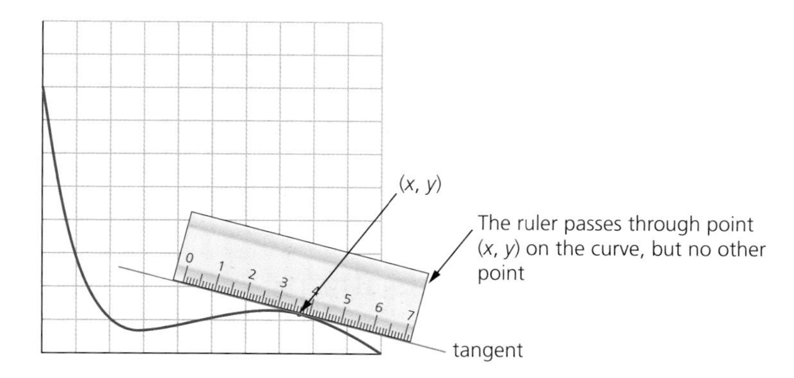

EXAMPLE 13

What is the gradient of the curve at point A?

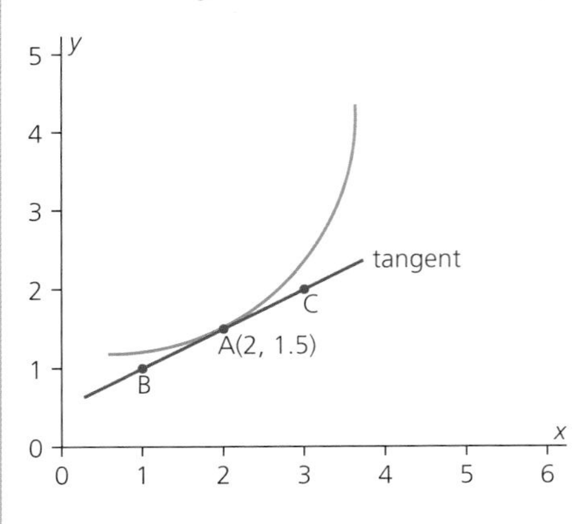

Answer

Draw a tangent to the curve at point A.

B has coordinates (x_1, y_1) (1,1) and C has coordinates (x_2, y_2) (3,2). The gradient of the curve at point A is equal to the slope of the straight line BC.

Calculate the gradient of BC:

$$m = \frac{y_2 - y_1}{x_2 - x_1} = \frac{2-1}{3-1} = \frac{1}{2} = 0.5$$

The gradient of the curve at point A is $\frac{1}{2}$, or 0.5.

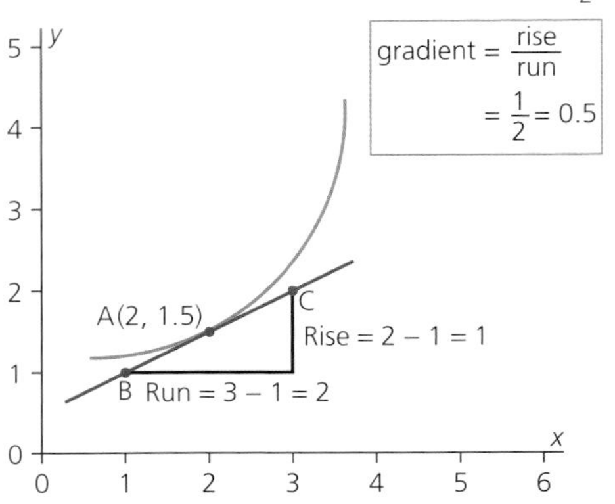

To find the slope of the curve at any other point, you would need to draw a tangent line at that point and then determine the slope of that tangent line. This is the method used to find the order of a reaction using the initial rates methods, see page 58.

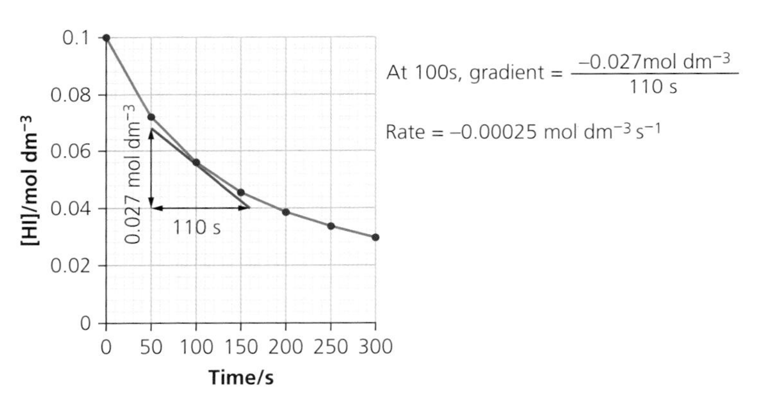

Figure 16.1 The graph shows how the rate of reaction at time 100 seconds is calculated, by drawing a tangent to the curve and calculating the gradient of the tangent.

Algebra

Logarithms

Large numbers are often complicated to deal with. By writing larger numbers in terms of their power to base ten, a smaller scale, called a log scale is generated which is often easier to comprehend. In chemistry, pH is defined in terms of a logarithmic scale. Long before calculators were invented, logarithm tables were used to simplify mathematical calculations. John Napier, a Scottish mathematician published the first logarithm tables in 1641.

Logarithms, or 'logs', express one number in terms of a base number, that is raised to a power.

TIP

There are two different buttons to calculate logarithms on most calculators. The 'log' button calculates logs to base ten and is the button which you should use. The 'ln' button calculates natural logs; *do not use this button*.

This is the log button you should use

Don't use this button – it does not calculate $\log_{10}$

For example, in the expression

$$100 = 10^2$$

10 is the base and 2 is the power or index. This expression can be written in an alternative way. In terms of logs it is written

$$\log_{10}100 = 2$$

This can be read as the 'log to base 10 of 100 is 2'. The relationship between the two expressions is shown below.

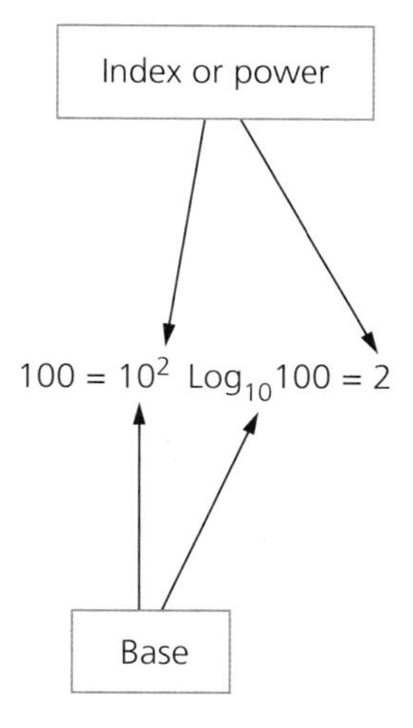

The base can be any positive number apart from 1, but in A-level Chemistry we will be using logs to base 10. Logarithms to base 10, $\log_{10}$, will be written simply as 'log', e.g. log100 = 2. You will use your calculator to find logs.

EXAMPLE 14

Check that you can use your calculator correctly by verifying that:

log 93 is 1.97

log 0.03 is −1.52

log 1.1×10^3 is 3.04

Two common equations involving logs which are used in chemistry are:

$$pH = -\log[H^+]$$

and

$$pK_a = -\log K_a$$

EXAMPLE 15

If a solution has $[H^+] = 0.50\ mol\,dm^{-3}$ find the pH of the solution using the equation $pH = -\log[H^+]$

Answer

$pH = -\log(0.50)$

To calculate this on your calculator, type in '–', then 'log', then '0.50'.

This should give an answer of 0.30

EXAMPLE 16

The K_a for ethanoic acid CH_3COOH is $1.74 \times 10^{-5}\ mol\,dm^{-3}$. Use the equation

$pK_a = -\log K_a$

to find the value of pK_a

Answer

$pK_a = -\log(1.74 \times 10^{-5})$

Type in '–', then 'log', then '1.74×10^{-5}'.

The answer should be 4.76

You will find many examples of pH and p*K*a calculations in Chapter 6.

Antilogs

You must be able to use your calculator to find logs of any positive number. In addition, sometimes you may be given the log of a number and must work backwards to find the number itself. This is called finding the **antilog** of the number. An antilog is the opposite of a log. To find the antilog on most simple scientific calculators:

1 press the second function (2ndF) inverse (inv) or shift button, then

2 type in the number, then

3 press the log button. It might also be labelled the 10^x button.

Press the shift (or 2ndF) button and then the log button to access the antilog function. The antilog function is 10^x as shown

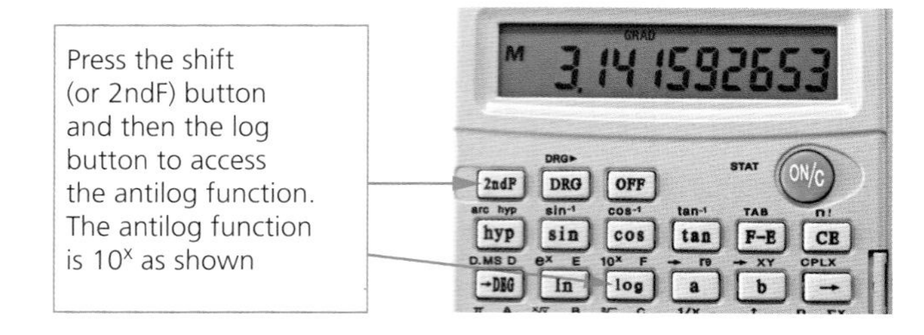

EXAMPLE 17

If $\log x = 2.3$. What is x?

Answer

To find x you need to find the antilog of 2.3.

On your calculator type in 'shift' (or 'inv' or '2ndF'), then 'log', then '2.3'

The answer should be 199.5.

EXAMPLE 18

If $\log b = -0.2$. what is b?

Answer

On your calculator find the antilog of −0.2. It should be 0.6 so $b = 0.6$

An antilog 'undoes' logs by raising the base to the log number.

For example:

$\log 100 = 2$, $\text{antilog} 2 = 100$

In finding the antilog, the calculator is performing the calculation 10^2.

EXAMPLE 19

Determine the concentration, in mol dm^{-3}, of nitric acid that has a pH of 0.65.

Answer

$pH = -\log[H^+]$

$0.65 = -\log[H^+]$

$-0.65 = \log[H^+]$

On your calculator, type in 'shift' (or '2ndF' or 'inv'), then 'log', then '−0.65,

This should give the value of $[H^+] = 0.22\ \text{mol dm}^{-3}$

An alternative way of answering this is to use the equation:

$[H^+] = 10^{-pH}$

$[H^+] = 10^{-0.65}$

This is calculated in the same way on your calculator.

EXAMPLE 20

Calculate the K_a of an acid which has a pK_a value of 2.89

Answer

$pK_a = -\log K_a$

$2.89 = -\log K_a$

$-2.89 = \log K_a$

On your calculator, type in 'shift' (or '2ndF or 'inv'), then 'log'. then '−2.89'. This should give the value of K_a of 1.29×10^{-3}

An alternative way of answering this question is by using the equation:

$K_a = 10^{-pKa}$

$K_a = 10^{-2.89}$

Logarithmic scales

Logs are a powerful way to reduce a set of numbers that range over many orders of magnitude to a smaller more manageable scale. pH is a logarithmic scale. The table below shows how taking the log of the hydrogen ion concentration reduces the range of numbers being dealt with. If values for $[H^+]$ range from 0.1 to 0.00000000000010 the pH can fit in the range 1 to 14!

pH	1	7	13	14
$[H^+]$/mol dm^{-3}	10^{-1}	10^{-7}	10^{-13}	10^{-14}

Presentation of data on a logarithmic scale can be helpful when drawing graphs. Logarithmic scales are used when the data:

- covers a large range of values; the use of logs reduces a wide range to a more manageable size which is easier to plot.
- may contain exponential or power laws since these will show up as straight lines. For example, for a second order reaction rate = $k[A]^2$, the graph of rate against concentration would give a curve (see page 59). If logs are taken of both sides, this gives:

  $\log(\text{rate}) = \log k + 2\log[A]$

  This has the same form as a straight line graph $y = mx + c$. Consequently, a plot of log rate against log[A] gives a straight line graph whose intercept is the value for log$k$ and the gradient is equal to the order of the reaction. This treatment is valid for any order values. The gradient (slope) of the graph line is always equal to the order.

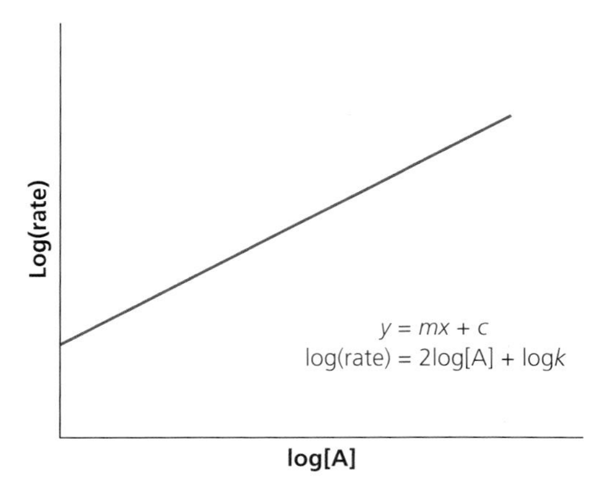

Sometimes you may be asked to plot a graph by calculating the log of some values, and then plotting the logs. Simply calculate the log values on your calculator and record them before plotting. An example of this type of graph is included with the activity in the kinetics chapter on page 62.

The Richter scale is used to measure the strength of earthquakes by taking the log of the amount of energy they release. It is a logarithmic scale. The largest earthquake ever recorded was in 1960 in Chile and reached 9.5 on the Richter scale.

Index

H

I

K

L

M

N

Free online resources

Answers for the following features found in this book are available online:

- Test yourself questions
- Activities

You'll also find Practical skills sheets and Data sheets. Additionally there is an Extended glossary to help you learn the key terms and formulae you'll need in your exam.

Scan the QR codes below for each chapter.

Alternatively, you can browse through all chapters at www.hoddereducation.co.uk/AQAChemistry2

How to use the QR codes

To use the QR codes you will need a QR code reader for your smartphone/tablet. There are many free readers available, depending on the smartphone/tablet you are using. We have supplied some suggestions below, but this is not an exhaustive list and you should only download software compatible with your device and operating system. We do not endorse any of the third-party products listed below and downloading them is at your own risk.

- for iPhone/iPad, search the App store for Qrafter
- for Android, search the Play store for QR Droid
- for Blackberry, search Blackberry World for QR Scanner Pro
- for Windows/Symbian, search the Store for Upcode

Once you have downloaded a QR code reader, simply open the reader app and use it to take a photo of the code. You will then see a menu of the free resources available for that topic.

1 Thermodynamics: Born-Haber cycles

4 Equilibrium constant K_p for homogeneous systems

2 Thermodynamics: Gibbs free energy change, ΔG, and entropy change, ΔS

5 Electrode potentials and cells

3 Rate equations

6 Acids and bases

7 Properties of Period 3 elements and their oxides

8 Transition metals

9 Transition metals: Variable oxidation states

10 Optical isomerism

11 The carbonyl group

12 Aromatic chemistry

13 Amines

14 Polymers, amino acids and DNA

15 Organic synthesis, NMR spectroscopy and chromatography

16 Maths for chemistry

17 Practical skills

18 Preparing for the examination